OPTICAL COHERENCE TOMOGRAPHY

OPTICAL COHERENCE TOMOGRAPHY

PRINCIPLES AND APPLICATIONS

Mark E. Brezinski MD, PhD

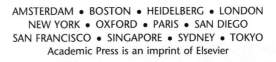

AMSTERDAM • BOSTON • HEIDELBERG • LONDON
NEW YORK • OXFORD • PARIS • SAN DIEGO
SAN FRANCISCO • SINGAPORE • SYDNEY • TOKYO
Academic Press is an imprint of Elsevier

Cover image courtesy of Brezinski, M.E., et.al. Circulation, 1996, 93:1206-1213.

Academic Press is an imprint of Elsevier
30 Corporate Drive, Suite 400, Burlington, MA 01803, USA
525 B Street, Suite 1900, San Diego, California 92101-4495, USA
84 Theobald's Road, London WC1X 8RR, UK

Library of Congress Cataloging-in-Publication Data
Brezinski, Mark E.
  Optical coherence tomography: principles and applications / Mark E. Brezinski.
       p. ; cm.
  Includes bibliographical references and index.
  ISBN-13: 978-0-12-133570-0 (hardcover : alk. paper)
  ISBN-10: 0-12-133570-4 (hardcover : alk. paper) 1. Optical tomography.
2. Optical tomography–Technological innovations. I. Title.
  QC449.5.B74 2006
  616.07'54–dc22

                                                                2006007869

British Library Cataloguing-in-Publication Data

A catalogue record for this book is available from the British Library.

ISBN 13: 978-0-12-133570-0
ISBN 10: 0-12-133570-4

For information on all Academic Press publications
visit our Web site at www.books.elsevier.com

Printed and bound in the United Kingdom
Transferred to Digital Printing, 2011

ABOUT THE AUTHOR

Dr. Mark Brezinski is an Associate Professor at Harvard Medical School and senior scientist at Brigham and Women's Hospital. His work, since 1994, has focused on the development of Optical Coherence Tomography (OCT), predominately in non-transparent tissue. He received his undergraduate training at King's College and his MD/PhD from Thomas Jefferson University. Dr. Brezinski did his residency at Brigham and Women's Hospital and postdoctoral training at Harvard Medical School. His cardiology fellowship was at Massachusetts General Hospital and Harvard Medical School, where he subsequently became a staff member in the Cardiac Unit, as well as a scientist in the Department of Electrical Engineering and Computer Science at Massachusetts Institute of Technology. In 2000, Dr. Brezinski moved to the Orthopedics Department at Brigham and Women's Hospital and Harvard Medical School, where he is currently a staff member. Dr. Brezinski has received many awards for his work, including his 1999 award from President Clinton for Scientists and Engineers. He is the author of over 200 publications and a recognized pioneer in OCT research. He was a co-founder of the Lightlab Corporation, which develops OCT technology. He was born in Nanticoke, Pennsylvania and currently resides in Marblehead, Massachusetts. Dr. Brezinski is currently the principal investigator on National Institutes of Health contracts NIH-RO1-AR44812, NIH R01 AR46996, NIH R01- EB02638, NIH-1-R01-HL55686, and NIH R01 EB000419.

CONTENTS

2   Light and Electromagnetic Waves   31

3   Light in Matter   57

4 Interference, Coherence, Diffraction, and Transfer Functions   71

SECTION II

5 Optical Coherence Tomography Theory   97

13 Other Technologies   353

16 OCT in the Musculoskeletal Disease 459

17 OCT in Oncology   493

Online companion site: http://books.elsevier.com/companions/0121335704

PREFACE

This book deals with the principles behind and the application of optical coherence tomography (OCT) for imaging in nontransparent tissue, which represents most of the tissue of the body. In brief, as will be seen in the text, OCT is a high resolution imaging technique that allows micron scale imaging of biological tissue over small distances. It uses ultrashort pulses or low coherence infrared light to perform high-resolution diagnostics.

Imaging in transparent tissue with OCT, specifically in the field of Ophthalmology, is addressed elsewhere (1). This book is also not a handbook on how to construct an OCT system. The following text is divided into three sections. The first section consists of four chapters intended for the nonphysical science. While most of the information is rudimentary to those in the physical sciences, it is intended to give the nonphysical scientist and clinician an appreciation of the fundamental mathematics and physics behind OCT. It is hoped that these sections will allow improved comprehension of OCT literature as well as chapters in section two of this text. In addition, references are provided of prominent textbooks in these areas for those with sufficient interest to pursue further study. The second section examines OCT theory and technology in considerable detail, attempting to be current with work as of March 2006. In many cases, extensive derivations are present, particularly when they are lacking in the literature, for several reasons, most importantly to delineate limitations and approximations involved in theories often taken as dogma. This section also attempts to review topics not comprehensively reviewed in the literature, from quantum noise sources to adjuvant OCT techniques. The third section has several aims. First, it reviews techniques useful in performing hypothesis driven research in both basic and clinical areas, as the research community demands. Second, this section reviews potential clinical applications not from a cursory or simplistic analysis. Instead, the goal is to provide a thorough understanding of the potential strengths and limitations of OCT for each clinical application, examining areas including epidemiology, competing technologies, pathophysiology, and economic barriers. Third, as accurately as possible, it attempts to define those subgroups of patients who could potentially benefit from OCT imaging, rather than making global claims of its applications to entire populations with each pathophysiology. The core portions of the book were written by a single author (myself), with the feedback and clerical assistance of colleagues (such as Bin Liu PhD, Alicia Goodwin, Diana Wells, Kathy Zheng, and Dorothy Branham), with the goal of providing synergy between the three sections. The exceptions are the chapters on clinical trials, image processing, and OCT Doppler, where it was felt these areas were best addressed by others.

The origins and development of OCT can be roughly divided into those of the 1980's, 1990's, and those in the new millennium. Ranging with interferometry has existed for over a century (2,3), but the 1980's (and to some degree the late 70's), saw the rise of two major advances, interest in low coherence interferometry (LCI) and technology development relevant to telecommunications. In particular, the latter saw the introduction of relatively inexpensive low coherent sources and fiber optics that were critical for OCT development. Some of the most important works are referenced below (4–10). These LCI techniques were subsequently applied primarily to imaging transparent tissue (11–15). The 1990's are arguably the period of greatest paradigm shifts toward advancing OCT. The 1990's initially saw the development of two dimensional low coherence imaging and its application to diagnosing pathology of the eye. In a *Science* publication in 1991, a group led by James Fujimoto PhD, with major contributions from Eric Swanson M.S., and David Huang, advanced low coherence ranging to two dimensions and coined the phrase "Optical Coherence Tomography" (OCT), the name the technology currently bears (16). Clinical studies in Ophthalmology continued through the 90's, focused primarily on retinal disease with Carmen Puliafito MD and Joel Schuman MD being major leaders (1,17,18). In 1994, we began work on developing OCT for imaging in nontransparent tissue, representing most of the tissue of the body, which would ultimately be published in *Circulation* (19). This work achieved several objectives. First, among other modifications, the median wavelength of the OCT beam was moved from 800 nm to the optical window at 1300 nm, allowing increased penetration in nontransparent tissue. Imaging was possible even through calcified tissue. Second, histologic correlations were performed on a micron scale, allowing effective hypothesis testing. Finally, the tissue chosen was 'vulnerable' plaques (which are discussed in chapter 15), demonstrating feasibility on a clinically important pathophysiology. Other work published on nontransparent tissue at this time included a review which was published in *Nature Medicine* based on data from the original *Circulation* paper, a brief report imaging a single human coronary artery with the corresponding histology, and a third that imaged human skin but without histologic correlation to confirm any structural detail (20–22).

For the remainder of the 1990's advances can be divided into four categories, identifying organ systems for which OCT demonstrated the most promise, technology development, commercialization, and early in vivo imaging studies.

We categorized those areas where OCT demonstrated its greatest potential in an article in *IEEE Journal of Quantum Electronics* and in general, this classification system still holds today and is discussed in more detail in chapter 12 (23). These areas are (1) when conventional tissue biopsy is not possible (ex: coronary artery or cartilage), (2) when false negatives occur (ex: screening for early cancer or assessing resected pathology), and (3) in guiding surgery and microsurgery. Specific areas where OCT demonstrated feasibility for clinical diagnostics include the cardiovascular system, joints, gastrointestinal tract, bladder, female reproductive tract, respiratory tract, surgical guidance, and dentistry (19,21,24–33). Among the most important technological advances that occurred in the 1990's included the development of improved delay lines, peripherals, femtosecond sources, and integrated OCT engines (34–41).

Commercialization of OCT technology began in the 1990's. For imaging of the eye, Humphrey Instruments[TM], a division of Carl Zeiss America[TM] has developed instruments for Ophthalmology (http://www.meditec.zeiss.com/). In 1998, Lightlab[TM] (formally Coherent Diagnostic Technologies) was formed primarily by myself, James G. Fujimoto PhD, Eric Swanson M.S., and the Carl Zeiss America[TM]. The company became a subsidiary of Goodman Co. Ltd. in 2002 (lightlabimaging.com).

While in vivo imaging was performed of human finger nails and skin, the most important work advancing the technology toward in vivo use was published in the journal *Science* (34). This study used in vivo catheter based imaging near real time in a synchronized system. Imaging was performed of the gastrointestinal and respiratory tracts of rabbits. This was extended to imaging of the cardiovascular system shortly afterward (42).

With the late 90's and through the new millennium, work moved toward adjuvant techniques, evaluating spectral based techniques, and in vivo human imaging.

A wide variety of adjuvants have been developed including Doppler OCT, single detector polarization sensitive OCT, dual detector polarization sensitive OCT, phase retrieval, entangled photon, image processing, elastography, dispersion analysis, and the use of contrast probes (25,43–63,76).

Spectral based interferometry has been around for over a century, but recently considerable interest has been focused on developing OCT systems that do not operate in the time domain (2,3,64–68). Spectral based OCT has the potential advantage of very high RF data acquisition rates. However, questions remain whether it can generate sufficient dynamic range for imaging in nontransparent tissue. These topics will be addressed in chapters 5 and 7.

In vivo human studies now have been performed in a wide range of organ systems (69–75). However, double-blinded clinical trials remain essentially nonexistent to this point.

OCT has undergone considerable advance over a relatively short period of time due the efforts and insights of many. While the technology has reached the stage of in vivo human studies, substantial basic work remains to be performed. Whether in some cases in vivo human studies are premature, and 'the cart is ahead of the horse', will be examined within the next several decades. Irrespective, we can look forward to exciting advances in this powerful developing technology.

The companion Web site containing color images and other information provided by the author to compliment the text can be found at http://books.elsevier.com.

REFERENCES

1. Schuman, J.S., C.A. Puliafito, and J.G. Fujimoto, 2004. Optical Coherence Tomography of Ocular Diseases. Slack Incorp.; 2nd ed. Thorofare, New Jersey.

2. Michelson, A.A., 1891. On the application of interference methods to spectroscopic measurements, *Phil. Mag. Ser.* 5(31), 338–341,

3. Rubens, H., and R.W. Wood, 1911, Focal isolation of long heat-waves. *Phil. Mag. Ser.* 6(21), 249–255.

4. Barnoski, M.K., and S.M. Jensen, 1976. Fiber waveguides: a novel; technique for investigating attenuation characteristics. *Applied Optics* 15(9), 2112–2115.

5. Wang, C.S., W.H. Cheng, C.J. Hwang, et.al., 1982. High power low divergence supperradiance diode. *Applied Physics Letters* 41(7), 587–589.

6. Ghafoori-Shiraz, and T. Okoski, 1986. Optical frequency domain reflectometry. *Optical and Quantum Electronic*s 18, 265–272.

7. Danielson B.L., and C.D. Whittenberg, 1987. Guided wave reflectometry with micrometer resolution. *Applied Optics* 26(14), 2836–2842.

8. Takada, K., I. Yokohama, K. Chida, and J. Noda, 1987. New measurement system for fault location in optical waveguide devices based on an interferometric technique. *Applied Optics* 26(9), 1603–1606.

9. Youngquist, R.C., S. Carr, and D.E.N. Davies, 1987. Optical Doherence domain reflectometry: a new optical evaluation technique. *Optics Letters* 12(3), 158–160.

10. Danielson, B.L., and C.Y. Bolsrobert, 1991. Absolute optical ranging using low coherence interferometry. *Applied Optics* 30, 2975–2979.

11. Fujimoto, J.G., S. Desilvestri, E.P. Ippen, C.A. Puliafito, R. Margolis, and A. Oseroff, 1986. Femtosecond optical ranging in biological systems. *Optics Letters* 11, 150–152.

12. Fercher, A.F., K. Mengedott, and W. Werner, 1988. Eye-length measurement by interferometry with partially coherent light. *Optics Letters* 13(3), 186–188.

13. Huang, D., J. Wang, C.P. Lin, et.al., 1991. Micron resolution ranging of cornea anterior chamber by optical reflectometry. *Lasers in Surgery and Medicine* 11, 419–425.

14. Hitzenbereger, C.K., 1992. Measurement of corneal thickness by low coherence interferometry. *Applied Optics* 31, 6637–6642.

15. Swanson, E.A., D. Huang, M.R. Hee, J.G. Fujimoto, C.P. Lin, and C.A. Puliafito, 1992. High speed optical coherence domain reflectometry. *Optics Letters* 17(1), 151–153.

16. Huang, D., E.A. Swanson, C.P. Lin, J.S. Schuman, W.G. Stinson, W. Chang, M.R. Hee, T. Flotte, K. Gregory, C.A. Puliafito, and J. G. Fujimoto, 1991. Optical Coherence Tomography. *Science* 254, 1178–1181.

17. Swanson, E.A., J.A. Izatt, M.R. Hee, D. Huang, C.P. Lin, J.S. Schuman, C.A. Puliafito, and J.G. Fujimoto, 1993. In vivo retinal imaging by optical coherence tomography. *Optics Letters* 18(21), 1864–1866.

18. Hee, M.R., J.A. Izatt, E.A. Swanson, D. Huang, J.S. Schuman, C.P. Lin, C.A. Puliafito, and J.G. Fujimoto, 1995. Optical coherence tomography of the human retina. *Arch. Ophthalm.* 113, 325–332.

19. Brezinski, M.E., G.J. Tearney, B.E. Bouma, J.A. Izatt, M.R. Hee, E.A.Swanson, J.F. Southern, and J.G. Fujimoto, 1996. Optical coherence tomography for optical biopsy. *Circulation* 93(6), 1206–1213.

20. Fujimoto, J.G., M.E. Brezinski, G.J. Tearney, S.A. Boppart, M.R. Hee, and E.A. Swanson, 1995. Optical biopsy and imaging using optical coherence tomography. *Nat. Med* 1, 970.

21. Brezinski, M.E., G.J. Tearney, B.E. Bouma, S.A. Boppart, M.R. Hee, E.A. Swanson, J.F. Southern, and J.G. Fujimoto, 1996. Imaging of coronary artery microstructure with optical coherence tomography. *Am. J. Card* 77(1), 92–93.

22. Schmitt, J.M., M.J. Yadlowski, and R.F. Bonner, 1995. Subsurface imaging of living skin with optical coherence microscopy. *Dermatology.* 191, 93–98.

23. Brezinski, M.E., and J.G. Fujimoto, 1999. Optical coherence tomography: high resolution imaging in nontransparent tissue. *IEEE Journal of Selected Topics in Quantum Electronics*, 5, 1185–1192.

24. Brezinski, M.E., G.J. Tearney, N.J. Weissman, S.A. Boppart, B.E. Bouma, M.R. Hee, A.E. Weyman, E.A. Swanson, J.F. Southern, and J.G. Fujimoto, 1997. Assessing atherosclerotic plaque morphology: comparison of optical coherence tomography and high frequency intravascular ultrasound. *Heart* 77, 397–404.

25. Herrmann, J.C., C. Pitris, B.E. Bouma, S.A. Boppart, J.G. Fujimoto, and M.E. Brezinski, 1999. High resolution imaging of normal and osteoarthritic cartilage with optical coherence tomography. *Journal of Rheumatology* 26(3), 627–635.

26. Tearney, G.J., M.E. Brezinski, J.F. Southern, B.E. Bouma, S.A. Boppart, and J.G. Fujimoto, 1997. Optical biopsy in human gastrointestinal tissue using optical coherence tomography. *Am. J. Gastro* 92, 1800–1804.

27. Pitris, C., M.E. Brezinski, B.E. Bouma, G.J. Tearney, J.F. Southern, and J.G. Fujimoto, 1998. High resolution imaging of the upper respiratory tract with optical coherence tomography. *American Journal of Respiratory and Critical Care Medicine* 157, 1640–1644.

28. Tearney, G.J., M.E. Brezinski, J.F. Southern, B.E. Bouma, S.A. Boppart, and J.G. Fujimoto, 1997. Optical biopsy in human urologic tissue using optical coherence tomography. *The Journal of Urology* 157.

29. Jesser, C.A., C. Pitris, D. Stamper, J.G. Fujimoto, M.E. Brezinski, 1999. Ultrahigh Resolution Imaging of Transitional Cell Carcinoma of the Bladder. *British Journal of Urology* 72, 1170–1176.

30. Pitris, C., A.K. Goodman, S.A. Boppart, J.J. Libus, J.G. Fujimoto, and M.E. Brezinski, 1999. High resolution imaging of cervical and uterine malignancies using optical coherence tomography. *Obstect. and Gyn* 93, 135–139.

31. Brezinski, M.E., G.J. Tearney, S.A. Boppart, E.A. Swanson, J.F. Southern, and J.G. Fujimoto, 1997. Optical biopsy with optical coherence tomography, feasibility for surgical diagnostics. *J. of Surg. Res.* 71, 32.

32. Colston, B.W., M. J. Everett, L. B. Da Silva, et.al., 1998. Imaging of hard and soft tissue structure in the oral cavity by optical coherence tomography. *Appl. Optics* 37, 3582–3585.

33. Wang, X.-J., T.E. Milner, J.F. de Boer, D.H. Pashley, and J.S. Nelson, 1999. Characterization of dentin and enamel by use of optical coherence tomography. 38(10), 2092–2096.

34. Tearney, G.J., M.E. Brezinski, B.E. Bouma, S.A. Boppart, C. Pitris, J.F. Southern, and J.G. Fujimoto, 1997. In vivo endoscopic optical biopsy with optical coherence tomography. *Science* 276, 2037–2039.

35. Tearney, G.J., B.E. Bouma, S.A. Bouma, B. Golubovic, E.A. Swanson, and J.G. Fujimoto, 1996. Rapid acquisition of in vivo biological images by use of optical coherence tomography. *Optics Letters* 21(17), 1408–1410.

36. Kwong, K.F., D. Yankelevich, K.C. Chu, J.P. Heritage, and A. Dienes, 1993. 400 Hz mechanical scanning optical delay line. *Optics Letters* 18(7), 558–560 (grating based delay line).

37. Tearney, G.J., B.E. Bouma, and J.G. Fujimoto, 1997. High speed phase- and group-delay scanning with a grating based phase control delay line. *Optics Letters* 22(23), 1811–1813.

38. Bouma, B.E., G.J. Tearney, S.A. Boppart, M.R. Hee, M.E. Brezinski, and J.G. Fujimoto, 1995. High-resolution optical coherence tomographic imaging using a mode-locked $Ti:Al_2O_3$ laser source. *Optics Letters* 20(13), 1486–1488.

39. Bouma, B.E., G.J. Tearney, I.P. Bilinski, B. Golubovic, and J.G. Fujimoto, 1996. Self phase modulated Kerr-lens mode locked Cr:forsterite laser source for optical coherence tomography. *Optics Letters* 22 (21), 1839-1841.

40. Tearney, G.J., S.A. Boppart, B.E. Bouma, M.E. Brezinski, N.J. Weissman, J.F. Southern, and J.G. Fujimoto, 1996. Single Mode Scanning Catheter/Endoscope for Optical Coherence Tomography. *Optics Letters* 21, 543–545.

41. Boppart, S.A., B.E. Bouma, C. Pitris, G.J. Tearney, J.G. Fujimoto, and M.E. Brezinski, 1998. Forward-scanning instruments for optical coherence tomographic imaging. *Optics Letters* 21(7), 543–545.

42. Fujimoto, J.G., S.A. Boppart, G.J. Tearney, B.E. Bouma, C. Pitris, and M.E. Brezinski, 1999. High resolution in vivo intraarterial imaging with optical coherence tomography. *Heart* 82, 128–133.

43. Izatt, J.A., M.D. Kulkarni, S. Yazdanfar, J.K. Barton, and A.J.Welch, 1997. In vivo bidirectional color Doppler flow imaging of picoliter blood volumes using optical coherence tomography. *Optics Letters* 22(18), 1439–1441.

44. Chen, Z., T.E. Milner, S. Srinivas, X. Wang, A. Malekafzali, M. van Gemert, and J.S. Nelson. Noninvasive imaging of in vivo blood flow velocity using optical Doppler tomography. *Optics Letters* 22(14), 1119–1121.

45. Kobayashi, M., H. Hanafusa, K. Takada, and J. Noda, 1991. Polarization independent interferometric optical time domain reflectometer. *J. Lightwave Technol.* 9, 623–628.

46. Hee, M.R., D. Huang, E.A. Swanson, and J.G. Fujimoto, 1992. Polarization sensitive low coherence reflectometer for birefringence characterization and ranging. *J. Opt. Soc. Amer. B.* 9(3), 903–908.

47. de Boer, J.F., T.E. Milner, M.J.C. van Gemert, and J. S, Nelson, 1997. "Two dimensional birefringence imaging in biological tissue using polarization sensitive optical coherence tomography. *Opt. Lett.* 15, 934–936,

48. Yao, G., and L.W. Wang, 1999. Two dimensional depth resolved Mueller matrix characterization of biological tissue by optical coherence tomography. *Optics Letters* 24, 537–539.

49. Drexler, W., D. Stamper, C. Jesser, X.D. Li, C. Pitris, K. Saunders, S. Martin, J.G. Fujimoto, and Mark E. Brezinski, 2001. "Correlation of collagen organization

with polarization sensitive imaging in cartilage: Implications for osteoarthritis." *J Rheum* 28, 1311–1318.

50. Liu, B., M. Harman, and M.E. Brezinski, 2005. Variables affecting polarization sensitive optical coherence tomography imaging examined through the modeling of birefringent phantoms. *J. Opt. Soc. Am. A.* 22, 262–271.

51. J.M. Schmitt, S.H. Xiang, and K.M. Yung, 1998. Differential absorption imaging with optical coherence tomography. *J. Opt. Soc. Am. A* 15, 2288–2296.

52. Rao, K., M.A. Choma, S. Yandanfar, et.al., 2003. Molecular contrast in optical coherence tomography by use of pump. *Opt. Lett.* 28, 340–342.

53. Xu, C., J. Ye, D.L. Marks, and S.A. Boppart, 2004. Near infrared dyes as contrast enhancing agents for spectroscopic optical coherence tomography. *Opt. Lett.* 29, 1647–1650.

54. Liu B, Macdonald EA, Stamper DL, Brezinski ME, 2004. Group velocity dispersion effects with water and lipid in 1.3 μm OCT system. *Phys Med Biol* 49, 1–8.

55. Schmitt, J.M., 1998. OCT elastography: imaging microscopic deformation and strain in tissue. *Opt Express* 3(6), 199–211.

56. Rogowska, J., N.A. Patel, J.G. Fujimoto, and M.E. Brezinski, 2004. Quantitative OCT elastography technique for measuring deformation and strain of the atherosclerotic tissues. *Heart* 90(5), 556–62.

57. Dave D., and T. Milner, 2000. Optical low coherence reflectometer for differential phase measurement. *Optics Letters* 25, 227–229.

58. Zhao, Y., Z. Chen, C. Saxer, et.al., 2000. Phase resolved OCT and ODT for imaging blood flow in human skin with fast scanning speed and high velocity sensitivity. *Optics Letters* 25, 114–116.

59. Nasr, M.B., B.E. Saleh, A.V. Sergienko, and M.C. Teich, et. al., 2003. Demonstration of dispersion cancellation quantum optical tomography. *Phys. Review Letters* 91, 083601-1–083601-4.

60. Kulkarni, M.D., C.W. Thomas, and J.A. Izatt, 1997. Image enhancement in optical coherence tomography using deconvolution. *Electron. Lett.* 33, 1365–1367

61. Schmitt, J.M., S.H. Xiang, and K.M. Yung, 1999. Speckle in optical coherence tomography. *J. Biomed. Optics* 4, 95–105.

62. Rogowska, J., and M.E. Brezinski, 2000. Evaluation of the adaptive speckle suppression filter for coronary optical coherence tomography imaging. IEEE Trans Med Imaging, 19(12), 1261–6.

63. Rogowska, J., and Mark E. Brezinski. Image processing techniques for noise removal, enhancement, and segmentation of cartilage images 2002. *Phys. Med. Biol.* 47, 641–655.

64. Loewenstein, E.V. (1966). The history and current status of Fourier transform spectroscopy. United States Air Force, Air Force Cambridge Research.

65. Fercher, A.F., C.K. Hitzenberger, G. Kamp, and S.Y. El-Zaiant, 1995. Measurement of intraocular distances by backscattering spectral interferometry. *Opt. Commun.* 117, 43–47.

66. Hausler, G., and M. Lindner, 1998. Coherence radar and spectral radar-dash; new tools for dermatological diagnosis. *J. Biomed. Opt.* 3, 21–31.

67. Chinn, S.R., E.A. Swanson, and J.G. Fujimoto, 1997. Optical coherence tomography using a frequency tunable optical source. *Optics Letters* 22(5), 340–342.

68. Lexer, F., C.K.H., and Fercher, A.F., 1997. Wavelength-tuning interferometry of intraocular distances." *Appl. Opt.* 36(25), 6548–6553.

69. H. Yabushita, B.E. Bouma, S.L. Houser, H.T. Aretz, I.-K. Jang, K.H. Schlendorf, C.R. Kauffman, M. Shishkov, D.-H. Kang, E.F. Halpern, and G.J. Tearney, 2002,

"Characterization of human atherosclerosis by optical coherence tomography," *Circulation* 106(13), 1640–1645.

70. Li, X., S.D. Martin, C. Pitris, R. Ghanta, D.L. Stamper, M. Harman, J.G. Fujimoto, and M.E. Brezinski, "High-resolution optical coherence tomographic imaging of osteoarthritic cartilage during open knee surgery", Arthritis Research & Therapy, 7, R318–323.

71. Li, X.D., S.A. Boppart, J. Van Dam, et.al., 2000. Optical coherence tomography: Advanced technology for the endoscopic imaging of Barrett's esophagus. *Endoscopy* 32, 921–930.

72. Jackle, S., N. Gladkova, F. Feldchtiech, et.al., 2000. In vivo endoscopic optical coherence tomography of esophagitis, Barrett's esophagus, and Adenocarcinoma of the Esophagus. *Endoscopy* 32: 750–755.

73. Zagaynova, E.V., O.S. Streltsova, N.D. Gladkova, et.al., 2002. In vivo optical coherence tomography feasibility for bladder disease. *J. Urol.* 167, 1492–1496.

74. Poneros, J.M., G.J. Tearney, M. Shiskov, et.al., 2002. Optical Coherence Tomography of the Biliary Tree during ERCP. *Gastrointestinal Endoscopy 55*, 84–88.

75. Escobar, P.F., J.L., A. White, et.al., 2004. Diagnostic efficacy of optical coherence tomography in the management of preinvasive and invasive cancer of uterine cervix and vulva. Intern. *J. Gynecol. Cancer.* 14, 470–474.

76. Giattina, S., B.K. Courtney, P.R. Herz, et.al., 2006. Measurement of coronary plaque collagen with polarization sensitive optical coherence tomography (PS-OCT). *Intern. J. Cardiol.* 107, 400–409.

ACKNOWLEDGMENTS

This book is dedicated to my mother and father, who have always loved me unconditionally, sacrificed so that I could succeed, and remain my heroes and role models. I am grateful to my family, particularly Buddy, Damian, and most especially Maria, for the support through the difficult times and the memories from the fun moments. And my friends, especially those in Nanticoke and Marblehead (the Muffin shop), I would like to thank for the great memories, past, present, and future. I would like to express my gratitude to my colleagues who I have learned from and worked with over the years, particularly Alan Lefer PhD, Charles Serhan PhD, Joseph Schmit PhD, Eric Swanson MS, and my former collaborator James Fujimoto PhD. Finally, I would like to thank the funding organizations, principally the NIAMS, NHLBI, and NIBIB, through whose support this text and other scientific efforts have been possible.

SECTION I

1 THE BASICS (MATH, WAVES, etc.) FOR THE NONPHYSICAL SCIENTIST

Chapter Contents

In spite of the large number of equations in this text, for the most part, the central points can be understood without following the equations in detail. The equations are predominately provided for those with a background in the field. This introductory chapter is not for the purpose of teaching the nonphysical scientist how to solve math problems with these tools discussed. Instead, the purpose is to allow the reader to have a sense of the concepts that are discussed with these derivations. For those interested in a more detailed understanding of the mathematics, reference texts are listed at the end of each chapter. Discussed in this chapter are the basic topics of general waves, trigonometry, exponentials, imaginary numbers, differentiation, integration, differential equations, vectors, waves, and Fourier mathematics. These topics are ultimately found throughout other chapters of the text.

1.1   GENERAL PROPERTIES OF WAVES AND COORDINATES

A light wave of a single wavelength can be thought of as a fluctuation or cyclical phenomenon that is constantly changing the electromagnetic field amplitude (defined in Chapter 2) between crests and troughs. It can therefore be described by cyclical mathematical functions, such as the trigonometric functions of cosine and sine or the exponential e^{ix}, which will also be explained in more detail shortly. Only a few trigonometric functions will be used in this text, to simplify the understanding of OCT. But before discussing trigonometric functions, exponentials, or waves, it is important to understand both the concept of a coordinate system (polar vs. rectangular coordinates) and how the position of the wave in the coordinate system is quantified (degrees vs. radians).

In Figure 1.1A and B, the principles of rectangular and polar coordinates are illustrated for what is known as a transverse wave, where the amplitude (oscillation) is perpendicular to the direction of propagation. The relative position in the cycle and amplitude can be quantified in either of these coordinate systems. Figure 1.1A is a rectangular coordinate system where one axis (x) is measuring a distance or time interval while the y axis measures the amplitude (a) of the wave. Light in a vacuum is a transverse wave. For the polar coordinates system in Figure 1.1B, amplitude is mapped as a function of angle or relative position in the cycle. Besides the coordinate system, the positions in the cycle are represented in Figure 1.1A and B, which are quantified in either units of degrees or radian. One complete cycle, where the wave returns to the same phase, can be described as either 360 degrees or 2π radians, which is seen in Figure 1.1A and B. The term for one complete cycle is the

A

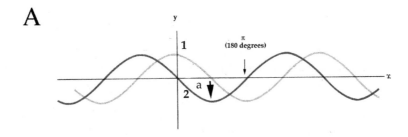

B

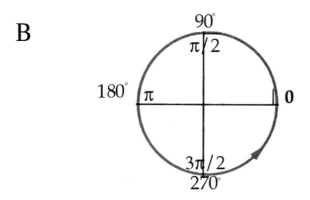

C

Figure 1.1 (A) This figure is a rectangular coordinate system where one axis (x) is measuring a distance or time interval while the y axis measures the amplitude (a) of the wave. The positions in the cycle are also represented here. The two waves are at different relative places in their cycle, or "out of phase." 1A (1) represents the sine function while 1A (2) is the cosine function. The value of the cosine function is one at zero degrees while the sine function is zero at zero degrees. (B) This is a polar coordinate system. Here amplitude is mapped as a function of angle or relative position in the cycle. The positions in the cycle are also represented. (C) This image demonstrates a propagating longitudinal wave along a spring, which is distinct from the transverse wave in 1A.

wavelength and is denoted by the symbol λ. The rate of wave fluctuation can also be described by the frequency (f), which is the number of complete cycles in a given time period. In this book, radians are generally used because they are easier to manipulate. It should also be noted that some waves move. Those waves have a speed that is determined by how quickly a given position on a wave moves in space or time. For example, the speed of light in a vacuum is 3×10^8 m/sec. It should be noted that along with transverse waves, there are also longitudinal waves, of which sound (in most media of interest) or vibrations of a spring are examples. Figure 1.1C demonstrates a propagating longitudinal wave along a spring. With a longitudinal wave, oscillations (amplitude) are in the same direction as the direction of propagation of the wave. In this text we will focus primarily on transverse waves, which is how light behaves in a vacuum.

When more than one wave is present, the position of each wave in the cycle can be different relative to the other. Figure 1.1A shows two waves that are at different relative places in their cycle or out of phase. In the example in Figure 1.1A, the peak of one wave is misaligned with the trough of the second wave, meaning the waves are 90 degrees or π/2 out of phase. If the waves had been aligned, they would have been described as in phase.

1.2 TRIGONOMETRY

Trigonometric functions are one way of describing cyclical functions in numerical terms, which is therefore important in the understanding of optical coherence tomography (OCT). Sine and cosine functions are repeating functions that differ by 90 degrees in the phase, which can be seen in Figure 1.1A, where 1.1A (1) is the representation of the sine function and 1.1A (2) is the cosine function. The value of the cosine function is one at zero degrees while the sine function is zero at zero degrees. These functions will be important in describing the properties of light and OCT.

As stated, trigonometric functions will primarily be used to describe cyclical functions, but it should be noted that they can be and will be used in analyzing certain other circumstances, such as the angles between vectors (described below) and lengths within geometric structures (equivalent to evaluating the lengths of the sides of a triangle) as shown in Figure 1.2. This type of application should become more clear in later chapters.

There are many rules for manipulating and using trigonometric functions, but for the most part they do not need to be discussed unless brought up specifically in the text. Three important examples will be listed here. First, the cosine is symmetrical about the axis (if you flip the wave about the axis, it looks the same) while the sine is not. Therefore, $\cos(-x) = \cos(x)$ {symmetric} and $\sin(x) = -\sin(-x)$. Second, since they are cyclical functions, $\cos(x + 2\pi) = \cos(x)$ and $\sin(x + 2\pi) = \sin(x)$. Third, $\sin^2 \phi + \cos^2 \phi = 1$. Besides the sine and cosine, additional trigonometric functions are listed in Table 1.1.

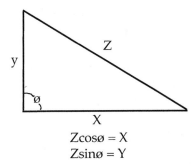

$$Z\cos\emptyset = X$$
$$Z\sin\emptyset = Y$$

Figure 1.2 This figure shows the lengths within geometric structures (equivalent to evaluating the lengths of the sides of a triangle). It is a simple illustration of functions of trigonometry, sine, and cosine.

Table 1.1 Trigonometry Identities

Reciprocal Identities

$\sin \Theta = \dfrac{1}{\csc \Theta}$	$\sec \Theta = \dfrac{1}{\cos \Theta}$	$\tan \Theta = \dfrac{1}{\cot \Theta}$
$\csc \Theta = \dfrac{1}{\csc \Theta}$	$\cos \Theta = \dfrac{1}{\sec \Theta}$	$\cot \Theta = \dfrac{1}{\tan \Theta}$

Quotient Identities

$\tan \Theta = \dfrac{\sin \Theta}{\cos \Theta}$	$\cot \Theta = \dfrac{\cos \Theta}{\sin \Theta}$

Pythagorean Identities

$\sin^2 \Theta + \cos^2 \Theta = 1$	$1 + \tan^2 \Theta = \sec^2 \Theta$	$1 + \cot^2 \Theta = \csc^2 \Theta$

1.3 IMAGINARY NUMBERS

Imaginary numbers will be used frequently throughout the text. To many people, it is concerning that numbers exist that are called imaginary. However, from a mathematical standpoint, it has been stated by many people that the concept of imaginary numbers is no stranger than negative numbers. In the early part of the last millennium, many had difficulty with negative numbers because the meaning of a negative object was not easy to comprehend. But today, few have difficulty comprehending what a negative balance on a credit card bill means, which is of course a negative number. Similarly, the imaginary number i (sometimes written as j) is just a mathematical tool to represent the square root of -1, which has no other method of description. Imaginary numbers are useful for manipulating many of the equations in physics. For our purposes, which will be clearer later, imaginary numbers are very useful for representing both cyclical phenomena and the phase of a wave(s).

A complex number is one that contains both real and imaginary components (the value of either component may be zero) and can be written in the form $z = x + iy$. In this equation, iy represents the term in the complex or imaginary domain while x is the "real" term. In general, these numbers may undergo the same operations as real numbers, with the example of addition shown here:

$$z = (2 + 4i) + (3 + 5i) = 5 + 9i = x + iy \tag{1.1}$$

An important concept with complex or imaginary numbers is the complex conjugate. If $y = (2 + 4i)$, then $y^* = (2 - 4i)$ is the complex conjugate. The multiplication of y by y^* yields a real rather than imaginary number $(2 + 8i - 8i + 16) = (18)$. This operation will be performed throughout the text to generate real from imaginary numbers.

1.4   THE EXPONENTIAL *e*

The number e is a powerful and important number in science and in describing the principles behind OCT. It is used extensively throughout this text predominately because it simplifies so many mathematical manipulations. In a large number of cases, it is used to replace the sine and cosine functions because in some cases it is easier for describing repetitive functions. While not particularly relevant to OCT physics, the actual value of e to eight significant digits is 2.7182818.

There are two reasons why e is useful to our study of OCT. First, it can be used to describe exponential growth or decay. In the formula, $y = e^{ax}$, if a is positive, it is exponential growth; if a is negative, it is exponential decay. As light penetrates through tissue, in general, it roughly undergoes exponential decay that needs to be compensated for in OCT image processing. Figure 1.3 demonstrates an example of exponential growth.

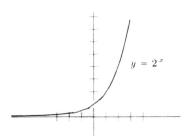

Figure 1.3   This demonstrates an example of exponential growth.

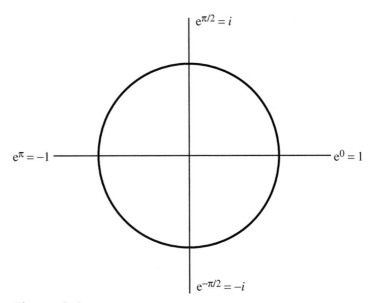

Figure 1.4 The second use of e, that is probably the most important with respect to this text, is its application in conjunction with imaginary numbers or more specifically the function $\boldsymbol{e}^{ix}$. This is a function that, like the sine and cosine, is cyclical in nature. Using radians for the value of x, $\boldsymbol{e}^{\pi i} = -1$, $\boldsymbol{e}^0 = \boldsymbol{e}^{2\pi i} = 1$, and $\boldsymbol{e}^{\pm i\pi/2} = \pm i$. This function can be plotted in polar coordinates as in Figure 1.4, so this again is a cyclic function and can be used similarly to the sine and cosine functions.

The second use of e, that is probably the most important with respect to this text, is its application in conjunction with imaginary numbers or more specifically the function e^{ix}. This is a function that, like the sine and cosine, is cyclical in nature. Using radians for the value of x, $e^{\pi i} = -1$, $e^0 = e^{2\pi i} = 1$, and $e^{\pm i\pi/2} = \pm i$. This function can be plotted in polar coordinates as in Figure 1.4, so this again is a cyclic function and can be used similarly to the sine and cosine functions. An example of this ease of manipulation is addition, where $e^{(Z1+Z2)} = e^{Z1}e^{Z2}$. A similar manipulation with trigonometric functions would be significantly more difficult. Not surprisingly, since the function e^{ix} is cyclical it can be related to the sine and cosine functions. One way they are related is through Euler's formula, $e^{i\Theta} = \cos\Theta + \sin\Theta$. This relationship is derived in Appendix 1-1. This appendix utilizes a power series, discussed in the next section, so that the reader may wait before reading Appendix 1-1. Through algebraic manipulation, Euler's equation may be rewritten in terms of the $\cos\Theta$ and $\sin\Theta$:

$$\cos\Theta = \frac{(e^{i\Theta} + e^{-i\Theta})}{2} \quad \sin\Theta = \frac{(e^{i\Theta} - e^{-i\Theta})}{2i} \tag{1.2}$$

1.5 INFINITE SERIES

Infinite series are frequently used in the physical sciences as well as pure mathematics and are literally the summation of an infinite number of terms. They can be used to evaluate functions including Fourier series, integrals, and differential equations. They have the general form of:

$$\sum_{n=0}^{\infty} c_n x = f(x)$$

A power series is perhaps the most important infinite series. With this series, the power sequentially increases with each variable and the equation has the general form:

$$\sum c_n x^n = c_0 + c_1 x + c_2 x^2 + c_3 x^3 + \cdots = f(x)$$

A power series may converge (approach a specific value) or diverge with any value of x. The domain of the function $f(x)$ is all x.

Example:

If $C_n = 1$.

$$\sum x^n = 1 + x + x^2 + x^3 + \cdots = 1/(1-x) = f(x)$$

This function converges when $-1 < x < 1$ and diverges when $|x| \geq 1$. This allows the $f(x)$ to be described in terms of a power series.

More generally, a power series has the form:

$$\sum c_n(x-a)^n = c_0 + c_1(x-a)^1 + c_2(x-a)^2 + c_3(x-a)^3 + \cdots$$

This is a power series in $(x-a)$ or a power series about a. Here, the power series specifies the value of a function at one point, x, in terms of the value of the function relative to a reference point. It is essentially the expansion in powers of the change in a variable ($\Delta x = x - a$). When $x = a$, all terms are equal to zero for $n \geq 1$ and so the power series always converges when $x = a$.

In many cases, the power series does not completely describe the function and the remainder term is required such that:

$$f(x) = P^n(x) + R_{N+1}(x)$$

This allows power series representation to be developed for a wide range of functions, where $P_n(x)$ is the power series and $R_{N+1}(x)$ is the remainder term. The remainder term may be represented by:

$$R_{N+1}(x) = (f^{n+1}(c)x^{n+1})/(n+1!)$$

where c has a value between 0 and x.

A Taylor series expansion, a type of power series, is very useful for defining and manipulating many functions. If f has a power series representation (expansion) at a, that is, if

$$f(x) = \sum c_n(x-a)^n \quad |x-a| < R$$

then its coefficients can be found by the formula:

$$c_n = f^{(n)}(a)/n!$$

Here $f^{(n)}(a)$ represents the nth derivative. The power series expansion can now be rewritten as:

$$f(x) = \sum \{f^{(n)}(a)/n!\}(x-a)^n = f(a) + \{f^1(a)/1!\}(x-a)^1 + \{f^2(a)/2!\}(x-a)^2$$
$$+ \{f^3(a)/3!\}(x-a)^3 + \cdots$$

This is the Taylor series expansion of the function f centered about a. Each coefficient can then be determined when each derivative has a value of $x = a$ or $f^{(n)}(0)$. This is best illustrated with the Taylor series expansion of e^x. The value of a here is equal to 0 for simplicity. Then:

$$f(0) = e^0 = 1$$
$$f^1(0) = xe^0 = x$$
$$f^2(0) = x^2e^0 = x^2 \ldots \text{etc.}$$

Then, substituting these coefficients we obtain:

$$e^x = 1 + x + x^2/2! + \cdots + x^{n-1}/(n-1)!$$

A wide range of functions can be represented by the power series.

Calculus

A large percentage of the mathematics used in this textbook involve integration or differentiation, which is the basis of calculus. The next two sections will try to give the nonphysical scientist a feel for the heart of calculus and the operations of differentiation and integration.

1.6　DIFFERENTIATION

Derivatives are very important in physics and in the understanding of the physical principles behind OCT. A derivative, for the purpose of this discussion, is a mathematical device for measuring the rate of change. But before going directly to the definition of a derivative, the concepts of a function and an incremental change must be discussed.

The term function will be used frequently in this text. In a relatively simple sense, a function can be described as a formula that tells you: If this is the value of one variable, how will the value of another variable be affected? In the following example, the variables x and y will represent distances.

$$y = 2x \tag{1.3}$$

For this function, when x is 2 meters (m), then it tells us that the position in the y direction is 4 m. When x is 3 m, y is 6 m. Throughout the text, we will face more complex functions, but the overall principles behind a function are the same.

A function does not need to contain single numbers but may also contain variables of change, including incremental change. By incremental change, we mean a change in a variable, such as position in the x direction (Δx), which is very small. So now, our function in Eq. 1.3 is looking at a change in each variable rather than identifying a specific position:

$$\Delta y = 2\Delta x \qquad (1.4)$$

So if the position in the x direction is changed by 3 m, the position in the y is changed by 6 m. Now, we will define the concept of the average rate of change of a variable, using here the change in the position (x) with time (t), which is expressed as:

$$\Delta x / \Delta t = \text{change in } x / \text{change in } t$$

Speed, a relatively familiar concept, is an example of a measure of rate of change. It is a measure of the change in distance with respect to time. So if an airplane has traveled 100 miles in 1 hour, the rate of change (Δ) or speed is 100 miles/hr.

The derivative is a special measure of rate of change. It is actually just the **instantaneous** (rather than incremental) rate of change of a variable over an infinitely small increment, rather than over a finite variable such as a minute.

Knowing that a derivative is related to an infinitely small rate of change, to more easily understand the derivative, the concept of limits is introduced. As an example such as $y = 1/x$, the function approaches a limit when x approaches infinity (α). The limit, the value of y, is 0 because $1/\alpha$ is zero.

This process can be generalized. If we use change in position (Δx) with respect to time as our variables, the limit as the time goes to zero (Δt) is:

$$\lim_{\Delta t \to 0} \Delta x / \Delta t = dx/dt$$

The term dx/dt is referred to as the derivative. This is more clearly explained in Figure 1.5. The value for the incremental or average change in region A is an average rate of change in that region. If we make A smaller and smaller until it becomes the infinitely small point B, we now have the instantaneous rate of change at that point or the derivative. The derivative can be expressed in a variety of ways, such as $df(x)/dx$, y', $f'(x)$, and dy/dx.

One of the difficult aspects of the derivative is that there are a multitude of rules for taking the derivative of functions. However, since the reader is not required to derive anything in this book, these rules will not be covered and can be found in any calculus textbook. We will, though, give an example of a derivative of a function for illustrative purposes.

In this example function, an object is moving in such a way that the relationship between the time the object travels (t) and position of the object (y) is:

$$y = t^2 \qquad (1.5)$$

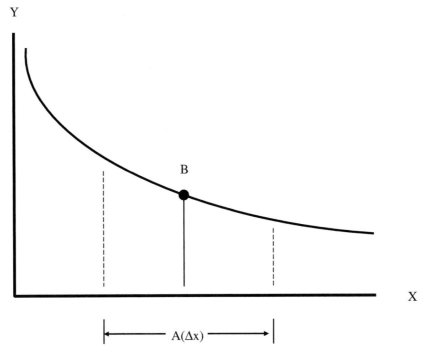

Figure 1.5 The term *dx/dt* is referred to as the derivative. This can be more clearly explained here. The value for the incremental change in region A is an average rate of change in that section. If we make A smaller and smaller until it becomes the infinitely small point B, we will have the instantaneous rate of change at that point or the derivative.

Then the velocity (v) or **rate of change** (derivative) of position is:

$$dy/dt = 2t = v \qquad (1.6)$$

This result is a rule of calculus where in this example, the derivative of a square is 2 times that of the variable. Because the resulting velocity here is not constant, but changes with time (it is a function of t rather than a constant), the object is therefore accelerating or increasing its velocity (a positive value) with time.

The derivative can be taken more than once. Therefore, if a second derivative is obtained, what is produced is the rate of change of the original rate of change. Among the common ways or symbols of expressing the second order derivative are:

$$d^2y/dx^2, f''(x), y'', d^2f(x)/dx^2$$

Going back to Eq. 1.5 for our moving object and taking the second derivative, we get:

$$d^2y/dt^2 = 2 \qquad (1.7)$$

In this case, the rate of change of the velocity in Eq. 1.6 is the acceleration. Because there are no variables left, just the constant 2, the object is moving at constant acceleration (but the velocity is not constant). If the second derivative had contained a variable, then the acceleration would not have been constant.

Again, there are a range of rules for performing derivatives, but they will not be discussed here, just the general principles. However, because the exponential will be used extensively throughout the text, an interesting point can be brought up with respect to this function. The exponential is the only function whose derivative is itself:

$$y = e^x \text{ then } dy/dx = e^x \tag{1.8}$$

1.7   INTEGRATION

Integration, to some degree, can be viewed as the inverse of differentiation. Although the integrals we will deal with are generally of a class known as definite, a brief discussion of indefinite integrals is in order as a foundation for definite integrals. In the previous discussion of differentiation, it was noted that if $y = x^2$, then $dy/dx = 2x$. The indefinite integration of $2x$ is:

$$y = \int 2x\,dx = x^2 + C \tag{1.9}$$

When the conditions of integration, the region over which the integration is performed, are not defined (indefinite integral), we end up with the reverse of differentiation plus a constant C, so it is not exactly a reverse of differentiation.

When the conditions of integrations are defined, for example between the physical points a and b, it is referred to as a definite integral. This is shown in Figure 1.6 and results in the loss of the constant C:

$$y = \int_a^b 2x\,dx = [b^2 + C] - [a^2 + C] = b^2 - a^2 \tag{1.10}$$

This, as it turns out, is the area under the curve between a and b in the Figure 1.6.

In general, with single integration, the results represent the area under a curve or line. However, unlike differentiation where the results generally always represent a rate of change, the results of integration are not always area (or volume), as we will see in an example with vector calculus. For example, we will see it used for measuring work performed along a given length later in the text.

Just like multiple differentiation, multiple integrations can be performed on a given function. An example of a double integral is given by the formula:

$$z = \int_c^d \int_a^b 1/xy\,dxdy \tag{1.11}$$

This particular formula is measuring the volume under a surface rather than area under a curve.

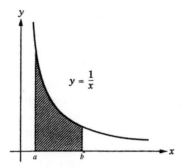

Figure 1.6 When the conditions of integrations are defined, for example between the physical points a and b, it is referred to as a definite integral. This is shown and results in the loss of the constant C, which was seen in the indefinite integral: $y = \int_b^a 2x\,dx = [b^2 + C] - [a^2 + C] = b^2 - a^2$.

1.8 DIFFERENTIAL EQUATIONS

The equations used in the text typically are more complex than these simple derivatives or integrals, but the principles are generally the same. Predominately, we are dealing with equations that contain differentials of multiple variables rather than single variables. They are referred to as differential equations. A differential equation is one that contains one or more derivatives, which will be more obvious in the following paragraphs. They are a very important class of equations that will be used extensively throughout the book and, for the most part, are used extensively throughout physics. A discussion on solving differential equations is well beyond the scope of this book, but this section is intended to give the reader a feel for these equations, making it easier to follow sections of this book. An example of a differential equation is the wave equation, which we will see again and which governs all kinds of physical phenomena including light waves, the motion of a spring, and the propagation of sound. This equation is:

$$\partial^2 y/\partial x^2 = (1/c^2)\partial^2 y/\partial t^2 \tag{1.12}$$

The equation contains two partial derivatives or different types of derivatives, $\partial^2 y/\partial x^2$ and $\partial^2 y/\partial t^2$. In other words, these are functions that are differentiated with respect to different variables, x or t, unlike the simple equations discussed up to this point. Classifying differential equations turns out to be very important in handling and understanding these equations. These equations are generally classified by three criteria: type, order, and degree.

The equation type is classified as either ordinary or partial. An ordinary equation contains only two variables, like x and y. A partial differential equation contains three or more variables. Most equations we will deal with are partial differential equations. Partial differential equations do not use the dx symbol but a ∂x symbol instead, which means that differentiation is occurring with respect to more than one variable. The wave equation shown above is an example of a partial differential equation, with derivatives both with respect to t and x.

The order of a differential equation means how many times the function is differentiated. The highest order term, say a second order differentiation, is used to classify the equation. In the wave equation, both terms are second order. This means that any given term is differentiated twice.

Some equations contain exponential terms such as squares or cubes. When describing the degree of a differential equation, we are defining it by the highest exponential term. If the equation contains a square as its highest exponential, then the degree is two. Because the wave equation has no exponential terms higher than 1, its degree is one.

An important term used in describing differential equations is the linearity. If the degree of each term, in other words its exponential, is either zero or one, the equation is said to be linear. A more detailed explanation of linearity is found in the references listed at the end of this chapter. The vast majority of equations dealt with in this book will be linear. This is an important point because nonlinear equations can be difficult, if not impossible, to solve.

The wave equation is again a good example for understanding how differential equations are used. Classical waves such as a sound wave in air or ripples on water can be represented mathematically. These waves in general must satisfy or be solutions of the wave equation. The process of satisfying the wave equation is best given by the equation for a standing or stationary wave on a string of fixed length [2]:

$$y_n = (A_n \cos w_n t + B_n W_n \cos w_n t) \operatorname{Sin} W_n x / c \qquad (1.13)$$

The actual meaning of the terms is not critical, but the coefficients in the formula are determined by the specific experimental setup (length of the string, tension, etc.).

Determining if Eq. 1.13 is a solution to the wave equation is as follows. First, we perform a second order differentiation with respect to $x (\partial^2 y / \partial x^2)$ (left side of the wave equation) on Eq. 1.13 and plug it into the wave equation. The second step is to perform second order differentiation with respect to time $(\partial^2 y / \partial t^2)$ and plug it into the right side of the wave equation. The two sides will be **equal** if Eq. 1.13 satisfies the wave equation. To study phenomena related to OCT, we frequently need to find solutions to the differential equations. The solutions will be stated, but from this section, hopefully the reader will now have an idea what satisfying a differential equation means.

1.9   VECTORS AND SCALARS

Throughout the text, the terms vector and scalar will be used. A scalar is a quantity that can be described with one number, the magnitude, such as speed, mass, or temperature. An example is a car with a speed of 5 miles per hour.

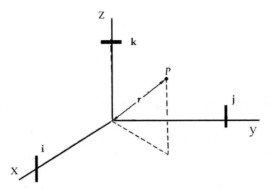

Figure 1.7 If we describe space in terms of *x*, *y*, and *z* components, then the velocity, a vector, has a magnitude, but a component or percentage of the total magnitude is found within each of the *x*, *y*, and *z* directions. This is shown here.

A vector is slightly more complex and requires defining both the magnitude and an addition quantity, usually direction. An example of a vector is velocity, as opposed to speed that is a scalar quantity. If we describe space in terms of *x*, *y*, and *z* components, then the velocity has a magnitude, but a component or percentage of the total magnitude is found within each of the *x*, *y*, and *z* directions. This is shown in Figure 1.7. If the particle is moving at 10 m/sec at a 45° angle in the *x*−*y* plane, its velocity in the *z* direction (v_z) is zero, while its velocity in the *x* and *y* directions will be 7.07 m/sec (using Pythagorean's theorem). If the particle is moving purely in the *x* direction, its velocities in the *y* and *z* directions will be zero, while its velocity in the *x* direction will be 10 m/sec. So once again, a vector does not just define a magnitude, but something in addition to the magnitude. In this case, it is the distribution in space of the magnitude.

1.10 UNIT VECTOR

An important tool for handling vectors is the unit vector. A unit vector is a vector that has unit magnitude or a value of one on each axis. This is easier to explain with the rectangular unit vectors *i*, *j*, and *k*. In Figure 1.7, we see these unit vectors plotted where each have a value of one. If we want to represent a given vector **F**, we can do it as a multiple of the individual unit vectors. Therefore:

$$\mathbf{F} = a_1 i + a_2 j + a_3 k \tag{1.14}$$

Here the *a*'s in front of the unit vectors are scalar terms by which the unit vector is multiplied. Together, they fully describe the vector. The value of the use of unit vectors may not be obvious at this point, but it will be seen that this formalism allows vectors in many instances to be more easily manipulated. The magnitude of vector **F** is given by:

$$\mathbf{F} = |F| = \sqrt{a_i^2 + a_2^2 + a_3^2} \tag{1.15}$$

1.11 SCALAR AND VECTOR FIELDS

A field is an abstract concept that is easier to explain with an example. If we plot the temperature over say the United States, at any given time we are assigning a number value (i.e., temperature) for each location. We are therefore plotting temperature as a function of its x and y position. This type of field is a special type known as a scalar field because at any given time, the field only has one value at any given location. Now, if we plotted the velocity of snow in an avalanche, each point has velocity, a vector quantity. In other words, at any given time, each point in the avalanche has defined both magnitude and direction. This type of field is known as a vector field. We will learn later that electric and magnetic fields are vector fields.

1.12 MATRICES AND LINEAR ALGEBRA

Again, linear equations are special equations that are dealt with throughout this text. As stated, these equations contain no squares or higher powers (for example, x^3). These equations have two important properties. First, the addition of two linear equations results in a new equation that is linear. Second, when each term is multiplied by a constant, the final sum is a multiple of the same constant:

$$z = x + y \text{ then}$$
$$az = ax + ay \tag{1.16}$$

Very often we will be dealing with several of these linear equations when we analyze a system. An example is the equations that describe how light propagates through a series of lenses (or mirror or grating), assuming the components are linear. As light propagates through the series of lenses, each lens can be represented by a given linear equation. Matrices allow these multiple equations to be analyzed all at once. If, for example, we have the following linear equations:

$$a = bx + cy$$
$$d = ex + fy \tag{1.17}$$

then in matrix form, the equations can be represented by:

$$\begin{vmatrix} a \\ b \end{vmatrix} = \begin{vmatrix} bc \\ ef \end{vmatrix} \begin{vmatrix} x \\ y \end{vmatrix} \tag{1.18}$$

As with integration and differentiation, the rules that guide the handling of these matrices are well described, but will not be discussed here. These matrix equations will be seen intermittently throughout the text, but the mechanisms by which they are solved is not necessary to understand. Just an understanding as to what a matrix represents should be sufficient.

1.13 WAVES

Light has an inherent wave nature, as discussed. Therefore, it is necessary to discuss waves to understand even the simplest properties of light. Waves are introduced here rather than the next chapter, even though this chapter focuses on mathematics. This is because an understanding of waves is needed to complete our introduction of the relevant OCT mathematics, particularly the concept of Fourier mathematics. One way to look at a classical wave is to view it as a disturbance which is self-sustained and can move through a medium, transmitting energy and momentum. Although this definition is sufficient for this book, the reader should understand that it is a simplified definition that is not applicable to all situations, particularly quantum mechanics. In the next chapter, we will learn that in a vacuum distant from the source, light is a transverse electromagnetic wave. As stated previously, transverse wave means that the medium (or in the case of light, the electromagnetic field) is displaced perpendicular to the direction of motion. Sound in air, on the other hand, is a longitudinal wave. This means the displacement is in the direction of motion as shown in Figure 1.1C.

An important point is that waves are not limited to pure sinusoidal or harmonic functions. A pulse can be considered a wave although we will see that it too can be represented as the summation of specific pure sinusoidal functions.

A one-dimension wave can be described by the formula $\psi(x, t) = F(x - vt)$, where ψ is a common symbol for a wave function, t is time, and v is velocity. $F(x - vt)$ is a general function that tells us the shape of the wave while x may represent the position in one dimension. A common example function of ψ, which is important in OCT imaging, is a Gaussian or bell-shape function $\psi(x, t) = e^{-(x-vt)2}$. This bell-shape disturbance is propagating at a velocity v.

To begin initially examining waves quantitatively, we will choose to start with sinusoidal or simple harmonic functions. For now, we will begin with the sine function. Amplitude, phase, frequency, velocity, and position need to be defined for the wave function, but these components will be introduced separately. The initial wave function will only be a function of position and will have the form:

$$\psi(x) = A \operatorname{Sin} kx \tag{1.19}$$

This function satisfies the wave equation. In the equation, A is the amplitude (the maximum) and k is the propagation constant. The propagation constant is related to the wavelength and its units are such that the product of kx has values of either degrees or radians. The spatial period is known as the wavelength and one complete cycle is 2π. Therefore, the propagation constant must have a value such that

$$k\lambda = 2\pi \quad \text{or} \quad k = 2\pi/\lambda \tag{1.20}$$

The wave is periodic in space. It should be noted that the wave we are describing currently is monochromatic, meaning that it consists of one wavelength and frequency. Soon, we will be introducing the mathematics to describe waves of multiple wavelengths. In reality, there are essentially no true monochromatic waves in the physical world. Even light from a laser contains a small range of wavelengths and this type of light generally is referred to as quasimonochromatic. The particular wave in Eq. 1.19 is a standing or stationary wave; it is not moving. To introduce the concept of movement, the equation is modified to include velocity:

$$\psi(x, t) = A \operatorname{Sin} k(x + vt) \tag{1.21}$$

So this wave is periodic in space and in time. The temporal period (T), the time for a complete cycle to pass a certain point, is described by:

$$kvT = 2\pi \quad \text{or} \quad T = \lambda/v \tag{1.22}$$

The equation can also be modified to use a term that is often more convenient, the angular frequency (ω). This is the number of radians per unit time and is defined as $2\pi/T$. It is the velocity multiplied by the propagation constant. To incorporate ω, the wave function can be modified to:

$$\psi(x, t) = A \operatorname{Sin} (kx + \omega t) \tag{1.23}$$

Equations 1.21 and 1.23 are of course equivalent. In the description of waves so far, we have ignored the possibility of a phase shift or taken into account where the wave begins. When talking about sine and cosine functions, it was stated that they were 90 degrees out of phase. This means their crests and troughs do not overlap. From Figure 1.1A, the sine wave is zero at position zero. This does not need to be the case if we introduce a phase factor, as a shift in the phase of the wave in either space or time can be introduced. The equation that includes phase is:

$$\psi(x, t) = A \operatorname{Sin} (kx + \omega t + \varepsilon) \tag{1.24}$$

If our phase factor is $-90°$ or $\pi/2$, the function will be equal to a cosine function, consistent with the definitions described above.

$$\psi(x, t) = A \operatorname{Sin} (kx + \omega t - \pi 2) = A \operatorname{Cos} (kx + \omega t) \tag{1.25}$$

We have now introduced terms into the sine function to control the cyclical waves place as a function of time or space and velocity, in addition to taking into account any phase shift. All the terms in the parentheses now give the position of the wave in the cycle at a specific location and time, which we will describe as the phase factor ($\varphi(x, t)$) (not to be confused with the phase).

$$\varphi(x, t) = kx + \omega t + \varepsilon \tag{1.26}$$

Now that we have defined the properties of a monochromatic wave, before we move on to polychromatic waves (waves of multiple wavelengths), several different derivatives will be introduced. The values of these derivatives for OCT will become more obvious as we move to polychromatic waves, because rate of change of some variables has significance to the technology. If we take the derivative of the phase factor ($\varphi(x, t)$) with respect to time at constant position, as we learned in the derivative section, we get:

$$|(\partial \varphi / \partial t)_x| = \omega \qquad (1.27)$$

A derivative, as stated, is a measure of the rate of change. This is the rate of change of phase factor with time at fixed position, which is the angular momentum (ω). Similarly, if we take the derivative with respect to position at fixed time, we get:

$$|(\partial \varphi / \partial x)_t| = k \qquad (1.28)$$

Therefore, the rate of change of phase factor with position is the propagation constant. With these two derivatives, a very important parameter can be defined, the phase velocity. Velocity is the change of position with time ($\partial x / \partial t)_\varepsilon$. From this, using straightforward algebra, the phase velocity is derived:

$$(\partial x / \partial t)_\varepsilon = (\partial \varphi / \partial t)_x / (\partial \varphi / \partial x)_t = +\omega / k = +v \qquad (1.29)$$

This is the speed of propagation of constant phase factor (for example, a peak) or the phase velocity, and is the angular velocity divided by the propagation constant.

Throughout the text, our representation of waves will not be limited to trigonometric functions. The exponential function, as stated, can also be used as a simple harmonic function taking into account position, velocity, wavelength, and initial phase. The complex representation, using an exponential function, is often easier to handle mathematically. The complex exponential representation can have the form of:

$$\psi(x, t) = e^{-i(kx + \omega t + \varepsilon)} \qquad (1.30)$$

It should be noted that so far the monochromatic waves described have been one dimensional. However, most waves are really three dimensional, and we will represent them with the formula:

$$\psi(\mathbf{r}, t) = A \, \mathrm{Sin} \, (\mathbf{\kappa} \cdot \mathbf{r} + \omega t + \varepsilon) \qquad (1.31)$$

where $\mathbf{r}$ now replaces x and represents the vector for the x, y, and z components. Therefore, we use $\mathbf{r}$ to represent a vector that takes into account the three different spatial components of the wave. A more detailed understanding of $\mathbf{r}$ is not necessary for the purposes of this text. The $\mathbf{r}$ is in bold because this is a standard representation for vectors (containing the three spatial components).

1.14 COMBINING WAVES OF TWO WAVELENGTHS

To this point, we have only been discussing monochromatic waves. All light contains more than one wavelength and is therefore polychromatic. OCT imaging uses a very wide range of wavelengths and is therefore often referred to as broadband light, a type of polychromatic wave. To study polychromatic waves, the discussion will begin by looking at light containing two different frequencies. Because the waves are linear, the waves can be added together and any combination of waves will be both linear and another solution to the wave equation. This is known as the principle of superposition.

We begin with two waves with different frequencies but the same amplitude and phase. The two waves are described by the wave functions $\psi_1(x, t)$ and $\psi_2(x, t)$, which have the formulas and summation described by:

$$\psi_1(x, t) = A \cos (k_1 x + \omega_1 t + \varepsilon_1) \tag{1.32}$$

$$\psi_2(x, t) = A \cos (k_2 x + \omega_2 t + \varepsilon_1) \tag{1.33}$$

$$\psi(x, t) = \psi_1(x, t) + \psi_2(x, t) = A \cos(k_1 x + \omega_1 t) + A \cos (k_2 x + \omega_2 t) \tag{1.34}$$

The cosine representation is used here rather than the sine term for convenience because it allows the use of a simple trigonometric entity. The rule is that:

$$\cos \Theta + \cos \alpha = 2 \cos \frac{(\Theta - \alpha)}{2} \cos \frac{(\Theta + \alpha)}{2} \tag{1.35}$$

Applying this rule to Eq. 1.34, we get:

$$\psi(x, t) = 2A \cos[(\omega_1 - \omega_2)t/2 - (k_1 - k_2)x/2] \\ \cos[(\omega_1 + \omega_2) t/2 - (k_1 + k_2)x/2] \tag{1.36}$$

Rearranging the equation in this fashion gives us a better understanding as to what happens when the two waves are combined. We will begin with the assumption that the frequencies of the two waves are similar but not exactly the same. The reason for selecting waves of similar frequencies will be discussed later. The combined wave function in Eq. 1.36 consists of two components, a slowly varying envelope, which is the first cosine term, and a rapidly varying component, the second cosine term. A plot of this function is shown in Figure 1.8. What the first term represents is amplitude modulation or modulation of the total intensity. It is the creation of beats (group) or an amplitude envelope. It is created because the crests and troughs of the waves at two different frequencies are adding at some points and canceling out at other points.

In addition to the slow amplitude modulation, there is a rapid oscillation in the signal, which is the second term in Eq. 1.36. Both the slow oscillation and the finer vibrations are seen in Figure 1.8.

Ultimately, with OCT, we will be concerned about the velocity of both the envelope (group velocity or beat velocity) and the velocity of the fine vibrations (phase velocities).

The phase velocity for the combination of two waves is:

$$v_p = \varpi/\mathbf{k} = [1/2](\omega_1 + \omega_2)/[1/2](k_1 + k_2) = \frac{(\omega_1 + \omega_2)}{(k_1 + k_2)} \tag{1.37}$$

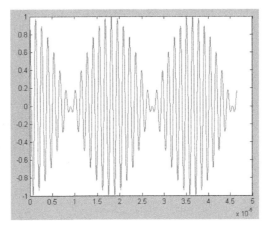

Figure 1.8 In addition to the slow amplitude modulation, there is a rapid oscillation in the signal that is the second term in Eq. 1.36. Both the slow oscillation "envelope" and the finer vibrations are seen here.

When the frequencies are almost equal, $w_1/k_1 \approx \omega_2/k_2 \approx v_p$. Therefore, through straightforward algebra:

$$\omega_1 + \omega_2/k_1 + k_2 = v_p(k_1 + k_2)/k_1 + k_2 = v_p \qquad (1.38)$$

The group velocity is given by:

$$v_g = 1/2(\omega_1 - \omega_2)/1/2(k_1 - k_2) = (\omega_1 - \omega_2)/(k_1 - k_2) = \frac{\Delta\omega}{\Delta k} \qquad (1.39)$$

The group velocity can be thought of as the velocity of the maximum amplitude of the group, so that it is the velocity with which the energy in the group is transmitted. The group velocity may be faster or slower than the phase velocity, which is a topic covered in Chapter 5.

1.15 FOURIER SERIES AND INTEGRALS

Now that the concept of polychromatic waves has been introduced, we will discuss the mathematics used to handle them, specifically Fourier mathematics. This is probably the most important area of mathematics used to describe the features of OCT.

In the previous section, we dealt with the addition of waves of two different frequencies that resulted in the formation of an AC component. To handle waves of many different frequencies and amplitudes, the analysis is not quite as simple and Fourier techniques are needed. This is because the crests and troughs add in a complex way. We begin with the example function, which is seen in Figure 1.9 and represents an infinite series of square pulses:

$$\psi(x) = A(\sin x - 1/3 \sin 3x + 1/5 \sin 5x - 1/7 \sin 7x \ldots) \qquad (1.40)$$

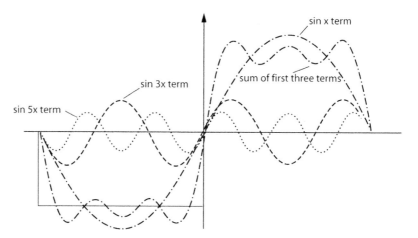

Figure 1.9 This figure demonstrates how adding waves of different frequencies individually progressively results in a function that more closely resembles the series of square pulses. The square pulses are the result of adding an infinite number of terms. It is appropriate to describe these results as a Fourier relationship between the amplitude distribution in the time domain and the frequency distribution in the frequency domain (Eq. 1.40).

For this function, we are adding different sinusoidal functions with different wavelengths and different weighting factors in front of the sine term to produce the pulse series. Figure 1.9 demonstrates how adding the factors individually progressively results in a function that more closely resembles the series of square pulses. Therefore, Eq. 1.40 can be viewed as consisting of a spectrum or series of different frequencies, which when added together form a specific pattern or function. In the case of Eq. 1.40, a repeating pattern of pulses is produced.

One way to describe these results is that a Fourier relationship exists between the amplitude distribution in the time domain (Figure 1.9) and the frequency distribution in the frequency domain (Eq. 1.40). This is actually a Fourier series that is discussed in detail in the next section. Fourier mathematics is describing how multiple frequencies at different amplitudes add up to give specific intensity shapes, again like a series of pulses.

Two different types of Fourier approaches exist, the Fourier series and the Fourier integral. Fourier series are used with periodic functions like the one in Figure 1.10. Fourier integrals are used with non-periodic functions, like a single pulse. Again, Fourier relationships are central to the understanding of the principles behind OCT, so we will discuss both.

1.15.1 Fourier Series

A statement of and the formula for a Fourier series (more fundamental than Eq. 1.40) will be presented but not proved. The statement is that virtually all periodic

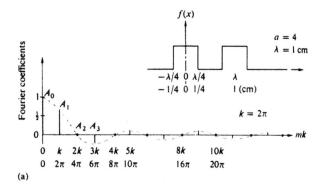

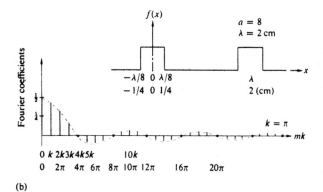

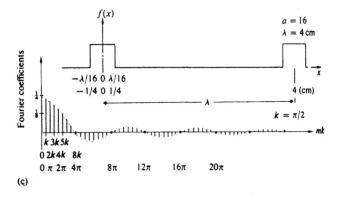

Figure 1.10 This figure demonstrates a series of square pulses and the corresponding frequency distribution in the Fourier domain. We see that as the frequencies within the beam get closer and closer, the square pulses get farther and farther apart. Therefore, if the frequencies summed up were spaced an infinitely small distance apart (i.e., no detectable separation), then we would ultimately get a single pulse. Of course, summing over an infinitely small distance is integration and is the basis of the Fourier integral. Courtesy of *Hecht Optics*, Fourth Edition, Figure 7.21, page 304 [2].

functions, like the one in Figure 1.9, for the purposes of this text, can be represented by the Fourier series:

$$\psi(x) = a_0/2 + a_n \cos nkx + b_n \sin nkx \qquad (1.41)$$

Here a_0, a_n, and b_n are coefficients which weigh the individual terms in the series that vary depending on the periodic function. The coefficients are appropriately weighed to produce the periodic function, some of which may have a value of zero. We see that Eq. 1.40 is a version of Eq. 1.41, where the value is zero for $a_0/2$, all the coefficients of the cosine terms, and every other coefficient of the sine terms.

In Figure 1.10, we see a series of square pulses and the corresponding frequency distribution. We see that as the frequencies within the beam get closer and closer, the square pulses get farther and farther apart. Therefore, if the frequencies summed were spaced an infinitely small distance apart (i.e., no detectable separation), then we would ultimately get a single pulse. Of course, summing over an infinitely small distance is integration and is the basis of the Fourier integral. Therefore, to describe in the frequency domain a non-repeating structure, such as a Gaussian pulse, integration is required.

1.15.2   Fourier Integral

The Fourier integral can be represented in a variety of ways. If you are looking at a pulse energy amplitude as a function of time, such as a femtosecond pulse, this is called analyzing in the time domain. If you are looking at the frequencies that add up to make the pulse, it is stated that you are analyzing in the frequency domain. Again, Fourier integrals will allow us to relate the frequency and time domains. The actual Fourier integrals can be represented in a range of ways. Different texts represent the integral in other ways, which are all equivalent. The representation used here, which will not be proved, is:

$$f(t) = \int_{-\infty}^{\infty} F(v)e^{i2}$$

\pivtdt

$$F(v) = \int_{-\infty}^{\infty} f(t)e^{-i2}$$

\pivtdt

$F(v)$ is the frequency distribution (i.e., amplitude of different frequencies) of a non-periodic function (for example, Gaussian pulse), whereas $f(t)$ is the time distribution of the amplitude of a non-periodic function or shape. In the first equation, $f(t)$ is referred to as the Fourier transform of $F(v)$. In the second equation, $F(v)$ is referred to as the inverse Fourier transform of $f(t)$. What the equation is saying is that the distribution in time of a non-periodic function $f(t)$ can be found from the integral of the various frequencies multiplied by an exponential. The second equation is basically the reverse process. Again, it should be reinforced that Eqs. 1.42 and 1.43 are written in slightly different ways in different textbooks and sometimes even in this text. This is because the π term and negative sign can be placed in a range of positions. All work equally as well, as long as the changes made in the two equations are consistent.

Why is Fourier math so important in understanding OCT? As an example, the sources used with OCT need to be broadband or contain a large number of frequencies. The reason is that the size of an appropriately shaped frequency distribution is proportional to the resolution of the system. This will be covered later in much greater detail.

1.16 DIRAC DELTA FUNCTION

An important function that we will be using throughout the text is the Dirac delta function. At first glance it will seem like a trivial function, but its significance becomes more obvious as it is actually put into use. The delta function is defined as:

$$\delta(a - a_0) = 0 \quad \text{when } a \neq a_0 \tag{1.44}$$

So this is a function that has a nonzero value only at one value, a_0. This will turn out to be a very useful mathematical device. The value of its area at a_0 is 1, while its height is infinite. One of its greatest features is the sifting property, where it is a method for assigning a number to a function at given points.

$$\int_{-\infty}^{\infty} f(a)\delta(a - a_0) = f(a_0) \tag{1.45}$$

This is important, for example, in the process of digitizing or dividing continuous data into segments. When a continuous signal is digitized, it is sampled or broken up at regular intervals. The mathematical procedure for doing this utilizes a series of Dirac delta functions.

1.17 MISCELLANEOUS BASIC PHYSICS

Before concluding this section, in addition to reviewing the relevant mathematics, some basic terms used in physics need to be defined including momentum, force, energy, power, work, and decibels. Momentum is the mass of an object times the velocity. Force, which has units here of Newtons, is defined in two ways. One is the change in momentum per unit time. In other words, an object's momentum remains constant unless acted on by a force. The second definition is that it is mass times the acceleration. They are equivalent definitions. Work can be defined as the force times the distance over which something is moved. Power is the change in work per unit time or how fast work is being done. Its units are watts or joules/second. So power is the rate at which work is being performed.

Energy is a little more difficult to define than power, work, or momentum and is one of the few things that is easier to envision on a quantum level than a classical one. The units for energy are joules or Newtons times meters. While momentum, for example, is a measurement of a changeable parameter, energy, like mass, is something that simply exists in the universe. Energy is something, in combination with mass (equivalent to energy through $e = mc^2$), which is conserved so that the total sum of mass and energy for the universe never changes even though it may change for a given system. Particles and fields possess energy (and can gain or lose it) in the form of either stored (potential) or active (kinetic) energy. The most concrete explanation for energy is that energy is what changes when work is performed on a system. (It should be noted that the concept that the mass and energy of the universe is constant is a belief but when taking into account the entire universe, this may not be completely true. Some theories describing the universe actually suggest it may not be constant. However, in our everyday world, it can be considered constant.) In subsequent chapters we will be concerned about the irradiance, which is the rate of flow of energy and is proportional to the square of the amplitude of the electric field.

With OCT, the maximal and minimal signal intensity measured is very large so that the decibel system, a method of reducing the number of digits used to describe data, is used for quantitating signal intensity. For an optical field, a decibel is:

$$dB = 10 \log (I/I_0) \tag{46}$$

Here, I_0 is the minimal signal detected by the system. For those who do not remember a log (base 10) from high school, it is the power 10 needs to be raised in order to get a certain number. So the log of 10 is 1, the log of 100 is 2. With OCT, the range of signal that can be detected by OCT is approximately 110 dB. This means that I is 10^{11} higher than I_o. This will be dealt with in more detail in Chapter 7.

REFERENCES

1. Goodman, J. W. (1996). *Introduction to Fourier Optics*, 2nd edition. McGraw-Hill, New York.
2. Hecht, E. (1998). *Hecht Optics*. Addison-Wesley Publishing Co., Reading, MA.
3. Maor, E. (1994). *e: The Story of a Number*. Princeton University Press, Princeton, NJ.
4. Pain, H. J. (1993). *The Physics of Vibrations and Waves*. Wiley, New York.
5. Thomas, G. B., Jr. (1972). *Calculus and Analytic Geometry*. Addison-Wesley Publishing Co., Reading, MA.

APPENDIX 1-1

DERIVING EULER'S FORMULA

A power series will be used to prove the relationship between the imaginary exponential and the sine/cosine terms through Euler's formula. A power series is a special type of infinite series. An infinite series is a common device in science and mathematics and is literally a summation of an infinite number of terms. The power series has the form:

$$\sum c_i x^i = c_0 + c_1 x + c_2 x^2 + c_3 x^3 + \cdots + c_n x^n \ldots$$

Functions like cosine and sine can be broken down into a power series or infinite series of numbers that have a specific relationship to one another. The power series representation of e, the sine function, and the cosine function are well known and can be found in most calculus textbooks. The power series for e^x is:

$$e^x = 1 + x + x^2/2! + x^3/3! + x^4/4! \ldots$$

The symbol ! means factorial and can best be described by an example: the term $4! = 4 \times 3 \times 2 \times 1 = 24$. The power series for e^{ix} is:

$$e^{ix} = 1 + ix - x^2/2! - ix^3/3! + x^4/4! \ldots$$

Notice that not all terms are imaginary and the signs are no longer all positive. This is because, for example, $-x^2/2!$, the square of i is -1. For $-ix^3/3!$, the cube of i is $-i$.

Now the imaginary and real terms can be grouped as follows:

$$e^{ix} = (1 - x^2/2! + x^4/4! \ldots) + i(x - x^3/3! + x^5/5! \ldots)$$

The power series in the first parentheses is that for the cosine function while the one in the second parentheses is the sine function multiplied by i. Therefore:

$$e^{ix} = \cos x + i \sin x$$

2   LIGHT AND ELECTROMAGNETIC WAVES

Chapter Contents

2.1 GENERAL

The focus of this chapter is electromagnetic radiation (EM) and how it behaves. Discussed in this chapter are some general comments on the nature of light, EM spectrum, reflection, refraction, propagation through optical components (lens and optical fibers), vector calculus, Maxwell's equations, and optical energy propagation.

The nature of light is complex and has been the source of debate for centuries. Conflicting experimental evidence suggested that it was either a particle or a wave. The history of this debate can be found in many textbooks and will not be discussed here. It has now been well established that light behavior is both particle and wave-like in nature, depending on what is being measured. The particle or graininess that makes up light is the photon. The energy of the photon is proportional to the frequency, $E = hf$, where h is Planck's constant with a value of 6.63×10^{-34} J/sec. For the most part light, when large numbers of photons are involved, behaves usually in a wave-like manner and can be treated as a classical wave. The reason for this, which will not be expanded upon in detail, is that photons behave *statistically* in a wave-like manner. The photon nature of light usually needs to be taken into account when dealing with small numbers of photons (low intensities), which includes areas such as the noise limits and the physics of lasers. In spite of centuries of research, many questions remain regarding the nature of light and extensive research on this topic is still ongoing.

The electromagnetic character of light is critical to understanding its behavior. In 1860, James Clark Maxwell published his mathematical theory of electromagnetism, which unified most electric and magnetic phenomena. These formulas established a wave equation for the propagation of electromagnetic waves. The velocity of this wave was the same as that for light, subject to the limitations of experimental measurement, and he successfully predicted that light was a propagating electromagnetic wave. The wave equation will be derived in Appendix 2-1 using Maxwell's equations, which are discussed below. However, a reader without previous experience with Maxwell's equations should read the remainder of the chapter before attempting to address Appendix 2-1. Qualitatively, the nature of light can be seen in Figure 2.1, where it is an oscillating electromagnetic field. From Maxwell's equations, a changing magnetic field produces an electric field whereas a changing electric field produces a magnetic field. Therefore, the two fields can recreate each other in a vacuum indefinitely, which explains the ability of light to propagate in free space. The electric and magnetic fields are perpendicular to each other and also perpendicular to the direction of propagation of the light. From Maxwell's equations, the speed of light in a vacuum is constant at 4×10^8 m/sec.

2.2 ELECTROMAGNETIC SPECTRUM

For an electromagnetic wave in a vacuum, the frequency and wavelength are related to the velocity via the formula $c = \lambda f$. It is through the frequency (or wavelength) that

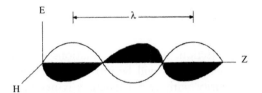

Figure 2.1 Qualitatively, the nature of light can be seen here, where it is an oscillating electromagnetic field.

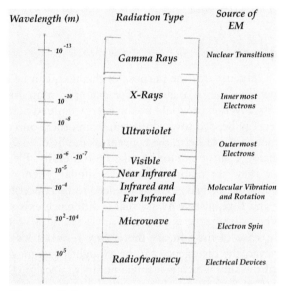

Figure 2.2 The electromagnetic (EM) spectrum or the classification based on wavelength and example sources.

an electromagnetic wave is classified. The term light is generally used to mean an electromagnetic wave with frequencies in what is known as the visible region, though the definition is extended here to include the near-infrared region where OCT imaging is performed. In Figure 2.2, the electromagnetic spectrum or the classification based on frequency is shown. The high energy region of the spectrum, which is the region of high frequency, consists of gamma and X-rays. The visible region is generally where the wavelengths range from approximately 4×10^{-7} m to 7×10^7 m. OCT imaging in nontransparent tissue is usually performed in the near-infrared region (for nontransparent tissue) at 1300 nm or 1.3×10^{-6} m. The low energy side of the spectrum or the longer wavelengths consist of radio waves and microwaves.

2.3   VECTOR CALCULUS

In the next section, Maxwell's equations that describe most of the properties of electromagnetic waves will be introduced. However, to appreciate these equations some familiarity with vector calculus is necessary. As with previous sections, this section is only intended for the reader to get a feel for this branch of mathematics and Maxwell's equations, not a thorough understanding. The three specific procedures of vector calculus that will be discussed are the divergence (or flux), curl, and gradient. These are differential equations that can be stated in either differential or integral forms, with the differential form as the focus here because, in general, this form is easier to work with.

2.3.1   Divergence

The divergence, for the purpose of this text, can be envisioned as a measurement of the net flow out of a surface. For example, if an imaginary sphere were around the sun, the divergence would be the energy emitted across the sphere by the sun, minus any energy headed toward the sun from other sources. In the case of the electric and magnetic field of light, the concept of the field flowing is sufficient for the purposes of analogy, but it cannot be taken too literally. Electric fields actually do not flow, rather they are associated with a similar phenomenon known as flux. For the purposes of this text, there is no need to be concerned about the difference between flow and flux, just be aware that a difference exists.

When divergence is measured, it is taking a vector and changing it to a scalar. In the case of the sun, all the energy released has both magnitude and direction. However, the divergence can be thought of as measuring the net total energy flow out of the sun, which is a single number, a scalar. Divergence can be symbolized as either Div $\mathbf{F}$ or $\nabla \cdot \mathbf{F}$. The formula for divergence is:

$$\text{Div } \mathbf{F} = \nabla \cdot \mathbf{F} = \partial \mathbf{F}x/\partial x + \partial \mathbf{F}y/\partial y + \partial \mathbf{F}z/\partial z \tag{2.1}$$

In Eq. 2.1, by taking the partial derivative with respect to x of the x component of the sun function (output) $\mathbf{F}(x, y, z)$, and doing the same for the y and z components, then adding them together the divergence is obtained.

2.3.2   Curl

The curl is typically used in describing work. If a rock is lifted 5 feet off the ground against gravity, work is being done. The curl can be thought of as a measure of how much work is done to move the rock along that line (whether it is curved or straight). The formula, although more complex in appearance then the gradient or divergence, is simply measuring this effect. The differential formula for the curl is:

$$\text{Curl } \mathbf{F} = \nabla X \mathbf{F} = \mathbf{i}(\partial \mathbf{F}z/\partial y - \partial \mathbf{F}y/\partial z) + \mathbf{j}(\partial \mathbf{F}x/\partial z - \partial \mathbf{F}z/\partial x)$$
$$+ \mathbf{k}(\partial \mathbf{F}y/\partial x - \partial \mathbf{F}x/\partial y) \tag{2.2}$$

We introduced the concept of the vector formula containing i, j, and k in the previous chapter and know these numbers cannot be added together. Therefore, our results will be represented by three numbers or a vector rather than a scalar. The three letters represent the x, y, and z components, respectively. Therefore, the value of the vector in the x direction, for example, is found from $(\partial \mathbf{F}z/\partial y - \partial \mathbf{F}y/\partial z)$.

2.3.3 Gradient

Most books like to use temperature to explain a gradient, so that tradition will be kept here. We begin with a plot of temperature over a plane in space. This would be a three-dimensional plot where for each x and y position, a specific value of the temperature exists. The temperature across the plane is not constant so a rate of change can be measured between different positions on the plane ($\Delta T/\Delta x$ and $\Delta T/\Delta y$). If the rate of change over an infinitesimally small region were measured, these would be partial derivatives and expressed as $\partial T/\partial x$ and $\partial T/\partial y$. At points where there is no change, obviously the rate of change would be zero. The gradient here is the measure of rate of change of temperature in two dimensions. The gradient is taking a scalar (the temperature) and converting it into a vector (change in temperature with respect to coordinates). The formula for the gradient in three spatial dimensions is:

$$\text{Grad } \mathbf{F} = \nabla \mathbf{F} = i\partial \mathbf{F}/\partial x + j\partial \mathbf{F}/\partial y + k\partial \mathbf{F}/\partial z \qquad (2.3)$$

2.4 MAXWELL'S EQUATIONS

Light, as stated, is an electromagnetic phenomenon. To study light, therefore, some understanding of electromagnetics and Maxwell's equations, which are based on vector calculus, is important. With these equations, many of the properties of light can be explored, from dipole radiation to light's speed. However, before going to these equations, some basic principles of electrical and magnetic fields need to be explored.

The universe consists of many particles that carry an electrical charge. Some have positive charge while others carry a negative charge. Alike charges repel one another while charges opposite in sign attract. Therefore, forces exist between the charges and an assumption is made that these forces are transmitted through a field. This field, the electric field, generates different amounts of force on a charge at different locations. The electric field ($\mathbf{E}$) is defined as $\mathbf{E}/q_0 = \mathbf{F_e}$, where $\mathbf{F_e}$ is the force experienced by q_0, the charge. The terms are in bold again because they are vector and not scalar quantities, with components in the x, y, and z directions. Similarly, there exists a field $\mathbf{B}$, the magnetic flux density, which is created by moving charges. There is some confusion in the literature over the name of $\mathbf{B}$. $\mathbf{B}$ is the name most people are referring to when they use the phrase magnetic field. However, historically, the term $\mathbf{H}$, which is described below, is the actual term bearing the name magnetic field. This text will follow the historic interpretation of $\mathbf{B}$, which is the most common usage.

Moving charges experience this force **B** such that the total force on the charge is $\mathbf{F} = \mathbf{E}/q_0 + q_0\mathbf{v}\mathrm{X}\mathbf{B}$. The X in the formula refers to a procedure performed on vectors (multiplication) known as the cross product. The term **v** is the velocity. It is not critical to understand the exact meaning of a cross product, and, it is sufficient to consider it the way vectors are multiplied. The direction of the new vector produced by the cross product is perpendicular to the other two original vectors.

In a vacuum, the only fields needed are the **E** and **B** fields. But since OCT imaging is not performed in a vacuum, the displacement vector **D** and the magnetic field **H** are introduced. These quantities deal with electromagnetic fields in situations other than a vacuum. The displacement vector **D** is given by the equation:

$$\mathbf{D} = \varepsilon\mathbf{E} + \mathbf{P} \tag{2.4}$$

The term **P** is the polarization vector, and it results from the ability of molecules and atoms to polarize or generate a field of their own when exposed to the applied field. The term ε is the dielectric constant or relative permittivity, which is a measure of how a material is affected by the influence of being placed in an electric field. Similarly, the magnetic field is described by:

$$\mathbf{H} = \mathbf{B}/\mu - \mathbf{M} \tag{2.5}$$

The magnetization vector **M** is analogous to **P** and is typically very small in circumstances dealt with for the subjects of this text and therefore will typically be ignored. The term μ is the permeability and is analogous to ε, except it is the induced magnetic field.

With the above information, the extremely important series of equations known as Maxwell's equations will be introduced. Maxwell's equations consist of four equations, Faraday's law, Ampere's circuit law, Gauss's law of electric fields, and Gauss's law of magnetic fields. When these formulas are applied in this text, it is assumed that the material is uniform (ε and μ are the same throughout the medium) and isotropic (ε and μ do not depend on direction). However, they remain applicable even if this were not the case.

The units used in these equations are usually a rationalized MKS system. For example, the unit for charge is the Coulomb or ampere-second and not, say, the ampere-minute. However, because historically some areas of electromagnetics do not use the MKS system, such as the study of scattering, there will be cases where other units will be required, specifically Gaussian units. This will result in some change in the appearance of the equations, specifically the constants.

As stated, Maxwell's equations can be written in what is referred to as either a differential or an integral form. Although most introductory texts use the integral form of these equations, because most nonphysical scientists lack a background in vector calculus, the differential forms are used predominately in this text because they are easier to comprehend.

2.4.1 Faraday's Law

Faraday's law basically states that when a magnetic field varies in time, it will result in the production of an electric field. In the example in Figure 2.3, by moving

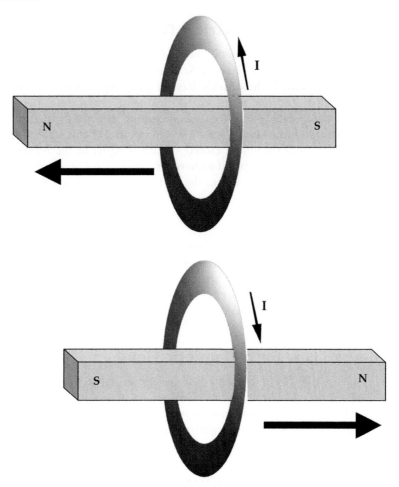

Figure 2.3 By moving a magnetic bar back and forth through a conducting loop of wire, electric current is produced in the wire, demonstrating Faraday's law. The current is I, north is N, and south is S.

a magnetic bar back and forth through a conducting loop of wire, electric current is produced in the wire. Mathematically, in the differential form, the equation that governs this phenomenon is:

$$\nabla x\, \mathbf{E} = -\partial \mathbf{B}/\partial t \qquad (2.6)$$

From the previous section on vector calculus, we know that the left side of the equation is the work necessary to move a charge along a line or a curl of the electric field. In the case of Figure 2.3, this is the work necessary to move electrons around the loop. From the previous chapter, the term on the right is a partial derivative **B** or rate of change of **B** with respect to time. So, once again, the equation is saying a changing **B** field results in an electric field, generating a force and resulting in a form of work.

2.4.2 Gauss's Law for Electrical Fields

To explain Gauss's law for both electrical and magnetic fields, it will be suggested that the electrical and magnetic fields flow. This is done for the purpose of analogy, since as stated, these fields have a property called flux rather than actual flow. However, for the purposes of this text, envisioning these fields as flowing is sufficient, since even flux, under current theories of quantum electrodynamics, is somewhat misleading. Gauss's law of electrical fields states that the net outward flux or flow of **D** through a closed surface is equal to the total charge enclosed by the surface. As an example, if a charge is at the center of an arbitrary sphere, the out flux through the sphere is due to the enclosed charge. This is described mathematically by the formula:

$$\nabla \cdot \mathbf{D} = \rho \tag{2.7}$$

From our discussion on vector calculus, the term on the left (divergence) represents flow/flux of **D** through the arbitrary surface and the term ρ on the right is the charge density (charge per unit volume). In the situations usually dealt with in this book, electromagnetic radiation propagating in space, with no free charges:

$$\nabla \cdot \mathbf{D} = 0 \tag{2.8}$$

2.4.3 Gauss's Law for a Magnetic Field

Gauss's law for a magnetic field is very similar to that of an electrical field, with one exception. Unlike an electrical charge, no magnetic charges have ever been identified. Magnetic fields come from moving electrical charges or fields. Therefore, there are no isolated positive or negative magnetic charges and the equation reads:

$$\nabla \cdot \mathbf{B} = 0 \tag{2.9}$$

This states that the divergence of the magnetic flux is always zero since there are no isolated magnetic changes.

As a side note for those interested, in a universe of electric charges but not magnetic charges, the magnetic field can be viewed as a relativistic correction to the electric field. That is why there is no relativistic correction per se in Maxwell's equations (i.e., they hold independent of the frame of reference). Therefore, if the speed of light was not finite or if all charges moved at relatively slow velocities, there would be no magnetic fields.

2.4.4 Ampere's Law

Ampere's law is in some ways the reverse of Faraday's law, but just slightly more complicated. The formula is stating that if an electric change is in motion or a changing electrical field is present, the result will be the development of a magnetic field. The equation that represents Ampere's law is:

$$\nabla x \, \mathbf{H} = \mathbf{J} + \partial \mathbf{D}/\partial t \tag{2.10}$$

The term on the left is straightforward and is analogous to the electric field term in Eq. 2.6. It is the curl of **H**. The first term on the right is the current density (**J**), the moving charge per unit volume. This term is generally zero for the situations we are examining in this text. This is because we are looking at electromagnetic waves rather than charges (such as an electron), so that **J** will usually be ignored. One important exception is when dealing with electric circuits where, for example, electrons are moving through a copper wire in a circuit. The second term on the right of Ampere's law is a change in the displacement current with changing time. The term is relevant to the generation of an electromagnetic field and ignoring **J** (since we generally do not deal with free charges), the equation can be reduced to:

$$\nabla x \, \mathbf{H} = \partial \mathbf{D}/\partial t \tag{2.11}$$

We now have equations that tell us a changing magnetic field leads to an electric field and a changing electric field leads to a magnetic field. In Appendix 2-1, we will derive the wave equations for electromagnetic radiation using Maxwell's equations. The equations describe constantly changing electric and magnetic fields. From this wave equation, it can ultimately be derived that the speed of a propagating electromagnetic wave is the same as that measured for the velocity of light. This is also derived in Appendix 2-1. The wave equation for the electric field portion is given by:

$$\nabla^2 \mathbf{E} = \varepsilon_0 \mu_0 \partial^2 \mathbf{E}/\partial t^2 \tag{2.12}$$

If we are interested in the spatial properties of a wave (amplitude and phase) and not the temporal properties, we can use the time-independent wave equation, known as the Helmholtz equation. This equation is also derived in Appendix 2-1 and will be used throughout the text.

Using these four equations, we will be able to derive the properties of electromagnetic radiation in a wide range of situations.

2.5 POLARIZATION

An important property of electromagnetic radiation is that it is polarized and polarization spectroscopy is an important technique used with OCT. We have already discussed that light can be represented as oscillating transverse electric and magnetic fields. If we envision the light propagating in the z direction and the transverse electric field is exclusively in the x direction, the electric field can be represented by:

$$\mathbf{E_x}\,(z, t) = iE_{0x} \cos(kz - \omega t) \tag{2.13}$$

For simplicity, a cosine is chosen here rather than an exponential because it is easier for most readers to envision in this case. If we have a similar wave, with its electric field exclusively in the y direction and at the same frequency but not necessarily in the same phase, the electric field can be given by:

$$\mathbf{E_y}\,(z, t) = jE_{0y} \cos(kz - \omega t + \varepsilon) \tag{2.14}$$

We will look at different ways Eqs. 2.13 and 2.14 can add together (i.e., polarization states).

First, if we assume that ε is zero or an integer multiple of 2π (which is equivalent), then the sum of the two fields is:

$$\begin{aligned} \mathbf{E}(z, t) = \mathbf{E_x}(z, t) + \mathbf{E_y}(z, t) &= \mathbf{i}E_{0x}\cos(kz - \omega t) + \mathbf{j}E_{0y}\cos(kz - \omega t) \\ &= (\mathbf{i}E_{0x} + \mathbf{j}E_{0y})\cos(kz - \omega t) \end{aligned} \qquad (2.15)$$

The electric field is therefore constant in orientation with the angle relative to the x axis, and from simple trigonometry the angle is given by:

$$\sin \o / \cos \o = \tan \o = (E_{0y}/E_{0x}) \qquad (2.16)$$

When $E_{0y} = E_{0x}$, the result is linearly polarized light, whose position from the axis is constant and is at an angle of 45 degrees from the x axis. Now, if we assume ε is equal to a multiple of $-\pi$, where $E_{0y} = E_{0x}$, since the second cosine is equal to negative 1, $\mathbf{E}(z, t)$ is given by:

$$\mathbf{E}(z, t) = (\mathbf{i}E_{0x} - \mathbf{j}E_{0y})\cos(kz - \omega t) \qquad (2.17)$$

This again is linearly polarized light with a different axis from that of Eq. 2.15. It is given by:

$$\tan \o = -E_{0y}/E_{0x} \qquad (2.18)$$

This is a linearly polarized wave at -45 degrees to the x axis if again, the intensities are equal. So the light is linearly polarized when the phase relationship (ε) between the two waves is 0 or a multiple of $\pm n\pi$. If the values of E_{0y} and E_{0x} are not equal, the light is still linearly polarized but the angle is no longer 45 or -45 degrees with respect to the x axis (Eqs. 2.16 and 2.18).

Now, if ε is $\pi/2$, the second cosine term becomes a sine term and Eq. 2.15 becomes:

$$\mathbf{E}(z, t) = \mathbf{i}E_{0x}\cos(kz - \omega t) + \mathbf{j}E_{0y}\sin(kz - \omega t) \qquad (2.19)$$

The peaks and troughs of the two waves are no longer aligned. If the peak intensities of the two waves are equal, $E_{0x} = E_{0y} = E$, the equation becomes:

$$\mathbf{E}(z, t) = E[\mathbf{i}\cos(kz - \omega t) + \mathbf{j}\sin(kz - \omega t)] \qquad (2.20)$$

This electric field is no longer at a constant angle with respect to the x axis. This is an electric field that is rotating through the x and y axes as it changes as a function of position on the z axis and time. When $(kz - \omega t)$ is equal to zero, the electric field is aligned with the x axis and has a magnitude of E along that axis. When $(kz - \omega t)$ is equal to $\pi/2$, the electric field is completely aligned with the y axis with a magnitude of E. If we are looking at the light coming from the source, the light appears to be rotating in a clockwise manner about the z axis and is usually referred to as right circularly polarized light.

If ε is $-\pi/2$, the electric field, for $E_{0x} = E_{0y} = E$, is given by:

$$\mathbf{E}(z, t) = E[\mathbf{i}\cos(kz - \omega t) - \mathbf{j}\sin(kz - \omega t)] \qquad (2.21)$$

This is circularly polarized light rotating counterclockwise and is referred to as left circularly polarized light. The light appears to be circular because no matter what angle the light is at a specific position, the magnitude is the same. When the waves are out of phase and $E_{0x} \neq E_{0y}$, which is the most common case, elliptically polarized light is produced. Elliptically polarized light means that as it rotates around the x and y axis, the irradiance varies. Therefore, as we look at the light coming from the source, it has an elliptical appearance with respect to rotation around the z axis.

Later in the text, we will be describing how both the OCT system and the tissue being imaged change the polarization state of the light. Polarization states, and changes in states, can be described quantitatively with one of two major approaches, either Jones or Mueller calculus. With Jones calculus, the state of light is defined by a vector known as the Jones vector. This defines the electric field in terms of the amplitude and phase of the x and y components. The Jones vector is written:

$$\mathbf{E} = \begin{vmatrix} E_x \\ E_y \end{vmatrix} = \begin{vmatrix} E_{0X} \mathrm{Cos}\ (kz - wt) \\ E_{0Y}\ \mathrm{Cos}\ (kz - wt + \varepsilon) \end{vmatrix} \tag{2.22}$$

This matrix is a description of the polarization state that the light is currently in. When polarization states are changed, for example by a filter or biological tissue, the change can be described by multiplying the Jones vector by a Jones matrix, which is a 2×2 matrix. An example is a Jones matrix of a linear polarization filter operating on the light:

$$\begin{vmatrix} 1 & 0 \\ 0 & 0 \end{vmatrix} \tag{2.23}$$

By performing matrix multiplication of the Jones vector (representing the light) by the 2×2 matrix (representing the polarization filter), the way light is altered by the filter is found.

An alternative to the use of Jones calculus is Mueller's calculus, where the polarization state is described in terms of the Stokes vector instead of the Jones vector. Unlike the Jones vector, the components of the Stokes vector represent empirically measured quantities. The Stokes vector is:

$$S = \begin{vmatrix} I \\ Q \\ U \\ V \end{vmatrix} = \{ I \quad Q \quad U \quad V \}, \tag{2.24}$$

where $I =$ the total intensity, $Q = I_0 - I_{90}$ (the difference in intensities between horizontal and vertical linearly polarized components), $U = I_{+45} - I_{-45}$ (the difference in intensities between linearly polarized components oriented at $+45°$ and $-45°$), and $V = I_{rcp} - I_{lcp}$ (the difference in intensities between right and left circularly polarized components). Transformation of polarization states by tissue or filter is then described in terms of a Mueller matrix, which is a 4×4 matrix. An example of a Mueller matrix is again a linear polarizer:

$$1/2 \begin{vmatrix} 1 & 1 & 0 & 0 \\ 1 & 1 & 0 & 0 \\ 0 & 0 & 0 & 0 \\ 0 & 0 & 0 & 0 \end{vmatrix} \tag{2.25}$$

The advantages of Mueller calculus are that it both uses easily measured quantities (such as total intensity), unlike the Jones vector that is more difficult to measure directly, and that it can be used to assess unpolarized light. Since many OCT sources are partially unpolarized, the latter is a significant advantage. The advantage of Jones calculus is that the calculations (but not measurements) are vastly simpler, particularly when many components are involved such as lenses, gratings, and fibers.

To this point, we have focused on the electromagnetic properties of light and the property of light known as polarization. In the following few sections, the topics of refraction and reflection will be dealt with in a very superficial manner. A more rigorous mathematical treatment of refraction and reflection are found in texts listed at the end of this chapter.

2.6   REFLECTION

Reflection occurs when there is a region of sharp refractive index mismatch. The refractive index will be explored in more detail in the next chapter, but for now it will be assumed to mean a measure of the velocity of light in a medium. Therefore, the velocity of light is not considered constant when it passes through different media. For illustrative purposes, light travels faster in the medium with low refractive index as compared to that of the high refractive index.

An example of external reflection, light going from a region of low to high refractive index, is shown in Figure 2.4A. This is a ray diagram where the direction of the light wave is represented by a line perpendicular to the wavefront, known as a ray. The refractive index of the external material where the ray is coming from, say air, is less than the refractive index of the other material the light is reflected off, which is when external reflection occurs. There are relatively basic laws that guide reflection. The Laws of Reflection state that (1) the incident angle ϕ_i must equal the reflected angle ϕ_r and (2) the incident and reflected light must be in the same plane. Reflection can be partial or total, with some to nearly all of the light being transmitted into the medium. The amount of reflection depends on the amount of mismatch, the angle, and the polarization of the incident light. However, an important principal, which needs to be taken into consideration, is that while reflection occurs because of this property of sharp refractive index mismatch, minimal reflection will occur if the change of refractive index is gradual, over the distance of a few wavelengths. This will be important in the design of certain optics to reduce reflection.

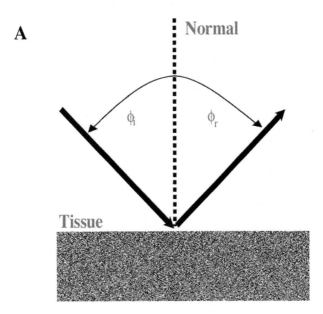

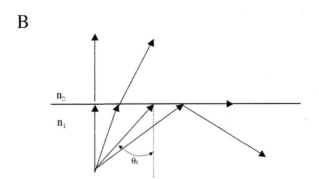

Figure 2.4 (A) External reflection, light going from a region of low to high refractive index. (B) Besides external reflection, there is also the important property of total internal reflection, which occurs when light goes from high to low refractive index. When light reaches a critical angle it is no longer transmitted.

Besides external reflection, there is also the important property of total internal reflection as shown in Figure 2.4B. Light traveling from a region of high refractive index to low refractive index may or may not penetrate the interface, depending on the angle. If the light is totally perpendicular to the surface, it will penetrate. However, there exists an angle beyond which, in this case of going from high to low refractive index, the light wave will be reflected internally rather than move through the surface. This is total internal reflection and is an important property for many optical devices, particularly optical fibers. This angle where total internal reflection occurs is the critical angle and is described by the formula:

$$\text{Sin } \Theta_c = n_2/n_1 \tag{2.26}$$

In this equation, Θ_c is the critical angle and is measured relative to the axis perpendicular to the surface. For the other two terms, n_2 is the refractive index of the external medium, and n_1 is the refractive index of the region light is initially propagating through. The larger the mismatch, the smaller the critical angle and the smaller the angle at which internal reflection will occur. While there exists an angle where light will be reflected with this type of mismatch is not covered here, but is present in any basic optics text, and as we will see, this is the basis of the physics of optical fibers.

2.7   REFRACTION

When light is transmitted from a medium of one refractive index to another of higher refractive index, the light is bent as shown in Figure 2.5. For illustrative purposes,

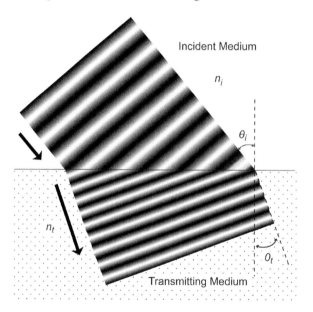

Figure 2.5   When light is transmitted from a medium of one refractive index to another of higher refractive index, the light is bent. For illustrative purposes, the left side of the wavefront that has entered the medium is now traveling slower than the light on the right. This is because the light on the right is still in a medium of lower refractive index and therefore traveling faster. Based on Figure 4.20, page 100, *Hecht Optics*, Third Edition [5].

in this figure, the left side of the wavefront that has entered the medium is now traveling slower than the light on the right. This is because the light on the right is still in the medium of lower refractive index and therefore traveling faster. This results in a relative bending of the light or change in the angle of propagation. The process is governed by Snell's law of refraction, which is:

$$n_i \sin \Theta_i = n_t \mathrm{Sin}\Theta_t \tag{2.27}$$

Here n_t is the refractive index of the medium from which the light is being reflected, n_i is the refractive index of the medium light is initially propagating through, Θ_t is the angle of transmitted light relative to the axis perpendicular to the surface, and Θ_i is the angle of incident light.

2.8 OPTICAL COMPONENTS

In this section, only those optical components and properties most relevant to the text will be discussed in detail, with other more specialized components introduced when they appear in the text. For example, the diffraction grating will be discussed after diffraction is introduced in Chapter 4. For the most part, the laws of refraction and reflection will provide the basics for guiding the propagation of light through many optical components. The description of light propagation through optical components (for example, lenses) can be broadly classified into two categories, geometric and physical optics. Geometric optics involves situations where the optical components have dimensions that are large relative to the wavelength of the light. This allows light to be represented as rays or straight lines passing through the device, and diffraction does not need to be taken into account. Physical optics, which will be touched on briefly in Chapters 4 and 5, involves situations where a significant amount of diffraction is occurring because the optical component is small in dimensions relative to the wavelength. For the most part, this means the light can no longer be treated with a simple ray diagram or as a line passing through the system.

2.9 LENS

A general understanding of geometric optics is important for building an OCT system, but is not necessary for appreciating the overall physics and application of OCT. However, several areas of geometric optics will be reviewed. Specifically, we will be discussing optical fibers, conventional lenses, and GRIN lenses.

Lenses are used extensively through OCT systems; some are fibers and some are free space. Most people are familiar with the general concept of a lens, like those in spectacles. They work primarily because the difference in the refractive index between the lens and its surrounding medium allows for refraction or bending of the light to adjust the direction of light propagation, in the case of spectacles to the eyes' focus. There is no need to review the properties of lenses in detail. Optical systems with lenses have properties such as aperture stop, pupils, and f-numbers, which can be read about in any standard optics textbook and again are not critical for the purposes of this textbook. However, those relevant to the discussion of OCT, such as the properties of a propagating Gaussian beam, are discussed in later chapters.

2.10   GRIN LENS

A graded index or GRIN lens, a special type of lens, is typically present in OCT catheters, arthroscopes, and endoscopes. GRIN lenses are usually fiber based and have advantages over free space lenses, particularly the ability to reduce reflection in fiber-based systems. The refractive index of this GRIN lens drops off radially (dark is the high refractive index) as shown in Figure 2.6. Most GRIN lenses have a parabolic refractive index profile that can be expressed as:

$$n_r = n_0(1 - ar^2/2) \tag{2.28}$$

where r is the radial position (distance) from the center of the fiber, n_0 is the maximum refractive index, n_r is the refractive index at a specific position, and a is a constant. Light can be focused with a GRIN lens just like a traditional lens, as shown in Figure 2.6. Among the advantages of the GRIN lens over a free space lens for OCT endoscopes, arthroscopes, and catheters is the fact that the surface is flat, in some cases eliminating lens-air interfaces (and therefore unwanted reflection).

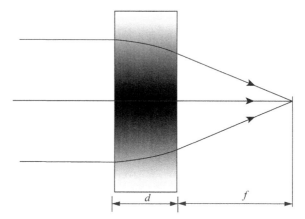

Figure 2.6   A graded index or GRIN lens is a special type of lens. It is typically present in OCT catheters, arthroscopes, and endoscopes. GRIN lenses are usually fiber based and have advantages over free space lenses, particularly their ability to reduce the reflection seen in fiber-based systems. With this type of lens, the refractive index drops off radially (dark is high refractive index). Light is focused with a GRIN lens just as it is with a traditional lens.

2.11 OPTICAL FIBER

Another important component to be discussed is the optical fiber, which in addition to being integral to OCT design, is used in a wide range of applications including telecommunications. Optical fibers can transmit light over long distances with minimal attenuation. The optical fiber is based on the concept of total internal reflection discussed above. Only one type of fiber will be discussed here: the step fiber which contains two important components, the core and the cladding, as shown in Figure 2.7. The refractive index of the cladding is less than the refractive index of the core, which is why total internal reflection can occur at the sharp interface (i.e., step). Most are made of fiberglass with impurities added to slightly alter the refractive indices. Not all light will propagate within the fiber and propagation depends on the angle between the light and the mismatch interface. Light must strike the core-cladding interface at or above the critical angle or else it will enter the cladding. This becomes clinically relevant, for example, with catheters and endoscopes because bending of the fiber as it passes through the body changes this angle and may lead to loss of light. This affects catheter and endoscope design. Light waves at angles less than the critical angles are transmitted through the fiber wall. The appropriate angles can be derived from simple geometry. The half acceptance angle Θ_i can be found from the critical angle as follows:

$$\text{Sin}\,\Theta_c = n_t/n_i \qquad (2.29)$$

where Θ_c is the critical angle between the cladding and the core, n_t is the refractive index of the cladding, and n_i is the refractive index of the core.

Figure 2.8 and the text below will be used to establish the angles light needs to hit the surface of the fiber at to propagate and not go through the cladding.

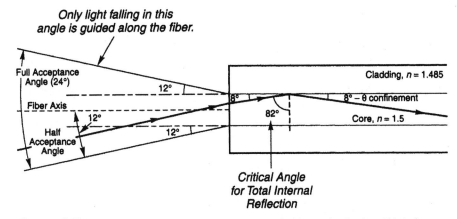

Figure 2.7   An optical fiber works on the concept of total internal reflection. This is the step fiber which contains two important components, the core and the cladding. Angular dependence is demonstrated. Based on Figure 3.30, page 63, *Understanding Fiber Optics* [6].

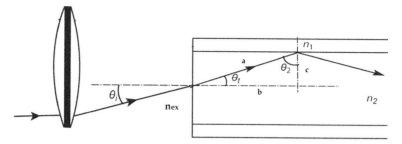

Figure 2.8   The angles at which light needs to hit the surface of the fiber in order to propagate and not go through the cladding are critical. Single mode fibers, as described in the text, have superior resolution to multimode fibers. However, single mode fibers make coupling more difficult and typically result in significant power losses. Still, because of their higher resolution, they are the fibers most often used when transmitting light in OCT systems.

From geometry we know that for the right triangle in this figure, $a^2 = b^2 + c^2$. Then, by dividing by a^2, we get:

$$1 = (b/a)^2 + (c/a)^2 \tag{2.30}$$

From our knowledge of trigonometry, we know that $b/a = \sin \Theta_c$ and using Eqs. 2.29 and 2.30, we get:

$$(c/a)^2 = 1 - (b/a)^2 = 1 - (n_t/n_i)^2 \tag{2.31}$$

Now, if we look at c/a from our triangle, it is the sin of the transmission angle Θ_a, through the face of the fiber. Therefore $c/a = (1 - (n_t/n_i)^2)^{1/2} = \sin \Theta_a$. So far, we have related the refractive index mismatch at the cladding-core interface to the transmission angle at the face of the fiber. We want to relate it to the incident angle on the fiber, but we have a formula for refraction. Therefore,

$$\sin \Theta_i = (n_i/n_{ex})(1 - (n_t/n_i)^2)^{1/2} \tag{2.32}$$

This formula tells us that the half angle of acceptance at the surface of the fiber is determined by the refractive index of the cladding, core, and space external to the fiber. For most fibers, the difference in the refractive index between the core and the cladding is on the order of 1% (Figure 2.7). This makes the critical angle on the order of 80°. A typical half acceptance angle is then 12°. Doubling the half acceptance angle gives the full acceptance angle. A property of optical fibers which should be clear from the above formulation is that the fibers are sensitive to sharp bends. The greater the bend in the fiber, the more likely it is that light will be lost into the cladding by going below the critical angle. This is why, with catheter and endoscope design, sharp angles need to be taken into account in the engineering.

Step fibers can be classified into single mode or multimode, which are seen in Figure 2.9. The propagation of light in a fiber will be examined initially with ray diagrams. Because the fiber is accepting light from a range of angles in a multimode fiber, light at a wide angle is traveling a longer distance than the light with a flat

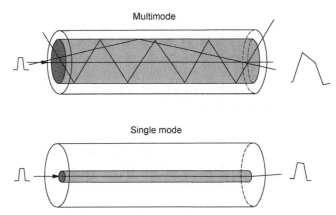

Figure 2.9 This figure illustrates how high resolution is achieved with single mode fibers using a ray diagram. It can be envisioned that in a multimode fiber, light in different modes travel different path lengths. When they combine at the end of the fiber, the signal is distorted by the different time of flights associated with each ray. In a single mode fiber, assuming only one polarization state, all photons have the same time of flight, leading to an output similar to the input.

angle. Each of the different paths are referred to as modes. This can result in modal dispersion, since the input pulses are taking different transit times to reach the end of the fiber. Because the modes are less in a single mode fiber compared to a multimode fiber, higher resolution can be maintained as in Figure 2.9.

When you are looking for high resolution, which you need with OCT, a fiber is used that is single mode, meaning only one path (assuming the light is linearly polarized) is taken by the light. To achieve single mode, a very small core diameter is used with the fiber. When imaging at 1300 nm, the fiber core diameter is typically 9 μm with a typical total fiber diameter of 124 μm. The core diameter (D) to achieve single mode can be found with the formula:

$$D \leq 2.4\lambda/\pi v \sqrt{n_i^2 - n_t^2} \tag{2.33}$$

From the formula it is obvious that the longer the wavelength, the larger the core diameter. Although the single mode fibers have superior resolution to multimode fibers, which is why they are used in OCT systems, single mode fibers make coupling more difficult and typically result in significant power losses (Figure 2.8).

Passing of light even through a single mode fiber can result in a loss in resolution. With the single mode fiber used with OCT, spectral dispersion (wavelength-dependent differences in refractive index), which will be discussed in more detail in a later chapter, can be the source of loss of resolution. However, the fibers chosen have minimal dispersion at 1310 nm, which is appropriate since we perform imaging at approximately 1300 nm. Although the attenuation from scattering within the fiber has a minimum at 1550 nm, which is significantly different from the 1300 nm light used with OCT, scattering is kept at a minimum by keeping fiber lengths short.

Besides dispersion and scattering, the fibers have a significant effect on the polarization of the light. When we are discussing single mode, we are actually talking about two modes, the two orthogonal polarization states of the light. The polarization state does not matter significantly in multimode fibers, since there is so much loss of the configuration of the original state, but this effect can have significant effects in single mode fibers. The light from the two polarization states can be envisioned as traveling both in the x and y states. There are two important polarization effects that occur in the single mode fibers. With the first, fibers are not ideal so that they may not be perfectly symmetrical with respect to refractive index. Therefore, light in one polarization state may travel a different distance than light in another polarization state. This results in the light from one state arriving at the end of the fiber at a different time than the light from the other state. The second polarization effect is that if the fiber is bent, the light in one polarization state will travel a different distance than light in another polarization state (because they travel in different axes of the fiber). Both these phenomena reduce resolution and are referred to as polarization mode dispersion. Some polarization sparing fibers do exist but are likely insufficient for general use with OCT.

For the sake of completeness, it should be noted that the ray diagrams used were just for illustrative purposes. In actuality, from Maxwell's equations, the energy is actually carried not just within the core, but slightly within the wall of the cladding due to what is called an evanescent wave, a quantum mechanical effect. Because of this, the effective core size is slightly larger than the 9 μm of the core. This is seen in Figure 2.10 where the wavefront is depicted rather than the ray diagram. Energy in the evanescent wave extends into the cladding, a finding with a simple ray diagram.

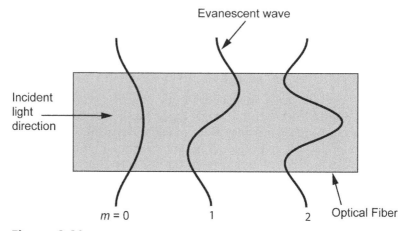

Figure 2.10   Propagation in an optical fiber using a wave front rather than ray diagram is shown in this figure. In this schematic, m is the mode of the wave. Zero mode is that of a single mode fiber. As the mode increases, more of the energy extends into the cladding in the form of an evanescent wave.

2.12 ENERGY PROPAGATION

Quantifying how the actual energy from light propagates is necessary to our study of OCT. The energy of the electromagnetic wave is shared equally by the electric and magnetic components, as described above. The magnitude of the power per unit area is given by the Poynting vector (S):

$$S = E \times B \tag{2.34}$$

At optical and near infrared frequencies, the light is cycling at a rapid rate, too fast to be measured by conventional detectors. Therefore, the measurement of energy with the light we are examining is generally described in terms of an average over time. The time averaged Poynting vector is known as the irradiance and is defined as the average energy per unit area per unit time. The term intensity is sometimes used to describe the magnitude of the electromagnetic radiation. However, since this term has several inconsistent definitions, use of irradiance is preferred when a quantitative definition is necessary. From time to time, the word intensity will show up in qualitative descriptions in this text, in which case the reader should view it as proportional to the irradiance. The irradiance is generally given by the formula:

$$I \equiv \langle S \rangle_T = (1/2)\varepsilon_0 n^2 \langle E^2 \rangle \tag{2.35}$$

This formula is sufficient for issues discussed within this text, which is averaged over small but finite regions of space. Outside these constraints, the formula is associated with ambiguity that requires the more exact integral which is described elsewhere (Born and Wolf).

2.13 DOPPLER SHIFTS

Doppler or phase shifts will be discussed in more detail in a later chapter; however, a brief description is appropriate here. Light, as we have said, has a frequency associated with it. If light interacts with a moving object, the moving object can actually change the frequency and wavelength of the light. An example would be a red blood cell moving away from the OCT beam. This frequency shift is known as a Doppler shift, which we will encounter throughout the text, and can increase or decrease the frequency of the original signal.

REFERENCES

1. Born, M. and Wolf, E. (1999). *Principles of Optics*. Cambridge University Press, Cambridge, UK.
2. Cheng, D. (1992). *Field and Wave Electromagnetics*. Addison-Wesley Publishing Co., Reading, MA.
3. Guenther, R. (1990). *Modern Optics*. John Wiley & Sons, New York.
4. Haurd, S. (1997). *Polarization of Light*. John Wiley & Sons, New York.
5. Hecht, E. (1998). *Hecht Optics*. Addison-Wesley Publishing Co., Reading, MA.
6. Hecht, J. (1999). *Understanding Fiber Optics*. Prentice Hall, Englewood Cliffs, NJ.

APPENDIX 2-1

ELECTROMAGNETIC WAVE EQUATIONS

Previously it was said that Maxwell was able to derive an equation for the propagation of an electromagnetic wave and calculated the velocity of this wave to be the same as that for the speed of light. In this appendix, we will make similar calculations. This simply requires using Maxwell's equations and a few rules of vector calculus to get to the answer. The equations used are first order differential equations in both $\mathbf{E}$ and $\mathbf{H}$. They will be used to derive one constant (speed of light) and second order equations. Using Faraday's law:

$$\nabla x \mathbf{E} = -\partial \mathbf{B}/\partial t \tag{1}$$

if we take the curl of both sides:

$$\nabla x (\nabla x \mathbf{E}) = -\nabla x (\partial \mathbf{B}/\partial t) \tag{2}$$

by using the identity $\mathbf{B} = \mu \mathbf{H}$ and rearranging:

$$\nabla x (\nabla x \mathbf{E}) = -\nabla x (\partial \mathbf{B}/\partial t) = -\mu_0 \partial/\partial t (\nabla x \mathbf{H}) \tag{3}$$

$\nabla x \mathbf{H}$ is the left side of Ampere's law, which equals $\partial \mathbf{D}/\partial t$. We therefore substitute:

$$\nabla x (\nabla x \mathbf{E}) = -\mu_0 \partial/\partial t (\partial \mathbf{D}/\partial t) = -\varepsilon_0 \mu_0 \partial/\partial t (\partial \mathbf{E}/\partial t) \tag{4}$$

So we have eliminated $\mathbf{H}$ and $\mathbf{B}$ from the equation. Now, we will use a rule not previously discussed (a vector identity):

$$\nabla x (\nabla x \mathbf{E}) = \nabla (\nabla \cdot \mathbf{E}) - \nabla^2 \mathbf{E} \tag{5}$$

So Eq. 2.4 now becomes:

$$\nabla (\nabla \cdot \mathbf{E}) - \nabla^2 \mathbf{E} = -\varepsilon_0 \mu_0 \partial/\partial t (\partial \mathbf{E}/\partial t) = \varepsilon_0 \mu_0 \partial^2 \mathbf{E}/\partial t^2 \tag{6}$$

Since $\nabla \cdot \mathbf{E}$ is zero by Gauss's law of electric fields equation (with no free charges):

$$\nabla^2 \mathbf{E} = \varepsilon_0 \mu_0 \partial^2 \mathbf{E}/\partial t^2 \tag{7}$$

A similar derivation can be made for $\mathbf{B}$:

$$\nabla^2 \mathbf{B} = \varepsilon_0 \mu_0 \partial^2 \mathbf{B}/\partial t^2 \tag{8}$$

These are wave equations of the form of the general wave equation described in Chapter 1. Therefore,

$$v = 1/\sqrt{\varepsilon_0 \mu_0} \tag{9}$$

Therefore, since ε_0 and μ_0 can be measured relatively easily, the speed of light can be calculated without actually measuring it.

HELMHOLTZ EQUATION

Potentials or fields, like the electric and magnetic fields, whose amplitude and phase vary with their position in space but are cyclic or sinusoidal with respect to time, can be represented by a wave equation which depends on space but not on time. Throughout the text, this equation will be very helpful in simplifying analysis and derivation when we seek only a spatial solution to an equation but not a temporal one. For the solution of the wave equation:

$$\mathbf{E(x, y, z, t)} = \mathbf{E(xyz)}e^{i\omega t} \tag{10a}$$

$$\mathbf{H(x, y, z, t)} = \mathbf{H(xyz)}e^{i\omega t} \tag{10b}$$

The quantities $\mathbf{E(xyz)}$ and $\mathbf{H(xyz)}$ are only dependent on position. Now a surface over which the field is constant is known as a wavefront. If the amplitude is constant over the wavefront, it is said to be homogeneous. The solution will be derived with Maxwell's equations assuming a simple, source free, nonconducting medium ($\rho = 0$, $\mathbf{J} = 0$, $\sigma = 0$) and the $\mathbf{P}$ and $\mathbf{M}$ can be ignored.

We begin eliminating the time dependence in Maxwell's equations. Faraday's law now becomes:

$$\nabla = -\partial \mathbf{B}/\partial t = -\mu \partial \mathbf{H}/\partial t \tag{11}$$

Differentiating Eq. 10b with respect to t

$$\partial \mathbf{H}/\partial t = -\mu i \omega \mathbf{H(xyz)} \tag{12}$$

Faraday's law now becomes:

$$\nabla x \mathbf{E} = -\mu i \omega \mathbf{H(xyz)} \tag{13}$$

Similarly, Ampere's law can be modified, so that

$$\nabla x \, \mathbf{H} = \mathbf{J} + \partial \mathbf{D}/\partial t = +\omega^2 \varepsilon \mathbf{E} \tag{14}$$

Now, if we apply the curl operation to Eq. 13, we obtain:

$$\nabla x (\nabla x \mathbf{E}) = -\mu i \omega \nabla x \mathbf{H(xyz)} \tag{15}$$

Differentiating Eq. 10a with respect to t and substituting into Eq. 15, we get:

$$\nabla x (\nabla x \mathbf{E}) = \mu \varepsilon \omega^2 \mathbf{E(xyz)} \tag{16}$$

Using the vector to identify $\nabla x (\nabla x \mathbf{E}) = \nabla (\nabla \cdot \mathbf{E}) - \nabla^2 \mathbf{E}$, and remembering that $\nabla (\nabla \cdot \mathbf{E}) = 0$ for no free charges, we get:

$$\nabla (\nabla \cdot \mathbf{E}) - \nabla^2 \mathbf{E} = \mu \varepsilon \omega^2 \mathbf{E(xyz)} = -\nabla^2 \mathbf{E} \tag{17}$$

Since $1/\mu\varepsilon = v^2$, and $\omega^2/c^2 = k^2$, the equation becomes:

$$(\nabla^2 + k^2)\mathbf{E}(\mathbf{x}, \mathbf{y}, \mathbf{z}) = 0 \tag{18}$$

More generally for any field of potential, we have the Helmholtz equation:

$$(\nabla^2 + k^2)\mathbf{F}(\mathbf{x}, \mathbf{y}, \mathbf{z}) = 0 \tag{19}$$

3   LIGHT IN MATTER

Chapter Contents

Until now, this book has focused, for the most part, on the properties of light in a vacuum. This chapter examines the properties of light within materials such as tissue. Specifically, it will examine the material properties of absorption, scattering, refractive index, and dispersion from a classical standpoint.

The interaction of light with atoms and molecules from a classical standpoint is best described by the entity known as the dipole moment (separation of opposite charges). Conveniently, the dipole moment can be understood through its analogy to a weight on a spring. Therefore, the discussion of the interaction of light with atoms and molecules will begin with the physics of a weight on a spring. Then, a more formal discussion of the dipole moment, where the weight is replaced by charged particles, will be undertaken.

The spring/mass in Figure 3.1A has a natural position of rest in which, unless the energy of the system is changed, it will remain at rest indefinitely. We will call this position x_0. In Figure 3.1B, energy has now been added to the system to displace the mass to x_1. Consistent with the terminology in Chapter 1, a force is applied over a distance $|x_1|$ (i.e., work). Because the overall work of the system has been changed, energy must have been added. Now, if the mass is not held in position it will move, and assuming for the moment that there are no other forces at work such as friction,

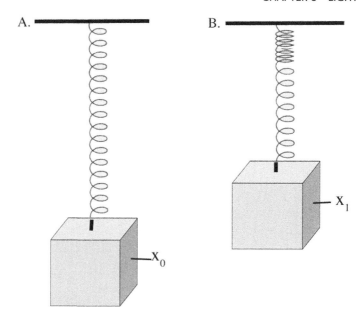

Figure 3.1   (A) Spring/mass has a natural position of rest in which, unless the energy of the system is changed, it will remain at rest indefinitely. We will call this position x_0. (B) Energy has now been added to the system to displace the mass to x_1. The energy is stored in the compressed spring.

the mass will begin to accelerate. Since potential energy is stored in the compressed spring, it will force the mass back in the direction of the equilibrium position (toward the x_0). The particular potential energy manifests itself as a restoring force (F_r):

$$F_r = -s|x_1| \tag{3.1}$$

Here, s will be referred to as the stiffness that is the restoring property of the string. The restoring force or potential energy is maximal when the mass is maximally displaced. Now, the mass begins to accelerate toward x_0. When the mass reaches x_0, the restoring force is now zero, but the energy added to the system has not disappeared. It is present in the form of kinetic energy. At x_0, the velocity and acceleration are at a maximum. The force at this point (kinetic energy) is given by:

$$F_a = ma = md^2x/dt^2 \tag{3.2}$$

In this formula, a is the acceleration and d^2x/dt^2 is its differential form in one dimension. The mass will move, reaching its maximum velocity at x_0 (the tension due to s is now zero), and will continue to move until the tension on the spring (stretched) results in the velocity going down to zero. At this point, all the force present is found in Eqs. 3.1 and 3.2. This constant oscillation will continue as the mass indefinitely goes back and forth until additional force/energy, such as friction, is introduced into the system. This constant oscillation occurs because energy had been added to the system during the initial compression. This exchange between kinetic and potential energy is simple harmonic motion. When the oscillating dipole moment is later introduced, which creates light (and other electromagnetic fields), it will be seen that oppositely charged particles, such as an electron and a proton, will behave like a mass on springs. Therefore, we will be looking at oscillating charges, atoms, or molecules in an analogous manner to the mass on the spring.

Now we were looking at a system from which energy was not added or subtracted (except during the initial compression), so that, combining Eqs. 3.1 and 3.2:

$$md^2x/dt^2 = -sx$$

or (3.3)

$$md^2x/dt^2 + sx = 0$$

Now, two types of force will be introduced, resistance and driving force. In the real world, as the mass moves back and forth, energy is lost as the mass interacts with air or other forces of nature resulting in resistance. The introduction of resistance can be described mathematically by the formula:

$$md^2x/dt^2 + rdx/dt + sx = 0$$

The new term takes into account constant resistance, which is dependent on the velocity (dx/dt) and r, and is assumed here to be a constant with dimensions of force per unit velocity. The resistance acts in a direction opposite the acceleration and therefore reduces the oscillation. A plot of the displacement as a function of time for damped oscillation is demonstrated in Figure 3.2. The dotted lines represent damped oscillations.

Finally, we will introduce the concept of forced oscillation. This topic will be addressed in considerable detail because of its relevance to the interaction of light with matter. When light interacts with matter, it drives the charges in the atoms and molecules, resulting in phenomena such as scattering and the refractive index.

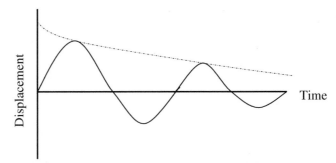

Figure 3.2 Resistance acts in a direction opposite of the acceleration and therefore reduces the oscillation as in Eq. 3.3. A plot of the displacement as a function of time for damped oscillation is demonstrated. The dotted line represents damped oscillations.

However, initially we will continue our analysis with the weight on the spring before moving to atoms.

In the case of a driving force, energy is constantly being introduced into the system, so Eq. 3.3 becomes:

$$md^2\,\mathbf{x}/dt^2 + rd\,\mathbf{x}/dt + s\,\mathbf{x} = F_D e^{i\omega t} \tag{3.4}$$

It should be noted that now $\mathbf{x}$ is bold. Previously we used bold to indicate that the quantity was a vector with x, y, and z components. In this case, the bold $\mathbf{x}$ represents a different kind of vector. While we are focusing exclusively on one dimension, the x dimension, the bold tells us what the phase of the oscillator (spring/mass) is, which may be, and typically is, distinct from that of the driving force.

Equation 3.4 is a differential equation. It will initially be assumed that the solution to this equation will be of the form $\mathbf{x} = \mathbf{A}e^{i\omega t}$. Here, $\mathbf{x}$ is the position of the weight due to the driving force. We are assuming that the frequency of oscillation of $\mathbf{x}$ is the same as the driving frequency, but amplitude and phase may not be. $\mathbf{A}$ may be real or complex and can be found by substituting this solution into Eq. 3.4. Therefore:

$$d\,\mathbf{x}/dt = i\omega\alpha\varepsilon^{i\omega t} = i\omega\,\mathbf{x} \tag{3.5}$$

$$d^2\,\mathbf{x}/dt^2 = -\omega^2\alpha\varepsilon^{i\omega t} = -\omega^2\,\mathbf{x}$$

Substituting this back into Eq. 3.4 yields:

$$-m\omega^2\alpha\varepsilon^{i\omega t} + ir\omega\alpha\varepsilon^{i\omega t} + s\,\mathbf{A}e^{i\omega t} = F_D e^{i\omega t} \tag{3.6}$$

Rearranging and solving for $\mathbf{A}$ we get:

$$\mathbf{A} = F_0/(-m\omega^2 + ir\omega + s) \tag{3.7}$$

Defining s/m as equal to ω_0, the natural angular frequency, we will divide the top and bottom by m to give:

$$\mathbf{A} = F_0/m(-\omega^2 + ir\omega/m + s/m) = F_0/m(-\omega^2 + i\gamma\omega + \omega_0{}^2) \tag{3.8}$$

Here γ is just a new constant related to the resistance. Now we will substitute $\mathbf{A}$ into $\mathbf{x} = \mathbf{A}e^{i\omega t}$ to find $\mathbf{x}$:

$$\mathbf{x} = F_0/m(i\gamma\omega + \omega_0{}^2 - \omega^2)e^{i\omega t} \tag{3.9}$$

The resistance phase delay φ induced by the force is given by the ratio of the imaginary to real component:

$$\text{imaginary/real} = \tan\varphi = \sin\varphi/\cos\varphi = -\gamma\omega/(\omega_0{}^2 - \omega^2) \tag{3.10}$$

We can say several things about the relationship between the phase of the driving force relative to the phase of the mass. First, from our knowledge of imaginary numbers in Chapter 1, the term $-i$ represents a phase shift between the mass and driving force of $-90°$, even if no resistance component is present. Second, an additional phase difference exists due to the ratio of the resistance to the resonance component.

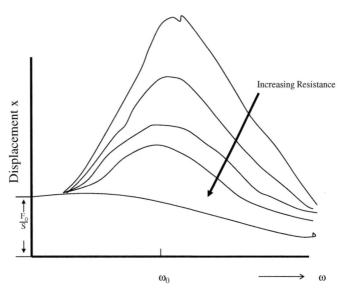

Figure 3.3 It can be seen that **x** is maximal at the frequency when ω^2 is equal to ω_0^2. This is known as resonance and the plot of displacement amplitude versus ω is shown at different resistance (r).

Returning to Eq. 3.9, it can be seen that **x** is maximal, assuming γ is constant, at the frequency when ω^2 is equal to ω_0^2. This is known as resonance and the plot of amplitude versus ω is shown in Figure 3.3. It should also be noted that as γ (or r) decreases, again a measure of resistance, the displacement (**x**) increases. If the resistance were zero, at resonance, the displacement would be infinite. With this background, we can move on to the oscillation dipole moment.

3.1 OSCILLATING DIPOLE MOMENT

The harmonic oscillator, combined with an understanding of the electrical dipole, allows us to use the Lorentz model to describe refractive index, absorption, and scattering. To illustrate the concept of an oscillating dipole, we will first use the example of a radio antenna. In our antenna, charge is initially separated and an alternating current is applied. This concept is best illustrated in Figure 3.4.

At maximal separation of charges, an electric field is produced in a manner analogous to a capacitor. As the charges come closer together, the electric field is changing (reducing). When the charges reach the point of overlap (Figure 3.4b), the electric field drops to zero. The changing electric field generates an alternating magnetic field according to Maxwell's equations (and vice versa).

Close to the dipole, the electric and magnetic fields are complicated in spatial orientation. However, at significant distances away from the dipole, a classical

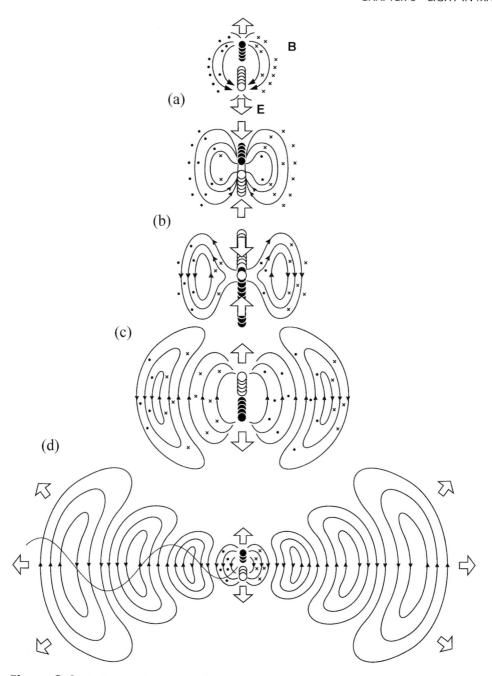

Figure 3.4 To illustrate the concept of an oscillating dipole, we will first use the example of a radio antenna. In our antenna, charge is initially separated and an alternating current is applied. The concept is well illustrated here. When the charges reach the point of overlap, the electric field drops to zero. (b) Propagating EM is seen in (c) and (d) arrows. Based on Figure 3.30, page 63, *Hecht Optics* [4].

electromagnetic wave is present. The irradiance distant from the source, which is not derived, is given by [2][4]:

$$I(\Theta) = (\mu_0{}^2\omega^4/32\pi^2c^2\varepsilon_0)(\sin^2\Theta/r^2) \tag{3.11}$$

Derivations of this equation can be found in most optics textbooks for those interested. Several important observations can be made when examining the equations. First, the irradiance of the electromagnetic radiation (EM) is proportional to the angular frequency to the fourth power. In the Lorentz model, we will see that this has implications toward the wavelength dependency of scattering. Second, irradiance falls off at a rate of $1/r^2$. Finally, the distribution of irradiance is doughnut-like around the dipole, with maximal irradiance perpendicular to the dipole movement and minimal in the line parallel to the dipole.

Thus an oscillating dipole generates a propagating electromagnetic wave. Therefore, when light interacts with an atom or molecule, the electrons present oscillate, leading to the generation of a secondary electromagnetic wave. The combining of these secondary waves with the primary wave will be the source of a new wave of different phase that will lead to the apparent velocity change in the light, the refractive index. When measuring the speed of light, one measures how long it takes for a portion of the wave cycle, say the peak, to go from one position to another. If the refractive index represents a relative change in velocity of light, what is really said is that a phase change results from the interaction of the incident light with the electrons and therefore the generation of a secondary wave. The phase change induced by combining the secondary wave results in a perceived change in velocity in the forward direction.

To further explain the interaction of light with matter, a model will be used that was originally developed by Lorentz in the late 19th century. The model combines results from Maxwell's equations with Newtonian mechanics. At that time, prior to any substantial understanding of the nature of the atom, the Lorentz model of the atom considered the electron (negative charge) to be held on to the nucleus (positive charge) by a spring-like interaction. Just like the spring/mass example, the electron has an equilibrium position about the nucleus. Although this is a simplistic model, it actually results in a reasonably good approximation.

When describing the interaction of light with matter we will be dealing with the electric field, and the magnetic field component of light can be ignored. This is because the electric field is far more influential and relativistic particle velocities are not involved. Therefore, the force driving the electrons will be the electric field. The electric field here will be a plane wave of the form:

$$E(z, t) = E_0 e^{-i(\omega t - kz)} \tag{3.12}$$

The force associated with the electric field (driving the electron) is $\mathbf{F}(z,t) = e\,E(z,t)$, were e is the charge on the electron. Equation 3.12 becomes:

$$F(z, t) = -\omega^2 + \omega^2\mathbf{x}_e + i\gamma\omega = eE_0 e^{-i(wt-kz)} = eE(z, t) \tag{3.13}$$

The solution of this differential equation is:

$$\mathbf{x}_e = eE_0 e^{-i(wt-kz)}/[m_e(i\gamma\omega + \omega_o{}^2 - \omega^2)] = e\mathbf{E}(z, t)/[m_e(i\gamma\omega + \omega_o{}^2 - \omega^2)] \tag{3.14}$$

where once again $\gamma\omega/(\omega_o^2 - \omega^2) = \text{Tan } \emptyset$ for the resistance induced phase delay. The greater the imaginary component relative to the resonance component, the greater the phase change in x_e. Also, it must be kept in mind that Eq. 3.14 describes the position and phase of the electron, not the generated electromagnetic wave. As discussed, the wave is produced by the changing position of the charges and not their absolute position. Because of this and the phase delays described above, it is apparent why the primary and secondary waves are generally of different phases.

We have described how the light leads to oscillations in the electrons (x_e) of the atom. Now we need to establish how it results in reradiation. To do this, we reintroduce the concept of a dipole moment. Atoms and molecules consist of positive and negative charges that can be separated by a distance x. The dipole moment (p) is the product of the charge times the distance it is separated ($p = ex$). The total dipole density in a medium is given by:

$$P(z, t) = Np = Nex \qquad (3.15)$$

P is the polarization and is the dipole moment per unit volume. N is the number of electrons per unit volume. We have seen how x_e changes after interacting with the light wave; then we get:

$$P(z, t) = [Ne^2 E_0/m(i\gamma\omega + \omega_0^2 - \omega^2)]e^{-i(wt-kz)} \qquad (3.16)$$

It was previously noted from Maxwell's equations that $(\varepsilon-\varepsilon_o)E = P$ for a polarized linear, isotropic, and homogeneous dielectric. This can be rearranged into $(\varepsilon-\varepsilon_o) = P/E$, which substituted into Eq. 3.16 gives:

$$\varepsilon = \varepsilon_o + Ne^2/m(i\gamma\omega + \omega_0^2 - \omega^2) \text{ or } \varepsilon/\varepsilon_o = 1 + Ne^2/\varepsilon_o m(i\gamma\omega + \omega_0^2 - \omega^2) \quad (3.17)$$

This equation can be used to generate the refractive index. In non-magnetic material the refractive index (n) is:

$$n \equiv c/v = \sqrt{e/e_o}$$

Equation 3.17 becomes:

$$\mathbf{n}^2 = \varepsilon/\varepsilon_o = 1 + Ne^2/[\varepsilon_o m(i\gamma\omega + \omega_0^2 - \omega^2)] \qquad (3.18)$$

This is the dispersion relationship and it tells us a few things. First, the refractive index is frequency dependent (ω), which will be expanded upon shortly. Second, because the mass and electric charges are constant, the refractive index at a given frequency for a specific material is dependent on ω_0, γ, and N, which is the number of contributing electrons per unit volume. Equation 3.18 can be made more general to take into account more than one type of electron from other molecules/atoms or from other locations within the atom.

$$n^2 = 1 + e^2/m\Sigma N_g/\varepsilon_{og}(i\gamma\omega + \omega_{0g}^2 - \omega^2) \qquad (3.19)$$

where N_g is the number of that species per unit volume. For gases, where the refractive index is close to 1, this can be expanded out {where $(1 + x^n) = 1 + mx + m(m-1)x^2/2 + \cdots$} to:

$$n \approx 1 + e^2/2m\Sigma N_g/\varepsilon_{og}(i\gamma\omega + \omega_{0g}^2 - \omega^2) \qquad (3.20)$$

For dense materials, the fields generated by other polarized atoms within the medium need to be taken into account. If we assume that the atoms are spherical and the material is isotropic, from Gauss's law on a dielectric spherical field then the new field the electron is exposed to is $P/3$. The derivation of the refractive index formula for a dense material can be found in the reference texts described later with the final result of [2]:

$$(n^2 - 1)/(n^2 + 2) = e^2/3m \Sigma N_g/[\varepsilon_{og}(i\gamma\omega + \omega_{0g}{}^2 - \omega^2)] \qquad (3.21)$$

This is known as the Clausius-Mossotti equation.

Referring back to any of the refractive index formulas, it can be seen that when ω is greater than ω_0, the refractive index is greater than one and the light behaves as if it is traveling slower than in a vacuum. When ω_0 is greater than ω, the refractive index is less than 1 and light behaves as if it were traveling faster than light in a vacuum.

3.2 DISPERSION

As stated, Eq. 3.26 tells us that the refractive index is frequency dependent. This means that as light passes through a medium, different wavelengths will travel at different speeds, which is called dispersion. This can be seen from Eq. 3.26. A prism is a device that works by producing dispersion. When ω_0 is much greater than ω, the refractive index increases with frequency, which is known as normal dispersion. When ω is much greater than ω_0, the refractive index decreases with increasing frequency and is referred to as anomalous dispersion.

Dispersion can be a good or bad thing. Dispersion can be used to intentionally disperse light into its different frequency components, as in the case of a prism. As light is directed at the prism at an angle, the angle at which light leaves the prism is dependent on the refractive index and therefore the frequency.

An example of where dispersion has a negative effect on imaging is with group velocity dispersion. We have previously discussed the fact that light of many wavelengths travels with a group velocity. We will see in the next chapter that these groups are used for ranging in an interferometer. However, when the light enters tissue rather than reflects off a mirror, for example, it undergoes dispersion. The different wavelengths return at slightly different times. The broader the bandwidth, the greater the dispersion. If the dispersion is too great, the reference and sample arm will no longer interfere in a way which does not allow ranging. Dispersion generally needs to be compensated for in OCT systems. A more rigorous treatment of dispersion will be presented in Chapter 5.

3.3 ABSORPTION

We will return to Eq. 3.14 to develop a better understanding of absorption. If we use the model of a mass on a spring, as the forced oscillation approaches the resonance ($\omega_0 = \omega$), the displacement of the mass increases. If γ is zero and the strength of the spring is finite, the spring will ultimately break, releasing the mass. Even though the harmonic oscillator is an excellent model for absorption, it is obvious that the electron is not held to the nucleus via a spring. From quantum mechanics, we can anticipate some modifications to the model. Here, an electron excited with a driving force at the resonance frequency (ω_{01}) can only be excited to another specific resonance state/energy level (ω_{0N}), dictated by the laws of quantum mechanics, or if the applied force is sufficient, released from the atom/molecule.

Without going into excessive detail, electrons exist in energy states about the nucleus. When the electron interacts with the light, it absorbs the energy of a photon of specific frequency and goes to a higher energy state. An electron from a given atom or molecule cannot receive photons of any energy but only of specific frequency or energy. This phenomenon is analogous to the resonance of the spring. This is why objects have color. If something is red, it is absorbing in the blue and green regions and scattering red light. If an object is black, it is absorbing all light. If an object is white, it is scattering all light.

As a general rule (but far from an absolute rule), when an electron absorbs light from the visible region, the electrons undergoing transitions are usually the ones in the outermost portion (shell) of the atom/molecule. With OCT, imaging is performed in the near-infrared region and generally involves electrons in bonds. These bonds have quantum mechanical vibrational or rotational states and the absorption of the photon results in the bond going to a higher energy state.

Absorbed photons are generally reradiated relatively rapidly, on the order of 10^{-8} to 10^{-9} seconds. From quantum mechanics, the mean radiation lifetime is given by the formula:

$$\tau = 3hc^3/2\pi\omega^3 D_o{}^2 \tag{3.22}$$

where the only new term is D_o which is the quantum mechanical dipole moment [3].

Absorption, dispersion, and refractive index are closely related. Figure 3.5 shows a plot of frequency versus real refractive index. At the resonance frequency, absorption will occur. However, as ω approaches the resonance frequency, it can be seen that the refractive index changes rapidly. This is because the energy is not absorbed but reradiated at a different phase. Furthermore, since the refractive index changes rapidly near the absorption peak, dispersion is particularly prominent. With OCT, a broadband source with a median wavelength near 1300 nm is used. Since a water absorption peak exists near 1380 nm, sources should be below 1300 nm to minimize dispersion.

3.4 SCATTERING

Light is scattered in tissue and is the basis of OCT imaging. Ultimately, it is related to the oscillating dipole, but it is more challenging to model than absorption or dispersion. Scattering will be treated at a superficial level in this section with minimal equations and the effects of polarization ignored. The scattering dealt with here is

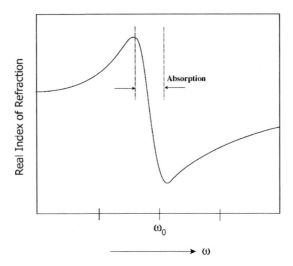

Figure 3.5 A plot of frequency versus real refractive index. At the resonance frequency, absorption will occur.

elastic scattering, meaning that the energy and frequency of the scattered light is the same as the light impinging on the scatterer. Inelastic scattering will be dealt with in Chapter 9. A more rigorous treatment of scattering will be included on the web page for this book (www.books.elsiever.com/companions).

Ultimately, we will find that most of the scattering we are concerned with will be from particles relatively large compared with the wavelength, composed of collections of oscillators, which have a different refractive index than the surrounding medium. It is the degree of mismatch, the size and shape of the scatter relative to the wavelength, and their relative density that determines light scattering.

To model scattering, we will begin with a gas consisting of a group of atoms that are separated from each other by a significant distance, more than $5\times$ their radius. The atoms are oscillating as per Eq. 3.26. The atoms then reradiate the radiation since they are oscillating dipoles. Since there is no type of organized pattern and the atoms are far apart, the arriving waves will be scattered in all directions, as shown in Figure 3.6. Therefore, the light will generate a uniform scattering pattern that is known as Rayleigh scattering. In general, Rayleigh scattering occurs when the particle is small relative to the wavelength, there is no organized, regular pattern, and the dipoles are relatively far apart. This is the situation in the upper atmosphere resulting in a blue sky because for Rayleigh scattering, the amplitude of scattering is $1/\lambda^4$. When light passes at an angle to the earth's surface, the blue light is scattered toward the earth while the red passes through. When the sun is just about to set, light travels its greatest distance through the atmosphere and most of the blue light has been scattered out, making the sun look red. OCT imaging is generally performed in situations where scattering is high and therefore not in the range of Rayleigh scattering.

When the atoms are closer together, say a half a wavelength apart, the light from one atom begins to predominately, destructively interfere with light from

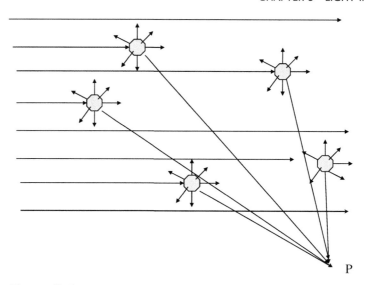

Figure 3.6 Since there is no type of organized pattern and the atoms are far apart, the phases of the arriving waves will be random in all directions or incoherent. However, wavelength dependency does exist as discussed in the text.

other atoms. Forward scattering begins to predominate since it is the only light which is in phase and constructively interfering. As the medium is packed even more densely and becomes well organized, the light passes through the medium almost exclusively in the forward direction. This is the origin of the refractive index in that the light is no longer angularly scattered. All that has occurred is that a phase shift has taken place as in Eqs. 3.18 and 3.21, which results in the phenomena behind the refractive index. This is the link behind the refractive index and scattering.

In general, OCT imaging is performed in a heterogeneous medium consisting of different atoms and molecules organized into particles. With OCT imaging, scattering is generally from particles much larger than the wavelength, meaning they are composed of a large number of oscillators. Mie's formula is a broader formula for determining the scattering of light, of which Rayleigh scattering is a special case. Mie's scattering tells us that light scattered from different sections of the same particle is now interfering with light from another section. The particle for the sake of this discussion is homogeneous but has a distinct refractive index from the environment. What becomes important now is the difference in refractive index between the inside and the outside of the particle, in addition to its size and shape. To a large degree, the particle becomes more and more forward scattering. With OCT this is problematic for several reasons. First, because OCT measures the intensity of backscattered light, this means a large majority of the signal will not be available for OCT imaging. Second, light that is scattered in any direction but backward can undergo multiple scattering, which is deleterious to OCT imaging for reasons discussed on the companion Web site.

3.5 SUMMARY

In this chapter we have modeled atoms and molecules as dipoles whose electrons (either in or out of bonds) can be induced to oscillate by a driving force, in particular, electromagnetic radiation. When the atom or molecule is relatively isolated, the oscillations result in the generation of a secondary wave that is uniform or ring shaped, but of different phase than the incident wave. However, when the impinging electromagnetic wave is near the resonance frequency, the energy is absorbed and the atom/molecule undergoes a transition in its energy state.

When the atoms/molecules are uniformly and densely packed, the resultant generated EM field cancels in all directions except the forward direction. Therefore, the EM propagates through the medium, but with a different phase than the incident light. This phase change of the forward directed light is the basis of the refractive index. When the medium becomes inhomogeneous, such as in the case of cells, the EM field from the dipoles composing the cells adds up in a complex manner. Now, the size, shape, and composition (refractive index) of the cell affect the scattering direction and intensity.

REFERENCES

1. Born, M. and Wolf, E. (1999). *Principles of Optics*. Cambridge University Press, Cambridge, UK.
2. Cheng, D. (1992). *Field and Wave Electromagnetics*. Addison-Wesley Publishing Co., Reading, MA.
3. French, A. P. and Taylor, E. F. (1978). *Introduction to Quantum Physics*. W.W. Norton and Company, New York, New York.
4. Hecht, E. (1998). *Hecht Optics*. Addison-Wesley Publishing Co., Reading, MA.
5. Milonni, P. and Eberly, J. (1988). *Lasers*. John Wiley & Sons, New York.

4 INTERFERENCE, COHERENCE, DIFFRACTION, AND TRANSFER FUNCTIONS

4.1 COHERENCE AND INTERFERENCE

This chapter covers the closely related topics of interference, coherence, diffraction, and transfer functions. Coherence, in particular, is important in understanding OCT. To adequately cover the topic, a significant amount of mathematics is required, but the principles can still be followed without a detailed mathematical analysis.

First order coherence (and second order interference), which involves fields, will be dealt with here. Second order coherence, which involves intensity correlations, is dealt with in later parts of this book.

4.1.1 Coherence

Coherence is a complex topic without a clear-cut definition. Its meaning will hopefully become clearer in later sections. This section is designed to give a mental picture of the phenomenon, but in later sections it will be addressed more rigorously. However, a definition taken from another source is a reasonable start:

> ...the amplitude and phase (of light) undergo fluctuations much too rapid for the eye or ordinary physical detectors to follow. If two beams originate in the same source, the fluctuations in the two beams are in general correlated, and the beams are said to be completely or partially coherent, depending on whether the correlation is complete or partial
>
> (Born and Wolf, *Principles of Optics*).

Although less accurate, coherence can also be envisioned as the degree of correlations that exist between the light fluctuations of two interfering beams.

Understanding light generation is useful in understanding coherence. Light sources in the real world are complicated. As has been stated, real sources are never truly monochromatic with even laser sources being quasimonochromatic. There is a range of reasons for this, which requires an understanding of how light is produced. In general, light is produced when an atom or molecule transitions from a higher energy state to a lower one. Energy has been absorbed to reach this higher state. The emission of light is not continuous from atoms but occurs over a period of $\sim 10^{-8}$ to 10^{-9} seconds. These short oscillatory pulses are referred to as wavetrains. One reason that light is not monochromatic is that, while the atoms are in the excited state, they are colliding. This generally results in the loss (or in some cases gain) of energy due to the transfer of energy from these collisions. Since the frequency of light is directly related to the energy ($E = hf$), these collisions result in changes in the frequency of the emitted light, therefore making the light polychromatic or containing multiple frequencies (wavelengths). Another mechanism by which the frequency range of emitted light is broadened is through the phenomenon of Doppler shifting. As stated in the previous chapter, the interaction of light with a moving object (or its release from) results in a Doppler shift. Since the atoms emitting light are moving in various directions, frequency shifts occur in the emitted light due to this movement. These are two reasons why light always contains more than one wavelength.

In Chapter 1, we discussed the properties of Fourier mathematics. From this discussion a wave can only be monochromatic if it is infinitely long. Specifically, the Fourier transform of a single frequency in the frequency domain (Dirac function) is spatially infinite in extent (spatial domain). Since light cannot be infinite in its extent or length, it can never be truly monochromatic; at best it can be only quasimonochromatic (small bandwidth).

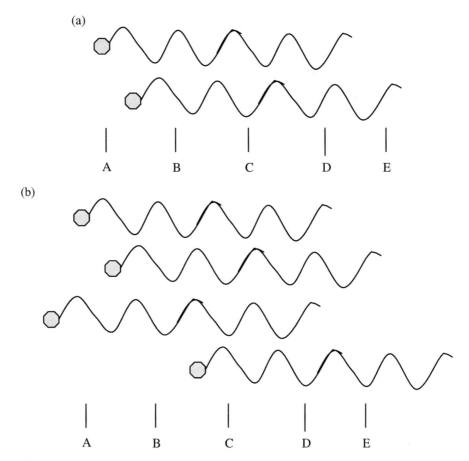

Figure 4.1 This is a simplified illustration of the principles behind coherence length. In (a), two overlapping of slightly different frequencies wavetrains in time are shown. Measurements between B and D yield identical results while those made between A and E do not. The distance between B and D is essentially the coherence length. In (b), four overlapping wavetrains, again all at slightly different frequencies, are shown but measurements are now only constant between C and D. This is because the 'coherence length' is shortened do to the increased bandwidth (range of wavelengths and phases).

To qualitatively appreciate coherence, envision the emission of two wavetrains. (This will be done for illustrative purposes but the reader should be aware that the analogy has some limitations, particularly at the quantum level.) The two wavetrains were emitted from two different atoms at slightly different times so that they have some overlap, but not completely as shown in Figure 4.1a. They are also at slightly different frequencies. If measurements were made at points B, C, and D, the results would be identical. However, measurements at A and E would yield different results. The time period over which the measurement is constant is known as the coherence time. This is an important term that will occur throughout the text. The distance light travels over that time, which is given by the coherence time × the speed of light, is known as the coherence length. Now envision four wavetrains being emitted as in Figure 4.1b. Again, the wavetrains are the same length but emitted at different times and all at slightly different frequencies. Now, the time period over which a measurement of the characteristics of the beam will yield the same result has decreased (just C and D). Therefore, it is said that the coherence time and length have shortened. In a real continuous wave (cw) source used with OCT, light is emitted by many atoms in such a way that the total average intensity (energy) remains relatively constant. Therefore, since the intensity is not varying dramatically, it is the amount of different frequencies and phases within the beam that influences the coherence length. A source with a small number of frequencies, like a laser, will have a very long coherence length/time, similar to the emissions from two atoms. A source with a wide range of frequencies, or broadband source, will have a shorter coherence length/time than one with a small range of frequencies. This again means the properties of the beam remain relatively constant only over the coherence length. This is addressed more rigorously below.

4.1.2   Interferometry and the Michelson Interferometer

OCT measures interference rather than backreflection through the use of an interferometer, which consists of a reference arm. The reference arm is needed since the backreflection intensity cannot be measured directly due to the high speed associated with the propagation of light, which is why OCT uses the intensity of interference to assess backreflection intensity indirectly. To introduce the concept of interferometry and ultimately coherence, a Michelson interferometer will be used. A schematic of the interferometer is shown in Figure 4.2. Light from the source is divided by a beam splitter. Half the light from the beam splitter is directed at a mirror in the sample arm while half is directed into a mirror that can move in the reference arm. Light reflects off both mirrors and is recombined at the beam splitter. Part of the light is then directed onto a detector. In the following discussion, first the light introduced into the interferometer will be of a single frequency (i.e., monochromatic). Then two monochromatic light waves of different frequencies will be examined in the interferometer. Finally, the light introduced will contain a wide range of continuous frequencies, as is the case with OCT. The different outcomes will be examined.

4.1.2.1 Interference of monochromatic waves with the same frequency

In the interferometer, light from the source, expressed in terms of the electric field E_{so}, is directed at the beam splitter. The beam splitter divides the light into E_r and E_s, which are the reference and sample arms, respectively. These two complex monochromatic plane waves have the same frequency, wavenumber, and phase as they are split. For the moment, we will ignore any losses occurring within the interferometer itself from scattering and other forms of attenuation. We will also assume that reflectivity off both mirrors is 100%. After reflecting off the two mirrors, the light recombines at the beam splitter, so that the electric field at the detector

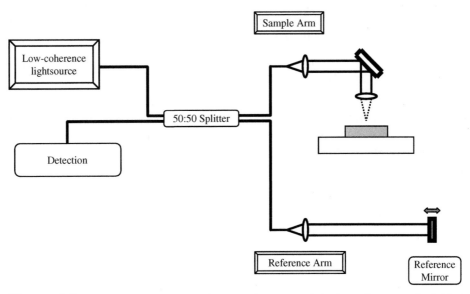

Figure 4.2 A simplified schematic of an interferometer. Light from the source is divided by a beam splitter. Half the light from the beam splitter is directed at a sample (which is a mirror here) while half is directed into a mirror that can move in the reference arm. Light reflects off both mirrors and is recombined at the beam splitter. Part of the light is then directed onto a detector.

is $E_D = (1/\sqrt{2})E_r + (1/\sqrt{2})E_s$. The reason it is $1/\sqrt{2}$ is that half the irradiance (and not the electric field) gets split back to the source in the beam splitter (while half goes to the detector). We will be examining what happens when monochromatic light in the two arms travels different distances and is then added together. Since $k = \omega/c$, we will represent these waves as:

$$E_r(x) = (1/\sqrt{2})E_{so}e^{i(\omega_r^{x/c})}$$
$$E_s(x) = (1/\sqrt{2})E_{so}e^{i(\omega_s^{x/c})}$$

(4.1)

It should be noted that we are ignoring phase changes induced by components of the interferometer to simplify calculations. These phase changes will be addressed in the next chapter. The only difference between the two waves is the distance they have traveled. So E_D equals:

$$E_D(x) = (1/\sqrt{2})E_r(x) + (1/\sqrt{2})E_g(x)$$
$$E_D(x) = (1/2)E_{so}e^{i(\omega_r^{x/c})} + (1/2)E_{so}e^{i(\omega_s^{x/c})}$$

(4.2)

Now what is generally measured at the detector is irradiance and not the electric field, which is given in Chapter 2 as the time average of the square of the electric field. This is represented by $I = \varepsilon v \langle EE^* \rangle_T$.

With the rules of complex numbers (ignoring the constants εv) this translates to:

$$I_D(x) \propto \langle E_D E_D^* \rangle_T = (1/2)\langle E_r E_r^* \rangle_T + (1/2)\langle E_s E_s^* \rangle_T + (1/2)\langle E_r^* E_s \rangle_T + (1/2)\langle E_s^* E_r \rangle_T$$

(4.3)

From Chapter 1, we know that $E_r{}^*$ and $E_s{}^*$ mean the complex conjugates of the electric field in the reference and sample arms, respectively. This becomes:

$$I_D(x) \propto \langle E_D E_D{}^* \rangle_T = (1/2)I_r(x) + (1/2)I_s(x) + (1/4)\langle E_{so}{}^* e^{-i\omega x_s/c}$$

$$\times E_{so} e^{-ix_r \omega/c} + E_{so} e^{i(\omega x_s/c)} E_{so}{}^* e^{-i(wx_s/c)} \rangle \tag{4.4}$$

In the right side of Eq. 4.4 the first two items represent the DC irradiance from sample and reference arms (which does not carry ranging information), respectively, as

$$I_r(x) \propto \langle E_r E_r{}^* \rangle_T \tag{4.4a}$$

$$I_s(x) \propto \langle E_s E_s{}^* \rangle_T \tag{4.4b}$$

We want Eq. 4.4 to be in the form:

$$I_D(x) = (1/2)I_r(x) + (1/2)I_s(x) + \langle \mathrm{Re}(1/2 E_{so} E_{so}{}^* e^{i(\omega x_r/c - \omega x_s/c)}) \rangle_T \tag{4.5}$$

To achieve this, we will use an identity from Chapter 1:

$$\cos\Theta = (e^{i\Theta} + e^{-i\Theta})/2 \text{ or } 2\cos\Theta = (e^{i\Theta} + e^{-i\Theta}) \tag{4.6}$$

Equation 4.4 now has the form:

$$I_D(x) = (1/2)I_r(x) + (1/2)I_s(x) + 1/4 \langle E_r E_s{}^* \rangle_T \cos\Theta \tag{4.7}$$

As a reminder, E_r and E_s are complex quantities. The terms $I_r(x)$ and $I_s(x)$ are DC terms of regular rapidly oscillating electrical fields of the original light waves, but the third term is the interference term. When Θ is 0 or a multiple of $\pm 2\pi$, the cosine's value is maximum at 1. The value of the interference term is then $1/4 \langle E_r E_s{}^* \rangle_T$. This is total constructive interference and the waves are said to be in phase (ignoring polarization effects). When the value of Θ is a multiple of $\pm\pi$, the value of the cosine is −1 and $I_D(x)$ is at a minimum. This situation is called total destructive interference. From Eq. 4.4a and 4.4b, it is known that $I_r(x) = I_s(x) = 1/2 \langle E_{so} E_{so}{}^* \rangle_T = (1/2)I_{so}$ (when reflectivity is equal in both arms). Equation 4.7 becomes:

$$I_D = (1/2)I_{so} + (1/2)I_{so}\cos\Theta = (1/2)I_{so}(1 + \cos\Theta) \tag{4.8}$$

When Θ equals 0, the intensity becomes I_{so}. When Θ is $\pm\pi$, the intensity is 0. A point that needs to be made here is the combined DC signal at the detector either will increase or decrease depending on the type of interference. However, there were no beats (or AC signal) as in the wave example in Chapter 1. The generation of beats requires at least two different frequencies.

4.1.2.2 Interference of monochromatic waves with two frequencies

With OCT, a Michelson interferometer is most commonly used with a broadband continuous source. Before moving on to interferometry with a broadband continuous source, we will take the intermediate step of looking at interferometry with the light of two frequencies. This will also allow us to introduce the concept of the autocorrelation function.

The most important aspect of the formulation is that unlike interference with waves of the same frequency, interference between waves of different frequencies results in *beats* or *amplitude modulation*. In using two frequencies, an assumption will be made that greatly simplifies the mathematics involved. The assumption is that, since the two frequencies are significantly far apart, the beat or amplitude modulation *within the source beam* (and not the beats from mismatch) is sufficiently rapid such that the detector cannot measure it. Therefore, beams will be treated initially like two independent beams whose detectable modulation will be due to the position of the reference arm relative to the sample arm. Then, how the detectable modulation of both beams adds together will be examined. Let us begin with Eq. 4.9:

$$I_D = 1/2 I_{so}(1 + \cos \Theta) \tag{4.9}$$

It is the relative position between the two arms that generates the significant beats, which we will see is embodied in the $\cos\Theta$ term. Again, the beats generated by the mixing of two wavelengths from the source cannot be detected because they are too rapid. Therefore, the two wavelengths can be treated independently and each represented by Eq. 4.9 (sample and reference arms of equal irradiance). They can be expressed as:

$$
\begin{aligned}
I_D = I_{D1} + I_{D2} &= 1/2 I_{so1}[1 + \cos(\omega_1 x/c)] + 1/2 I_{so2}[1 + \cos(\omega_2 x/c)] \\
&= 1/2 I_{so1} + 1/2 I_{so2} + 1/2[I_{so1}\cos(\omega_1 x/c) + I_{so2}\cos(\omega_2 x/c)]
\end{aligned} \tag{4.10}
$$

where subscripts *so*1 and *so*2 denote the two frequency components from the source and subscripts *D*1 and *D*2 denote the two frequency components at the detector. If $I_{D0} = 1/2 I_{so1} + 1/2 I_{so2}$, which are constant or DC terms, then Eq. 4.10 can be rewritten as:

$$I_D = I_{D0}(1 + \gamma_u(x)) \text{ where } \gamma_u(x) \text{ equals :}$$
$$\gamma_u(x) = 1/2[(I_{so1}/I_{D0})\cos(\omega_1 x/c) + (I_{so2}/I_{D0})\cos(\omega_2 x/c)] \tag{4.11}$$

This normalized function $\gamma(x)$ is called the degree of partial coherence or complex degree of coherence and is a measure of both interference and coherence. When $\gamma_u(x)$ equals 1, I_D is then the maximum with a value of 2 I_{D0}. This is complete coherence of the two light beams. When $\gamma_u(x) = 0$, there is no interference and I_D is at a minimum (DC signal only) with a value of I_{D0}. This is complete incoherence of the two light beams. Any value in between is known as partial coherence. The relation between ω_1 and ω_2 is critical for establishing $\gamma_u(x)$ or the degree of coherence. Assuming that $I_{so1} = I_{so2}$, then Eq. 4.11 can be rewritten through a trigonometric identity (Chapter 1) as:

$$\gamma_u(x) = [(I_{so2}/I_{D0})\cos(\omega_1 + \omega_2)x/(2c)]\cos[(\omega_1 - \omega_2)x/(2c)] \tag{4.12}$$

The first cosine term never equals zero (because there are no negative frequencies) and is a very fast oscillation. The second cosine term is the slow oscillating term, as in Chapter 1, and when it equals zero, $\gamma(x)$ equals zero. It is the AC oscillation or envelope of the first cosine term. This second term equals zero when:

$$(\omega_1 - \omega_2)x/(2c) = \pm\pi/2 \tag{4.13}$$

Again, the key feature here is that, by using two light waves of different frequencies, we develop amplitude modulation or beats, distinct from the outcome in the previous section with monochromatic waves.

4.1.3 Partial Coherence Using a Source with Continuous Frequencies

From our understanding of Fourier mathematics in the first chapter, it is not surprising that by using a finite number of frequencies, as in the previous section, the amplitude fluctuated in a periodic manner. Furthermore, this suggests that if a continuous spectrum were used, we would get a non-periodic or finite function. This is the case and is the subject of this section. In the discussions that follow, it is assumed that all quantities are stationary. Stationary means that the time average is independent of the time of origin chosen. So we are now going to modify Eq. 4.12 so that it contains an infinite number of wavelengths separated from each other by an infinitely small intervals.

$$I_D = (1/2)\int_0^\infty I_{so}(k)(1 + \cos\Theta)dk \tag{4.14}$$

Because ω and k are proportional to one another, for convenience, we have switched back to the spatial domain (k) since this is typically how OCT is described in the literature. Using a different cosine identity (Eq. 4.6) and letting Θ be equal to kx where x is the pathlength difference, Eq. 4.14 can be broken up into

$$\begin{aligned}
I_D &= 1/2\int_0^\infty I_{so}(k)dk + 1/4\int_0^\infty I_{so}(k)(e^{ikx} + e^{-ikx})dk \\
&= 1/2I_{so} + (1/4)\int_{-\infty}^\infty I_{so}(k)e^{ikx}dk
\end{aligned} \tag{4.15}$$

where $I_{so}(k)$ is the source's power at a given value of k and total power I_{so} equals $\int_0^\infty I(k)_{so}$. The integral $\int_{-\infty}^\infty I_{so}(k)e^{ikx}dk$ is referred to as the autocorrelation function (Γ_{11}). OCT measures the autocorrelation function and uses it to represent backreflection. The autocorrelation function can also equivalently be represented by:

$$\Gamma_{11} = \langle E_r(t)E_s{}^*(t + \tau)\rangle_T$$

where τ is the time delay between the two interferometer arms. Its relationship to the complex degree of coherence is:

$$\gamma_{11}(x) = \Gamma_{rs}/\sqrt{I_r I_s}$$

A relationship exists between the autocorrelation function and the power spectrum or spectral density function that is derived in Appendix 4-1. The power spectrum is the frequency-dependent spectrum (frequency distribution within beam) of the light that would be measured by an ideal spectroscope. The relationship is that, for second order coherence, the autocorrelation function of a stationary random process and the power spectrum of the process is a Fourier transform pair.

$$\Gamma_{11}(x) = \int P(\omega)e^{-i\omega x/c} d\omega \qquad (4.16)$$

It is nice that we have a function $\Gamma_{11}(x)$ that gives us an idea of the degree of coherence, but what does this mean from a practical standpoint? $\Gamma_{11}(x)$ can be looked at with respect to the interferometer in two ways: (1) the Fourier transform of the power spectrum and (2) how well two distinct portions of the light wave correlate.

The envelope of the correlation term can be written in terms of the degree of partial coherence or complex degree of coherence $\gamma_{11}(x)$ by multiplying the autocorrelation function by $\sqrt{I_r I_s}$. Equation 4.15 becomes:

$$I_D = (1/2)I_r + (1/2)I_s + Re\gamma(x)_{11}\sqrt{I_r I_s} \qquad (4.17)$$

This is a normalized function so that the real part of it has values from 0 to 1. The values of the degree of coherence are classified as follows:

$$|\gamma(x)_{11}| = 1 \text{ coherent limit}$$
$$|\gamma(x)_{11}| = 0 \text{ incoherent limit}$$
$$0 < |\gamma(x)_{11}| < 1 \text{ partial coherence}$$

How does it relate to the differences in pathlength in the interferometer? Returning to the concept of coherence time, it was stated that light remains constant to measurements for a period of time, then undergoes an abrupt change in phase (Figure 4.1). This period of constant composition is known as the coherence time. The distance light travels during the coherence time is referred to as the coherence length (l_c) and is found by multiplying the coherence time by the speed of light. The relation between the coherence length and $|\gamma(x)_{11}|$ for a linear decay is:

$$|\gamma(x)_{11}| = 1 - l/l_c \qquad (4.18)$$

when $l < l_c$ and l is the difference in pathlength between the two mirrors. When $l > l_c$, the value of the function is zero. What this tells us is that when the distance traveled in each arm is different by an amount less than the coherence length, we get an AC component in addition to the DC component at the detector. When the distance is greater than coherence length, only the DC signal is noted. Now let us say we have a way of measuring the AC component. We can see now that if we know the distance light travels in the reference arm, we can determine that the mirror in the sample is the same distance to within the coherence length. To within the coherence length means that, if the coherence length were 30 μm, then when we get the AC signal, the two mirrors are aligned (optical path length equal) to within 30 μm. This is ranging. We are using the reference arm to tell us how far away something is in the sample arm. If we decreased our coherence length to 10 μm (and we would do that by increasing the bandwidth as per Eq. 4.16), we would increase our ability to tell where the mirror is, or resolution, by a factor of three.

Two closing points will be discussed before moving on to diffraction. The first is that we made the assumption that there were no losses in the interferometer. This is not entirely true so for correctness, constants C_1 and C_2 should be in front of all the intensity and electric field terms. This is done in the next chapter. Second, polarization effects were ignored and will be discussed in later chapters.

4.2 DIFFRACTION

A detailed discussion of diffraction is not warranted because it is a broad area, and we will limit ourselves to those topics directly related to OCT. These include two important areas, diffraction limited imaging and the diffraction grating. The diffraction grating will be introduced later on in the text, so that the focus here will be on the meaning of diffraction limited imaging. As diffraction is a complex topic, it is again not discussed in detail, but references are listed at the end of the chapter for those interested.

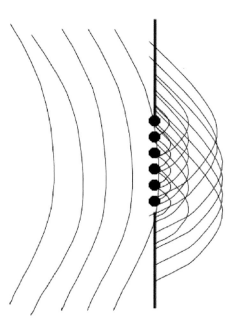

Figure 4.3 Huygens principle. This is best described by envisioning a water wave approaching a wall with an opening. With this simple illustration the points along the opening now serve as sources of wavefronts. However, the Huygens principle does not take into account the phase difference among points.

4.2.1 Huygens and Huygens-Fresnel Principle

A simple beginning to the understanding of diffraction is through the presentation of the Huygens principle. A simple statement of the Huygens principle is that every point on a propagating wavefront serves as the source of new spherical secondary wavelets. The wavefront at this time consists of these new wavelets and the new wavefront is the summation of these wavelets. This is best described by envisioning a water wave approaching a wall with an opening. Basically, this means that as the wave in Figure 4.3 engages the wall with the opening, the wave in the opening behaves as if it consists of an array of new sources. The problem with this model is that it does not take into account frequency or wavelength of the wave, which is known to have a critical effect on diffraction. The Huygens-Fresnel principle was then developed which states that[3]

> every unobstructed point of a wavefront, at any given instant, serves as a source of spherical secondary wavelets (with the same frequency as the primary wave). The amplitudes of the optical field at any point beyond are the superposition of all these wavelets (considering their amplitudes and relative phases).

This is shown in Figure 4.4 where it can be seen that as the wavelength gets longer, the diffraction pattern or bending gets wider. This wavelength dependency

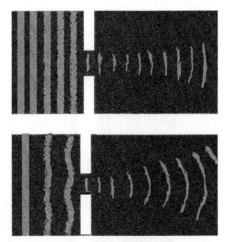

Figure 4.4 This figure illustrates the Huygens-Fresnel principle which states that "every unobstructed point of a wavefront, at any given instant, serves as a source of spherical secondary wavelets (with the same frequency as the primary wave). The amplitudes of the optical field at any point beyond are the superposition of all these wavelets (considering their amplitudes and relative phases)." As the wavelength gets longer, the diffraction pattern or bending gets wider.

is one of the most important characteristics of diffraction. The second characteristic is, the smaller the opening, the greater the diffraction. This is the advantage of the Huygens-Fresnel model over the Huygens model. It should be no surprise that the relationship between the opening and the width of the diffraction pattern is a Fourier relationship. A rigorous mathematical description of diffraction will not be considered here, but reference texts that address the issue are discussed below.

4.2.2   Fraunhofer and Fresnel Diffraction

We can artificially divide diffraction into two general extremes, either Fraunhofer (far-field) diffraction or Fresnel (near-field) diffraction. OCT imaging is basically always in the far-field but both types will be defined here. For Fraunhofer diffraction, if the source and point of observation are sufficiently far from the opening, then the diffraction pattern will be spread out, contain multiple fringes, and will not resemble that of the shape of the aperture or opening. This is seen in Figure 4.5. If, on the other hand, the point of observation is near the aperture, then Fresnel diffraction will dominate, which most closely resembles the aperture, as shown in Figure 4.6. Near and far are clearly relative, but will depend on, among other things, the size of the aperture and the wavelength. Also, some authors describe the Fresnel region as both the near- and far-field. Readers should therefore be aware, when reading other texts, to clarify how the Fresnel region is defined.

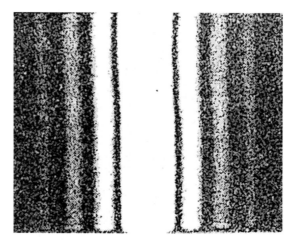

Figure 4.5   For Fraunhofer diffraction, if the source and point of observation are sufficiently far from the opening, then the diffraction pattern will be spread out, contain multiple fringes, and will not resemble that of the shape of the aperture or opening.

Fresnel Diffraction

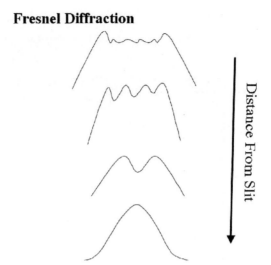

Distance From Slit

Figure 4.6 If, on the other hand, the point of observation is near the aperture, then Fresnel diffraction will dominate, which most closely resembles the aperture.

4.2.3 Diffraction Limited

A well-focused, aberration-free converging lens never focuses light to a single point but always has some sort of diffraction pattern associated with it. The smaller the focus or longer the wavelength, the worse the diffraction effect. This is particularly important since the lateral resolution with OCT is essentially dependent on the lens used. Since most of the lenses dealt with here are circular, this section will focus on diffraction of a circular aperture. The formula for a circular aperture will not be derived but merely stated; for those interested, the derivation can be found from the referenced sources. The formula is important because it allows us to understand the ultimate limit to resolution in the far-field. The equation, which is derived elsewhere, is:

$$I(\Theta) = I(0)[2J_1(ka \sin \Theta)/ka \sin \Theta]^2 \qquad (4.19)$$

The equation is plotted in Figure 4.7 and tells us how the intensity of light is redistributed as a function of the angle (Θ) and a distance measurement (a) (what specific angle and distance is not critical to the discussion). In this formula, $I(0)$ is the irradiance at the center and k is the wave number. The term J_1 is known as a Bessel function of the first kind (or sometimes referred to as just a Bessel function) and shows up frequently in many physics applications and this text. This function is the solution of a second order differential equation known as Bessel's differential equation. Since this differential equation is second order, there is a second independent solution referred to as a Bessel function of the

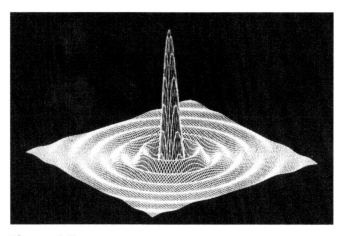

Figure 4.7   Equation 4.19 is plotted and shows how the intensity of light is redistributed as a function of the angle (Θ) and a distance measurement (a). The specific angle and distance is not critical to the discussion. Courtesy of *Hecht Optics*, Figure 10.28, page 462 [3].

second kind or Neumann function. Because this second function yields physically impossible solutions in situations we are interested in, then the Bessel function of the first kind is deemed the real solution. The general formula for a Bessel function of the first order is

$$J_n(t) = (1/2\pi) \int_0^{2\pi} e^{it \sin \o - in \o} d\o \tag{4.20}$$

Although a strange looking equation that many are unfamiliar with, it is not really any more difficult to utilize than the trigonometric functions we have discussed previously. In general, a function should be suspected as a Bessel function when the exponential contains a sine or cosine.

The center of irradiance surrounded by the dark ring is the Airy spot. It has a finite width and contains 84% of the total irradiance, which is shown in Figure 4.7. The Airy spot is the carrier of relevant information transmitted through the lens. If we focus a perfect circular lens onto a point source, the Airy disc represents the source light.

Resolution, in this case, can be thought of as the ability to discriminate between two incoherent sources, as seen in Figure 4.8. According to Rayleigh's criterion, two Airy discs are just barely resolvable when the central peak of one Airy disc coincides with the first minimum of the other (Figure 4.8). Therefore, the wider the Airy spot, the lower the resolution. The central disc or Airy pattern is surrounded by a dark band which occurs when $I(\Theta)$ equals zero. From charts on Bessel functions, $J_1(ka \sin \Theta)$ equals zero when $ka \sin \Theta$ equals 3.83. Knowing this, for a point source at the focal length of the lens, Eq. 4.22 can be rearranged to give the maximum of the Airy disc:

$$I_m(\Theta) = 1.22 f\lambda/D \tag{4.21}$$

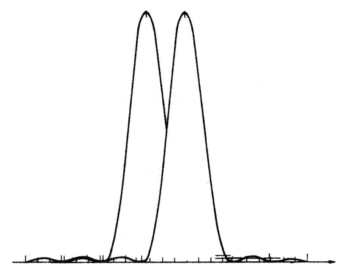

Figure 4.8 Resolution, in this case, can be thought of as the ability to discriminate between two incoherent sources, as seen in this figure. According to Rayleigh's criterion, two Airy discs are just barely resolvable when the central peak of one Airy disc coincides with the first minimum of the other.

where D is the diameter of the circular aperture. This tells us that a larger diameter lens and shorter wavelengths result in smaller Airy discs and therefore higher resolutions. Then, a theoretical perfect lens is said to be diffraction limited because, at a given f, λ, and D, the Airy spot will have a finite width that determines the resolution. This is illustrated in Figure 4.9. In the top half of the figure is the object imaged. It consists of structures that are extremely thin (i.e., delta functions). The bottom of the figure is how the structures are resolved.

Since the system can only resolve the structures to the width of the Airy function, the ability to image the object is limited. For the two structures at A, they can barely be resolved while for C, the three structures cannot be resolved and appear as a single structure. Current imaging systems are typically diffraction limited since components, such as lenses, are not perfect.

4.3 CONVOLUTION AND TRANSFER FUNCTIONS

The system can also be defined in terms of the convolution and transfer function. The convolution function is an important mathematical concept used to understand the combination of two functions, in our case the object and the imaging system. It involves taking one function, sweeping it across another, and taking the area of overlap at every point to produce a new function. It essentially "blends" the two functions together. Looking at it another way, when performing

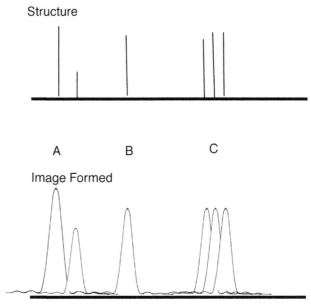

Figure 4.9 Then, a theoretical perfect lens is said to be diffraction limited because, at a given f, λ, and D, the Airy spot will have a finite width that determines the resolution. The top half contains the object imaged. It consists of structures that are extremely thin (i.e., delta functions). The bottom shows how the structures are resolved. Since the system can only resolve the structures to the width of the Airy function, the ability to image the object is limited.

a convolution, an image is obtained by establishing an appropriately weighted point spread function (PSF) (such as the Airy function in perfect circular aperture) to the location of each point on the image then adding them up (assuming for now an incoherent source). Figure 4.10 shows two identical square waves, $h(x)$ and $f(x)$. $h(x)$ is swept across $f(x)$ and the enclosed area is taken at each point. The areas at each point are summed to form the new function $g(x)$. $h(x)$ is said to be *convolved* with $f(x)$, or vice versa. The corresponding expression describing the convolution of the two functions $f(x)$ and $g(x)$ is:

$$h^*f = h \otimes f = g(x) = \int (h(X)f(x - X)dX \qquad (4.22)$$

where h^*f denotes the convolution of $h(x)$ and $f(x)$ and is sometimes denoted by $h \otimes f$. The integral is generally taken to infinity but one could set limits to isolate a desired area. Even though the integral is carried to infinity, the resultant function is finite because $g(x)$ is nonzero only when the curves overlap, that is $g(x-\tau)$ is nonzero. It is useful to visualize a convolution using different types of functions such as the Dirac delta function or a Gaussian function (see Figure 4.12). Here $G(x)$ is the performance of the input, $F(x)$ is a Gaussian function, and $H(x)$ is a Dirac

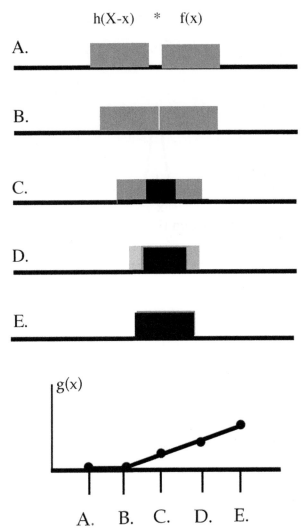

Figure 4.10 This demonstrates the principles behind the convolution. It shows two identical square waves, $f(x)$ and $g(x)$. $f(x)$ is swept across $h(x)$ and the enclosed area is taken at each point. The areas at each point are summed to form the new function $g(x)$. $f(x)$ is said to be *convolved* with $h(x)$, or vice versa.

function which are how each system analyses an input. The respective responses are $R(x)$. Notice how the spread of each function affects $G(x)$. The convolution of an OCT system will be described in the next chapter.

There are several important properties of the convolution that need to be discussed. It is evident that $f*g = g*f$ (substituting $-x$ for x and watching the limits of integration). Of even greater interest is the convolution theorem which states that

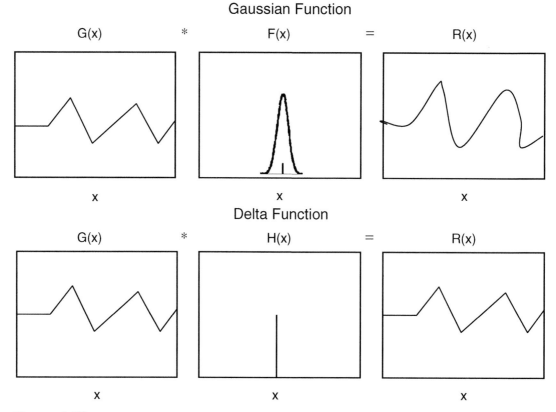

Figure 4.11   It is useful to visualize a convolution using different types of functions such as the Dirac delta function, $H(x)$, or a Gaussian function, $F(x)$. The Gaussian function smoothes out the input signal, $G(x)$, while the Dirac delta function has the same output as input.

the Fourier transform of the convolution equals the product of the individual Fourier transforms:

$$F\{h\} = F\{f^*g\} = F\{f\} \cdot F\{g\} \tag{4.23}$$

This can be useful in solving certain problems. Very often, it is easier to perform the Fourier transform on the two functions, multiply them, and then performing the inverse Fourier transform on the product.

4.3.1   Transfer Function

In generating an image with an optical system, the final image $I_M(x, y)$ can be viewed as the convolution between the light from the imaged object $I_0(x, y)$ and the PSF, $h(x, y)$.

$$I_M(x, y) = h(x, y) \otimes I_0(x, y)$$

As we discussed in the previous section, it can often be easier and more informative to analyze the system by performing the Fourier transform on $I_0(x, y)$ and $h(x, y)$ and then multiplying them.

$$F\{I_M(x, y)\} = F\{h(x, y)\} \cdot F\{I_0(x, y)\}$$

Here, $F\{h(x, y)\}$ is the transfer function of the system.

This initial discussion of transfer functions will be for an incoherent source (i.e., we do not need to consider interference). Coherent sources will be addressed in subsequent chapters. The PSF is useful for determining the maximal resolution of the system and above we represented the image as overlapping PSF. However, this is not the only parameter that is useful to know when studying an optical system. Higher resolution does not necessarily mean greater performance, which is why transfer functions become useful. An example that does not involve an optical system, but illustrates the point, is a quality stereo system that responds to high frequencies. The response to high frequencies is analogous to the high resolution of our optical system, but this does not mean that a system with a lower frequency response does not perform better at lower frequencies. Therefore, by exampling the spatial frequency in the Fourier domain, additional insight can be provided into the system performance.

To better understand our system's performance, we will introduce the concepts of contrast or modulation. One way of defining contrast or modulation is with the formula:

$$\text{contrast} \equiv I_{max} - I_{min}/(I_{max} + I_{min}) \tag{4.24}$$

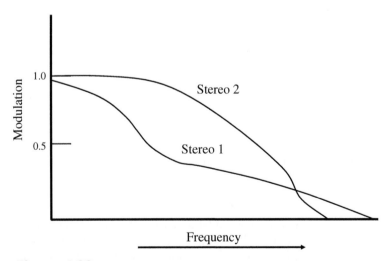

Figure 4.12 In this figure, stereo 1 has the higher frequency response but stereo 2 has better performance at lower frequencies. This is equivalent to a theoretical OCT system with lower resolutions having better performance than another high resolution OCT system when examining structures significantly larger than the coherence length. Based on concepts from [3].

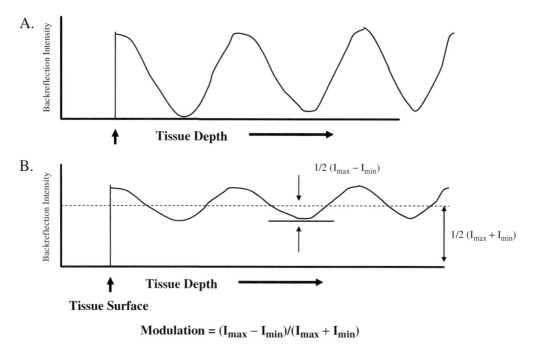

Modulation = $(I_{max} - I_{min})/(I_{max} + I_{min})$

Figure 4.13   This figure further illustrates the principles behind modulation. In addition to resolution, it is important to measure contrast or modulation. The modulation transfer function (MTF) is measured using tissue that has a backreflection intensity that is sinusoidal in nature with depth, as shown in A. In B, the output is demonstrated. Contrast has been reduced when passing through the optical system.

In later chapters on image processing, other definitions of contrast will be examined. In Figure 4.12, stereo 1 has the higher frequency response but stereo 2 has better performance at lower frequencies. This is equivalent to an OCT system with lower resolutions having better performance than another high-resolution OCT system when examining structure significantly larger than the coherence length.

Transfer functions can give additional information in addition to the PSF, including the contrast and how the system is performing on objects significantly larger than the PSF. When dealing with multiple spatial frequencies, the ratio of the image modulation to the object frequencies at all frequencies is referred to as the modulation transfer function. In Figure 4.13A, the modulation transfer function (MTF) is measured using tissue that has a backreflection intensity which is sinusoidal in nature. When looking at Figure 4.13B, the measured response, contrast has been reduced on passing through the optical system. Here $I_{max} - I_{min}$ is the total fluctuation from peak to trough and one-half $I_{max} + I_{min}$ is the mean.

With an incoherent source, there is no consistent relationship between the phases within the light beam. Therefore, the intensities can be added directly (i.e., interference is not a concern). However, with a coherent source, phase differences will occur within different sections of the image. This results in interference phenomena in addition to the pattern that would be produced by incoherent light. Among the interference phenomena is the granularity that is referred to as speckle.

References

1. Born, M. and Wolf, E. (1999). *Principles of Optics*. Cambridge University Press, Cambridge, UK.
2. Cheng, D. (1992). *Field and Wave Electromagnetics*. Addison-Wesley Publishing Co., Reading, MA.
3. Hecht, E. (1998). *Hecht Optics*. Addison-Wesley Publishing Co., Reading, MA.
4. Teich, M. and Saleh, B. (1991). *Fundamentals of Photonics*. John Wiley & Sons, Inc., New York.

APPENDIX 4-1

WEINER KHINCHIN THEOREM

The Weiner Khinchin (sometimes spelled Khintchine) theorem states that a Fourier relationship exists between the autocorrelation function of a stationary, random process and the power spectrum of the process. A relationship had previously been known between random processes and the power spectrum prior to the landmark works of Weiner and Khinchin. This included work occurring in a range of fields from economic cycle and marriage rates to a recently found paper by Albert Einstein.[1-4] Therefore, some controversy remains about the origin of the theorem. However, it is clear that Weiner and Khinchin brought the topic to the forefront.[5]

An electric field can be represented as the integral sum of its individual frequency components.

$$E(\omega) = \int_{-\infty}^{\infty} E(t)e^{i2\pi\omega t} dt \tag{1}$$

The time average of the electric field is given by:

$$|E(\omega)|^2 = \int\int_{-\infty}^{\infty} E^{\#}(t)E(t')e^{i2\pi\omega(t'-t)} dt dt' = \int\int E^*(t)E(t+\tau)e^{2\pi i\omega\tau} dt d\tau \tag{2}$$

where $\tau = t' - t$. The electric field autocorrelation function for the period T that is large but not infinite can be represented by:

$$\Gamma_{11}(t) = lim(1/T)\int_0^T E^*(t)E(t+\tau)dt$$
$$T \rightarrow \infty$$

It is assumed that the statistical properties of the electric field are time independent or stationary. Therefore, although the process is averaged over a finite period, it is considered representative of integration over an infinite period. The autocorrelation function can then be substituted into Eq. 2 and it becomes:

$$P(\omega) = |E(\omega)|^2 = \int_{-\infty}^{\infty} \Gamma_{11}(t)e^{i2\pi\omega\tau} d\tau \tag{4}$$

This can be recognized as a Fourier integral with P(ω) and $\Gamma_{11}(t)$ representing the Fourier pair. The power spectrum (or spectral density) is a Fourier transform pair with the autocorrelation function of a stationary random process.

REFERENCES

1. Einstein, A. (1914). Method for the determination of the statistical values of observations concerning quantities subject to irregular fluctuations. *Arch. Sci. Nat.* 37, 254–256.

2. Hooker, R. H. (1901). Correlation of the marriage-rate with trade. *J. R. Statist. Soc.* 64, 485–492.
3. Moore, H. L. (1914). *Economic Cycles: Their Law and Cause,* Macmillan, New York.
4. Clayton, H. H. (1917). Effect of short period variation of solar radiation on the earth's atmosphere. *Smithsonian Miscellaneous Collections,* 68, Pub. 2446.
5. Mandel, L. and Wolf, E. (1995). *Optical Coherence and Quantum Optics.* Cambridge University Press, New York.

SECTION II

5 OPTICAL COHERENCE TOMOGRAPHY THEORY

Chapter Contents

5.1   GENERAL

This chapter examines the major physical principles behind optical coherence tomography (OCT) imaging. These principles will be derived in considerable detail but for those only interested in the final results, skip the intermediate steps. The extensive derivations are provided to emphasize assumptions and approximations made so that the limitations of mathematical relationships presented in the literature can be understood. They are also provided for those in training who often find equations in the literature whose origins are not obvious.

Time domain OCT (TD-OCT) with a changing optical group delay in the reference arm will be examined first focusing on mechanisms generating high axial resolutions. It will initially be described in terms of the axial point spread function (PSF) and later in terms of the transfer function, as both are used in the literature. Dispersion and principles determining lateral resolution will then be discussed. Finally, this chapter concludes by examining spectral domain OCT (SD-OCT), which has recently generated considerable attention. Other aspects of the technology, such as signal to noise ratio (SNR) and polarization properties, will be discussed in later chapters. Throughout, the focus will be on the physics and application of OCT, not on details relevant to constructing the device.

OCT is analogous to ultrasound, measuring the backreflection intensity of infrared light rather than sound. However, unlike ultrasound, the backreflection intensity cannot be measured electronically due to the high speed associated with the propagation of light. Therefore, a technique known as low coherence interferometry is used. OCT requires an interferometer and will be described here with a Michelson interferometer, shown in Figure 5.1, the most common embodiment and similar to that in Chapter 4. In TD-OCT, light from the source is divided evenly by the beam splitter, half toward the sample and half toward a moving mirror (Figure 5.1A). Light reflects off the mirror and from within the sample. The light is recombined by the beam splitter and directed at the detector. If the pathlengths match to within a coherence length, interference will occur. OCT measures the intensity of interference and uses it to represent backreflection intensity. Figure 5.1B and C are schematics for Fourier domain (FD-OCT) and swept source (SS-OCT), respectively. They will be discussed later in the chapter.

5.2   TIME DOMAIN OCT (TD-OCT)

5.2.1   Interferometry with a Monochromatic Source

The performance of OCT imaging can be described mathematically using an approach similar to that used in Chapter 4, starting initially with a monochromatic wave. However, in this section, the interferometer will be described in more detail, in terms of the electrical field rather than the intensity, so that both amplitude and phase components can be followed. The mathematical manipulations are slightly more complicated since now the phase is introduced through the imaginary

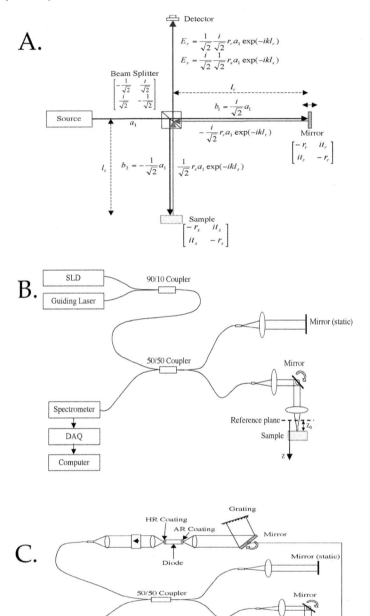

Figure 5.1 Image A is a schematic of time domain (TD-OCT). The embodiment includes a moving mirror in the reference arm. The schematic includes alterations in phase due to beam splitters and mirrors as well as reductions in intensities at different stages. Images B and C are schematics of Fourier domain (FD-OCT) and swept source (SS-OCT), respectively.

term i and negative signs, where we will see that the former represents a 180-degree change in phase while the latter is a 90-degree phase change. In subsequent sections we will generally ignore these phase relationships for simplicity, but the reader can refer back to this section when necessary although these particular phase changes. We will initially begin with a monochromatic complex plane wave of the form[1]:

$$E_{so} = E_0 e^{-ik(\omega)z} \tag{5.1}$$

Once again, $k = 2\pi/\lambda$.

In the interferometer, beam splitters and mirrors are represented by the general matrix:

$$\begin{bmatrix} -r & it \\ it & -r \end{bmatrix} \tag{5.2}$$

where r is the reflectivity and t is the transmission. The negative sign represents a 180-degree phase shift while i is a 90-degree phase shift. For the 50/50 beam splitter the ABCD matrix is:

$$\begin{bmatrix} -1/\sqrt{2} & i/\sqrt{2} \\ i/\sqrt{2} & -1/\sqrt{2} \end{bmatrix} \tag{5.3}$$

The two light beams entering each arm are represented by:

$$E_r = -(1/\sqrt{2})E_{so} \quad \text{and} \quad E_s = (i/\sqrt{2})E_{so}$$

The two light beams from the beam splitter are directed onto mirrors in the reference and sample arms (later a sample with backreflection as a function of depth will be introduced). The matrix for the mirrors is:

$$\begin{bmatrix} -r & 0 \\ 0 & -r \end{bmatrix} \tag{5.4}$$

After reflecting off the mirrors and entering the beam splitter, the electric fields become:

$$E_{r2} = r_r(1/\sqrt{2})E_{so}e^{-2kl_r} \quad \text{and} \quad E_{s_2} = -r_s(i/\sqrt{2})E_{so}e^{-2kl_s} \tag{5.5}$$

A phase change of 180 degrees occurred from the mirrors represented by the change in sign. Also, the pathlength in both arms is represented by $2kl_r$ and $2kl_s$. Repassing through the beam splitter, the field in the detector arm becomes:

$$E_D = (i/\sqrt{2})E_{r2} + (-1/\sqrt{2})E_{s2} = r_r(i/\sqrt{2})(1/\sqrt{2})E_{so}e^{-i2kl_r} + r_s(i/\sqrt{2})(1/\sqrt{2})E_{so}e^{-i2kl_s}$$

$$= E_R e^{-i20kl_r} + E_S e^{-i2kl_s} \tag{5.6}$$

We will be using E_R and E_S (field between the splitter and the detector exit) rather than E_r and E_s for simplicity. Dispersion, backscattering from more than one depth, and sample birefringence will be examined later. The detector measures the irradiance and not the electric field, so phase information in general is lost (we will see ways of retrieving phase information in later chapters). The irradiance approaching the detector arm is obtained by squaring both sides of Eq. 5.6. Performing both the squaring and utilizing the normalized function $\gamma(x)$ complex degree of coherence as described in Chapter 4, the following is obtained:

$$I_D = |E_D|^{\sim 2} = \langle E_D E_D^* \rangle = \langle E_R E_R^* \rangle + \langle E_D E_D^* \rangle + E_R E_S^* e^{-i2kl_r + i2kl_s} + E_S E_R^* e^{-i2kl_r + i2kl_s}$$

$$= I_R + I_S + \mathrm{Re}\gamma(z)_{11} \sqrt{I_r I_s} [e^{+i2k(l_r - l_s)} + e^{+i2k(l_r - l_s)}] \tag{5.7}$$

I_D represents the combine signal approaching the detector.

An identity from trigonometry is now reintroduced so that we have a better understanding of the physical implications of Eq. 5.7. Since $\cos\Theta = (e^{i\Theta} + e^{-i\Theta})/2$ or $2\cos\Theta = (e^{i\Theta} + e^{-i\Theta})$, then it becomes:

$$I_D = I_R + I_S + 2\mathrm{Re}\gamma(z)_{11} \sqrt{I_r I_s} \cos(2k\Delta l). \tag{5.8}$$

Therefore, again as in Chapter 4, the irradiance has DC components (I_R, I_S) and an amplitude modulated as a function of the pathlength difference. Remembering that $k = 2\pi/\lambda$, the argument of the cosine becomes $(2\pi\Delta l)/(\lambda/2)$, so that the period of the interference is half the wavelength relative to the pathlength mismatch. Depending on the pathlength mismatch, constructive or destructive interference is occurring in the detector arm. So at a given mismatch, the interference term is a constant (no envelope modulation) and shifts the baseline of the DC signal. Since I_{so}, the total intensity of the source is constant when I_D increases from constructive interference; the irradiance directed in the source arm is decreased due to conservation of energy. When destructive interference is occurring in the sample arm, irradiance in the source arm is increased.

With TD-OCT, the mirror in the reference arm is moving with a velocity v_M and therefore a Doppler or frequency shift occurs in the signal. The Doppler shift is given by $f_D = 2v_M/\lambda$. Therefore, the argument becomes $2\pi f_D t$ and Eq. 5.8 becomes:

$$I_D = I_R + I_S + 2\mathrm{Re}\gamma(z)_{11} \sqrt{I_r I_s} \cos(2\pi f_D t) \tag{5.9}$$

5.2.2 Low Coherence Interferometry with OCT

Equation 5.9 gives a general formula for the Michelson interferometer, using a monochromatic light source, with a 50/50 beam splitter and a moving mirror in the reference arm for light. To introduce low coherence interferometry, using a broad bandwidth source, this formula will be slightly modified. To simplify our understanding of the underlying physics, phase and amplitude alterations by the beam splitter and mirror will not be included (replaced by proportionality sign). Also, the DC terms will be treated as constants. We will focus on light in the detector arm consisting of components from the sample $\{E_S(\omega)\}$ and reference $\{E_R(\omega)\}$, which are given by:

$$E_R(\omega)\ \alpha\ E_r(\omega)e^{(-ikr(\omega)2l_r-\omega t)}$$

$$E_S(\omega)\ \alpha\ E_s(\omega)e^{(-iks(\omega)2l_s-\omega t)} \tag{5.10}$$

The irradiance in the detector arm is therefore proportional to where E_R and E_S are the field entering the reference and sample arm, respectively:

$$I_D\ \alpha\ \text{real}\left\{\int_{-\infty}^{\infty}\langle E_S(\omega)E_R(\omega)^*\rangle d\omega\right\} \tag{5.11}$$

Based on our discussion in previous sections, it is also proportional to the power spectrum of the source $\{G(\omega)\}$:

$$I_D\ \alpha\ \text{real}\left\{\int_{-\infty}^{\infty}G(\omega)e^{-i\Delta\phi(\omega)}d\omega\right\} \tag{5.12}$$

where $\phi(\omega)$ equals $2\ k_S(\omega)l_S - 2\ k_R(\omega)\ l_R$.

The light source is band-limited with a central frequency of $\omega_{\bar{0}}$. Now, assuming a *uniform, linear, and non-dispersive medium*, k is the propagation constant in each arm and can be considered equal. We will approximate that it can be rewritten as a first order Taylor expansion (Chapter 1) so that the phase mismatch is solely dependent on the pathlength difference:

$$k_s(\omega) = k_R(\omega) = k(\omega_0) + k'(\omega_0)(\omega - \omega_0) \tag{5.13}$$

The wavenumber is now described with respect to ω_0. Here $\Delta\phi(\omega) = k\ (\omega_0)2\Delta l + k'$ $(\omega_0)\ (\omega - \omega_0)\ 2\Delta l$ and $\Delta l = l_S - l_R$, and Eq. 5.12 can be written as:

$$I\alpha\ \text{real}\left\{\exp[-i2\Delta lk(\omega_0)]\int_{-\infty}^{\infty}G(\omega-\omega_0)\exp[-i(\omega-\omega_0)k'(\omega_0)(2\Delta l)]d(\omega-\omega_0)/2\pi\right\} \tag{5.14}$$

To introduce the concept of phase and group delay, the equations can be rewritten in the form:

$$I\alpha\,\text{real}\left\{\exp[-i2\Delta l\omega_0k(\omega)/\omega_0]\int_{-\infty}^{\infty}G(\omega-\omega_0)\exp[-i(\omega-\omega_0)k'(\omega_0)(2\Delta l)]d(\omega-\omega_0)/2\pi\right\}$$

If the phase ($\Delta\tau_p$) and group ($\Delta\tau_g$) are defined as:

$$\Delta\tau_p = (k(\omega_0)/\omega_0)(2\Delta l) = (2\Delta l)/v_p \quad v_p = \omega_0/k(\omega_0) \tag{5.15}$$

$$\Delta\tau_g = k'(\omega_0)\ (2\Delta l) = (2\Delta l)/v_g \quad v_g = 1/k'(\omega_0) = 1/(\partial k/\partial\omega(\omega_0))$$

where v_p and v_g are the phase and group velocity, respectively. Then Eq. 5.14 becomes:

$$I\alpha\,\text{real}\left\{\exp[-j\omega_0\Delta\tau_p]\int_{-\infty}^{\infty}G(\omega-\omega_0)\exp[-j(\omega-\omega_0)\Delta\tau_g]\frac{d(\omega-\omega_0)}{2\pi}\right\} \tag{5.16}$$

Now, with OCT it is optimal to have a source with a Gaussian spectrum to optimize resolution since the autocorrelation function of a Gaussian source is Gaussian. A Gaussian spectrum can be written in the form:

$$G(\omega - \omega_0) = \left(\frac{2\pi}{\sigma_\omega^2}\right)^{1/2} \exp\left[-\frac{(\omega - \omega_0)^2}{2\sigma_\omega^2}\right] \qquad (5.17)$$

where σ_ω is the standard deviation of the angular frequency spectrum. We want to rewrite Eq. 5.16 when the equation is Gaussian. Substituting into equation 5.17, we obtain:

$$I \alpha \operatorname{real}\left\{\exp[-j\omega_0\Delta\tau_p]\int_{-\infty}^{\infty}\left(\frac{2\pi}{\sigma_\omega^2}\right)^{1/2}\exp\left[-\frac{(\omega-\omega_0)^2}{2\sigma_\omega^2}\right]\exp[-j(\omega-\omega_0)\Delta\tau_g]\frac{d(\omega-\omega_0)}{2\pi}\right\}$$

$$\alpha \operatorname{real}\left\{\exp[-j\omega_0\Delta\tau_p]\left(\frac{2\pi}{\sigma_\omega^2}\right)^{1/2}\int_{-\infty}^{\infty}\exp\left[-\frac{(\omega-\omega_0)^2}{2\sigma_\omega^2}\right]\exp[-j(\omega-\omega_0)\Delta\tau_g]\frac{d(\omega-\omega_0)}{2\pi}\right\}$$

$$(5.18)$$

The following integral formula can be used to reduce equation:

$$\int_{-\infty}^{\infty}\exp(-ax^2)\exp(-j2\pi kx)dx$$
$$= \int_{-\infty}^{\infty}\exp\left[-\frac{a(2\pi x)^2}{4\pi^2}\right]\exp(-j(2\pi x)k)\frac{d(2\pi x)}{2\pi} \qquad (5.19)$$
$$= \sqrt{\frac{\pi}{a}}\exp\left(-\frac{\pi^2 k^2}{a}\right)$$

Here, x represents a dummy variable and does not represent a unit of distance. To solve Eq. 5.18 with Eq. 5.19, it is useful to identify the following correspondents between components of the integral:

$$\frac{d(\omega-\omega_0)}{2\pi} \leftrightarrow \frac{d(2\pi x)}{2\pi}$$
$$\exp[-j(\omega-\omega_0)\Delta\tau_g] \leftrightarrow \exp(-j(2\pi x)k)$$
$$\exp\left[-\frac{(\omega-\omega_0)^2}{2\sigma_\omega^2}\right] \leftrightarrow \exp\left[-\frac{a(2\pi x)^2}{4\pi^2}\right]$$
$$\Delta\tau_g \leftrightarrow k$$
$$\frac{4\pi^2}{2\sigma_\omega^2} \leftrightarrow a$$
$$x^2 \leftrightarrow (\omega-\omega_0)^2/4\pi^2$$
$$x \leftrightarrow (\omega-\omega_0)/2\pi \qquad (5.20)$$

By using Eq. 5.19 and the above correspondence, Eq. 5.18 becomes:

$$I \propto \text{real}\left\{\exp[-j\omega_0\Delta\tau_p]\left(\frac{2\pi}{\sigma_\omega^2}\right)^{1/2}\left(\frac{\pi}{\frac{4\pi^2}{2\sigma_\omega^2}}\right)^{1/2}\exp\left(-\frac{\pi^2\Delta\tau_g^2}{\frac{4\pi^2}{2\sigma_\omega^2}}\right)\right\}$$

$$I \propto \text{real}\left\{\exp[-j\omega_0\Delta\tau_p]\exp\left(-\frac{\sigma_\omega^2\Delta\tau_g^2}{2}\right)\right\} \tag{5.21}$$

This represents the autocorrelation term for a Gaussian source. The equation contains two oscillatory terms. The first oscillatory term is a rapid oscillatory term or the phase modulation. The second oscillatory term is much slower, is Gaussian in shape, and is the envelope of the autocorrelation function as in Figure 5.2. It is this second term that primarily carries ranging information. The autocorrelation function is the inverse Fourier transform of the source power spectrum as per the Wiener-Khinchine theorem (Appendix 4-1). Therefore, a source with a Gaussian frequency spectrum ideally yields a Gaussian autocorrelation function, which is critical in ranging and performing high resolution imaging. Examples of the Fourier transform (linear and logarithmic) of the three non-Gaussian spectrums are also

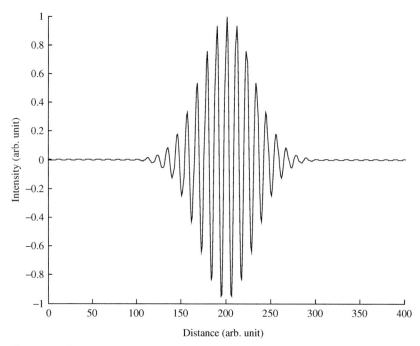

Figure 5.2   The components of the autocorrelation function are shown. The second oscillatory term of Eq. 5.21 is much slower, here is Gaussian in shape, and is the envelope of the autocorrelation function. This is what primarily carries ranging information. The high speed oscillations represent the phase and are used, for example, in Doppler or dispersion assessments.

shown in Figure 5.3. In Figure 5.3A, a Gaussian spectrum is shown which has a Gaussian FFT. In Figure 5.3B, a Gaussian spectrum with noise is shown where the log Fourier transform is significantly distorted. In Figure 5.3C, a sinc function is shown and in this case, the logarithmic Fourier transform is essentially unusable. In Figure 5.3D, a Gaussian-like function with side lobes is shown where again the Fourier transform is severely distorted.

Blindness is a particular problem when the autocorrelation function is not Gaussian. Blindness basically occurs when the autocorrelation plot designated for one pixel spreads into the adjacent pixel. This results when either side lobes are present or the tail is relatively large, and the intensity of that autocorrelation function

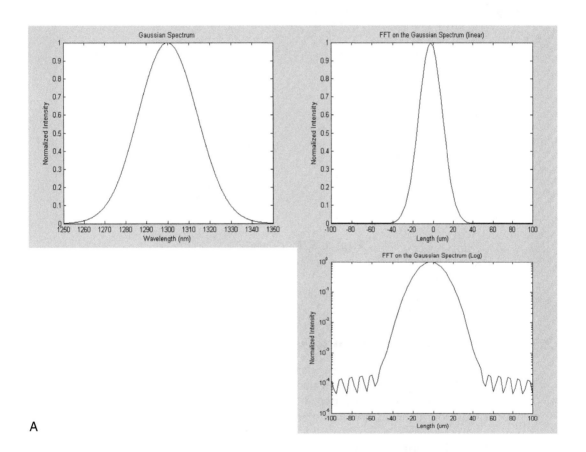

A

Figure 5.3 Source spectrum and their corresponding linear and logarithmic FFT. In A, an ideal Gaussian sources is demonstrated, which has a Gaussian FFT. In B, a Gaussian source with noise is examined where the Fourier transform is clearly distorted when compared to A. In Figure C, a sinc function is shown and in this case, the logarithmic Fourier transform is essentially unusable. In D, a Gaussian like function with side lobes is shown where again the Fourier transform is severely distorted.

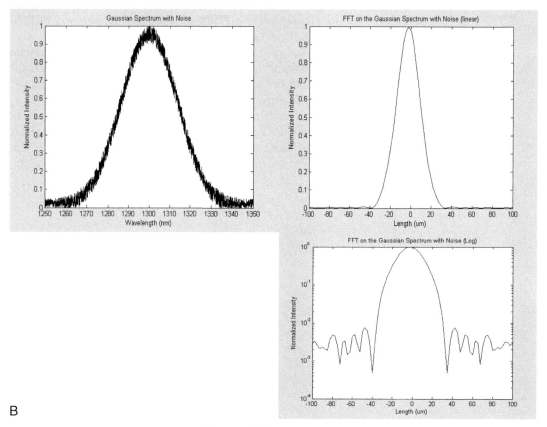

B

Figure 5.3 Continued.

is much greater than that of the adjacent pixel. This is why Gaussian autocorrelation functions are optimal.

5.2.3 Determination of Axial Resolution

In the previous section, it was demonstrated how OCT with a broadband source resulted in an AC envelope in the interference profile which will be used for high resolution ranging. Below two different approaches to calculating the coherence length from Eq. 5.21 are illustrated.

5.2.3.1 Approach 1

In the first approach, we will rewrite Eq. 5.21 in terms of frequency f whose units here are Hz.

$$I \propto \text{real}\left\{ \exp\left[-j2\pi f_0 \Delta\tau_p\right] \int_{-\infty}^{\infty} G(f - f_0) \exp\left[-j2\pi(f - f_0)\Delta\tau_g\right] d(f - f_0) \right\} \quad (5.22)$$

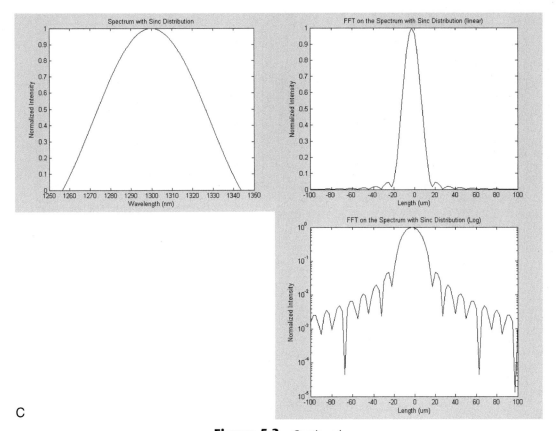

C

Figure 5.3 Continued.

And the corresponding Gaussian function is:

$$G(f - f_0) = \left(\frac{1}{2\pi\sigma_f^2}\right)^{1/2} \exp\left[-\frac{(f - f_0)^2}{2\sigma_f^2}\right] \qquad (5.23)$$

where σ_f is the standard deviation of the frequency spectrum. The envelope of the isolated autocorrelation function, which carries ranging information, is given by the slowly oscillating term of the solution to Eq. 5.21 and is:

$$I_{AC} = \exp\left(-2\pi^2\sigma_f^2\Delta\tau_g^2\right) \qquad (5.24)$$

Usually the full width half maximum (FWHM) width of I_{AC} is used as the parameter to characterize the axial resolution. So the FWHM value (i.e., half maximum) can be calculated by (the $1/e$ value can be calculated in a similar manner):

$$\exp\left(-2\pi^2\sigma_f^2\Delta\tau_g^2\right) = \frac{1}{2} \qquad (5.25)$$

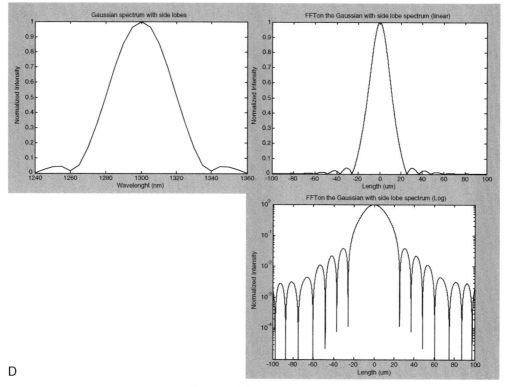

D

Figure 5.3   Continued.

Then through simple algebra we can solve for the group delay:

$$\exp\left(-2\pi^2\sigma_f^2\Delta\tau_g^2\right) = \exp(-\ln 2)$$
$$2\pi^2\sigma_f^2\Delta\tau_g^2 = \ln 2 \tag{5.26}$$

Isolating the optical group delay:

$$\Delta\tau_g = \frac{\sqrt{\ln 2}}{\sqrt{2\pi}\sigma_f} \tag{5.27}$$

since the standard deviation of the frequency can be expressed in terms of the wavelength:

$$\sigma_f = \frac{c\sigma_\lambda}{\lambda_0^2} = \frac{c\Delta_\lambda}{2\lambda_0^2\sqrt{2\ln}} \tag{5.28}$$

The σ_λ (standard deviation) can be expressed as the full width half maximum of the bandwidth of source spectrum:

$$\Delta_\lambda = 2\sqrt{2\ln 2}\sigma_\lambda \approx 2.3548\sigma_\lambda \tag{5.29}$$

which yields:

$$\Delta\tau_g = \frac{2\left(\sqrt{\ln 2}\right)\left(2\sqrt{2\ln 2}\right)\lambda_0^2}{\sqrt{2\pi}c\Delta_\lambda}$$
$$= \frac{4\ln 2\lambda_0^2}{\pi c\Delta_\lambda} \tag{5.30}$$

Then the difference in optical group delay can be converted to the FWHM of the coherence length:

$$\Delta\tau_g = \frac{2\Delta l}{c} \tag{5.31}$$

So

$$\Delta l = \frac{2ln2\lambda_0^2}{\pi\Delta_\lambda}\text{(Axial Resolution)} \tag{5.32}$$

This is the standard equation for defining the axial resolution for OCT imaging in free space. Therefore, the resolution depends on the square of the wavelength and the bandwidth of the source. Since the wavelength values used for imaging in nontransparent tissue do not differ by large amounts (1250 to 1350 nm), the resolution depends primarily on the bandwidth of the source. The femtosecond lasers have the highest bandwidths; this will be discussed in the next chapter. An example image at 4-μm resolution of a *Xenopus* organism is demonstrated in Figure 5.4.[2] Some studies are performed at wavelengths of approximately 850 nm, which is the wavelength used for imaging the transparent tissue of the eye. However, the penetration is so limited that it is of questionable value in assessing pathology in nontransparent tissue.

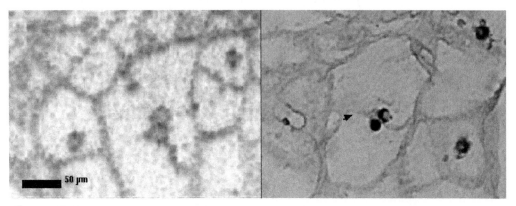

Figure 5.4 True subcellular imaging can be achieved with OCT using ultra-high bandwidth lasers. Femtosecond lasers currently have the highest bandwidths, which will be discussed in the next chapter. An example image at 4 μm resolution of a *Xenopus* organism is demonstrated. Not only are individual cells identified by their subcellular structures as nuclei. However, the high resolutions are associated with short confocal parameter as will be discussed later in the chapter. Image courtesy of Boppart, S.A., et.al., Proc. Natl. Acad. Sci., 1997. 94: 4256-4261 [2].

5.2.3.2 Approach 2

Again rewriting Eq. 5.21 in terms of frequency f the axial resolution can also be found more formally from the following.

$$I \propto real\left\{\exp[-j2\pi f_0 \Delta \tau_p] \int_{-\infty}^{\infty} G(f - f_0) \exp[-j2\pi(f - f_0)\Delta \tau_g]d(f - f_0)\right\} \quad (5.33)$$

$$G(f - f_0) = \left(\frac{1}{2\pi\sigma_f^2}\right)^{1/2} \exp\left[-\frac{(f - f_0)^2}{2\sigma_f^2}\right] \quad (5.34)$$

Equation 5.34 is the power spectrum and total power here is one

$$\int_{-\infty}^{\infty} G(f - f_0)df = 1 \quad (5.35)$$

Substituted with Eq. 5.35, Eq. 5.34 is derived as:

$$I \propto real\left\{\exp[-j2\pi f_0 \Delta \tau_p] \int_{-\infty}^{\infty} \left(\frac{1}{2\pi\sigma_f^2}\right)^{1/2} \exp\left[\frac{-(f - f_0)^2}{2\sigma_f^2}\right]\right.$$
$$\left. \times \exp[-j2\pi(f - f_0)\Delta \tau_g]d(f - f_0)\right\}$$
$$\propto real\left\{\exp[-j2\pi f_0 \Delta \tau_p]\left(\frac{1}{2\pi\sigma_f^2}\right)^{1/2} \int_{-\infty}^{\infty} \exp\left[-\frac{(f - f_0)^2}{2\sigma_f^2}\right]\right.$$
$$\left. \times \exp[-j2\pi(f - f_0)\Delta \tau_g]d(f - f_0)\right\} \quad (5.36)$$

According to the integration formula:

$$\int_{-\infty}^{\infty} \exp(-ax^2) \exp(-j2\pi kx)dx = \sqrt{\frac{\pi}{a}}\exp\left(-\frac{\pi^2 k^2}{a}\right) \quad (5.37)$$

The items in the integral correspond to:

$$d(f - f_0) \leftrightarrow dx$$
$$\Delta \tau_g \leftrightarrow k$$
$$\frac{1}{2\sigma_f^2} \leftrightarrow a \quad (5.38)$$

Equation (5.36) can be further derived as:

$$I \propto real\left\{\exp[-j2\pi f_0 \Delta \tau_p] \int_{-\infty}^{\infty} G(f - f_0) \exp[-j2\pi(f - f_0)\Delta \tau_g]d(f - f_0)\right\}$$
$$\propto real\left\{\exp[-j2\pi f_0 \Delta \tau_p]\left(\frac{1}{2\pi\sigma_f^2}\right)^{1/2}\left(\frac{\pi}{\frac{1}{2\sigma_f^2}}\right)^{1/2} \exp\left(-\frac{\pi^2 \Delta \tau_g^2}{\frac{1}{2\sigma_f^2}}\right)\right\}$$
$$\propto real\left\{\exp[-j2\pi f_0 \Delta \tau_p] \exp(-2\pi^2 \sigma_f^2 \Delta \tau_g^2)\right\} \quad (5.39)$$

According to the Heisenberg uncertainty principle, for light with a Gaussian spectrum, the standard deviations of the time-frequency product satisfies:

$$\sigma_t \sigma_f = \frac{1}{4\pi} \tag{5.40}$$

Equation 5.39 can be simplified as:

$$I \propto real\{\exp\left(-\frac{\Delta\tau_g^2}{8\sigma_t^2}\right)\exp[-j2\pi f_0 \Delta\tau_p]\} \tag{5.41}$$

The envelope of the normalized OCT signal I is:

$$I_{AC} = \exp\left(-\frac{\Delta\tau_g^2}{8\sigma_t^2}\right) = \exp(-2\pi^2 \sigma_f^2 \Delta\tau_g^2) \tag{5.42}$$

The relation between standard deviation in frequency σ_f and wavelength σ_λ is:

$$\sigma_f = \frac{c\sigma_\lambda}{\lambda_0^2} \tag{5.43}$$

So the envelope expression becomes:

$$I_{AC} = \exp\left(-\frac{2\pi^2 c^2 \sigma_\lambda^2}{\lambda_0^4}\Delta\tau_g^2\right) \tag{5.44}$$

Usually we use the FWHM value to replace the standard deviation.

$$\Delta_\lambda = 2\sigma_\lambda \sqrt{2\ln 2} \approx 2.3548\sigma_\lambda \tag{5.45}$$

So

$$\begin{aligned}
I_{AC} &= \exp(\frac{-\pi^2 c^2 \Delta_\lambda^2}{4\ln 2\lambda_0^4}\Delta\tau_g^2) \\
&= \exp[-\frac{\pi^2 \Delta_\lambda^2}{4\ln 2\lambda_0^4}(c\Delta\tau_g)^2]
\end{aligned} \tag{5.46}$$

Since the optical path is double pass,

$$\begin{aligned}
I &= \exp[-\frac{\pi^2 \Delta_\lambda^2}{4\ln 2\lambda_0^4}(2\Delta L)^2] \\
&= \exp[-\frac{\pi^2 \Delta_\lambda^2}{\ln 2\lambda_0^4}\Delta L^2] \\
&= \exp[-\frac{4\ln 2\Delta L^2}{(\frac{2\ln 2\lambda_0^2}{\pi\Delta_\lambda})^2}
\end{aligned} \tag{5.47}$$

ΔL is geometrical mismatch, so the FWHM width of the coherence length is;

$$\Delta l = \frac{2ln2\lambda_0^2}{\pi\Delta_\lambda} \tag{5.48}$$

This is identical to Eq. 5.31 and is the PSF of the system.

Detection and filtering of the signal is discussed in Chapters 6 and 7. Most commonly, though, after the signal is detected, it is bandpass filtered, demodulated, low pass filtered, and then converted to a digital signal.

5.2.4   Transfer Function

In this section, OCT will be described in terms of the transfer function. Some groups have done this through an inverse scattering assessment, which will not be done here.[3] However, Appendix 5-1 includes summaries of Green's theorem and Green's functions for those interested in evaluating these works. In addition, Hilbert's transforms, analytical signals, and Cauchy's principal value are discussed and are used throughout the text. The incident light field at the coupler can be described as a Fourier integral[1]:

$$E_0(t) = \int_{-\infty}^{\infty} a_0(f) \exp\{i[\phi_0(f) - 2\pi ft]\}df \tag{5.49}$$

$a_0(f)$ represents the amplitude spectrum of the incident light and $\phi_0(f)$ represents the phase spectrum. The bandwidth of the incident light used is finite around a certain optical frequency f_0. Apart from an inessential constant, the observable intensity of the incident light is described as:

$$I_0 = 2\langle E_0^2(t)\rangle = 2\int_{-\infty}^{\infty} G(f)df \tag{5.50}$$

where $G(f)$ is referred to as the spectral density or the spectrum of the source with electromagnetic radiation, whereas it is called the power spectrum of the random process. $G(f)$ includes the negative frequency components.

Ideally, the incident light beam is split evenly into the sample and the reference beam by the coupler. The light field at the exit of the interferometer, which is reflected by the mirror in the reference arm whose reflection coefficient is p_R, can be expressed as

$$E_R(t) = \frac{1}{2}\int_{-\infty}^{\infty} a_0(f)p_R \exp\left\{i\left[\phi_0(f) + \delta + 4\pi\frac{f}{c}z_R - 2\pi ft\right]\right\}df \tag{5.51}$$

Here $E_R(t)$ is the field from the reference arm after it has entered the detector arm.

It is assumed the origin of coordinates O_S in the sample arm is set up at the sample surface. The origin of coordinates in the reference arm O_R can be chosen at a point from where the optical group delay to the coupler is matching with that between

O_S and the coupler in the sample arm. δ in Eq. 5.51 represents a fixed phase delay. The mirror position is z_R off the O_R, while c is the light speed in a vacuum. Assuming no additional attenuation in the reference arm, the intensity of the returned reference beam will be as:

$$I_R = 2\langle E_R^2(t)\rangle = \frac{p_R^2}{4} \times 2 \int_{-\infty}^{\infty} G(f)df = \frac{p_R^2 I_0}{4} \qquad (5.52)$$

In Eq. 5.52, the square of the reflection coefficient represents the reflectivity of the mirror (p_R).

In the sample arm, the light beam encounters scattering and reflection in a sample. Defining $p(z)$ as the amplitude backscattering coefficient distribution in the tissue, where z is measured from O_S, the light field at the exit which is contributed by a scatter in the sample arm, at position z, can be written as:

$$E_S(t,z) = \frac{1}{2} \int_{-\infty}^{\infty} a_0(f)p(z)\exp\{i[\phi_0(f) + \delta + 4\pi\frac{n_S(f)f}{c}z - 2\pi ft]\}df \qquad (5.53)$$

E_S (t,z) is the field in the detector arm from the reference arm. In Eq. 5.53, $n_S(f)$ represents the refractive index of the sample and $p(z)$ carries the diagnostic information.

At the exit of the interferometer, the two return beams from the reference and the sample arm recombine coherently, described as:

$$E_D(t,z) = E_R(t) + \int_0^z E_S(t,z')dz' \qquad (5.54)$$

The intensity of the light field at the exit will be:

$$I_D = 2\langle E_D^2(t,z')\rangle = 2\langle E_R^2(t)\rangle + 2\left\langle\int_0^z E_S^2(t,z')dz'\right\rangle + 4\int_0^z \langle E_S(t,z')E_R(t)\rangle dz' \qquad (5.55)$$

The first term in Eq. 5.55 corresponds to the intensity of the reflected reference beam I_R. The second term corresponds to the total intensity of the returned sample beam which is contributed by all the scatters, and the mutual interference of all scattering waves which was called the autocorrelation function of scattering or a parasitic term.[4,5] If no strong reflection exits in the sample, the second term can be approximated to the incoherent intensity sum in the returned sample beam for TD-OCT, as will be discussed later. This may not be the case for SD-OCT. It is then:

$$I_S = 2\int_0^z \langle E_S^2(t,z')\rangle dz' = \frac{I_0}{4}\int_0^z p^2(z')dz' \qquad (5.56)$$

where $p^2(z)$ represents the scattering coefficient in terms of intensity.

Then Eq. 5.55 can be further derived as:

$$I_D \approx I_R + I_S + \int_0^z \text{Re}\left\{\int_{-\infty}^{\infty} G(f)p(z')p_R \exp\{i4\pi\frac{f}{c}[n_S(f)z' - z_R]\}df\right\}dz' \qquad (5.57)$$

Eq. 5.57 describes the light intensity perturbation at the exit of the interferometer. The third item is the interferometric signal, in which the diagnostic information $p(z)$ is encoded. The intensity perturbation at the exit with TD-OCT is detected either by a single detector or dual balanced detection.

For simplification on the derivation, the intensity spectrum of the light source, including negative frequencies, is assumed to be a Gaussian distribution as:

$$G(f) = \frac{I_0}{4} \times \frac{1}{\sigma_f \sqrt{2\pi}} \{\exp[-\frac{(f-f_0)^2}{2\sigma_f^2}] + \exp[-\frac{(f+f_0)^2}{2\sigma_f^2}]\} \quad (5.58)$$

The intensity spectrum includes the negative frequency components and making the function even. Δ_f represents its standard deviation bandwidth. Assuming that no group velocity dispersion exits in this OCT system, the refractive index becomes frequency independent as $n_s(f_0)$. Substituting Eq. 5.58 into Eq. 5.57 it becomes:

$$I_D = I_R + I_S + \frac{I_0}{2}\int_0^z \{p(z')p_R \cos[2\pi f_0(\frac{n_s(f_0)2z'}{c} - \frac{2z_R}{c})]\exp[-2\pi^2\sigma_f^2(\frac{2n_s(f_0)z'}{c} - \frac{2z_R}{c})^2]\}dz' \quad (5.59)$$

since the Fourier transform of the real Gaussian function is still Gaussian and real.

Equation 5.59 also can be rewritten using the convolution theorem, using wavenumber k instead of angular frequency f.

$$I_D = I_R + I_S + \frac{I_0}{2}\int_0^z p(z')p_R \cos[4\pi k_0(n_S(k_0)z' - z_R)]\exp[-8\pi^2\sigma_k^2(n_S(\omega_0)z' - z_R)^2]\}dz' \quad (5.60)$$

The OCT system in terms of the convolution is then:

$$I_D(z_R) = I_R + I_S + \frac{I_0}{2}p_R p(\frac{z_R}{n_S(k_0)}) \otimes [\cos(4\pi k_0 z_R)\exp(-8\pi^2\sigma_k^2 z_R^2)] \quad (5.61)$$

Now expressing it in terms of the coherence length, we use the uncertainty relationship:

$$\sigma_k \sigma_l \geq \frac{1}{4\pi} \quad (5.62)$$

The relationship between the coherence length and the standard deviation of the width of the autocorrelation function is:

$$\Delta_l = 2\sqrt{2\ln 2}\sigma_l \quad (5.63)$$

So the convolution can be expressed in terms of the coherence length:

$$I_D(z_R) = I_R + I_S + \frac{I_0}{2}p_R p(\frac{z_R}{n_S(k_0)}) \otimes \{\cos[4\pi k_0 z_R]\exp[-\frac{4\ln 2 z_R^2}{\Delta_l^2}]\}$$

$$\Delta_l = \frac{2\ln 2\lambda_0^2}{\pi\Delta_\lambda} \quad (5.64)$$

The axial resolution is now expressed in terms of the FWHM coherence length of light source and Δ_λ is its FWHM bandwidth. k_0 is the central wavenumber in a vacuum. Equation 5.64 gives the description of the TD-OCT signal and its dependence on the mirror position in the reference arm. The diagnostic information $p(z)$ backreflection intensity as a function of depth is encoded in the AC portion of this signal, which is detected by a spatially linear system as the impulse response or transfer function and is given by:

$$H(K) = F|h(z_R)| = \left| \cos[4\pi k_0 z_R] \exp[-\frac{4\ln 2 z_R^2}{\Delta_l^2}] \right| \tag{5.65}$$

The optical transfer function (OTF) is proportional to spectrum function $G(f)$, but the argument becomes k, the spatial frequency. If the spectrum of light source is Gaussian-type, the OTF is also a Gaussian function. Its central spatial frequency and its standard deviation can be written as:

$$k_0 = \frac{1}{\lambda_0};$$

$$\sigma_k = \frac{\sigma_\lambda}{\lambda_0^2} \tag{5.66}$$

where λ_0 is the central wavelength of the probing beam in the object and σ is the standard deviation of spectral bandwidth. Figure 5.5 gives a typical Gaussian-type OTF curve of an OCT system and helps us understand some aspects of OCT not always obvious from the PSF.[6] Figure 5.5 and Eq. 5.65 describe the OCT system as a bandpass spatial filter with zero phase shift. In theory structures in the order of half wavelength have maximum contrast. Tissue constituents of various sizes

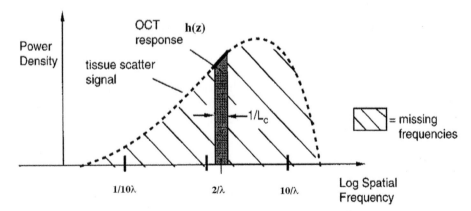

Figure 5.5 A typical Gaussian OTF curve of an OCT system is shown that illustrates some aspects of OCT not always obvious from the PSF6. The OCT system is shown as a bandpass spatial filter with zero phase shift. In theory, structures in the order of half wavelengths have maximum contrast. Tissue constituents of various sizes with refractive index mismatches are present. Structures with spatial frequencies significantly above or below those with inverse coherence length are not readily detected, although the reason that this is of only limited theoretical relevance is outlined in the text. Courtesy of Schmitt, J. M., et al., 1999. *J. Biomed. Optics* 4; 95–105 [6].

with refractive index mismatches are present. According to Figure 5.5, structures with spatial frequencies significantly above or below those with the inverse coherence length are not readily detected by OCT. This seems to suggest that large structures with low spatial frequencies such as lipid collection will not be detected. However, most biological structures are not particularly homogeneous but consist of regional variations in refractive index. This makes the lack of sensitivity to low spatial frequencies not likely to result in significant loss of detection. Also, because impulse signal sequence may inherently manipulate the spatial spectrum to the passing band, the spatial filtering characteristics of an OCT system also imply that any deconvolution procession on the OCT signal is not necessarily robust.

5.3   DISPERSION

In Chapter 3, dispersion was described qualitatively to provide a better understanding of tissue interaction with light. In this section, scalar light wave analysis is applied to OCT imaging, which is relevant to theoretically modeling and quantifying dispersion. Dispersion is important both for its ability to distort OCT images and its potential to be used diagnostically. In view of the importance of dispersion in OCT imaging, relevant equations are derived in considerable detail so that an understanding of assumptions and approximations are made clear. Again, the perturbation of the incident light field at the coupler is described as a Fourier integral:

$$E_0(t) = \int_{-\infty}^{\infty} a_0(f) \exp\{i[\phi_0(f) - 2\pi ft]\} df \qquad (5.67)$$

$a_0(f)$ represents the amplitude spectrum of incident light and $\phi_0(f)$ represents its phase spectrum. The power spectrum includes symmetrical components in both positive and negative frequencies and is finite around f_0, with certain intensity

$$I_0 = \langle E_0(t) E_0^*(t) \rangle = 2 \int_{-\infty}^{\infty} G(f) df \qquad (5.68)$$

In the sample arm, the light beam encounters scattering in the sample. Defining $\chi(f, z)$ as the absorption and scattering characteristics of the sample, z is measured from O_S, and the backscattered light from one scattering spot can be written as

$$E_S(t, z) = \frac{1}{2} \int_{-\infty}^{\infty} a_0(f) |\chi(f,z)| \exp\{i[\phi_0(f) + \phi_S(f,z) - 2\pi ft]\} df \qquad (5.69)$$

In Eq. 5.69, $\phi_S(f, z)$ represents the phase shift in the sample arm. As we know, the distribution $\chi(f, z)$ carries the diagnostic information and the absorption portion is assumed constant throughout the sample. Scattering events are assumed single scattering. The total intensity of the returned scattering light contributed by all the scatters within the tissue then can be expressed as

$$I_S = \int_0^z \langle E_S(t, z') E_S^*(t, z') \rangle dz' = \frac{I_0}{4} \int_0^z \left| \chi(f,z') \right|^2 dz' \qquad (5.70)$$

At the exit of the interferometer, the two return beams from the reference arm and the sample arm recombine coherently, described as

$$E_D(t,z) = \int_0^z [E_S(t,z') + E_R(t,z_R)]dz' \tag{5.71}$$

The intensity of the light field at the exit of the interferometer is

$$
\begin{aligned}
I_D &= \int_0^z \langle E_D(t,z')E_D^*(t,z')\rangle dz' \\
&= \langle E_R(t)E_R^*(t)\rangle + \int_0^z \langle E_S(t,z')E_S^*(t,z')\rangle dz' \\
&\quad + \int_0^z [\langle E_S(t,z')E_R^*(t)\rangle + \langle E_S^*(t,z')E_R(t)\rangle]dz' \tag{5.72}
\end{aligned}
$$

Assuming the extinction coefficients in the reference arm and sample arm (other than sample) are 1, the mirror is a perfect reflector, and since only the AC portion of the interference signal is used to construct OCT images, Eq. 5.72 can be simplified to the interference portion:

$$I_D \propto \int_0^z \text{Re}\left\{ \int_{-\infty}^{\infty} G(f)\left| \chi(f,z')\right| \exp\{i[\phi_S(f,z') - \phi_R(f,z_R)]\}df \right\}dz' \tag{5.73}$$

Polarization effects are ignored at this point. Since the spectrum of the light source is finite around f_0, Eq. 5.73 can be approximated in terms of the second order Taylor series/expansion of the phase shift:

$$
\begin{aligned}
I_D \propto \int_0^z &\left(G(f)\left| \chi(f,z')\right| \exp\{i\{[\phi_S(f_0,z') - \phi_R(f,z_R)] \right. \\
&\times \text{Re}\left\{ \int_{-\infty}^{\infty}
\begin{array}{l}
+[\phi_S'(f_0,z') - \phi_R'(f_0,z_R)](f - f_0) \\
+\frac{1}{2}[\phi_S'(f_0,z') - \phi_R'(f_0,z_R)](f - f_0)^2\}df
\end{array}
\right\}dz' \tag{5.74}
\end{aligned}
$$

In Eq. 5.74, the first derivative of the phase shift represents the group delay of the beam in respective arms. The second derivative of the phase shift is proportional to the relative group velocity dispersion (GVD). The terms in Eq. 5.74 can be rewritten as in terms of:

$$
\begin{aligned}
\Delta\tau_p(z,z_R) &= \frac{\phi_S(f_0,z) - \phi_R(f_0,z_R)}{2\pi\omega_0}, \\
\Delta\tau_g(z,z_R) &= \frac{\phi_R'(f_0,z) - \phi_S'(f_0,z_R)}{2\pi}, \\
\Delta\tau_c^2(z,z_R) &= \frac{\phi_R''(f_0,z) - \phi_S''(f_0,z_R)}{2} \tag{5.75}
\end{aligned}
$$

where $\Delta\tau_p$ and $\Delta\tau_g$ refer to the phase delay difference and group delay difference between the reference and sample arm. The term $\Delta\tau_c$ characterizes the relative difference in GVD between the two arms. We can then replace the spectral items in Eq. 5.74 with a new variable $s = f - f_0$ for simplicity. Equation 5.74 can therefore be written as:

$$I_D \propto \mathrm{Re}\{\exp(i2\pi\Delta\tau_p f_0)$$

$$\times \int_{-\infty}^{\infty} |\chi_S(s)| G(s) \exp[i2\pi(\Delta\tau_g s + \Delta\tau_c^2 s^2)]ds\}; \qquad (5.76)$$

The integral portion of the interference signal is the amplitude modulation (AM) signal of the envelope. As seen previously, FWHM of the envelope in terms of the coherence length is the PSF of the OCT system when light is reflected off a mirror.

If the dispersive length (in terms of the physical distance) of a sample is defined as L_S, $\Delta\tau_c^2$ can be represented as:

$$\Delta\tau_c^2 = -k_S''(f_0)L_S \qquad (5.77)$$

k_s is, as previously described, the wavenumber in the sample arm. To solve for $k_s''(\omega_0)$, we rewrite k_s as:

$$k_s = \frac{n_s(f)f}{c} \qquad (5.78)$$

Then the first derivative is:

$$k_s' = \frac{dk_S(f)}{d\omega} = \frac{[n_S'(f)f + n_S(f)]}{c}$$

$$= [-\lambda \frac{dn_S(\lambda)}{d\lambda} + n_S(\lambda)]/c \qquad (5.79)$$

The second derivative is then:

$$k_s'' = \frac{d^2 k_S(f)}{df^2} = \frac{d[-\lambda \frac{dn_S(\lambda)}{d\lambda} + n_S(\lambda)]}{d\lambda} \times \frac{d\lambda}{cdf}$$

$$= \frac{d^2 n_S(\lambda)}{d\lambda^2} \times \frac{\lambda^3}{c^2} \qquad (5.80)$$

The items in Eq. 5.75 can be calculated by Eq. 5.80.

$$\Delta tp = \frac{2n_S(f_0)}{c}z - \frac{2n_R(f_0)}{c}z_R$$

$$\Delta tg = \frac{2[n_S'(f_0)f_0 + n_S(f_0)]}{c}z - \frac{2[n_R'(f_0)f_0 + n_R(f_0)]}{c}z_R$$

$$\Delta t^2 c = \frac{n_S''(f_0)f_0 + 2n_S'(f_0)]}{2c}z - \frac{n_R''(f_0)f_0 + 2n_R'(f_0)}{2c}z_R \qquad (5.81)$$

Equation 5.75 can also be rewritten in terms of wavelength λ_0.

$$\Delta tp = \frac{2n_S(\lambda_0)}{c}z - \frac{2n_R(\lambda_0)}{c}z_R$$

$$\Delta tg = \frac{2[-\lambda_0 \frac{dn_S}{d\lambda}(\lambda_0) + n_S(\lambda_0)]}{c}z - \frac{2[-\lambda_0 \frac{dn_R}{d\lambda}(\lambda_0) + n_R(\lambda_0)]}{c}z_R$$

$$\Delta t^2 c = 2\pi \frac{d^2 n_S}{d\lambda^2}(\lambda_0)\frac{\lambda_0^3}{c^2}z - 2\pi \frac{d^2 n_R}{d\lambda^2}(\lambda_0)\frac{\lambda_0^3}{c^2}z_R \qquad (5.82)$$

These three terms represent the phase delay, group delay, and GVD factor. Defining group indices n_g and group velocity dispersion parameter D as[7]:

$$\Delta tp = \frac{2n_S(\lambda_0)}{c}z - \frac{2n_R(\lambda_0)}{c}z_R$$

$$\Delta tg = \frac{2n_S^g(\lambda_0)]}{c}z - \frac{2n_R^g(\lambda_0)}{c}z_R$$

$$\Delta t^2 c = 2\pi D_S(\lambda_0)\frac{\lambda_0^2}{c^2}z - 2\pi D_R(\lambda_0)\frac{\lambda_0^2}{c^2}z_R \tag{5.83}$$

where:

$$\Delta\tau_c^2 = \frac{\lambda_0^2}{c^2}D(\lambda_0)L_S \tag{5.84}$$

The spectrum of the source is assumed to be Gaussian:

$$G(f)|\chi(f,z)| = \frac{I_0}{4} \times \frac{p(z)}{\sigma_f\sqrt{2\pi}}\exp[-\frac{(f-f_0)^2}{2\sigma_f^2}] \tag{5.85}$$

Then the AC portion of Eq. 5.81 can be rewritten as:

$$\frac{1}{2}\int_{-\infty}^{\infty}\sqrt{\gamma_R(f)\gamma_S(f)}G(f)\exp\{i2\pi[\Delta\tau_g(f-f_0)+\Delta\tau_c^2(f-f_0)^2]\}df$$

$$= \frac{1}{\sigma_f 2\sqrt{2\pi}}\int_{-\infty}^{\infty}\exp[(-\frac{1}{2\sigma_f^2}+i\Delta\tau_c^2)(f-f_0)^2]\exp[i2\pi\Delta\tau_g(f-f_0)]d(f-f_0) \tag{5.86}$$

using the identity:

$$\int_{-\infty}^{\infty}\exp(-ax^2)\exp(-i2\pi kx) = \sqrt{\frac{\pi}{a}}\exp(-\frac{\pi^2 k^2}{a}) \tag{5.87}$$

where a corresponds to the item in Eq. 5.86:

$$a = \frac{1}{2\sigma_f^2} - i\Delta\tau_c^2 \tag{5.88}$$

Solving Eq. 5.86 according to Eqs. 5.87 and 5.88, the resolution modified by dispersion is now given by:

$$\sigma_\tau' = \left|\sqrt{\frac{1}{2\pi^2}(\frac{1}{2\sigma_f^2}-j\Delta\tau_c^2)}\right| = \frac{1}{\pi\sqrt{2}}(\frac{1}{4\sigma_f^4}+\Delta\tau_c^4)^{\frac{1}{4}}$$

$$= \frac{1}{2\pi\sigma_\omega}[1+4(\sigma_f\Delta\tau_c)^4]^{\frac{1}{4}} \tag{5.89a}$$

$$\sigma_\tau = \frac{1}{2\pi\sigma_f} \tag{5.89b}$$

$$\sigma_\tau' = \sigma_\tau[1+\frac{1}{4\pi^4}(\frac{\Delta\tau_c}{\sigma_\tau})^4]^{\frac{1}{4}} \tag{5.89c}$$

Simplifying Eq. 5.89c with Eq. 5.84, we get:

$$\sigma'_\tau = \sigma_\tau[1 + \frac{1}{4\pi^4}\frac{\lambda_0^4}{c^4}D_S^2(\lambda_0)L_S^2(\frac{1}{\sigma_\tau})^4]^{\frac{1}{4}} \tag{5.90}$$

The relationship between the standard deviation of frequency σ_f and wavelength σ_λ is given by:

$$\sigma_f = \frac{c\sigma_\lambda}{\lambda_0^2} \tag{5.91}$$

Then the modified FWHM coherence length σ'_l and coherence length σ_l of the light source are given by:

$$\sigma'_l = 2\sqrt{2\ln 2}\,c\sigma'_\tau$$
$$\sigma_l = 2\sqrt{2\ln 2}\,c\sigma_\tau \tag{5.92}$$

The σ'_l or the new resolution can be given by:

$$\sigma'_l = \sigma_l[1 + 4(\frac{\sigma_\lambda^2}{\lambda_0^2}D_S(\lambda_0)L_S)^2]^{\frac{1}{4}} \tag{5.93}$$

expressed in terms of the coherence length.

Now expressing it in terms of the coherence length, we use the uncertainty relationship

$$\sigma_k\sigma_l \geq \frac{1}{4\pi} \tag{5.94}$$

The coherence length is then expressed in terms of the standard deviation

$$\Delta_l = 2\sqrt{2\ln 2}\,\sigma_l \tag{5.95}$$

where Δ_l represents the equivalent FWHM width of the PSF in terms of

$$\Delta'_l = \Delta_l\left[1 + \frac{c^4(\ln 2)^2\Delta\tau_c^4(z_R)}{\pi^2\Delta_l^4}\right]^{\frac{1}{2}}$$
$$= \Delta_l\left\{1 + \frac{\Delta_\lambda^4}{4(\ln 2)^2\lambda_0^4}[D_S(\lambda_0)L_S]^2\right\}^{\frac{1}{2}} \tag{5.96}$$

The PSF now expressed in terms of D_S and L_S in Eq. 5.96 represents the GVD parameter and equivalent dispersion length. Equation 5.96 implies several things. First, that increasing D_S leads in general to broadening of the PSF. Second, D_S or L_S induces a nonlinear oscillation term and the envelope of the PSF is modulated by this nonlinear oscillation.

Figure 5.6 is a simple drawing of a Gaussian pulse before and after passing through a dispersive medium. The pulse has become widened and the frequency distribution is no longer symmetric. In Figure 5.7, data obtained from the

Without Dispersion

With Dispersion

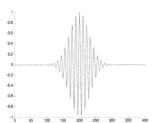 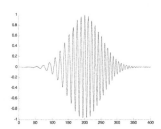

Figure 5.6 A simple drawing of a Gaussian pulse before and after passing through a dispersive medium. The pulse has become widespread and the frequency distribution is no longer symmetric after dispersion has occurred.

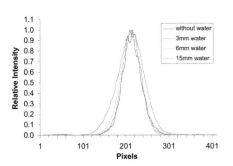 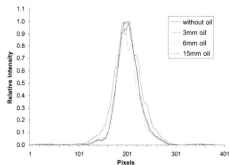

Figure 5.7 Data obtained from the autocorrelation function off a mirror after light has passed through different amounts of water and oil. The water is shown on left; the oil on the right. It can be seen that water results in a substantially greater broadening of the PSF even at depths as small as 5 mm. Courtesy Liu, B., et al., 2004. *Phys. Med. Biol.* 49, 1-8 [5].

autocorrelation function of a mirror after light has passed through different amounts of water and oil are shown. The PSF in air is 12 μm. It can be seen that water, with an absorption peak near 1380 nm, demonstrates GVD even after passing through as little as 6 mm of water with only minimal effect from oil.[8] In Figure 5.8, two sides of the inner tubing are imaged. In Figure 5.8A, two thin reflections are noted when no dispersion is introduced. However, as the dispersion is increased in Figure 5.8B, the reflections appear wider. As they begin to overlap, due to the asymmetry of the frequencies, high frequency vibrations appear in what is now a single large peak (both peaks overlap). These high frequency oscillations are referred to as chirping. It should be noted that while the dispersion of the autocorrelation is dependent on the physical distance light travels, chirping is dependent on the optical distance between two scatters.

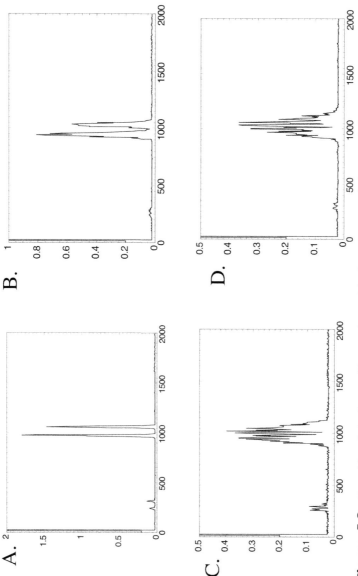

Figure 5.8 Two sides of the inner tubing are imaged in the presence and absence of dispersion. In A., two thin reflections, are noted when no dispersion is introduced. However, as the dispersion is increased in B, the reflections appear wider. As they begin to overlap, due to the asymmetry of the PSF, high frequency oscillations occur and are referred to as chirping (C and D).

Dispersion is compensated for through a variety of mechanisms. With TD-OCT, this includes focusing in the grating based delay line,[9] acoustical-optic modulator (Appendix 6-1),[10] introduction of transparent material in the reference arm,[11] and autofocus algorithms.[12]

5.4 LATERAL RESOLUTION

5.4.1 Light Beam

The lateral or transverse resolution of OCT is determined primarily by the ability to focus the infrared "beam."[27,28] In other words, it is determined by how light is directed and collected from the sample. Maxwell's equations are utilized to model beam focusing. However, due to approximations made in the derivations, equations generated are not exact solutions.

The derivations are straightforward albeit lengthy. Therefore, we will leave the step-by-step derivations to other sources and will focus on the final conclusions and their limitations.[27] The discussion here will be about incoherent light propagation for simplicity, but spatially coherent relationships still hold with OCT as described in the beginning of this chapter which contribute to phenomena such as speckle which is discussed in subsequent sections. For the analysis, we can confine ourselves to the spatial components of the beam, in other words solutions of the Helmholtz and paraxial Helmholtz equation.

The easiest way to model propagation would be with a plane wave, which is a solution of the Helmholtz equation. It has the form of $\psi(z) = A_p e^{ik.z}$, where all terms have been previously described, with the direction of propagation in the z direction. However, a plane wave has infinite extent and therefore energy and so is not physically realizable. The next possible solution would be a spherical wave $\psi(r) = (A_s/r)e^{ikr}$, where A_s is a measure of the initial energy. This radiates from a single point and has finite energy, with its irradiance obeying the inverse square law. However, it is not directed (i.e., not a beam), a requirement for describing lateral resolution with OCT.

The equations of propagation we will be looking for are solutions of the paraxial Helmholtz equation, which will be described shortly, and include the paraboloidal wave (in between a spherical and planar wave) and a Gaussian beam. For a paraxial wave, the energy is concentrated around the line of propagation (unlike a spherical or plane wave) and the wavefront normals propagate approximately in the direction of the center of the beam.

The paraboloidal wave, which is our first paraxial wave, can be approximated by the Fresnel approximation of a spherical wave, which will have an origin at $r = 0$. We will be examining points $r = (x, y, z)$ sufficiently close to the z axis but far from the origin so that $(x^2 + y^2)^{0.5} < z$ (i.e., paraxial or beam-like). Defining θ as the angle between the average direction of propagation and position of interest off axis, we have $\theta^2 = (x^2 + y^2)/z^2 < 1$. So r equals:

$$r = (x^2 + y^2 + z^2)^{0.5} = z(1 + \theta^2)^{0.5} \qquad (5.97)$$

Performing a Taylor series expansion we obtain:

$$r = z(1 + \theta^2/2 - \theta^4/8 + \ldots) \approx$$

$$z(1 + \theta^2/2) = z + (x^2 + y^2)/2z \qquad (5.98)$$

This approximation for r will be substituted into the spherical wave equation, $\psi(r) = (A_S/r)e^{ikr}$. For the phase of the paraboloidal, $r = z + (x^2 + y^2)/2z$ while for the magnitude we will drop the second term so that $r = z$. The resulting Fresnel approximation is:

$$\psi(r) = (A/z)\exp(-ikz)\exp\{-ik(x^2 + y^2/2z)\}\exp - i\phi(z) \qquad (5.99)$$

This function has beam-like properties, but we will see that a closely related function, the Gaussian beam, is a better approximation to the light behavior being modeled.

5.4.2   Paraxial Wave Equation

One way of constructing our beam is to modulate the complex envelope of a plane wave, making it slowly varying with position. We will do this with a Gaussian beam. The Gaussian beam must satisfy the paraxial wave equation. The paraxial wave equation will be derived here to point out approximations made, identifying limits to the approach. Specifically, Gaussian beams do not represent rigorous solutions of Maxwell's equations, but approximations that hold only if the minimal beam diameter (waist) is large in comparison to wavelength (i.e., beam diverges only slightly). For the paraxial wave to satisfy the Helmholtz equation, it must satisfy the paraxial Helmholtz equation.

We begin with the Helmholtz equation, which has been described in Appendix 2-1 and is time independent. The equation is given by:

$$(\nabla^2 + k^2)E(x, y, z) = 0$$

$$\nabla^2 = \frac{\partial^2}{\partial x^2} + \frac{\partial^2}{\partial y^2} + \frac{\partial^2}{\partial z^2} \qquad (5.100)$$

Let E be the space-dependent part of a monochromatic scalar wave. By the paraxial approximation, we initially assume a plane wave is propagating parallel to the z direction, whose complex amplitude is described by:

$$E(x, y, z) = \psi(x, y, z)e^{-ikz} \qquad (5.101)$$

where $\psi(x, y, z)$ is the complex envelope. Performing the derivatives in Eq. 5.100 on Eq. 5.101, the following terms are derived:

$$\frac{\partial E}{\partial z} = \frac{\partial \psi}{\partial z}e^{-ikz} - ik\psi e^{-kz} = (\frac{\partial \psi}{\partial z} - ik\psi)e^{-ikz}$$

$$\frac{\partial^2 E}{\partial z^2} = (\frac{\partial^2 \psi}{\partial z^2} - ik\frac{\partial \psi}{\partial z})e^{-ikz} - ik(\frac{\partial \psi}{\partial z} - ik\psi)e^{-ikz}$$

$$= \frac{\partial^2 \psi}{\partial z^2}e^{-ikz} - 2ik\frac{\partial \psi}{\partial z}e^{-ikz} - k^2\psi e^{-ikz} \qquad (5.102)$$

Substituting these equations into the Helmholtz equation:

$$(\frac{\partial^2 \psi}{\partial x^2} + \frac{\partial^2 \psi}{\partial y^2} + \frac{\partial^2 \psi}{\partial z^2})e^{-ikz} + k^2 \psi e^{-ikz} - 2ik\frac{\partial \psi}{\partial z}e^{-ikz} - k^2 \psi e^{-ikz}$$

$$= 0$$

$$\Downarrow$$

$$[\frac{\partial^2 \psi}{\partial x^2} + \frac{\partial^2 \psi}{\partial y^2} + (\frac{\partial^2 \psi}{\partial z^2} - 2ik\frac{\partial \psi}{\partial z})]e^{-ikz} = 0 \tag{5.103}$$

Under the *important assumption* that k is a very large number:

$$\frac{\partial^2 \psi}{\partial z^2} \langle\langle 2ik\frac{\partial \psi}{\partial z} \tag{5.104}$$

Equation 5.103 becomes the paraxial Helmholtz equation:

$$\frac{\partial^2 \psi}{\partial x^2} + \frac{\partial^2 \psi}{\partial y^2} - 2ik\frac{\partial \psi}{\partial z} = 0 \tag{5.105}$$

This is the slow varying envelope approximation of the Helmholtz equation. This solution resembles the time-dependent Schrodinger's equation. The simplest solution is paraboloidal, but the most important, we will see, is the fundamental or zero order Gaussian beam.

5.4.3 Relationship between the Paraboloidal and Gaussian Wave

Before moving on to analyzing the detailed form of Eq. 5.105, for a Gaussian function, the relationship between the complex envelope of a paraboloidal and Gaussian wave shall be expanded on. The complex envelope of a paraboloidal wave is given by the general formula:

$$\psi(r) = (\alpha_1/z)\exp\{-ik(x^2 + y^2)/2z\}\exp - i\phi(z) \tag{5.106}$$

where $\exp\{-ik(x^2 + y^2)/2z\}$ modulates the envelope. For a Gaussian function, we replace z by $q(z)$ where $q(z) = z + iz_0$ and z_0 is a constant. The equation modulating our complex envelope is now:

$$\psi(r) = (\alpha_1/q(z))\exp\{-ik(x^2 + y^2)/2q(z)\}\exp - i\phi(z) \tag{5.107}$$

The shift by iz_0, we will see, has very significant implications.

5.4.4 The Gaussian Wave

A Gaussian beam is basically a spherical wave field whose amplitude in the transverse direction has a Gaussian pattern. It, like other solutions of the paraxial wave equation, has the form of Eq. 5.107. Deriving the equations behind a Gaussian wave is straightforward but tedious. These derivations can be found in the referenced sources at the end of the chapter.[27,28] This discussion will focus on those properties of the Gaussian waves important in describing their behavior with OCT and the underlying assumptions and associated limitations. The equation used to describe the Gaussian wave here is:

$$\psi = (W_0/W(z)) \exp\{-[x^2 + y^2/W^2(z)]\} \exp\{-ik(x^2 + y^2)/2R(z)\} \exp - i\phi(z)$$

$$(5.108)$$

The equation can be divided into components of the amplitude $((W_0/W(z)) \exp\{-(x^2+y^2/W^2(z)\})$, a phase component $\{\exp -i\phi(z)\}$, and modulation or paraxial component $\exp\{-ik(x^2+y^2)/2R(z))\}$. The meaning of this equation will become clearer but for now, speaking in generalities, $R(z)$ represents the radius of curvature, $\phi(z)$ is the phase difference between the an ideal plane wave and the Gaussian wave, W_0 is the radius of the minimum width of the Gaussian wave, and $W(z)$ is the width of the beam at an arbitrary point z (spot size). A drawing of a Gaussian beam is shown in Figure 5.9. It has a minimal radius at the beam waist (W_0), the beam behaves relatively collimated between $\pm z_R$, and it then diverges rapidly afterwards. This deviation from the pattern of a spherical or paraboloidal wave results from the substitution of the imaginary displacement in Eq. 5.107.

5.4.5   Properties of a Gaussian Wave

Important properties of the Gaussian beam are its minimal radius (beam waist), Rayleigh range/confocal parameter, intensity profile, power, and radius of curvature.

The Gaussian beam has a minimal radius at the beam waist, W_0. The beam waist is given by:

$$W_0^2 = \lambda z_R/n\pi \tag{5.109}$$

Here z_R is known as the Rayleigh range as shown in Figure 5.9. Two times z_R is known as the confocal parameter or depth of focus. This is the range over which the

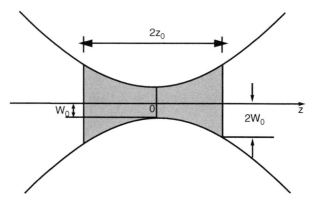

Figure 5.9   The profile of the Gaussian beam, demonstrating the lateral intensity distribution. W0 represents the spot size while 2z0 represents the Rayleigh range. Drawn by Karen Aneshansley.

beam is relatively collimated. At z_R the beam radius is a $\sqrt{2}$ larger than at the waist, it has 2 times the area, and the intensity at the beam axis is one-half the peak intensity. In addition, the radius of curvature of the wavefront is the smallest so that the wavefront has the greatest curvature ($R = 2z_0$).

The intensity profile of the beam in general is given by:

$$I(x, y, z) = I_0(W_0/W(z))^2 \exp\{-2[x^2 + y^2]/W^2(z)\} \tag{5.110}$$

Within a transverse plane, the beam drops to $1/e$ radially at $r = \omega(z)$, and the power $1/e^2$. Assuming no sources of energy loss, the power is constant at any value of r and is given by:

$$P = (1/2)I_0(\pi W_0^2) \tag{5.111}$$

Approximately 86% of the power exists within the circle of radius $W(z)$. The diameter at which the intensity drops to $1/e^2$ is generally used to refer to the diameter of the beam.

The Gaussian beam has its greatest radius of curvature $R(z)$ at z_R as shown in Figure 5.11A. $R(z)$ is given by:

$$R(z) = z[1 + (z_R/z)^2] \tag{5.112}$$

If $z < z_R$, $R(z) \approx z_R^2/z$ and the wavefront is planar in nature. If $z > z_0$, $R(z) \approx z$ and therefore behaves more like a spherical wave.

5.4.6 Important Points of a Gaussian Beam Relevant to OCT

1. The smaller the beam waist, the shorter the confocal parameter or distance over which the beam is relatively collimated. This is illustrated in Figure 5.10. When the beam waist approaches 5 μm, the confocal parameter drops to approximately 5 μm. Therefore, when small beam waists are required, such as in conjunction with an extreme broad band source, additional techniques are required such as dynamic focus tracking in a manner analogous to confocal microscopy, C-mode scanning imaging, or the use of special lenses.[14–16]

2. The paraxial approximation (Eq. 5.107) tells us that the beam waist must be large relative to the wavelength (angular deviation is small). As high resolution OCT goes beyond 5 μm using a 1300-nm source, these parameters are of the same order of magnitude, raising concerns that the paraxial equation may begin to break down.

3. The derivations above are for a monochromatic wave. With OCT, we are dealing with a broadband source. Since the focal length of lenses are wavelength dependent, chromatic aberration remains an issue, particularly with a broad bandwidth source. Therefore, for high resolution OCT, appropriately designed achromatic lenses are indicated.

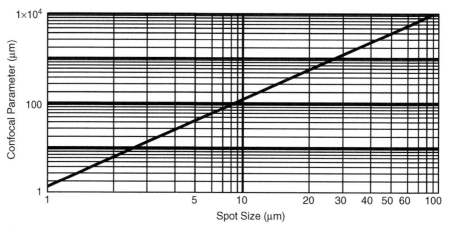

Figure 5.10   The smaller the beam waist, the shorter the confocal aprameter or distance over which the beam is relatively collimated. When the beam waist approaches 5 çm, the confocal parameter drops to approximately 5 μm.

4. The power within the beam waist or spot size is defined at a radius of $1/e^2$ that contains 86% of the total power. The remaining 14% needs to be taken into account when directing Gaussian beams, such as into single mode fibers.

5. The derived equations were done for the zero or fundamental Gaussian mode. Some of the equations will vary when examining higher orders. However, as most OCT propagation occurs through single mode fibers, higher modes can largely be ignored.

6. When propagating through optical components, it is important to maximize the numeric aperture to minimize spatial filtering described under the transfer function section.

7. The derivation of Gaussian propagation was done with incoherent waves. However, spatial coherence remains important and contributes to such phenomena as speckle which is discussed in the following section.

8. Although not dealt with here in detail, the greatest radius of curvature generally is not at the beam waist and the position not constant with z, as shown in Figure 5.11A. This has the consequence that the minimum spot size on the target of interest is not necessarily achieved by having the beam waist at the target. In general, minimal spot size is achieved with the beam waist between the target and source, which is discussed elsewhere in detail and in Figure 5.11B.[11] However, this positioning of the beam waist in front of the target reduces the depth of focus so that the two phenomena need to be balanced.

9. A Gaussian beam propagating through optical components cannot generally be described by simple lens maker formulas. Generally the ABCD law for Gaussian beam is used which relates the input and output of optical systems.

10. Different waves are defined by different parameters. A planar wave can be characterized by its amplitude and direction. A spherical wave is

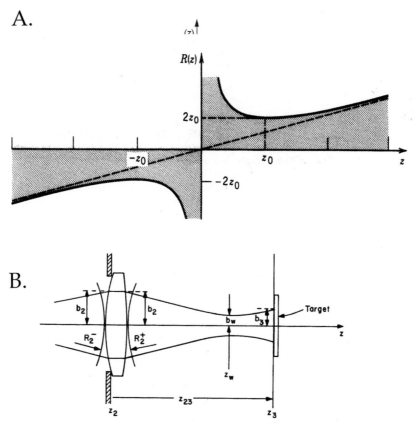

Figure 5.11 Properties of a Gaussian beam. In A, the effect of distance on the radius of curvature is noted. The point of minium radius of curvature is not at the beam waist but at z_0. Courtesy of Teich, M. and Saleh, B. *Fundametals of Photonics*, figure 3.1-6 [28]. In B, it can be seen that because of this radius of curvature, the minimal spot size on a target is not necessarily at the beam waist. In general, the minimal spot size is between the target and source, as discussed in reference 17. However, this positioning of the beam waist in front of the target reduces the depth of focus so that the two phenomena need to be balanced. Image courtesy of Gaskill, J. D. *Linear Systems, Fourier Transforms, and Optics*, figure 10-31 [17].

characterized by its amplitude and origin. A Gaussian wave, however, needs to be defined by additional parameters. It is characterized by four parameters: direction, peak amplitude, the location of the waist, and one additional parameter, like the waist radius or the confocal parameter.

5.4.7 Speckle

Until now we have treated the lateral resolution via an incoherent approach. However, OCT is a technology based on the coherence of backreflected light.

Figure 5.12 The phenomena of speckle. Two or more sites separated by a distance of an odd multiple of one-half the wavelength, but within the coherence length, result in speckle (constructive and destructive interference of the wavefront). The graininess produced in the image deteriorates the image quality, but as will be seen in chapter 11, it is also a mechanism for producing contrast.

This issue will be partially addressed via the optics of speckle. Speckle has risen to importance with the development of coherent sources, particularly the laser. As we will see in Chapter 11, speckle is both signal carrying and degrading.[6]

The signal degrading component alters the shape of the backreflected wavefront through localized regions of constructive and destructive interference. Optimal performance with OCT is achieved when only single backreflected light is detected. However, multiply scattered light from out-of-focus sites results in speckle. Two or more sites separated by a distance of odd multiples of one-half the wavelength, but within the coherence length, result in speckle (constructive and destructive interference of the wavefront). Therefore, speckle is dependent on the coherence length, multiple scattering, and the aperture of the detector. The graininess produced in the image deteriorates the image quality, as shown in Figure 5.12. However, if all the waves backreflect from the sample volume, interfere constructively, and produce contrast, it improves the OCT image.

Therefore, when considering the image-resolving ability of OCT, both spatial and temporal coherence need to be taken into account.

5.5 SPECTRAL RADAR OR FOURIER DOMAIN OCT (FD-OCT)

In addition to the more commonly used time domain TD-OCT described in the initial portions of this chapter, spectral domain techniques (SD-OCT) have recently become of interest including Fourier domain OCT (FD-OCT) and swept source OCT (SS-OCT) (see Figure 5.1B and C). FD spectroscopy is a technique that, in part, dates back to the time of Michelson at the turn of the previous century.[18,19,24] It has been applied to low coherence reflectometry and OCT over the last decade.[20–23] With FD-OCT, a broadband source is used similar to TD-OCT, but the mirror in the reference arm typically does not move. Depth information is obtained by evaluating the spectrum of the processed interferogram by directing the returning light onto a Charged Coupled Device (CCD). Less common, a dispersive element is placed in the detector arm, which is scanned through the bandwidth, and a single detector is used. The Fourier transform of the spectrum provides a backreflection profile as a function of depth. Depth information is obtained by the different interference profiles from different pathlengths in both arms. Larger differences in the optical pathlength between the sample and reference arm result in higher frequency interference signals. FD-OCT is described using an approach common in the literature, including radar and holography work.[22,23] When we examine dynamic range and SNR in Chapter 7, a more detailed approach FD-OCT will be used.

The signal in the sample arm comes from different depths within the sample and is combined with the signal in the reference arm to give the interferogram. At the detector, the different wavenumbers result from the combined signal of both arms. The total interference signal $I(k)$ is given by the spectral intensity distribution of the light source times the square of the sum of the two backreflected signals.

$$I(k) = G(k)|a_R \exp{(i2kr)} + \int_{z_0}^{\infty} a(z)\exp{\{i2kn(z)(r+z)dz|} \tag{5.113}$$

Here the first term within the square is the signal from the reference arm while the second term comes from the sample arm. $G(k)$ is the spectral intensity distribution of the light source, a_R is the reflection amplitude of the coefficient reference arm, $a(z)$ is the backscattering coefficient of the object signal, with regard to the offset z_0, n is the refractive index, z_0 is the offset of the reference plane and object surface, $2r$ is the pathlength in the reference arm, $2(r+z)$ is the pathlength in the object arm, and $2z$ is the difference in pathlength between the sample and reference arm. The reflection coefficient in the reference arm, a_R, will arbitrarily be set at 1 and since we are only interested in the pathlength difference between both arms (ignoring dispersive and polarization effects), we will define r as zero. Using the relationship:

$$|y+z|^2 = (y+z)(y^*+z^*) = |y|^2 + 2\text{Re}[yz^*] + |z|^2 \tag{5.114}$$

The components of Eq. 5.113 become:

$$|y|^2 = [\exp{(i2kr)}](a_R)^2 \exp{(-i2kr)} = (a_R)^2 \tag{5.115a}$$

$$2\,\text{Re}[yz^*] = 2\,\text{Re}\{[\exp{(i2kr)}] \int_{z_0}^{\infty} a(z)\exp[\{-i2kn(z)\,(r+z)\}dz]\}$$

$$= 2\int_{z_0}^{\infty} a(z)\,\cos[2kn(z)z]dz \tag{5.115b}$$

$$|z|^2 = |\int_{z_0}^{\infty} a(z)\,\exp{i2kn(z)\,(r+z)dz}|^2$$

$$= \int_{z_0}^{\infty} a(z)\,\exp{i2kn(z)\,(r+z')dz'} \int_{z_0}^{\infty} a(z)\,\exp\{i2kn(z)\,(r+z)\}dz \tag{5.115c}$$

Now $I(k)$ can be written:

$$I(k) = G(k)\left(1 + 2\int_{z_0}^{\infty} a(z)\cos\{2knz\}dz + \int_{z_0}^{\infty}\int_{z_0}^{\infty} a(z')a(z)\exp\{i2kn(z-z')\}dzdz''\right)$$

$$\tag{5.116}$$

The first term is a DC term $\{G(k)\}$. The second term encodes the depth information of the object. The backreflection intensity is found in $a(z)$ while the corresponding optical pathlength difference is found in the argument of the cosine term. The third term describes the mutual interference of all elementary waves. We must make the assumption that $a(z)$ is symmetric with respect to $z = 0$ (since $a(z) = 0$ for $z < z_0$) so that we can obtain $a(z)$ from the Fourier transform of $I(k)$. Therefore:

$$A(z) = 1/2 \; a(z) + 1/2 a(-z) \tag{5.117}$$

Now Eq. 5.116 becomes:

$$I(k) = G(k)\left(1 + \int_{-\infty}^{+\infty} A(z)e^{\{-i2knz\}}dz + (1/4)\int_{-\infty}^{+\infty} AC[A(z')]\exp^{\{i2kn(z)z\}} dz\right) \tag{5.118}$$

Here, we have replaced the cosine function by an exponential (so that it is the form of a Fourier transform) by using Euler's formula $e^{i\Theta} = \cos\Theta + i\sin\Theta$. We are integrating over $-\infty$ to $+\infty$ such that $\int_{-\infty}^{\infty} a(z)a(z + z_0)dz$ where $z_0 = z' - z$. Expressing Eq. 5.118 in terms of the Fourier transform (F) of the last two terms,

$$I(k) = G(k)(1 + 1/2F\{A(z)\} + (1/8)F\{AC[A(z')]\}) \tag{5.119}$$

Taking the inverse Fourier transform of Eq.5.119, we obtain:

$$F^{-1}I(k) = F^{-1}\{G(k)\} \otimes (\delta(z) + 1/2A(z) + (1/8)AC)[A(z')]) = A \otimes (B + C + D) \tag{5.120}$$

The equation is valid for a single scattering event and constant refractive index. We are interested in the information contained within $A \otimes C$. The convolution of A (source intensity distribution) with B, the Dirac delta function $\delta(z)$, is the DC signal. We can separate it from C by shortening the optical pathlength in the reference arm compared to the sample arm by a few hundred millimeters. Increasing separation between arms results in increasing frequency in the interferogram while $A \otimes B$ remains around zero. This is shown in Figure 5.13. Similarly, we need to remove $A \otimes D$, which represents the mutual interference among scatters in the sample arm. This is achieved in two ways.

First, in strongly scattering medium, the $A \otimes C$ is generally much larger than the $A \otimes D$ term. Second, displacements using the same reference are used to separate B, and the $A \otimes D$ term tends to remain around $z = 0$. However, high backscattering deep within tissue can lead to overlap of the terms which can result in artifacts in the data obtained.

As stated, the greater the mismatch between the two arms, the higher the frequency content of the detected signal. For a depth of only a few millimeters the frequency is rather high. Therefore, the ability of the CCD to detect the high frequencies determines the penetration depth.

While the axial resolution is governed by the same principles as TD-OCT, given by Eq. 5.30,

$$\Delta l = (2\ln2/\pi)(\lambda_0^2/\Delta\lambda) \tag{5.121}$$

but the resolution decreases with depth (higher mismatch).

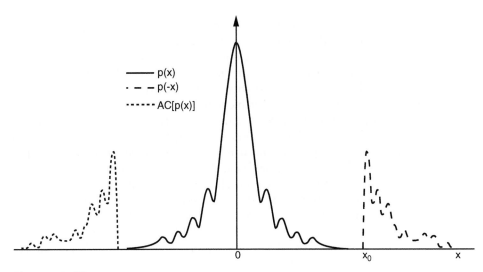

Figure 5.13 Interferogram recorded with spectral radar or SD-OCT. Diagnostic information is contained in either $p(x)$ or $p(-x)$, the latter is symmetrically identical to the former. The region around zero (AC) contains minimal diagnostic information and is, therefore, filtered from the final digitized profile. Drawn by Karen Aneshansley.

The CCD has a limited spectral range $\Delta\zeta$. If $\Delta\zeta$ is too small, not capturing the full available spectrum, the effective axial resolution will decrease. If $\Delta\zeta$ is too large, the pixel spacing is reduced, resulting in a reduced measurement range without improving axial resolution. The pixel spacing is chosen as $\Delta l/2$ (Nyquist condition), which corresponds to one-half the axial resolution. The spectral range is therefore given by:

$$\Delta\zeta = (\pi/2\ln2)(\Delta\lambda) \qquad (5.122)$$

If the detector has N pixels, then Z_{max}, the axial measurement range equals:

$$Z_{max} = (\Delta l/2)(N/2) \qquad (5.123)$$

N is divided by two because the Fourier transform of the real spectrum has conjugate symmetry about the zero delay. This equation demonstrates that for a given source bandwidth or coherence length, the axial scan range is determined by the number of pixels.

The data collected by the CCD are sampled at equal wavelength intervals. The output is typically digitized using a four-channel data acquisition board with 12-bit resolution and continually transferred to computer memory. The Fourier transform, however, links z and k space. Because of the nonlinear relationship between k and λ the Fourier transform of the data acquired directly from the spectrometer is improper. Therefore, the signal is unevenly sampled in k space but evenly sampled in λ space, which results in broadening of the PSF function (as well as the amplitude) at higher mismatches. Preprocessing of the data is therefore needed, which is typically done with a nonlinear scaling algorithm. Also, there is a decrease in amplitude with optical group mismatches due to the limited spectral resolution of the spectrometer, which requires further correction. It results from the sensitivity of the interpolation process to small errors at high fringe frequencies, which can be minimized by utilizing more pixels in the CCD.

Dispersion compensation can be performed in the software with spectral OCT (Appendix 5-1). This involves canceling the frequency-dependent nonlinear phase induced by dispersion mismatch between the reference and sample arm. This is achieved by the following. First, the reference arm spectrum is measured separately and subtracted from the interference spectrum. Second, rescaling is performed as described above. Third, a Hilbert transform is performed to generate the imaginary part of the complex analytical signal (Appendix 5-1). Fourth, the complex analytical signal is constructed. Fifth, phase correction is performed, the most arbitrary part of the procedure, unless an isolated specular reflection within the tissue is noted. Finally, a Fourier transform is performed to obtain the backscattering profile as a function of depth.[30]

Motion artifacts are a serious problem with spectral domain techniques. With SD-OCT, the integration process of the CCD further enhances motion artifacts. If a phase drift over more than π occurs during a single A-scan, there is degradation of the image. In one study, the axial motion of a sample relative to the probe beam results in decreased SNR due to fringe washout.[25] Transverse sample motion does not cause fringe washout, but results in degradation in transverse resolution and SNR. In addition, fringe washout may potentially occur if the beam is transverse scanned across tissue structure with spatially varying depth.[25] Recently, the use of a pulsed source or a cw wavelength swept source has been proposed to reduce these artifacts.[26]

5.6   SWEPT SOURCE (SS-OCT)

With SS-OCT, the light beam from the source is again evenly split by a 2×2 coupler and comes through and back in the reference and the sample arm, then recombines at one of the ports of the coupler in terms of the exit of the interferometer. The light intensity perturbation at the exit can be described as

$$\left\{ I_D = I_R + I_S + \frac{1}{2} \int_0^z \int_0^\infty G(f) p(z') p_R \cos\{\frac{4\pi f}{c}[n_S(f)z' - z_R]\} df \right\} dz' \qquad (5.124)$$

As shown in Figure 5.1C, the origin of coordinates O_S in the sample arm is usually set up at the sample surface. Consequently the origin of coordinates in the reference arm O_R will be chosen at a point from where the optical group delay to the coupler matches that between O_S and the coupler in the sample arm. In Eq. 5.124, the function $p(z)$ represents the backscattering coefficient distribution in the sample, and z is measured from O_S. The mirror position is z_R off the O_R. Hereafter, the integration of $G(f)$ on the total positive frequency is defined as the intensity I_0.

Function $n_s(f)$ represents the refractive index of the sample, assuming no dispersion. Constant c is the light velocity in a vacuum.

The first term in Eq. 5.124 corresponds to the intensity of the reflected reference beam I_R. The second term corresponds to the total intensity of the returned sample beam which is contributed by all the scatters, and the mutual interference of all scattering waves, which was called the autocorrelation function of scattering[4] or a parasitic term.[5] If strong reflections exist relatively deep within the tissue, they may be difficult to remove with SD-OCT techniques.

Using a swept/tunable laser source instead of the wide band source, the spectral interferogram can be obtained with a single detector at the exit of the interferometer. The axial detection performance of SS-OCT can be conducted on a same base of signal evolution as TD-OCT. From Eq. 5.124, assuming an ideal spectrum is captured by a detector, the spectral interferogram can be described as

$$I_D(f) = \frac{1}{4} G(f) \left[p_R^2 + \int_0^z p_S^2(z')dz' + 2\text{Re}\left\{ \int_0^z p(z')p_R \exp\{j4\pi \frac{f}{c}[n_S(f_0)z' - z_R]\}dz' \right\} \right]$$

(5.125)

It can also be written in terms of wavenumber k:

$$I(k) = \frac{1}{4} G(k) \left[p_R^2 + \int_0^z p_S^2(z')dz' + 2\text{Re}\left\{ \int_0^z p(z')p_R \exp\{j4\pi k[n_S(k_0)z' - z_R]\}dz' \right\} \right]$$

(5.126)

where $G(k)$ is the intensity spectrum of the light beam. From Eq. 5.126, the detected spectrogram can be described as

$$I'_D(k) = \frac{1}{4}\{G(k)[\exp(-\frac{k^2}{2\sigma_k^2}) \otimes \text{comb}(\frac{k}{\delta'_k})]\}$$

$$\{p_R^2 + \int_0^z p_S^2(z')dz' + 2\text{Re}\{\int_0^z p(z')p_R \exp\{j4\pi k[n_S(k_0)z' - z_R]\}dz'\}\} \qquad (5.127)$$

The spectrum of the light source is described as a discrete spectral intensity distribution with a line width σ_k (standard deviation) and tuning mode-hop δ_k'. Tuning mode-hops result from the inability to compensate for changes in cavity length. So the retrieved signal can be expressed as

$$F^{-1}\{I_D(k)\} = \frac{1}{4} F^{-1}[G(k)] \otimes [\exp(-2\sigma_k^2 z''^2)\text{comb}(\delta'_k z'')] \otimes$$

$$\{F^{-1}[p_R^2 + \int_0^z p_S^2(z')dz'] + p_R p(\frac{z'' + 2z_R}{2n_S(k_0)}) + p_R p(\frac{-z'' + 2z_R}{2n_S(k_0)})\} \qquad (5.128)$$

From Eq. 5.128, it is seen that the mode-hop will limit the A-scan range and the laser line width will degrade the A-scan profile.

In SS-OCT, as in FD-OCT, no moving parts are required for axial scan (ignoring the tuning mechanism in the laser source).

In an SD-OCT, the spectral interferogram of a single reflector located at z_0 off the O_S can be expressed according to Eq. 5.129 as

$$I_D(k) = \frac{1}{4} G(k) \{p_R^2 + p^2 + 2pp_R \cos[4\pi(z_0 - z_R)k]\} \tag{5.129}$$

z_R is the mirror position off the O_R. p is defined as the backscattering coefficient of the single scatter.

5.6.1 The Resolution

Applying the Fourier transform, the retrieved a-scan can be described as:

$$i_D(z'') = \frac{1}{4} F^{-1}[G(k)] \otimes \{[p_R^2 + p^2]\delta(z'') + pp_R[\delta(\frac{z'' - 2z_0 + 2z_R}{2}) + \delta(\frac{-z'' - 2z_0 + 2z_R}{2})]\} \tag{5.130}$$

As Eq. 5.130 indicates, the designated signal $p(z'')$ is convoluted by the reverse Fourier transform of the spectrum $G(k)$. The power spectrum is assumed to be Gaussian distribution in k-space as

$$G(k) = \frac{I_0}{4} \times \frac{1}{\sigma_k \sqrt{2\pi}} \{\exp[-\frac{(k - k_0)^2}{2\sigma_k^2}] + \exp[-\frac{(k + k_0)^2}{2\sigma_k^2}]\} \tag{5.131}$$

where σ_k is the standard deviation of the spectrum. The Fourier transform will be

$$F^{-1}\{G(k)\} = I_0 \cos(2\pi k_0 z'') \exp[-2\pi^2 \sigma_k^2 (z'')^2] \tag{5.132}$$

The spectrum width in k-space has a linear relationship in frequency domain as

$$\sigma_k = \frac{\sigma_f}{c} \tag{5.133}$$

Consider the reciprocity relation between σ_f and coherence time σ_t as

$$\sigma_t \sigma_f = \frac{1}{4\pi} \tag{5.134}$$

and the coherence length known as

$$\sigma_l = c\sigma_t \tag{5.135}$$

Substituting Eqs. 5.133, 5.134, and 5.135 into Eq. 5.132, we can get

$$F^{-1}\{G(k)\} = I_0 \cos(2\pi k_0 z'') \exp[-\frac{(z'')^2}{8\sigma_l^2}] \tag{5.136}$$

Use the FWHM width instead of the standard deviation as

$$\Delta_l = 2\sqrt{2\ln 2}\sigma_l \tag{5.137}$$

where Δ_l is the FWHM width of the PSF of TD-OCT as

$$\Delta_l = \frac{2\ln 2\lambda_0^2}{\pi\Delta_\lambda} \tag{5.138}$$

Thus Eq. 5.136 can be rewritten in terms of Δ_l

$$F^{-1}\{G(k)\} = I_0 \cos(2\pi k_0 z'') \exp[-\frac{4\ln 2(z'')^2}{(2\Delta_l)^2}] \tag{5.139}$$

According to the linear system theory, the axial resolution of SS-OCT is determined by the system response function as Eq. 5.139 demonstrates. Analogous to TD-OCT, the FWHM width of the response function determines the axial resolution. The FWHM width in FD-OCT is twice the coherence length Δ_l that is defined in TD-OCT as axial resolution. But it does not mean that axial resolution in SS-OCT is better than in TD-OCT, because the rescaling factor in FD-OCT is twice the one in SS-OCT. Theoretically both FD-OCT and TD-OCT have the same axial resolution that is in the order of coherence length of the light source.

As we know, any spectrum measurement has a certain spectral resolution Δ_k. According to the discrete Fourier transform theory, the axial measuring range of FD-OCT will be

$$z''_{max} = \frac{1}{2\Delta_k} \tag{5.140}$$

It is measured in free space. We can rescale it to the geometrical size of the measured sample and use wavelength resolution Δ_λ to replace wavenumber resolution Δ_k. Taking into account the group index of the sample, the maximum detectable range will be

$$z_{max} = \frac{\lambda_0^2}{2n(\lambda_0)\Delta_\lambda} \tag{5.141}$$

REFERENCES

1. Born, M. and Wolf, E. (1999). *Principles of Optics*, 7th edition (expanded). Cambridge University Press, UK.
2. Boppart, S. A., Tearney, G. J., Bouma, B. E., Southern, J. F., Brezinski, M. E., and Fujimoto, J. G. (1997). Noninvasive assessment of the developing Xenopus cardiovascular system using optical coherence tomography. *Proc. Natl. Acad. Sci.* 94, 4256–4261.
3. Hellmuth, T. (1997). Contrast and resolution in optical coherence tomography. *Proc. SPIE* 2926, 228–232.
4. Fercher, A. F., Hitzenberger, C. K., Kamp, G., and Elzaiat, S. Y. (1995). Measurement of intraocular distances by backscattering spectral interferometry. *Opt. Commun.* 117(1–2), 43–48.

5. Wojtkowski, M., Kowalczyk, A., Leitgeb, R., and Fercher, A. F. (2002). Full range complex spectral optical coherence tomography technique in eye imaging. *Opt. Lett.* 27(16), 1415–1417.

6. Schmitt, J. M., Xiang, S. H., and Yung, K. M. (1999). Speckle in optical coherence tomography. *J. Biomed. Opt.* 4, 95–105.

7. Neumann, E. G. (1988). *Single Mode Fiber*. Springer-Verlag, Berlin.

8. Liu, B., Macdonald, E. A., Stamper, D. L., and Brezinski, M. E. (2004). Group velocity dispersion effects with water and lipid in 1.3 μm OCT system. *Phys. Med. Biol.* 49, 1–8.

9. Niblack, W. K., Schenk, J., Liu, B., and Brezinski, M. E. (2003). Dispersion in a gruting based delay line for OCT. *Applies Optics* 42(19) 4115–4118.

10. Chen, Y. and Li, X. (2004). Dispersion management up to third order for real time optical coherence tomography involving a phase or frequency modulator. *Opt. Express* 12, 5968–5978.

11. Drexel, W., Morgner, U., Kartner, F. X. et al. (1999). *In vivo* ultrahigh resolution optical coherence tomography. *Opt. Lett.* 24, 1221–1223.

12. Marks, D., Oldenburg, A. L., Joshua Reynolds, J. et al. (2003). Autofocus algorithm for dispersion correction in optical coherence tomography. *Appl. Opt.* 42, 3038–3046.

13. Kogelnik, H. and Li, T. (1966). Laser beams and resonators. *Proc. IEEE,* Vol. 54, pp. 1312–1329.

14. Schmitt, J. M., Lee, S. L., and Yung, K. M. (1997). An optical coherence microscope with enhanced resolving power in thick tissue. *Opt. Commun.* 142, 203–207.

15. Ding, Z., Ren, H., Zhao, Y. et al. (2002). High resolution optical coherence tomography over a large depth range with an axicon lens. *Opt. Lett.* 27, 243–246.

16. Lexer, F., Hiizenberger, C. K., Drexler, W. et al. (1999). Dynamic coherent focus OCT depth-independent transversal resolution. *J. Mod. Opt.* 46, 541–553.

17. Gaskill, J. D. (1978). *Linear Systems, Fourier Transforms, and Optics*. John Wiley & Sons, New York, pp. 435–441.

18. Loewenstein, E. V. (1966). The history and current status of Fourier transform spectroscopy. United States Air Force, Air Force Cambridge Research Laboratories.

19. Michelson, A. A. (1891). *Phil. Mag. Ser.* 5(31), 338–341.

20. Tokunaga, E., Terasaki, A., and Kobayashi, T. (1993). Induced phase modulations of chirped continuum pulses studied with a femtosecond frequency domain interferometer. *Opt. Lett.* 18, 370–372.

21. Fercher, A. F., Hitzenberger, C. K., Kamp, G., and El-Zaiant, S. Y. (1995). Measurement of intraocular distances by backscattering spectral interferometry. *Opt. Commun.* 117, 43–48.

22. Hausler, G. and Lindner, M. (1998). Coherence radar and spectral radar — new tools for dermatological diagnosis. *J. Biomed. Opt.* 3, 21–31.

23. Wolf, E. (1969). Three dimensional structure determination of semi-transparent objects from holographic data. *Opt. Commun.* 1, 153–156.

24. Rubens, H. and Wood, R.W. (1911). *Phil. Mag. Ser.* 6(21), 249–255.

25. Yun, S. H., Tearney, G. J., de Boer, J. F., and Bouma, B. E. (2004). Motion artifacts in optical coherence tomography with frequency domain ranging. *Opt. Express* 12, 2977–2998.

26. Yun, S. H., Tearney, G. J., de Boer, J. F., and Bouma, B. E. (2004). Pulsed-source and swept source spectral domain optical coherence tomography with reduced motion artifacts *Opt. Express* 12, 5614–5624.

27. Guenther, R. (1990). *Modern Optics*. John Wiley & Sons, New York.

28. Teich, M. and Saleh, B. (1991). *Fundamentals of Photonics*. John Wiley & Sons, New York.

29. M. Wojtkowski, V. J. Srinivasan, T. Ko, et al. (2004). Ultrahigh resolution, high speed Fourier domain optical coherence tomography and methods for dispersion compensation. *Opt. Express* 12, 2404–2422.

BIBLIOGRAPHY

Kowalczyk, A. and M. W. (2004). Ultrahigh sensitive imaging of the eye by spectral optical coherence tomography. *Opt. Security Safety* (5566), 79–83.

Fercher, A. F., C. K. H., and Sticker, M. (2001). Numerical dispersion compensation for partial coherence interferometry and optical coherence tomography. *Opt. Express* 9(12), 610–822.

Wax, A., C. Y., and Izatt, J. (2003). Fourier-domain low-coherence interferometry for light-scattering spectroscopy. *Opt. Lett.* 28(14), 1230–1232.

Rollins, A. M. and J. A. I. (1999). Optimal interferometer designs for optical coherence tomography. *Opt. Lett.* 24(21), 1484–1486.

Goubovic, B., B. E. B., and Tearney, G. J. (1997). Optical frequency-domain reflectometry using rapid wavelength tuning of a Cr^4 +:forsterite laser. *Opt. Lett.* 22(22), 1704–1706.

Cense, B. and N. N. (2004). Ultra high-resolution high-speed retinal imaging using spectral-domain optical coherence tomography. *Opt. Express* 12(11), 2435–2447.

Chen, N. G. and Zhu, Q. (2002). Rotary mirror array for high-speed optical coherence tomography. *Opt. Lett.* 27, 607–609.

Deepak, U. and B. C. (1985). Precision time domain reflectometry in optical fiber systems using a frequency modulated continuous wave ranging technique. *J. Lightwave Technol.* LT-3(5), 971–977.

Dominik H., C. K., and Rupp, F. (2000). High resolution imaging with advanced chirp optical coherence tomography. Coherence domain optical methods in biomed science and clinical applications IV 3915, 83–89.

Lexer, F., C. K. H., and Fercher, A. F. (1997). Wavelength-tuning interferometry of intraocular distances. *Appl. Opt.* 36(25), 6548–6553.

Coquin, G. A. and K. W. C. (1988). Electronically tunable external cavity semiconductor laser. *Electron. Lett.* 24, 599–603.

Gianotti, J. B., C. R. Walti, et al. (1997). Rapid and scalable scans at 21m/s in optical low-coherence reflectometry. *Opt. Lett.* 22, 757–759.

Hsiung, P.-L., X. Li, et al. (2003). High-speed path-length scanning with a multiple-pass cavity delay line. *Appl. Opt.* 42, 640–648.

Kowalczyk, M. W. A. A. (2002). Full range complex spectral optical coherence tomography technique in eye imaging. *Opt. Express* 27(16), 1415–1417.

Bail, M., G. H., and Linder, J. M. Optical coherence tomography with the "Spectral Radar" — Fast optical analysis in volume scatterers by short coherence interferometry. *SPIE* 2925, 298–303.

Wojtkowski, M., R. L., and Kowalczyk, A. (2002). *In vivo* human retinal imaging by Fourier domain optical coherence tomography. *J. Biomed Opt.* 7(3), 457–463.

Wojtkowski, M., T. B., and Targowski, P. (2004). Real-time and static *in vivo* ophthalmic imaging by spectral optical coherence tomography. *Ophthal. Technol. XIV* 5314, 126–131.

Choma, M. A., M. V. S., and Yang, C. (2003). Sensitivity advantage of swept source and Fourier domain optical coherence tomography. *Opt. Express* 11(18), 2183–2189.

Nassif, N. A., B. C., and Park, B. H. (2004). *In vivo* high-resolution video-rate spectral-domain optical coherence tomography of the human retina and optic nerve. *Opt. Express* 12(3), 367–376.

Nassif, N. and B. C. (2004). *In vivo* human retinal imaging by ultrahigh-speed spectral domain optical coherence tomography. *Opt. Lett.* 29(5), 480–482.

Passy, R., N. G., and von der Weid, J. P. (1994). Experimental and theoretical investigations of coherent OFDR with semiconductor laser sources. *J. Lightwave Technol.* 12(9), 1622–1630.

Leitgeb, R. A., W. D., and Unterhuber, A. (2004). Ultra-high resolution Fourier domain optical coherence tomography. *Opt. Express* 12(10), 2156–2165.

Rainer, A., Leitgeb, R. A., C. K. H., and Fercher, A. F. (2003). Phase-shifting algorithm to achieve high-speed long-depth-range probing by frequency-domain optical coherence tomography. *Opt. Lett.* 28(22), 2201–2203.

Yun, S. H., C. B. et al. (2003). High-speed wavelength-swept semiconductor laser with a polygon-scanner-based wavelength filter. *Opt. Soc. Am.* 28, 1881–1883.

Chinn, S. R., E. A. S. et al. (1997). Optical coherence tomography using a frequency-tunable optical source. *Opt. Lett.* 22, 340–342.

Yun, S. H., G. T., and Bouma, B. (2003). High-speed spectral domain optical coherence tomography at 1.3 μm wavelength. *Opt. Express* 11, 3598–3599.

Sorin, W. V. (1990). *IEEE Photon. Technol. Lett.* 2, 902–906.

Su, C. B. (1997). Achieving variation of the optical path length by a few millimeters at milliseconds rates for imaging of turbid media and optical interferometry: A new technique. *Optics Lett.* 22, 665–667.

Haberland, U. W. R. et al. (1995). Continuously tuned external cavity semiconductor laser. *Proc. SPIE* 2389, 503–507.

Eikhoff, W. and R. U. (1981). Optical frequency domain reflectometry in single-mode fiber. *Appl. Phys. Lett.* 39(9), 693–695.

Trutna, W. R. J., and Stokes, L. F. (1993). Investigation of highly scattering media using near infrared continuous wave turnable semiconductor laser. *Lightwave Technol.* 11, 1279–1283.

Guan, A., Lambsdorff, N., Kuhl, J. and Wu, B. C. (1988). Fast scanning autocorrelator with 1-ns scanning range for characterization of mode locked ion lasers. *Review of Scientific Instruments* 59, 2088–2090.

Zvyagin, A. V., A. D., Smith, J. et al. (2003). Delay and dispersion characteristics of frequency-domain optical delay for scanning interferometry. *J. Opt. Soc. Am. A* 20, 333–341.

APPENDIX 5-1

General:

In this appendix, Hilbert transforms, analytical functions, Green's function, and Cauchy's Principal Value will be discussed since they arise in different portions of the next few chapters.

Cauchy's Principle Value:

The Cauchy Principal Value (CPV) of a divergent integral is the limit of a function as a circular region around a singularity as the radius goes to zero. In other words, the region of the singularity is being excluded from the integral. For example, the equation $y = x + 1/x$ is divergent between -1 to 2 with a singularity at $x = 0$. With $f(x)$ being the CPV:

$$f(x) = \lim_{R \to 0}[\int_{-1}^{-r}(x + 1/x)dx + \int_{r}^{2}(x + 1/x)dx] =$$
$$= \lim_{R \to 0}[x^2 + \ln(|x|)]|_{-1}^{-r} + [x^2 + \ln(|x|)]|_{r}^{2}$$
$$= 3 + \ln 2 \qquad (1)$$

The singularity is therefore removed from the calculations. Of note, the letter P is often placed in front of the integral to identify it as a CPV.

Hilbert transforms and Analytical Signals

The Hilbert transform and the related Kronig-Kramers relationship link the real and imaginary parts of the transfer function of a linear shift invariant causal system. A causal system is one that, when $t < 0$, the impulse response vanishes. In other words, there cannot be a response prior to the input.

The causal system is asymmetric so that the Fourier transform of the impulse response must be complex. In addition, if the impulse response is real, the Fourier transform must be symmetric.

If all frequencies of one part of the transfer function are known (such as the real portion), then the other can be determined completely (imaginary).

The Hilbert Transform of a real valued function $s(t)$ is obtained by convoluting the signal, ($S(t)$ with $1/\pi t$ to obtain $S(t)$). The impulse response is therefore $1/\pi t$. Specifically:

$$S(t) = H[S(t)] = S(t) * h(t) = (1/\pi)P\int_{-\infty}^{+\infty} s(x)/(t - x)dx \qquad (2)$$

where $h(t) = 1/\pi t$ and:

$$H(\omega) = F[h(t)] = -i \, \text{sign}(\omega) \qquad (3)$$

which equals $+i$ for $\omega < 0$ and $-i$ for $\omega > 0$. The Hilbert transform therefore has the effect of shifting the negative frequency components of $s(t)$ by $+90$ degrees and the positive frequency components by −90 degrees. It then represents a filter where ideally the amplitude of the spectral components are left unchanged but the phases are altered by $\pi/2$. The integral in equation 2 is considered a Cauchy Principal Value (P) around $x = t$, where a singularity exists. It is dealt with as described above.

Again, because of causality, the real and imaginary parts of a function can be related. If $F\{h(t)\}$ is the Fourier transform of a transfer function, and $R(f)$ and $X(f)$ are the Hilbert transforms of the real and imaginary portions respectively, then:

$$F\{h(t)\} = R(f) + iX(f) \tag{4}$$

Then the real and imaginary components can be related through, often referred to as the dispersion relationship:

$$X(f) = -(1/\pi)P\int_{-\infty}^{+\infty} R(y)/(t-y)dy \tag{5}$$

$$R(f) = -(1/\pi)P\int_{-\infty}^{+\infty} X(y)/(t-y)dy$$

When the integrals are rewritten such that the intervals are from 0 to ∞, the equations are known as the Kramers-Kronig relations.

Very often, it is useful to express a signal in terms of its positive frequencies, such as in demodulation. A signal that contains no negative frequencies is referred to as an analytical signal. For any complicated function signals which are expressible as the sum of many sinusoids, a filter can be constructed which shifts each component by a quarter cycle, which is a Hilbert transform filter, and ideally keeps the magnitude constant. Let $S_a(t)$ be the analytical function of $S(t)$ and $S_i(t)$ be the Hilbert transform of $S(t)$. Then:

$$S_a(t) = S(t) + iS_i(t) \tag{6}$$

Again, positive frequencies are shifted $-\pi/2$ and by $\pi/2$ for the negative frequencies. How this results in removal of negative frequencies is as follows. The original function $S(t)$ is broken into its positive and negative components, $S_+(t) = e^{i\omega t}$ and $S_-(t) = e^{-i\omega t}$. Now adding a −90 degree phase shift to the positive frequencies and +90 degrees to the negative frequencies:

$$S_{i+}(t) = (e^{-i\pi/2})(e^{i\omega t}) = -ie^{i\omega t} \tag{7}$$

$$S_{i-}(t) = (e^{i\pi/2})(e^{-i\omega t}) = ie^{-i\omega t}$$

Now the positive and negative frequencies of $S(t)$ are:

$$S_a+(t) = e^{i\omega t} - i^2 e^{i\omega t} = 2e^{i\omega t} \tag{8}$$

$$S_a-(t) = e^{i\omega t} + i^2 e^{-i\omega t} = 0$$

It is clear that the negative frequencies have been removed to produce the analytical signal.

GREEN'S THEOREM AND GREEN'S FUNCTIONS

Green's function is used with OCT both for scattering theory and in the evaluation of the system transform function. Green's function here will be introduced through Green's theorem. Green's theorem is a corollary of Gauss's theorem, a commonly used theorem is vector calculus, which states that the volume integral of the divergence of a vector bounded by a closed surface equals the surface integral of that vector over the closed surface. Mathematically, Gauss's theorem states that

$$\int_S \mathbf{V} \cdot d\sigma = \int_V \nabla \cdot \mathbf{V} d\tau \tag{9}$$

Another way of looking at Gauss's theorem is by merely recalling the definition of the divergence of a vector. Divergence is analogous to net outflow, so the divergence within a volume with a closed surface by definition would be a measurement of the vector field over just the surface, which is where the "outflow" is occurring.

Green's theorem, which can be derived from Gauss's theorem, is:

$$\int_V \left(u\nabla^2 v - v\nabla^2 u \right) \cdot d\tau = \int_S (u\nabla v - v\nabla u) \cdot d\sigma \tag{10}$$

which again takes a more complicated volume integral and reduces to a simpler surface integral. Here, u and v are two scalar functions, where $\mathbf{V} = u\nabla v - v\nabla u$.

Among the most important applications of Green's theorem is with differential equations, where Green's function can be used to solve second order inhomogeneous partial differential equations. For a given second order linear inhomogeneous differential equation, the Green's function is a solution that yields the effect of a point source, which mathematically is a Dirac delta function. Expressed formally, for a linear differential operator of the form

$$L[y(\mathbf{r})] = f(\mathbf{r}) \tag{11}$$

the Green's function is the solution of

$$L[G(\mathbf{r},\mathbf{r}')] = \delta(\mathbf{r} - \mathbf{r}') \tag{12}$$

Therefore, the Green's function can be taken as a function that gives the effect at $\mathbf{r}$ of a source element located at $\mathbf{r}'$. An example with electrostatic potentials will be used for illustrative purposes. Specifically, Poisson's inhomogeneous equation:

$$\nabla^2 \psi = -\frac{\rho}{\varepsilon} \tag{13}$$

will be solved. Here, ρ is the charge and ε is the electric permitivity of the field. Thus, according to equation 13, the corresponding Green's function for equation 14 would be the solution to

$$\nabla^2 G = -\delta(\mathbf{r} - \mathbf{r}') \tag{14}$$

Thus, G can be viewed as the electrostatic potential at $\mathbf{r}$ corresponding to a point source at $\mathbf{r}'$. The Green's function is derived by Green's theorem. Green's theorem now becomes:

$$\int_V \left(\psi \nabla^2 G - G \nabla^2 \psi\right) \cdot d\tau = \int_S (\psi \nabla G - G \nabla \psi) \cdot d\sigma \tag{15}$$

This differential equation can be reduced by making the volume extremely large so that the surface integral becomes negligible and disappears, yielding:

$$\int_V \psi \nabla^2 G \cdot d\tau = \int_V G \nabla^2 \psi \cdot d\tau \tag{16}$$

Substituting in equation 15, equation 17 becomes:

$$\int_V \psi(\mathbf{r}')\delta(\mathbf{r} - \mathbf{r}') \cdot d\tau = \int_V \frac{G(\mathbf{r},\mathbf{r}')\rho(\mathbf{r}')}{\varepsilon} \cdot d\tau \tag{17}$$

Integrating with the extracting property of the Dirac delta function yields

$$\psi(\mathbf{r}) = \int_V \frac{G(\mathbf{r},\mathbf{r}')\rho(\mathbf{r}')}{\varepsilon} \cdot d\tau \tag{18}$$

The complex second order differential equation has now been simplified to the form:

$$y(\mathbf{r}) = \int G(\mathbf{r},\mathbf{r}')f(\mathbf{r}')d\tau \tag{19}$$

and to solve it we now only need to derive the appropriate Green's function and integrate its product with the inhomogeneous part of the equation, f. Typically, the appropriate Green's function can either be found with a table or derived by using knowledge of the physical situation being represented by the differential equation. In the case of Poisson's inhomogeneous equation, the appropriate Green's function is:

$$G(\mathbf{r},\mathbf{r}') = \frac{1}{4\pi|\mathbf{r} - \mathbf{r}'|} \tag{20}$$

Thus, after plugging the Green's function back into the appropriate integral, we get the solution for Poisson's inhomogeneous equation,

$$\psi(\mathbf{r}) = \frac{1}{4\pi\varepsilon} \int \frac{\rho(\mathbf{r}')}{|\mathbf{r} - \mathbf{r}'|} \cdot d\tau \tag{21}$$

which is consistent with the established formula for electrostatic potential. Although Poisson's inhomogeneous equation is relatively simple and straightforward, this method can be used to solve many second order inhomogeneous partial differential equations, including equations as complicated as light and particle scattering.

REFERENCE TEXTS

1. Born, M. and Wolf, E. (1999). *Principles of Optics*, Cambridge University Press, Cambridge, UK.
2. Cheng, D. (1992). *Field and Wave Electromagnetics*, Addison Wesley Publishing Co., Reading, MA.
3. Bracewell, R. (1999). *The Fourier Tranform and its Application*, third edition, McGraw Hill, New York.
4. Arfker, G. B. and Walker, H. J. (2001). *Mathematical Methods for Physicists*, fifth edition, Academic Press, Boston.

6 OPTOELECTRONICS AND OPTICAL COMPONENTS

Chapter Contents

6.1   GENERAL

This book is not designed to provide a blueprint to the reader on how to build an OCT system. This can be found in other sources.[1] The emphasis here is on the physical principles and applications of the technology. However, an understanding of certain optical and electronic components is necessary both to compare embodiments and to understand some of the strengths and limitations of the technology. These include the sources, delay line, interferometer design, detector, electronic filter, and beam directing devices.

6.2   SOURCES

The ideal source for OCT, whether swept or not, would typically have a center wavelength around 1300 nm, a broad spectral bandwidth, a Gaussian beam profile, high output power, low noise, compact design, be inexpensive, and be simple to integrate into a medical imaging device. Semiconductor and femtosecond sources represent the sources that are currently more commonly used.

6.2.1   Semiconductor Sources

Semiconductor source theory will not be reviewed here as it is found in basic optics and optoelectronics textbooks. These continuous wave sources include the super-luminescent diode and quantum well sources. As a simple model of a light emitting diode (LED), in a forward-biased pn junction, a draft electron falls into a hole in a process known as recombination. Analogous to electron transition, a photon is generated. If a large density of forward biasing occurs in the junction, as in a superluminescent diode (SLD), a large density of electrons and holes are formed, and the power outputs can be raised to the order of 1 mW. However, current cannot be raised indefinitely or radiation amplification exceeds the loss. Lasing will inevitably occur, which can destroy the broad spectrum characteristics. These sources have the advantage of a relatively low cost, small size, Gaussian profile, and output modulated by varying the current. Two SLDs with slightly differing wavelengths are frequently combined to produce even broader bandwidths. However, the power of these sources (under 1 mW) is not sufficient for *in vivo* imaging, so their usage has generally been limited to slower *in vitro* systems.

Multiple quantum well semiconductor sources (semiconductor optical amplifiers, SOA) are the most widely used for high speed imaging. They are compact, "turn key," have a Gaussian profile, respectable power levels, approximately 10-μm resolution, and appropriate center frequency. SOAs use very thin active layers stacked alternatively with thin inactive regions. The quantum well regions are the areas of minimal potential for electrons in the conduction band and holes in the valence band. The varying dimensions of the stacked layers vary the potentials, which results in the broad spectrum (analogous to Bragg diffraction in Appendix 6-1), in addition to lower thresholds and higher gain. They are relatively inexpensive (typically between \$5,000–\$15,000), have a near Gaussian spectrum, and a broad bandwidth (up to 80 nm). Disadvantages include the fact that the power, while large compared with the SLD (up to 25 mW), is at the limit for high speed imaging, and bandwidth that is small relative to crystal lasers. In addition, they are at least partially unpolarized.

6.2.2 Femtosecond Lasers

Two solid state crystal, Kerr-lens modelocked (KLM) lasers (Appendix 6-1) have become among the most important sources in OCT, the titanium sapphire (Ti:Al$_2$0$_3$) and the chromium forsterite (Cr^{4+}Mg$_2$SiO$_4$) KLM lasers, particularly when high resolutions or source power is required.[9,15] They are pulsed sources with durations in the femtosecond range, achieved with KLM. These paramagnetic lasers contain lanthanide and actinide ions that are used to dope host material such as glass or crystals. The vibrational interaction between the host and doped ion leads to the broad bandwidth of light absorption and emission. The lasers are optically activated to population inversion with pump lasers. These lasers allow high powers and broad bandwidths but are generally complex and expensive. Due to source instability, dual balanced detection is typically required.

For the Ti:Al$_2$0$_3$, the crystal is generally pumped with a cw argon laser. The peak power is approximately 400 mW, a center wavelength in the range of 800 nm, and bandwidth approaching 200 nm (this is variable depending on the laboratory). Free space resolutions below 2 μm have been achieved with this source. However, as the optimal wavelength for imaging in nontransparent tissue is about 1300 nm, this source is predominately of value when used in transparent tissue (i.e., the eye) or over distances of less than 500 μm.

The Cr^{4+}Mg$_2$SiO$_4$ offers superior penetration in nontransparent tissue compared with the Ti:Al$_2$0$_3$. The laser is usually pumped by a diode-pumped, cw Nd:YAG laser at approximately 1 μm. The center wavelength is at 1280 nm with a source power of 300 mW. The bandwidth is 200 nm allowing resolutions below 4 μm. Because of the lower noise level, dual balanced detection is less critical. Also, since the median wavelength is near the optical window for scattering in biological tissue, the source is more practical for imaging human pathology. In combination with its high power, it is particularly useful as a source for *in vivo* imaging. An example image generated of a *Xenopus* organism with this source using a TD-OCT system is shown in Figure 5.4. Not only are cells seen, but their individual nuclei can be identified.

6.2.3 Doped Fiber Amplifiers

Rare earth doped fibers can be used as amplified sources when they are pumped with a laser having a high powered, continuous output at an optical frequency slightly higher than that of the desired fiber output.[2] The most common doping agents are erbium, neodymium, and ytterbium. The pumping energy is stored as a population inversion which results in amplification of the guided fluorescence. Angle cleaving of the fiber suppresses lasing and produces a broad bandwidth. The pump lasers of interest typically operate at a wavelength near 1480 nm. Spectral filtering is also performed to suppress gain narrowing. It does not require optical-electrical conversion or electrical amplification, an additional attractive feature. However, there are several disadvantages that have prevented its widespread use. First, Gaussian-like spectrums are generally difficult to obtain. Second, the requirement for a high power pumping source is impractical. Third, the power levels must be carefully controlled to prevent saturation.

Examples of newer sources that are under investigation include a high power Raman continuum light source and an ultra-high resolution compact femtosecond Nd:glass laser, which is spectrally broadened in a high numerical aperture single mode fiber.[3,4] It should be noted that Raman fiber sources are similar to rare earth doped fibers in that stimulated emission is produced through Raman scattering rather than fluorescence. To date, their output has been in the range of 1700 nm.

6.2.4 Wavelength Scanning

Sweeping the frequencies of a broadband Gaussian source is important for SSOCT. Ideally, the spectral line width will be small, the sweep rate fast, mode-hopping minimized, and the sweep rate linear. A variety of frequency sweeping approaches have been performed using a broadband source plus tunable filter, tunable laser, or polygonal scanner. These include wavelength scanning such as acousto-optic tuning (poor wavelength resolution), electro-optic tuning, rotating mirror grating filter (limited sweep rate), and polygon-scanner-based wavelength filter. The polygon-scanner-based wavelength filter semiconductor laser has recently received considerable attention and consists of a diffraction grating, an afocal telescope, and a polygonal scanner. The polygon reflects back only a small portion of the dispersed spectrum normal to the front mirror facet of the polygon. Tuning rates of up to 1150 nm/ms are achievable and a 70-nm wavelength span, a line width of less than 0.1 nm, and a 9-mW cw output.[5] Most of the external cavity tunable laser techniques have the general limitation of mode-hopping, which makes it hard to compensate for cavity length changes.

6.3 INTERFEROMETERS

The Michelson interferometer with a 50/50 beam splitter represented the original configuration of OCT systems, and this configuration probably remains the most common. In this configuration, the reference arm power is partially attenuated to maximize/optimize performance, but the optimal attenuation remains controversial. Most current designs use dual balanced detection, which should be standard at this point for TD-OCT systems and possibly SS-OCT systems to reduce excess noise (discussed in the next chapter). Some groups use unbalanced primary beam splitters, such as 90/10 splitters, where most of the power is directed toward the sample arm. This is done since sample backreflection is substantially less than reference arm signal.

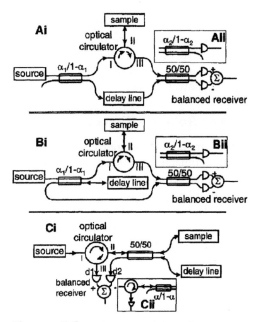

Figure 6.1 Schematic of three less common interferometer designs. All three designs use optical circulators based on the Faraday Isolator in Appendix 6-1. The first two use Mach-Zehnder interferometers while the third is based on Michelson interferometer. Dual balanced detection is examined in chapter 7. Courtesy Rollins, A.M. and Izatt, J.A. (1999). *Optics Letters.* 24, 484–486. [7]

More complex designs have been proposed.[7,8] Three power-maximizing designs have been proposed by Dr. Izatt et al. and are shown in Figure 6.1.[7] The first two use a Mach-Zehnder interferometer, which generally utilizes two beam splitters and two mirrors to divide and recombine the beam. This interferometer allows the easy introduction of modulators, here Faraday rotators (see Appendix 6-1). The third design is based on a modification of the Michelson interferometer. The literature is filled with subtle variations on the traditional Michelson designs, such as the inclusion of modulators (Appendix 6-1) to remove mechanical vibrations, but in general most systems closely resemble the original configuration.

6.4 DELAY LINES

Changing the optical group delay in the reference arm is a requirement of TD-OCTs. While many delay line configurations exist, the grating based delay line probably is the most common high speed imaging system, which is discussed in detail. Also presented are three other common embodiments used over the last

10 years: galvanometer retroreflector, fiber stretcher, and rotating elements. Although many parameters are important in building a delay line such as a duty cycle, linearity, and range, we will focus on parameters most relevant to users of the technology such as overall scan rate, group delay, phase delay, polarization variability, and power losses.

6.4.1 Galvonometer Retroreflector

The simplest version of an optical delay line is the mirror placed on a linearly scanning galvanometer in the reference arm. The galvanometer typically scans at under 30 mm/sec. In our early systems, this corresponded to the generation of an image in 30 seconds (approximate scan depth of 3mm) when the pixel size is 300×300. This is insufficient for preventing motion artifacts and the general need to acquire a large number of images over very short durations. However, as many experimental questions remain about OCT imaging (e.g., which pathology it recognizes and the mechanisms behind scattering), a place still remains for this original embodiment. Of note, the induced Doppler shift is:

$$f_d = 2v/\lambda_0 \tag{6.1}$$

with a resulting bandwidth of:

$$\Delta f_d = 2v\Delta\lambda/\lambda_0^2 = f_d\Delta\lambda/\lambda_0 \tag{6.2}$$

The group delay and phase delay change in parallel, there is no polarization instability, and there are no significant power losses associated with the mirror or any aberrant reflections.

6.4.2 Piezoelectric Fiber Stretcher

Due to the speed limitations associated with the galvanometer, fiber stretching was introduced as a method to increase acquisition rate, which allowed acquisition rates on the order of 4 frames per second (250×250 pixels).[10] With this approach, the single mode fiber in the reference arm is wrapped around piezoelectric transducers, which are typically polycrystalline ceramics such as lead-zirconate-titanate.[11] A piezoelectric transducer (PZT) is a multicrystalline ceramic which, when an electric charge is applied, results in expansion of the crystal. The centers of the crystal elements undergo realignment resulting in a state of the entire crystal that is no longer at the minimal volume, resulting in crystal expansion as a function of voltage.

The PZTs are displaced in proportion to the voltage applied, stretching the fiber and increasing the optical group delay. The PZTs can be stacked to increase their effect but this can result in friction and heating, which can lead to breakdown of the PZTs. PZTs have resonance frequencies that are dependent on the mass and elastic properties of the crystal, as well as electrical susceptibility. While driving the PZT, the driving frequency, which is usually a triangular waveform, must not couple into the resonance frequency of the PZT or oscillations will result that distort the desired waveform and potentially damage the ceramic. A common design consists of 40 m of fiber wound around the stacks and driven by a 600 Hz triangle waveform. However, 400 W of power were used to drive the stack.

Several significant problems are associated with the PZT approach: hysteresis, polarization mismatch, polarization mode dispersion, and high voltages.

The PZT Stretcher does not behave as a homogeneous unit but as if it consists of microdomains. It can be envisioned then that as the domains move in response to the field, friction develops between them, dampening the response. Therefore, the response to the applied voltage is not linear so a triangle waveform does not lead to a triangular response. At the rate of 4 frames per second, first order correction is possible, but at higher data acquisition rates correcting for hysteresis can be quite difficult.

The PZT is also subject to breakdown at temperatures near 80°C. The increased temperature is due to power dissipation and friction, which is worsened by the use of a triangular waveform or large numbers of stacks. The PZT Strecher is also plagued with three polarization problems. The first is fixed polarization mismatch due to the winding of the fibers. As has been stated earlier, bending the fiber generally results in changes in the polarization state of the light in the fiber. The fiber is wrapped around the PZT many times, resulting in alterations in the polarization state. This phenomenon can be partially compensated for by winding the fiber in the sample arm around an inactive PZT the same number of times. The second polarization problem is slow drift that results from the gradual increase in temperature of the PZT. The slow drift is controlled by careful temperature control. Finally, polarization mode dispersion (PMD) occurs due to the active stretching of the fiber (different polarization states at different expansions of the crystal). PMD is partially compensated for through the use of Faraday rotators placed in each arm, which are described in Appendix 6-1. An alternative approach for compensating for PMD, which has not been applied to OCT but has been used in other delay lines, is to use two stacks rather than one with one orthogonal to the other.

6.4.3 Grating Based Delay Lines

Although some groups continue to use delay lines based on galvanometers or fiber stretching, most groups have moved on to a grating based delay line. The delay line was originally developed for femtosecond pulse shaping and was introduced in other delay lines in the early 1990s.[12,13] Largely through the efforts of G. Tearney and B. Bouma, it was developed as an OCT delay line.[14] The delay line will initially be described qualitatively. However, in view of its widespread use, this will be followed by a more rigorous quantitative analysis. Qualitatively, light from the reference arm is directed onto a grating. The dispersed light is then focused on a tilted mirror, as shown in Figure 6.2. A phase shift of reflected light off the mirror (when the mirror is tilted) occurs as a function of the position on the mirror and therefore, since the light is dispersed, as a function of frequency. The linear phase ramp in the frequency domain results in an optical group delay in the time domain.

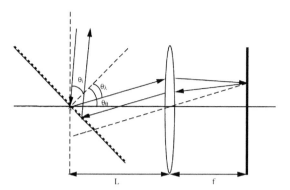

Figure 6.2 Schematic of the reference arm of the grating based delay line. The principles are discussed in the text.

The light that reflects off the mirror is redirected onto the grating, which results in an inverse Fourier transform. The light then reenters the reference arm fiber. By rotating the angle of the mirror, different group delays are introduced.

This embodiment has advantages in addition to speed. First, the groups and phase delays can be controlled separately. The optical group delay, at fixed lens, fiber, and grating positions, is dependent on the angle of the mirror. The phase delay is determined by where the center frequency in the beam is relative to the axis of the mirror (and how fast the mirror is moving). If the center frequency was on the axis of the mirror, no Doppler shift would occur. The higher the dispersed light is off the mirror axis, the greater the Doppler shift. The second advantage of the grating-based technique is that dispersion can be compensated for without the need for additional optical components. If dispersion occurs in the tissue, it can be compensated for by altering the position of the mirror/lens relative to the grating.

More formally, the single-pass delay line is shown in Figure 6.2, but the actual configuration is a folded double-pass design for more ease of alignment. The phase delay $\{\emptyset(\lambda)\}$ at a given wavelength can be given by simple geometry in Eq. 6.3[15,16]

$$\emptyset(\lambda) = -2k\Delta z = -2kx \tan \gamma \tag{6.3}$$

Δz is the distance up the mirror, x is the distance of the axis of the beam, and γ is the angle of tilted mirror. For now, we will assume the grating and the mirror are both at the focal length of the mirror ($f = L$). Describing the phase change in terms of the incident light, once again through geometric considerations:

$$x = L \tan (\Theta_0 - \Theta(\lambda)) = f \tan (\Theta_0 - \Theta(\lambda)) \tag{6.4}$$

where $\emptyset(\lambda)$ is the angle between the incident light on the grating and the diffractive light while Θ_0 is the angle between the incident light and the axis of the beam. Multiplying both sides by $-2k \tan \gamma$ to get $\emptyset$ gives:

$$\emptyset(\lambda) = -2kx \tan \gamma = 2kf \tan \gamma \tan (\Theta_0 - \Theta(\lambda)) \tag{6.5}$$

From the grating equation which relates the angle dependence as a function of wavelength:

$$\sin \Theta_i + \sin \Theta_r = m\lambda_r/d \tag{6.6}$$

where d is the spacing in the grating, m is the order, Θ_i is the incident angle on the grating, and Θ_r is the angle reflected off the grating. This can be rewritten as:

$$\Theta_r = \sin^{-1}(m\lambda_r/d - \sin \Theta_i) \tag{6.7}$$

Substituting this into Eq. 6.5 for $\Theta(\lambda)$, it becomes:

$$\text{ø}(\lambda) = -2kx\tan \gamma = 2kf\tan \gamma \tan(\Theta_0 - \sin^{-1}(m\lambda_r/d - \sin \Theta_i)) \tag{6.8}$$

The group delay is the infinitesimal change in phase with angular frequency. Therefore, Eq. 6.8 will be expressed in terms of ω:

$$\text{ø}(\lambda) = -(2\omega f \tan \gamma/c)\gamma \tan(\Theta_0 - \sin^{-1}(2m\pi c/\omega d - \sin \Theta_i)) \tag{6.9}$$

The group delay (τ_g) is now given by:

$$\tau_g = 4\pi f \tan \gamma/\omega_0 d\cos \Theta_0 = 2f\lambda_0 \tan \gamma/cd\cos \Theta_0 \tag{6.10}$$

The length of the group delay, l_g, is τ_g times c and is given by:

$$l_g = 2\pi\lambda_0 \tan \gamma/d\cos \Theta_0 \tag{6.11}$$

Assuming that the angle deviations are small, $\tan \gamma \approx \gamma$, the length is given by:

$$l_g = 2\pi\lambda_0\gamma/d\cos \Theta_0 \tag{6.12}$$

The phase delay, and importantly the phase velocity, are distinct from the group delay. The phase again is given by Eq. 6.3 ($\text{ø} = -2k\Delta z$) so that the change of phase with time is:

$$\text{ø}(\lambda) = -2kx_0 \tan \gamma = -4\pi x_0 \tan \gamma/\lambda_0 \tag{6.13}$$

The angle γ is changing at a rate of $\omega_\gamma t$ so that:

$$d\text{ø}/dt = 4\pi x_0 \tan \omega_\gamma t/\lambda_0 = 4\pi x_0\omega_\gamma t/\lambda_0 \tag{6.14}$$

for small γ. Therefore, the phase velocity is a function of x_0, the position of the center frequency on the mirror. If the center frequency is on the axis of the mirror, the phase velocity is zero. The higher up the mirror the center frequency is, the greater the phase velocity.

From Chapter 5, the modulation of the intensity envelope is given by $\cos(2\pi f_D t)$ so that f_D is given by:

$$f_D = 2\omega_2 tx_0/\lambda_0 \tag{6.15}$$

The delay line from our group can image at 8 frames per second at 256×256 pixels (2000 scans per second). Disadvantages include its power loss associated with the grating and chromatic dispersion with a broad band source. Dispersion relationships have been derived in detail in previous work so only the final results are included.[16–18] The PSF modified by reference arm induced dispersion is now described by:

$$PSF = (4[L - f] m^2 \lambda_0 \Delta\lambda)/d^2 \cos^2 \Theta_0 \qquad (6.16)$$

where f is the focal length of the lengths and L is the distance between the incident point on the grating and the lens. Also in the equation, d is the grating spacing and m is the diffraction order. The group velocity dispersion (GVD) or second order dispersion is given by:

$$GVD = -2m^2 \lambda_0^3 (L - f)/\pi c^2 d^2 \cos^2 \Theta_0 \qquad (6.17)$$

A large number of hardware and software modifications of the above design have been introduced, but the base design is the most widely used. A particularly interesting modification is the introduction of an electro-optic phase modulator in the delay line to stabilize phase variations, which may reduce the mechanical noise described in the next chapter.[19]

6.4.4 Rotating Elements

A variety of rotating elements have been proposed for varying the optical group delay in the reference arm including a rotating mirror, cube, and polygon. These embodiments are typically associated with a nonlinear phase velocity that requires correction.[20–22] The most interesting is the CAM,[23] which is similar to the circular involute reflector[24] for non-OCT use. The CAM is similar to a rotating polygon mirror, but instead of a flat surface, it consists of a series of spiral sections around its circumference whose changing angle allows for a constant phase velocity. The advantages include high acquisition rates, a single rotating part with constant phase velocity (at constant rotational frequency) but varying group delay, and no power losses such as those induced by the grating. Disadvantages include no direct dispersion compensation mechanism.

6.5 DETECTORS

TD-OCT and SS-OCT use single PIN diodes for detecting the optical signal whereas FD-OCT typically uses a CCD. FD-OCT can be performed with a single detector and grating (or similar device), but this is rarely done. The CCD collects photoelectron charges during the exposure time where PIN diodes produce a continuous photo-electron current. Therefore, energy is measured rather than power. With the single diode, dual balanced detection is generally used to suppress excess noise. The physics of PIN diodes can be found in most optics textbooks.[38]

6.5.1 CCD

CCD physics is briefly reviewed here because its limitations are important to the function of spectral radar. Light is focused onto an array of detectors and the output

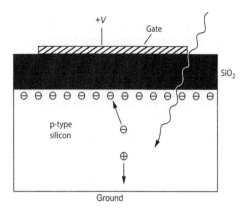

Figure 6.3 Schematic of an element of a CCD. The principles behind the MOS capacitor structure are discussed briefly in this chapter and the following. Based on concepts from Wilson and Hawkes, *Optoelectronics*, third edition. [38]

of each detector must be read out sequentially. A schematic of one of the basic units of the CCD is the metal oxide semiconductor (MOS) capacitor shown in Figure 6.3. A layer of silicon dioxide is formed on p-type silicon. The gate is a metal electrode that is positively biased with respect to the silicon. Electron-hole pairs are formed by incoming photons. Electrons are attracted to the silicon as long as a positive voltage is maintained while holes move in the opposite direction. These electrons are trapped under the metal by a potential well. Electrons are accumulated over the measurement or integration period.

Now, the series of MOS capacitors have to be read out sequentially. Once a suitable integration time has been achieved, the charge can be moved down the chain of MOS via a repeated sequence of potentials down the gates, thereby shifting the effective quantum well. A new light integration cannot be performed until the charges have been moved down the length of the array. To increase scanning rate, a second shielded CCD—called the transport register—lays alongside the field and is shielded from the incoming light. When the charge is transferred to the transport registry, it can be read out sequentially while the original CCD is accumulating the next image. When multiple arrays are present the read out electrons are transferred to the read out registration. As will be seen in the next chapter, the increased complexity of the CCD can both improve performance and lead to significant noise sources.

6.6 DETECTION ELECTRONICS

Most of the detection electronics will not be discussed, particularly circuit design, as the focus of this book is understanding the physical principles. However, an appreciation of certain aspects of the electronics and optical components is necessary for understanding the strengths and limitations of various OCT embodiments.

6.6.1 TD-OCT

With TD-OCT, after the signal is detected, it is passed through a bandpass filter to remove low frequency noise (see Figure 6.4A). After bandpass filtration, the signal is demodulated (usually log based 10 for reasons discussed below) by either synchronous demodulation (mixing) or asynchronous demodulation (envelope detection) (see Figure 6.4B). The primary purpose of demodulation is to remove the carrier frequency (Doppler shift). With TD-OCT, the signal passes through demodulation logarithmic amplifiers which compress the data and make the large SNR signal more assessable. This is an advantage over FD-OCT techniques, which are described in the next chapter. The carrier frequency is then removed and A-D conversion performed.

With both mixing and envelope detection, the carrier frequency must be much greater than the modulation (information) frequency. With mixing, demodulation is performed by mixing the measured signal with a signal with a frequency identical to the carrier frequency through a lock-in amplifier. It is essential that the phase of the carrier and added signal are locked in. This results in shifting the spectrum of the modulation signal (autocorrelation function) to its original position on the frequency axis. The signal is then low pass filtered to remove the carrier signal (see Figure 6.4C). With asynchronous demodulation, as the name suggests, it avoids the need for synchronization (phase and frequency) between the input signal and the demodulation signal. It depends on the fact that the autocorrelation function is positive and its angular frequency (the envelope) varies slowly when compared with the carrier. However, there is a trade-off between power efficiency in the output and the quality of the demodulation signal. Also, asynchronous demodulation requires more transmitting power than synchronous.

6.6.2 FD-OCT

With FD-OCT, the signal is detected by a photodiode array, followed by digitization usually with a 12- or 16-bit A-D converter (data acquisition board). The data are then transferred to continuous computer memory. Prior to discrete Fourier transform (DFT), the points in the spectrum should be evenly distributed in k-space mapped from λ-space. A DFT is performed on the digitized data. It is filtered to remove fixed pattern noise, then spectral averaging is performed through averaging the background spectrum. The image undergoes processing as described in the previous chapter, which includes dispersion compensation, compensation for different frequency sensitivity of the diodes, and logarithmic compression (although post-DFT).[25]

With SS-OCT, only a single diode is used and the camera output is then digitized using a data acquisition board and then the data stream is sent to computer continuous memory. The data processing typically involves zero paddling, interpolation,

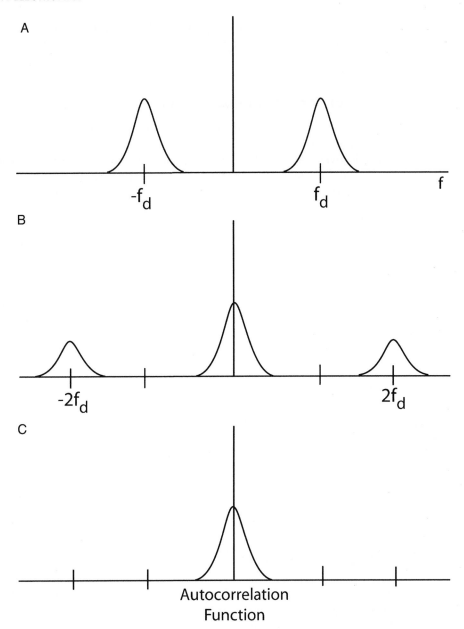

Figure 6.4 Detection electronics in most TD-OCT systems. In A, the detected signal has undergone bandpass filtration to remove low frequency noise (Chapter 7). In B, demodulation is performed to separate the autocorrelation function from the carrier. In C, low pass filtration is performed to remove the carrier prior to A-D conversion.

and mapping to linear k-space prior to fast Fourier transform. Logarithm processing can be performed after the FFT. In addition, other processing may be required as discussed for SD-OCT and Chapter 7.

6.6.3   A-D Conversion/Frame Grabbers

A-D conversion is obviously required for digital monitoring. The A-D converter samples at 2^n where n is the number of bits in the A-D converter and the dynamic range is given by $10 \log(2^N)$. Some authors define dB as $20 \log(2^N)$ so that knowing the definition used is critical in interpreting published performances. Using the first definition, 12 bit corresponds to 36 dB. When a logarithmic demodulate is used (TD-OCT), the dynamic range is given by $10(2^N) - \log(90)$, which can be as high as 60–80 dB. The four most common sources of noise in the A-D converter are offset error, scale error, nonlinearity, and non-monotonicity.[26,27] Quantization and the associated error is particularly significant with FD-OCT techniques because the FFT is performed of the signal after A-D conversion and not on the analog signal.

A frame grabber is used for high speed imaging and consists of an A-D converter, programmable pixel clock (which is for calibration and fires at a constant set rate), acquisition control unit, and a frame buffer. It requires, in addition to the video signals, line synchronization (which synchronizes with the A-scan) and frame synchronization (which synchronizes with the lateral scan).

All digital OCT systems are currently under investigation that have the advantage of having the ability to alter the bandpass filter through computer control. This is particularly useful for Doppler OCT. For the realization of a fully digital OCT system, a high sampling rate as well as a high dynamic range of the A-D converter is required. The sampling rate of the A-D conversion must be high enough to digitalize the highest frequency of the interference signal. Theoretically, by the Nyquist theorem, a sampling rate of at least twice the highest sampled frequency is required. To be well above the Nyquist frequency sampling is at four times the heterodyne frequency.[28]

6.7   LIGHT DELIVERY DEVICES

A variety of devices are used to perform imaging with OCT including rotating catheters/guidewires, linearly translating catheters, forward imaging devices, and Micro Electro-Mechanical System (MEMS) based technology.

6.7.1   Rotational Catheters/Guidewires

OCT imaging catheters have a wide range of applications including use in the cardiovascular system, gastrointestinal tract, urogenital tract, and respiratory tract.

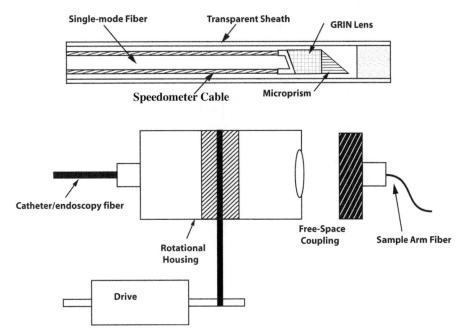

Figure 6.5 Common OCT catheter design. The top is the distal end of catheter while the bottom is the free space proximal coupling. Drawn by Karen Aneshansley.

The catheters generally consist of a GRIN lens (for focusing), prism or mirror (for directing light), single mode optical fiber, outer casing, and an inner sheath such as a speedometer cable for rotating all but the outer casing (some devices rotate the fiber within the casing without the speedometer cable). The proximal end has a free space that allows the catheter to rotate while keeping the proximal end fixed (Figure 6.5).

Several limitations with the OCT catheter design exist that need to be considered. These include internal reflection, fixed focus, nonlinear rotations, catheter fragility, polarization artifacts, cross-sectional diameter of the catheter, and the proximal coupling.

Internal reflections can occur when refractive index mismatches exist within the catheter and therefore need to be minimized. If too high, they not only can lead to power losses, but if high enough, can lead to detector saturation. Generally, the largest back reflections occur at the fiber GRIN interface. Care must be taken while designing to minimize these internal reflections.

The catheter or guidewire generally has a fixed focal length. Paraxial wave propagation is used to calculate the beam properties. The refractive index profile of the GRIN lens is given here by:

$$n = n_0(1 - \{x^2 + y^2\}/2h^2)$$

where n_0 is the index at the center of the lens and $1/h$ is a measure of the rate of decrease of the index as we move away from the axis. The matrix for the lens is then given by[15,29]:

$$\begin{bmatrix} A & B \\ C & D \end{bmatrix} = \begin{bmatrix} \cos\left(\dfrac{l}{h}\right) & \dfrac{h}{n_0}\sin\left(\dfrac{l}{h}\right) \\ -\dfrac{n_0}{h}\sin\left(\dfrac{l}{h}\right) & \cos\left(\dfrac{l}{h}\right) \end{bmatrix}$$

where l is the length of the lens. The $ABCD$ matrix cascade for the design in Figure 6.5 has been derived elsewhere and is given by:[15,29]

$$\begin{bmatrix} A & B \\ C & D \end{bmatrix} = \begin{bmatrix} 1 & l_w \\ 0 & 1 \end{bmatrix}\begin{bmatrix} 1 & \frac{l_p}{n_p} \\ 0 & 1 \end{bmatrix}\begin{bmatrix} \cos\left(\frac{l_G}{h}\right) & \frac{h}{n_G}\sin\left(\frac{l_G}{h}\right) \\ n_G\sin\left(\frac{l_G}{h}\right) & \cos\left(\frac{l_G}{h}\right) \end{bmatrix}\begin{bmatrix} 1 & \frac{l_c}{n_c} \\ 0 & 1 \end{bmatrix}$$

where l_G is the length of the GRIN lens, l_c is the distance between the lens and fiber, l_w is the distance between the prism and point of focus, and l_p is the physical distance light travels in the prism. The beam waist and confocal parameter are therefore determined by the fixed physical properties of the GRIN lens and the length of the prism, along with the controllable distance between the fiber and lens. Although varying the distance between the fiber and lens would give some flexibility to alter the focus, this is currently not implemented with OCT catheters. Therefore, catheters are designed with fixed focal distance and confocal parameter, which makes it difficult to image lumens of widely different diameters or perform imaging when the catheter is not centered. This is minimized with the use a confocal parameter of a few millimeters.

If light is propagated as a ray, then total internal reflection would occur at the prism directing all the light to the sample and collecting it. However, as the beam enters the prism it is diverging. Therefore, not all light is at or below the critical angle and therefore passes through the prism face, which results in spatial filtering. Therefore, it is optimal to coat the prism with metal.

Another difficulty associated with rotational imaging is tied into the fact that penetration depth is limited to only a few millimeters. When dealing with organs with large lumens, such as the esophagus and stomach, a rotational image will appear predominately as the empty space of the large lumen surrounded by a thin tissue layer (Figure 6.6). Therefore, in these organs, forward or translational imaging is preferred.

The fiber optics within the catheters and guidewires are relatively fragile. Therefore, they need to be designed to maximize mechanical stability, particularly if engineered to very small cross-sectional diameters.

The typical OCT catheter and guidewire are rotated from the proximal end. In typical clinical scenarios, the catheter goes through a significant amount of bending. This frequently results in nonlinear rotation of the optics, distorting the image. For this reason, MEMS technology (discussed below) is examined to overcome this limitation.

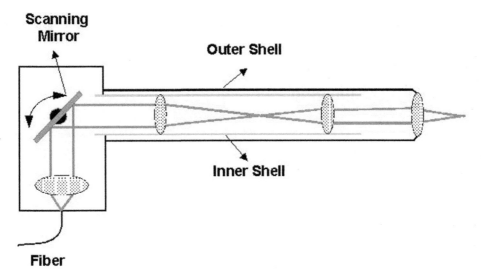

Figure 6.6 Schematic of a second generation OCT hand held probe. Courtesy Li, X. et.al. (2005). *Arthritis Research and Therapy. 7*, R318–323. [31]

As stated, the OCT catheters and guidewires consist of a single mode fiber. With catheter bending or compressions (automated pullback), the birefringence of the optical fiber changes. This results in artifacts in the image. This topic will be discussed in more detail in Chapter 8.

Small cross-sectional diameter devices are being pursued, particularly an OCT imaging guidewire. The current smallest imaging catheter is the OCT imaging wire from LightLab Imaging which has a cross-sectional diameter of 0.017". Guidewires are important for intravascular imaging, as apposed to a large diameter catheter, because they allow interventional and flushing catheters to pass over them and be switched while the guidewire remains. However, to obtain clinical utility, the catheters will need to have the mechanical properties similar to current interventional guidewires.

The free space coupling at the proximal end of the catheter has some significant complexity associated with it. The clinician will need to connect catheters and guidewires quickly. Therefore, the procedure should be almost turnkey. However, quick alignment across the coupling and minimizing power losses can be an engineering challenge.

6.7.2 Translational Devices

A translational device is one that scans along the length of the catheter. Uses include imaging joints and organs with large lumens, such as the esophagus. The optics are usually the same as those for the rotational catheters as well as the same paraxial wave equations. Most of the limitations and engineering challenges that exist are described above. In particular, birefringence artifacts from the push-pull action are a significant problem. However, these catheters do not suffer from nonlinear rotation and the proximal coupling does not need to be free space because no rotation is involved.

6.7.3 Forward Imaging

Forward imaging devices are important to OCT including translational stages, OCT microscope, hand-held probes, flexible endoscope, and the rigid endoscope. OCT systems where the specimen is placed on a moveable stage with submicron resolutions are commonly used for *in vitro* studies. Where they have no direct clinical use, they have been instrumental in both technology development and assessing the capabilities of OCT.

Hand-held probes are useful particularly during open surgical procedures or imaging within the mouth. Various probes have been developed which displace the fiber through such methods as a piezoelectric controlled cantilever or electromagnetic displacement of the fiber.[30] Our groups and others currently use a device where the light beam is redirected by a galvanometer driven mirror and controlled with a series of relay lenses as shown in Figure 6.6.[31] In the design of these devices, factors that need to be considered are lateral scanning rate, linearity of scanning, working distance, and spot size. In addition, the device needs to be constructed to minimize profile in the operative field, contain no electrical sources that could potentially endanger the patient/surgeon, and be relatively easy to sterilize.

Surgical OCT microscopes have been produced for various applications. They are used in clinical medicine during open surgical procedures. They allow colinear alignment with the viewing axis of a surgical microscope, which allows high magnification visualization of scan location during imaging. In addition, these high numerical aperture microscopes provide significant working distances that allow manipulation of tissue while imaging. To date, they have not been widely used for any specific clinical applications, but could easily be envisioned as guiding micro-surgical procedures such as neurosurgery, microvascular repair, nerve repair, or even micromanipulation of the prostate to prevent impotence.

Both rigid and flexible endoscopes can be integrated with OCT systems. There are various ways this can be approached. One technique is to use a fiber bundle (of single mode fibers) for forward imaging and the image captured with a CCD. However, cross talk between different optical fibers limits its performance. Another technique would be to use either a series of relay lenses or a GRIN image relay rod in a rigid endoscope. For flexible endoscopy a single mode fiber runs down the length of a shaft and the tip scanned through techniques similar to those used with a hand-held probe.[32]

6.7.4 MEMS

To overcome some of the limitations of the devices described above, MEMS has recently been pursued. MEMS refers to mechanical components of micrometer size and includes 3D lithographic features of various geometries. It is of interest particularly because it can be used to direct light at the end of light-delivery devices, scanning in a rotational, translational, or forward direction. The ultimate objective is to develop these devices to reduce catheter/endoscope size and nonlinear rotation. They are typically manufactured using planar processing similar to semiconductor processes such as surface micromachining and/or bulk micromachining. Size varies from micrometers to millimeters. Due to the large area to volume ratio, surface effects such as electrostatics and wetting generally dominate volume effects such as inertia or thermal mass. They are fabricated using modified silicon fabrication technology, molding, plating, wet etching, dry etching, and electro-discharge machining. Several prototypes have been published for OCT, but they have yet to reach the stage of clinical utility.[33-37]

REFERENCES

1. Bouma, B., and Tearney, G. 2002. *Handbook of Optical Coherence Tomography*. Marcel Dekker Inc., New York, New York.
2. Bouma, B., Nelson, L.Y., Tearney, G.J., et.al., 1998. Optical coherence tomographic imaging of human tissue at 1.55 μm and 1.81 μm using Er- and Tm- doped fiber sources. *J. Biomedical Optics* 3: 76–79.
3. Bourquin, S., Aquirre, A.D., Hartl, I., et.al., 2003. Ultrahigh resolution real time OCT imaging using a compact femtosecond Nd: Glass laser and nonlinear fiber. *Optics Express* 11: 329–339.
4. Hsiung, P., Y. Chen, T. Ko, et.al., 2004. Optical coherence tomography using a continuous-wave, high-power, Raman continuum light source. *Optics Express* 12: 5287–5295.
5. Yun, S.H., C. Boudoux, G.J. Tearney, et.al., 2003. High-speed wavelength-swept semiconductor laser with a polygon-scanner based wavelength filter. *Opt. Lett.* 28: 1981–1983.
6. Yun, S.H., G.J. Tearney, J.F. deBoer, and B.E. Bouma, 2004. Motion artifacts in optical coherence tomography with frequency-domain ranging. *Optics Express* 12: 2977–2998.
7. Rollins, A.M. and J.A. Izatt, 1999. Optimal interferometer designs for optical coherence tomography. *Opt. Lett.* 24: 484–486.
8. Bouma, B.E. and Tearney, 1999. Power efficient nonreciprocal interferometer and linear scanning fiber-optic catheter for optical coherence tomography. *Opt. Lett.* 24: 531–533.
9. Bouma, B.E., et al., 1995. High resolution optical coherence tomography using a mode locked Ti: Al_2O_3. *Opt. Lett.* 21: 1839–1841.
10. Tearney, G.J., B.E. Bouma, S.A. Boppart, et al., 1996. Rapid acquisition of an in vivo biological image by use of optical coherence tomography. *Opt. Lett.* 21: 1408–1410.
11. The Peizo Book, Burleigh Instruments, Inc. Fisher, N.Y.
12. Fork, R.L., C.H. Brito, P.C. Becker, and C.V. Shank, 1987. Compression of optical pulses to six femtoseconds by using cubic phase compensation. *Opt. Lett.* 12: 483–487.
13. Kwong, K.F., D. Yankelevich, K.C. Chu, et.al., 1993. 400-Hz mechanical scanning optical delay line. *Opt. Lett.* 18: 558–560.
14. Tearney, G.J., B.E. Bouma, J.G. Fujimoto, 1997. High speed phase and group delay scanning with a grating based phase control delay line. *Opt. Lett.* 22: 1811–1813.

15. Tearney, G.J. 1997. Optical Biopsy of In Vivo Tissue Using Optical Coherence Tomography: Doctoral Thesis, MIT, Cambridge, Mass.
16. Niblack, W.K., J.O. Schenk, B. Liu, M.E. Brezinski, 2003. Dispersion in a grating-based optical delay line for optical coherence tomography. *Appl. Opt.* 42(19): 4115–4118.
17. Liu, B., E.A. Macdonald, D.L. Stamper, and M.E. Brezinski, 2004. Group velocity dispersion effects with water and lipid in 1.3 μm OCT system, *Physics in Medicine and Biology*, 49: 923–930.
18. Chen, Y. and X. Li, 2004. Dispersion management up to third order for real time optical coherence tomography involving a phase or frequency modulator. *Optics Express* 12: 5968–5978.
19. de Boer, J.F., C.E. Saxer and J.S. Nelson, 2001. Stable carrier generation and phase-resolved digital data processing in optical coherence tomography. *Applied Opt.* 40: 5787–5790.
20. Windecker, R., M. Fleischer, B. Franze and H.J. Tiziani, 1997. Two methods for fast optical coherence tomography and topography *J. Mod. Opt.* 44: 967–977.
21. Su, C.B., 1997. Achieving variation of the optical path length by a few millimeters at millisecond rates for imaging of turbid media and optical interferometry. *Opt. Lett.* 22: 665–667.
22. Szydlo, J., N. Delachenal, R. Gianotti, et.al., 1998. Air turbine driven optical coherence reflectometry at 28.6 kHz scan repetition rate. *Opt. Commun.* 154: 1–4.
23. Swanson, et.al., U.S. patent 6,191,862.
24. Hecht, U.S. patent 3,776,637.
25. Nassid, N.A., B. Cense, B.H. Park, M.C. Pierce, S.H. Yun, B.E. Bouma, G.J. Tearney, T.C. Chen and J.F. de Boer, 2004. In vivo high resolution video-rate spectral-domain optical coherence tomography of the retina and optic nerve. Substantially reduced real time acquisition rate. *Optics Express* 12: 367–376.
26. Courtesy National Semiconductor Corp.
27. Horowitz, P. and W. Hill, 1997. *The Art of Electronics.* 2nd edition. Cambridge University Press, Cambridge, UK, p. 615.
28. Izatt, J.A., M.D. Kulkarni, S. Yazdanfar, J.K. Barton and A.J. Welsh, 1997. In vivo bidirectional color doppler flow imaging of picoliter blood volumes using optical coherence tomography. *Opt. Lett.* 22(18): 1439–1441.
29. Tearney, G.J., S.A. Boppart, B.E. Bouma, et.al., 1996. Scanning single mode fiber optic-endoscope for optical coherence tomography imaging. *Opt. Lett.* 21: 543–545.
30. Boppart, S.A., B.E. Bouma, C. Pitris, et.al., 1997. Forward imaging instruments for optical coherence tomography. *Opt. Lett.* 22: 1618–1620.
31. Li, X, S.D. Martin, C. Pitris, R. Ghanta, D.L. Stamper, M. Harman, J.G. Fujimoto, M.E. Brezinski, 2005. High-resolution optical coherence tomographic imaging of osteoarthritic cartilage during open knee surgery. *Arthritis Research & Therapy* 7: R318–R323.
32. Sergeev, A.M., V.M. Gelikonov, G.V. Gelikonov, et.al., 1997. In vivo endoscopic optical biopsy with optical coherence tomography. *Optics Express* 1: 443–440.
33. Tran, P.H., D. Mukai, et.al., 2004. In vivo endoscopic optical coherence tomography by use of a rotational microelectromechanical system probe. *Opt. Lett.* 29: 1236–1238.
34. Zara, J.M., S. Yazdanfar, K.D. Rao, et.al., 2003. Electrostatic micromachine scanning mirror for optical coherence tomography. *Opt. Lett.* 28: 628–630.
35. Xie, H., Y. Pan, and G.K. Fedder, 2003. Endoscopic optical coherence tomography imaging with a CMOS-MEMS micromirror. *Sensors and Actuators A.* 103: 237–241.

36. Jain, A., A. Kopa, Y. Pan, et.al., 2004. A two-axis electrothermal micromirror for endoscopic optical coherence tomography. IEEE *J. of Quantum Electronics*. 10: 636–642.

37. Xie, T., H. Xie, G. Fedder, et.al., 2003. Endoscopic optical coherence tomography with new MEMs mirror. *Electronics Letters* 39(21): 1531–1536.

38. Wilson, J. and J. Hawkes, 1998. *Optoelectronics, Third Edition*, Prentice Hall Europe, London, UK.

APPENDIX 6-1

6A-1 MAGNETO-OPTICAL EFFECT

Until now the properties of the material have for the most part been assumed to be isotropic, meaning the physical properties of interest do not change with direction. Some materials rotate the plane of polarization and are referred to as optically active. For these materials, the dielectric constant is complex and the imaginary part is represented by an antisymmetric tensor.

In 1845, Michael Faraday demonstrated that, in the presence of a magnetic field, some materials can behave as if they are optically active. In particular, he discovered that the plane of light polarization passing through glass rotates when a magnetic field is applied in the direction of light propagation. However, unlike optically active material, where the effect depends on the direction of propagation of light, the Faraday effect depends on the direction of the magnetic field.

How the magnetic field induces the change in polarization can be modeled via a modification of classical dispersion theory or the harmonic oscillatory described in Chapter 3. To do so, we envision, for illustrative purposes, that the magnetic field decomposes the linear polarized light into circular right- and left-handed components. We return to the concept that the electron is held to an atom via a string-like spring constant. The rotating electric fields from each component result in circular rotation in the electrons of the atom. The magnetic field now modifies the spring constant, such that it is decreasing in one direction and increasing in another, which leads to two new resonance frequencies and therefore refractive indicies.

Figure 6A.1 shows the Faraday isolator. The magnetic field is in the direction of light propagation. The light, as it passes through the medium, is rotating. The rotation is generally given by the formula:

$$\varphi = VBL \tag{1}$$

where φ is the angle of rotation (in minutes of arc), B is the magnetic flux density, V is the Verdet constant, and L is the distance transversed through the medium. V is characteristic for a given medium and is dependent on the temperature and frequency. An important property is that, as the light passes back through the medium after reflecting off the mirror, it is rotated the same amount *in the same direction* rather than rotating back to the original position, as would occur with optically active material. The net result of a double pass through a Faraday rotator is two times the rotation of single pass.

A Faraday circulator would work as follows. The linearly polarized light passing through the linear polarizer on the left is rotated 45 degrees as it passes through the rotator. (After passing through the rotator for the second time, the light is now orthogonal to the original beam. The light therefore cannot pass through filter one and is reflected.)

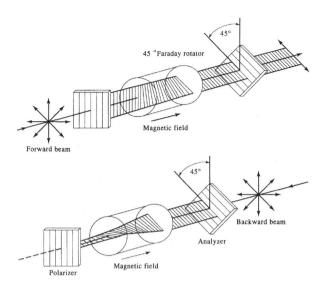

Figure 6A.1 Schematic of a faraday isolator. The top image represents the incoming light while the bottom is the backreflected. The principles are described in the text. Courtesy of Yariv, A. (1997). *Optical Electronics in Modern Communications.* [6]

6A-2   ELECTRO-OPTICS

An electro-optic material has a refractive index that is a function of the applied electric field. Since the index only slightly varies with the electric field, it can be expanded as a Taylor series about $E_0 = 0$:

$$n(E) = n(0) + \alpha_1 E + (1/2)\alpha_2 E^2 + \cdots \tag{2}$$

where $\alpha_1 = dn/dE$ and $\alpha_2 = d^2\,n/dE^2$. As pointed out elsewhere it is convenient to express the electro-optic effect in terms of the electrical impermeability (η)[1]:

$$\eta = \varepsilon_0/\varepsilon = 1/n^2 \tag{3}$$

so Eq. 2 becomes:

$$\eta(E) \approx \eta + \gamma E + \xi E^2 \tag{4}$$

where $\gamma = 2a_1/n^3$ and $\xi = -a_2/n^3$.

The values of these coefficients are a function of the field and the polarization.

Pockels Effect

If the induced birefringence changes linearly with the electric field, the third term in Eq. 2 can be ignored and it becomes:

$$\eta(E) \approx \eta + \gamma E \tag{5}$$

The effect generally only occurs in crystals that lack a center of symmetry. There are 20 crystal symmetries with this property, the same ones which exhibit the piezoelectric effect. Among the common media used are potassium dihydrogen phosphate (KDP) and potassium dideuterium phosphate (KD*P).

Pocket Cells

A pocket cell consists of a crystal between two electrode plates (either perpendicular or parallel to the direction of light propagation), and orthogonal linear polarizers at the beam entrance and exit points. They have a response time less than 10 ns (1×10^9 Hz). Pocket cells are used for generating ultra-short pulses, phase/polarization modulation, or amplitude modulation, depending on their configuration.

Kerr Effect

The Kerr effect is based on the fact that optically transparent substances, typically nitrobenzene or carbon disulfide, become birefringent when placed in an electric field. The crystals are symmetrical and invariant under field reversal. Therefore, the γE in Eq. 4 can be eliminated. The optical axis corresponds to the applied field. The Kerr effect is then proportional to the square of the field:

$$\eta(E) \approx \eta + \xi E^2 \qquad (6)$$

The Kerr effect is often referred to as the quadratic electro-optic effect.

Kerr Cell

The voltage across the Kerr cell is such that when linearly polarized light is passed through, circularly polarized light is produced. After the light reflects off a mirror and passes back through a crystal, it is linearly polarized but orthogonal to the original light and exit. When the voltage is switched to zero, the light is parallel and a large pulse is produced. The great advantage is that it can respond to frequencies faster than 10^{10} Hz and therefore can serve as a Q-switch for pulsed lasers. Kerr cells typically require 10 to 20 kV while pocket cells can operate at significantly less voltage.

6A-3 ACOUSTO-OPTICS

Acousto-optics is, in general, the manipulation of light by sound.[2] It can be distinguished from photo-acoustics, which is the generation of sound by light. In one form of acousto-optics, imaging depends on the presence of a light absorber in tissue.[3] An ultrasound beam is passed perpendicular to the light, and in regions were light absorption is occurring, the detected signal is reduced. Bragg and Raman-Nath scattering are postulated as the mechanism for the frequency shifts, although other mechanisms, such as vibrations of cells and organelles, are possible.[4] With Bragg and

Raman-Nath scattering, the column of an acoustical wave behaves like a diffraction grating and diffracts a perpendicular beam of light. The acoustic wave produces a spatial-density variation in the density of the medium. The spatial-density variation produces, in turn, a concomitant perturbation in the refractive index. The spatial variations in index of refraction produces at any given instant a phase grating, which then diffracts the incident light beam into one or more directions.

The process of diffraction from an acoustical column is, of course, more complicated than this picture for a variety of reasons. One reason is that the index of refraction changes produced by the elastic wave are tensor quantities having off-diagonal elements. A transverse or shear acoustical wave causes the diffraction of a light beam without a corresponding change in density owing to the complexity of the elasto-optical interaction. Other reasons are dealt with below.

On the quantum level, the event that is occurring is a photon annihilating a phonon within the solid. Due to conservation of energy, the total vector energy of the photons must be the initial photon energy plus the energy of the annihilated phonon. Since $E = hv$, this means that the phonon annihilation corresponds to a change in frequency of the reflected light.

I will start with a simplified explanation of the process in the far-field. Polarization changes will not be taken into account for now. When the frequency of the light (ω) is much greater than the frequency of the sound (Ω), and the width of the ultrasound beam (L) is large compared to the width of the beam, Bragg scattering results. This is seen as two deflected beams in Figure 6A.2. When the frequency of sound is high and the beam width is small, Raman-Nath scattering occurs which leads to multiple orders, as shown in Figure 6A.3. In general, the regions between the two diffraction patterns can be described by the Klein-Cook parameter.[5]

The Klein-Cook parameter is

$$Q = 2\pi\lambda_L L/n_o\Lambda^2 \tag{7}$$

where n_o is the refractive index without acoustical stimulation and Λ is the acoustical wavelength. In general, when $Q < 1$, Raman-Nath diffraction occurs and when $Q > 1$ Bragg diffraction occurs. Another useful parameter is the Raman-Nath parameter:

$$v = 2\pi(\Delta n)L/\lambda_L = 2\pi\kappa p L/\lambda_L \tag{8}$$

where Δn is the refractive index change, L is the width of the ultrasound beam, λ_L is the wavelength of light, κ is the piezoelectric coefficient ($\partial n/\partial p$), and p is the ultrasound pressure. The propagation of the modulated electric field from the sample can be described with this parameter:

$$E_r = (E_o/\sqrt{2})\exp i[(\omega_L)t + v(x,t) + \varphi_s) \tag{9}$$

where ω_l is the initial frequency and φ_s is the phase constant associated with the sample arm.

Acoustical-optic devices include modulations, spectrum analyzer, scanners, filters, and isolators. A frequency modulator shifts the frequency by an amount equal to

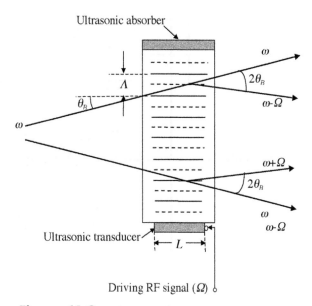

Figure 6A.2   Schematic of Bragg scattering. The ultrasound signal acts as a diffraction grating creating the original signal and a signal with shifted frequency. Based on concepts from Korpel, A. (1998). *Acousto-Optics.* [2]

the acoustical frequency. To be used as a modulator, a wide bandwidth is optimal. The bandwidth over which the frequency can be modified is determined by the range over which the Bragg condition is satisfied, which is:

$$\sin \Theta_B = \lambda/2\lambda_s \tag{10}$$

where Θ_B is the angle of the incident light wave relative to the angle perpendicular to the sound wave, λ is the wavelength of the incident light, and λ_s is the wavelength of sound. Under the Bragg condition, the incident and diffracted light is the same and only one frequency shift has occurred (i.e., no Raman-Nath diffraction).

The use of an acoustical-optic device as a scanner is based on the fact that an approximately linear relationship exists between the angle of deflection and the sound frequency. The relationship is basically,

$$2\Theta \approx \lambda f/v_s \tag{11}$$

where f is the sound frequency and v_s is the velocity of sound. However, the technique requires rotating the incident angle as the sound frequency is changed, which makes it inefficient.

Another major usage is as a tunable spatial filter. As stated:

$$\text{Sin}\Theta_B = \lambda/2\lambda_s \tag{12}$$

Therefore, reflection can occur only for a single optical wavelength at a given $\text{Sin}\Theta_B$ or λ_s. This can be used to isolate a small frequency distribution from a broadband source through changing the angle or sound frequency.

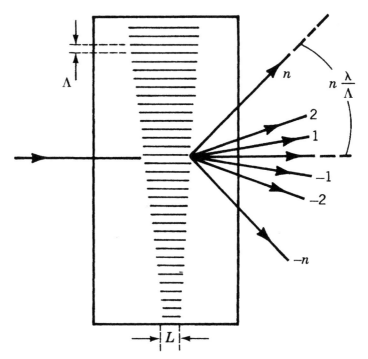

Figure 6A.3 Schematic of Raman-Nath scattering. The ultrasound beam, at a higher frequency, is now diffracting the OCT signal into multiple orders. Based on concepts from Saleh, B. E. and Teich, M. C. (1991). *Fundamentals of Photonics.* [1]

REFERENCES

1. Saleh, B. E. and Teich, M. C. (1991). *Fundamentals of Photonics.* Wiley-Interscience, New York.
2. Korpel, A. (1998). *Acousto-Optics.* Marcel Dekker, Inc., New York.
3. Wang, L. and Zhao, X. (1997). Ultrasound-modulated optical tomography of absorbing objects buried in tissue-simulating turbid media. *Appl. Opt.* 36(28), 7277–7282.
4. Raman, C.V. and Nath, N. S. (1935). Diffraction of light by high frequency sound waves. *Proc. Indian Acad. Sci.* 2A, 406–412, 1935; Klein, W. R. and Cook, B. D. (1967). Unified approach to ultrasonic light diffraction. *IEEE Trans. Sonics Ultrasonics* 14, 123–134.
5. Singh, V. R. (1990). Instrumentation techniques for acousto-optic studies in complex materials. *Appl. Acoustics* 29, 289–304.
6. Variv, A. (1997). *Optical Electronics in Modern Communications*, Fifth Edition. Oxford Press, New York.

7 NOISE AND SYSTEM PERFORMANCE WITH TD-OCT AND SD-OCT

Chapter Contents

7.1 INTRODUCTION

Maximizing the performance of OCT in terms of its signal to noise ratio (SNR) and dynamic range is critical for *in vivo* imaging.[2–6] This is seen in Figure 7.1,

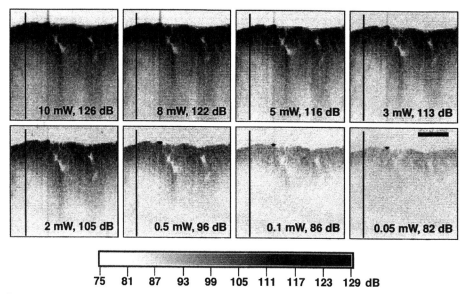

Figure 7.1 Effect of SNR on image penetration. It can be seen that increasing SNR results in increased imaging penetration. Courtesy of Boppart, S.A. (1999). Doctoral Thesis, MIT. [39]

where SNR in the range 126 dB is vastly superior to those in the range of 82 dB.[39] Traditionally, the theoretical maximum SNR quoted for OCT is:

$$SNR = 10 \log \eta P_s / 2 h v N E B \tag{7.1}$$

However, we will see in this chapter that the actual dynamic range and SNR are more complex than this suggests. With the different embodiments and their corresponding noise sources, a more accurate understanding of the system performance is critical. In particular, noise sources, which vary depending on the embodiments, and other factors reducing dynamic range, such as limitations of the optoelectronics, becomes critical to understand.

Comparing the performance of embodiments such as TD-OCT and FD-OCT based on currently available data is difficult for several reasons.[8-19] First, this is a complex issue, with many sources of performance reduction that are often currently poorly quantified. Second, many works claim superior performance of certain embodiments based on quantum noise limited detection, which may not be possible to implement. Third, most studies only consider a very limited number of noise sources. Fourth, data are often not reported in sufficient detail to make accurate comparisons. For example, it is often unclear whether the data are real time or post extensive processing and whether the data are obtained from the image or the raw RF signal. Fifth, it is also important to note that higher SNR does not automatically mean a larger dynamic range, as there is not a one to one correspondence.

This chapter focuses on identifying a wide range of image-degrading sources in OCT systems and how they compare between three OCT embodiments (TD-OCT, FD-OCT, and SS-OCT). With all the variability of terminology in the literature, this chapter will define the usage of many terms, of which the reader may find some unfamiliar. Therefore, references are included frequently for the sources of these definitions. This chapter attempts to integrate, based on the most current data on noise sources, signal loss, SNR, and dynamic range in the field of OCT. For theoretically characterizing the SNR and dynamic range of TD-OCT or SD-OCT, a traditional fiber-optic Michelson interferometer will be used, assuming polarization or dispersive effects in the interferometer are similar in the different embodiments and therefore not considered in their comparisons. Similarly, the systems will be assumed to be optimally aligned and appropriate optoelectronics used. Many of these issues have been recently reviewed.[40]

7.2 DEFINITIONS

Because different definitions exist in the literature for dynamic range, sensitivity, and SNR, the definitions used in this chapter will be defined to prevent ambiguity.[40] The different definitions used by investigators may result in some of the varying performance data described for modalities in the literature. It should also be remembered that equations may be different depending if the optical or electronic values are being measured.

Dynamic range — In a system or device, the ratio of a specified maximum level of a parameter, such as power, current, voltage, or frequency to the minimum detectable value of that parameter, which is usually expressed in dB. In this chapter the focus will be in the dynamic range of the final digitized image.

SNR — The definition of the SNR used here is the ratio of the amplitude of the maximal signal to the amplitude of noise signals at a given point in time. This chapter will define SNR in decibels and in terms of peak values for impulse noise and root-mean-square values for random noise off the RF signal. In defining or specifying the SNR, the definition of both the signal and noise should be characterized, e.g., 'peak' signal and 'peak noise' peak, in order to avoid ambiguity. Photo current is the detected parameter in TD-OCT and SS-OCT whereas photo-generated charge numbers are the detected parameters in FD-OCT. The different parameters do not affect the SNR as long as parameters are appropriately defined. Most SNRs describe the analog or RF signal prior to digitization. Some groups extrapolate back the optical SNR from the RF measurements, of which the reader should also be aware.

Sensitivity — The sensitivity is defined here as the minimum input signal required to produce a specified output signal having a specified SNR. With OCT, the SNR is often chosen as 1. So the minimum input signal, usually defined as the minimum sample reflection intensity, equals the amount of signal that can produce the same amount of output as noise.

Decibel — We define decibel as

$$decibels = 10 \log x \qquad (7.2)$$

which is what is commonly used in the literature. However, this can be confusing since some authors use a definition found in the electronics literature:

$$decibels = 20 \log x \qquad (7.3)$$

This second definition may be more appropriate in view of what data are being presented, but since the former has dominated the literature for over a decade, this chapter will stay with it. This has recently become important as with SD-OCT, investigators are often using the second, essentially doubling SNR and dynamic range compared with previous TD-OCT work.

7.3 NOISE

The focus will be on three sources of noise, those generated by the (1) light source, (2) interferometer, and (3) detector/electronics. In some instances, sources of signal loss (e.g., A-D conversion) will be referred to as noise for the convenience of grouping, which is a liberal use of the term.

7.3.1 Overview

7.3.1.1 Source

The major types of noise from light sources are photon shot noise and photon excess noise. The photon shot noise is intrinsic to the quantum nature of the source. The sources used in OCT, superluminescent diodes (SLD), quantum well devices (below the lasing threshold), and broadband lasers typically follow Bose-Einstein statistics whereas nearly monochromatic lasers follow Poisson statistics. However, at low photon counts, the Poisson and Bose-Einstein statistics are very similar, so the distinction is more critical at higher intensities. Photon excess noise is minimal in the setting of a source with a Poisson distribution. Sources following Bose-Einstein statistics produce photon excess noise, which is in excess of photon shot noise and can be on the order of 19 dB.[21,22] An important source of photon excess noise is second order correlations or photon bunching, often referred to as Brown-Twiss correlations.[23] Since the arrival times of bunched photons (Bosons) in a balanced dual detector OCT system will occur almost simultaneously, this technique can be used to remove significant amounts of excess noise.[24] We will see that dual balanced detection allows for substantial reduction of noise for TD-OCT and possibly some SS-OCT systems, depending on the source. However, while FD-OCT may be susceptible to high excess noise, dual balanced detection has not been implemented in this embodiment.

Two major sources of photon shot noise are relevant to OCT.[20] The first is spontaneous fluctuations (position-momentum uncertainty) at optical frequencies of electrons localized in the atomic or crystalline field of the source. The second source is field fluctuations caused by the quantum-mechanical uncertainty of electric and magnetic fields.

7.3.1.2 Interferometer

There are several major sources of interferometer noise that include photon pressure, mechanical vibrations, and vacuum fluctuations, in addition to the DC offset which is not technically noise but can significantly alter performance. Mechanical noise, from moving parts in the interferometer, such as from small scale nonlinear movement of a moving mirror, will be discussed in the section TD-OCT (7.6.1). A DC component is produced in the interferometer as part of the interference effect, as described in Chapter 5. The DC offset needs to be filtered out for the system to approach the quantum noise limit. With the low pass filter design of some SD-OCT designs, complete filtration of the DC offset may not be possible.

Although this fact is rarely addressed, the light field in the interferometer is not classical, and ideally should be treated with second order quantization. This essentially treats the optical field in a vacuum as a collection of quantum harmonic oscillators of varying energy. Most investigators typically treat the light as classical while the detection process may or may not be treated as quantum mechanical. For assessment of system performance, both should be treated quantum mechanically, which is the approach taken here (second order quantization). For second order quantization, the quantum harmonic oscillator has a lowest energy state of $(1/2)h\omega$. The nature of vacuum noise (as opposed to vacuum fluctuations) can then be interpreted as the light field being represented by one mode with average occupation number N and the remaining modes (vacuum) at $(1/2)h\omega$. The vacuum fluctuations per se are not sensed by the detector. However, as the modes impinge on the detector, beat signals are generated at the frequency differences between the light and vacuum modes.[25] In other words, the quantum noise is the result of vacuum fluctuations in all other modes beating with one strongly occupied mode. It therefore becomes easily understandable that this noise source is proportional to the square root of the source power. Although vacuum fluctuations have been examined in depth in other fields, such as technologies looking for gravitational waves and in telecommunications, they are largely unrecognized in OCT work.[25,26] One group noted its presence when an interface of high specular reflection was present in the fiber optics, but the influence of vacuum fluctuations extends well beyond this effect.[27] It will become particularly relevant to SS-OCT where higher instantaneous intensities at single frequencies are involved. It should be remembered that when the AC signal approaches the noise floor in the detector arm (near sensitivity limit), back-action or energy directed back to the source arm is greatest, leading to increased vacuum beating in the source arm which propagates through the interferometer. However, it has been suggested that, particularly when the AC signal is well above the noise floor, the major source of vacuum fluctuations is from the exit port.[25,28] These fluctuations can be imagined to originate at the dissipation associated with the photodetector. They then bounce off the detector and interfere with light emerging from the interferometer. It should also be noted that vacuum fluctuations are influenced by external potentials (e.g., the Aharonov-Bohm effect) in addition to the light wave field, suggesting that these potentials need to be considered in interferometer design.

Photon pressure results because DC photons impinging on the reference mirror carry momentum and are a function of the total intensity. Therefore, photon pressure likely creates its greatest noise problem with SS-OCT.

7.3.1.3 Electronic noise sources

Both classical and nonclassical electronic noise need to be considered when comparing technologies. This includes $1/f$ noise, thermal noise (Johnson noise and dark current), and preamplifier noise. In addition, noise specific to the CCD and signal loss differences in A to D conversion will be discussed in subsequent sections.

At any junction, including metal-to-metal, metal-to-semiconductor, and semiconductor-to-semiconductor, conductivity fluctuations occur from $1/f$ noise. The causes of these are still not completely understood. The rms $1/f$ noise current is given by[29]:

$$\sigma_{nf} \propto I_d \left(\frac{\Delta f}{f}\right)^{\frac{1}{2}} \tag{7.4}$$

The shunt resistance R_{sh}, in a photodiode, due to voltage across a dissipative circuit element, has a Johnson or Nyquist noise associated with it. These fluctuations are most often caused by the thermal motion of the charge carriers. The magnitude of this generated current noise is[30]:

$$\sigma_{nT} = \left(\frac{4kT\Delta f}{R_{sh}}\right)^{\frac{1}{2}} \tag{7.5}$$

Dark current, which is often confused with current shot noise, occurs in the absence of an irradiance field and follows predominately classical thermodynamic principles. It results when random electron-hole pairs are excited by sufficient thermal energy to enter the conduction band.[31]

Preamplifier cascaded to the photodiode does contribute to the noise characteristics of the system, but it contributes less than the noise from the photodiode. Therefore, it is not likely to be significant with TD-OCT.

A nonclassical noise source is current shot noise, which is due to momentum-position uncertainty of the electrons in the current.[20,32] In the second quantization description, the wave function of the system is described in terms of the occupation numbers of one electron state where the occupation numbers can take on values of 0 or 1, in accordance with the exclusion principle. The alternative is to describe the many-electron wave function of the system as a Slater determinant of one-electron wave functions.

Contrary to many misconceptions, the electron shot noise of the detector is not noise produced by photon shot noise, thermal noise, or vacuum fluctuations. However, photon shot noise and vacuum fluctuations entering the detector can be modeled for simplicity as annihilation (photons)-creation (electrons) operators at the surface of the detector. Since the annihilation-creation operators are incompatible operators, fluctuations at the detector surface fall out naturally. Current shot noise occurs once electrons are generated. A small amount of current shot noise will also result from quantum mechanical tunneling. But at the temperatures used during most OCT experiments, electron shot noise due to momentum-position uncertainty is far in excess of that generated by tunneling.

7.3.1.4 Quantum and shot noise

Quantum mechanical sources of noise are likely very important in OCT imaging but are rarely discussed. When examining noise within the system, the phrase "shot noise limit" is frequently used as the ultimate objective of minimal signal limit. As previously alluded to, this represents a term that is used very differently depending on the author. The term shot noise dates back to a misunderstanding of the physics of the cathode ray tube, where the incident field on the detector was viewed as classical rather than taking into account quantization of the field. As indicated above, we will limit our discussion of shot noise as the quantum noise limit in the source, interferometer, and detector. Most quantum noise sources result from measurements limited by the uncertainty principle. The quantum noise limit here will be defined as when photon shot noise, vacuum fluctuations, photon pressure, and/or current shot noise predominate in the system over other noise sources. Quantum mechanical tunneling, Raman scattering (because of the short length of the fibers), and phonon interactions will not be a focus as they are only a minority of quantum noise sources. Quantifying the quantum mechanical effects in different OCT systems is in its infancy. Some quantitative relationships will be discussed here, but comprehensive analysis of quantum effects can not be made at this time as insufficient experimental data are available. Furthermore, a complete description will likely depend on the exact configuration of the system.

Quantitating photon shot noise, which again is determined primarily by uncertainty of the field and lattice, depends heavily on the nature of the source, intensity of the light beam, and method for inducing excitation, in addition to other factors. Therefore, quantitation of photon shot noise (i.e., from the source) is beyond the scope of this chapter. However, the topic has been extensively reviewed elsewhere by Henry and Kazarinov.[20] The photon shot noise depends on the type of source but one derivation for a laser is given by:

$$\Delta \hat{r} = \sqrt{\omega/Q_e}\,\Delta\hat{A} - \frac{1}{2}\left[\hat{f}e^{i\Delta\hat{\phi}} + e^{-i\Delta\hat{\phi}}\hat{f}^{+}\right]$$

where f represents the zero point fluctuations, Q_e is the Q value due to the output coupling through the output mirror of the laser cavity, and ΔA is the expectation value of the internal field operator.

The Hamiltonian for the vacuum fluctuations is given by $H = \hbar\omega_k(a_k^{+}a_k + 1/2)/2\pi$, where a_k^{+} and a_k are the creation and annihilation operators, respectively.[33] As previously stated, vacuum fluctuations are not directly measured by the detector.

Instead, their interaction with the optical field results in beats that are detectable and a source of noise. The light electric field operator $E_k(t)$ for the mode k at the point r is given by:

$$E_k(t) = i\left(\frac{\eta\omega_k}{2\varepsilon_0 V}\right)^{\frac{1}{2}}\left\{a_k e^{-i(\omega_k t - k \cdot r)} - a_k^+ e^{i(\omega_k t - k \cdot r)}\right\} \tag{7.6}$$

where V is the quantization volume and ε_0 is the permittivity of free space.

The shot noise current, described above, is due principally to position-momentum uncertainty of the electron and is given $\Delta x\, \Delta p \geq h/4\pi$. It is distinct from current due to thermodynamic effects in that it cannot be deduced through classical techniques. Phenomena due to the restrictions of the uncertainty principle can be optimized for particular measurements through nonclassical approaches. For example, the quantum noise limit refers to the minimum value of the combination of amplitude and phase. Techniques exist which decrease amplitude uncertainty at the expense of phase uncertainty, such as squeezing and quantum non-demolution, which may ultimately be applied to OCT imaging as they are currently examined in telecommunications.[25–26]

An additional potential source of quantum mechanical noise is photon pressure, particularly at the mirrors. The photon pressure is given by[25]:

$$(\Delta z)_{rp} \approx (bh\omega\tau/2\pi mc)(a^2 e^{2\gamma} + \sinh^2\gamma)^{\frac{1}{2}} \tag{7.7}$$

where b is the number of bounce events at each end mirror, τ is the duration of measurement, and γ is the squeezing factor. If photon pressure has a significant effect on OCT imaging, it is likely when DC intensities are relatively high.

7.4 NOISE IN CCD VERSUS PHOTODIODE

Detector noise plays an important role in OCT, whether evaluating a photodiode array or CCD. Photodiodes have classical noise sources that can be suppressed more readily in TD-OCT but not easily in SD-OCT. These include $1/f$ noise, thermal noise in resistive elements, dark noise, and preamplifier noise. The diodes also have current shot noise that cannot be suppressed by classical methods.

Noise in a CCD is typically separated into two types of noise above those of a PIN diode: random noise and pattern noise.[34] Pattern noise is a kind of spatial noise that is induced mainly by the non-uniformity of the pixels' responsiveness, and fixed deviations of performance between pixels in the absence of illumination. Here we focus on the random noise in individual pixels and the array. Unlike the random noise, pattern noise may slightly alter the form of the autocorrelation function, but not significantly alter the SNR and dynamic range with high quality CCD.

The photo-sensing element in each pixel of CCD is typically a photodiode. So any noise sources discussed above for the photodiode are present in CCD.

Additional noise sources include read out noise and reset noise will contribute significantly in a CCD. CCD employs an electronic network including many capacitors, transistors for signal integration, signal transferring, and final output of the signal. This infrastructure is commonly named readout stage in a CCD, as discussed in Chapter 6. Noise is generated in this readout portion of the CCD. Readout noise includes additional thermal noise and $1/f$ noise in the CCD. $1/f$ noise arises mainly in transistor circuits where there are numerous junctions, which implies the critical role of $1/f$ noise in the noise of a CCD. CCD employed in FD-OCT cannot suppress such a noise simply, which will deteriorate the detecting performance more in FD-OCT than TD-OCT.

Each photodiode in the array has to be reset through a MOSFET during the interframe period before restarting the signal integration. Effectively, this is a capacitance charged through the resistance of the MOSFET channel. The reset noise is an uncertainty about the voltage on the capacitor and can be described as rms voltage $\sqrt{kT/C}$. k is the Boltzmann constant and T is the absolute temperature.

Time integration is a method to improve SNR in a CCD. A detected signal $s(t)$ with the presence of additive noise $n(t)$ can be described as a random process $X(t)$. Assuming the signal intensity is S and the rms intensity of noise is σ, the SNR of this process will be

$$SNR = \frac{S}{\sigma} \tag{7.8}$$

If we measure such random processes N times, N samples $x_1(t)$, $x_2(t)$, ..., and $x_N(t)$ will be obtained. The sum of these samples will be a new random process that has the signal intensity $N \times S$, while the rms intensity of noise only increases $\sqrt{N}$ time as $\sqrt{N} \times \sigma$. The consequence of such a process is that the SNR is increased as

$$SNR = \sqrt{N}\frac{S}{\sigma} \tag{7.9}$$

This mechanism is used in a CCD for potentially improving SNR. Based on the operating mode of a CCD, the method of addition mentioned above is realized specifically by integration. Assuming the photocurrent of each individual photodiode in the array is I_p, the noise in rms current is σ_n. If the photocurrent of each photodiode is read out time-sequentially by a multiplexer or $x - y$ addressing, it will only be equal to a single photodiode. The readout current has not been integrated with the SNR as

$$SNR = \frac{I_p}{\sigma_n} \tag{7.10}$$

Typically in the CCD used with OCT, the photocurrent of each photodiode is used to charge a potential well (capacitor) during a period ΔT. The accumulated charge packet represents the signal and will be transferred out. The fluctuation of the charge number is the noise. According to the statistics, the mean value of the charge number (signal) increases proportionally to ΔT, while the rms fluctuation of the charge number (noise) is only proportional to $\sqrt{\Delta T}$. Hence the SNR of the readout signal will be as

$$SNR = \frac{I_p \Delta T}{\sigma_n \sqrt{\Delta T}} = \sqrt{\Delta T} \frac{I_p}{\sigma_n}$$ (7.11)

From Eqs. 7.10 and 7.11, the SNR is obviously improved by integration. That might partially compensate for the deterioration of SNR in FD-OCT, which is induced by low frequency noise sources previously described. However, it should be pointed out that an additional disadvantage of performing OCT with signal integration is that any vibration in the system during the integration period will cause distortion in detecting performance. Therefore, its applicability to moving tissue such as coronary arteries may be limited. This has been recently addressed by one group using a swept source, but its ultimate effectiveness needs to be assessed.[38]

7.5   A-D CONVERSION IN SD-OCT AND TD-OCT

Signal loss from analog to digital (A-D) conversion is important in determining performance. Typically, the analog SNR ranges from 80 to 130 dB with OCT. But the displayed image has a substantially reduced dynamic range due to the signal loss associated with the limitations of A-D conversion. For A-D conversion we will temporarily assume the interval Δ between the discrete levels is always uniform, which determines the quantization noise (i.e., signal loss) whose rms intensity is proportional to Δ. The maximum level of the uniform quantizer is $2M\Delta$ where M is the bits of the A-D converter. Thus the maximum image dynamic range is limited by M. A 14-bit A-D converter has an ~40 dB dynamic range for SD-OCT. From Chapter 6, A-D converters are not ideal and Δ is not completely uniform but has periodic bit errors. This is particularly important when there is extensive post processing after A-D conversion.

The application of A-D conversion and its respective signal loss are different for TD-OCT and SD-OCT. In TD-OCT, digital processing is applied to the analog autocorrelation signal after electronic filtering, which allows prior $\log_{10}$ demodulation to maintain a high dynamic range. In a SD-OCT, the digital Fourier transform is conducted to calculate the A-scan signal from the quantized spectral interferogram, prior to signal processing. The dynamic range of the calculated A-scan signal in the SD-OCT will therefore inevitably be reduced because the logarithmic amplification of the output signal is not performed or is performed after A-D conversion.

7.6   EMBODIMENT AND THEORY

7.6.1   TD-OCT

With TD-OCT, in the reference arm, a mirror is scanned providing a low noise a-scan primarily through a heterodyne detection process. Again, the light beam from the source is evenly split by a 2×2 coupler and comes through and back in the reference and the sample arm, then recombines at one of the ports of the coupler in terms of the exit of the interferometer. The light intensity perturbation at the exit can be described as

$$I_D = I_R + I_S + \frac{1}{2} \int_0^z \int_0^\infty G(f)p(z')p_R \cos\left\{\frac{4\pi f}{c}[n_s(f)z' - z_R]\right\}df\}dz' \qquad (7.12)$$

The coordinates have already been described in Chapter 5.

The first term in Eq. 7.12 corresponds to the intensity of the reflected reference beam I_R. The second term corresponds to the total intensity of the returned sample beam I_S, which is a DC source contributed by all the scatters. It is important to reiterate that mutual interference of backscattering in the sample occurs, which is called the self-correlation function of scattering, or a parasitic term.[7,35–36] This phenomenon is likely to have limited relevance to system performance. For TD-OCT, if no strong reflections exit in the sample, the second term can be approximated to the total incoherent intensity of the returned sample beam (DC signal). This will not be the case for SD-OCT: Thus I_R and I_S can be, respectively, represented as a portion of I_0.

$$I_R = \frac{p_R^2 I_0}{4} \qquad (7.13)$$

$$I_S = \frac{I_0}{4} \int_0^z p^2(z')dz' \qquad (7.14)$$

where the coefficient p_R $(0 \leq p_R \leq 1)$ represents the variable attenuations introduced in the reference arm. The third item in Eq. 7.12 is the interferometric term, the actual information carrying the OCT signal (AC term), in which the diagnostic information $p(z)$ is encoded. As stated, the unfiltered DC signal represents image-degrading energy unless removed.

For time domain operation, the light intensity perturbation is converted into an electronic signal (usually the photocurrent) by either a single optical detector or a dual balanced detector approach; the latter is used to remove excess noise.

For simplification, the intensity spectrum of the light source is assumed to have a Gaussian distribution with a full width half maximum (FWHM) bandwidth Δ_λ, and it is assumed that the mirror in the reference scans in a constant velocity V_R and no polarization or dispersion alterations occur. Apart from a constant that is related to the light beam size, the time-sequentially generated signal can be described in terms of the convolution as

$$i_D(t) = \alpha\left\{I_R + I_S + \frac{I_0}{2}p_R p\left(\frac{V_R t}{n_s(\lambda_0)}\right) \otimes \left\{\cos\left[\frac{4\pi V_R t}{(\lambda_0)}\right]\exp\left[-\frac{4\ln 2(V_R t)^2}{\Delta_l^2}\right]\right\}\right\} \qquad (7.15)$$

$$\Delta_l = \frac{2\ln 2\lambda_0^2}{\pi\Delta_\lambda} \qquad (7.16)$$

where α represents the responsiveness of the detector, which is defined as the electronic signal intensity per unit light power. Δ_l represents the FWHM coherence length of light source, and λ_0 is the central wavelength of the light source in a vacuum. The refractive index of the sample is constant as $n_s(\lambda_0)$. Equation 7.15 indicates that two major parts compose the detector output, an amplitude modulation (AM) OCT signal with a certain central Doppler frequency $f_c = 2V_R/\lambda_0$ and FWHM bandwidth $\Delta f_{TD} = 2V_R\Delta_\lambda/\lambda_0^2$, and a DC offset caused by I_R and I_S.

The use of a mechanically induced optical group delay has parameters that need to be considered to prevent signal loss. These include optical power loss, polarization effects, dispersion effects, nonlinear motion, and microvibrations. Optical power loss in a properly aligned OCT system does occur when the reference arm configuration is a grating based delay line. However, in most systems this is not a significant issue as reference arm power is at least partially attenuated for optimal performance. TD-OCT systems using a fiber stretcher delay line have significant problems in maintaining polarization, but this design is no longer commonly used. The most common embodiments do not suffer from this limitation. Although a small amount of dispersion does occur in the OCT system, this is readily compensated for as described in Chapters 5 and 6. Nonlinear motion of the mirror in the reference arm is typically partially compensated for through modification of the driving waveform. Even though incomplete compensation can result in distortion of the image, it will not significantly affect SNR or dynamic range. Microvibrations are typically low frequency and can be filtered with the bandpass filter. If they do enter the bandpass filter, algorithms exist for their correction, as well as through the use of optical modulators.[37]

TD-OCT offers several advantages in approaching high SNR and near quantum noise detection. First, dual balance detection offers a mechanism for removing excess noise. Second, the Doppler shift induced by the mechanical movement in the reference arm offers several advantages for noise reduction. This includes allowing bandpass filtration to be performed that reduces the DC signal offset, $1/f$ noise, and detector noise. Third, it has certain advantages over using a CCD, as described above. Fourth, logarithmic demodulation is performed before A-D conversion rather than after, reducing data loss. Fifth, the level of post processing used with SD-OCT, such as nonlinear scaling algorithm, dispersion compensation, and correction for decreasing amplitude with increasing optical group mismatches. Finally, in theory, photon shot noise, photon pressure, and vacuum fluctuations should be reduced relative to SS-OCT since the power in a given frequency at one time is lower.

7.6.2 SD-OCT

SD-OCT does not require a moving mirror. From Eq. 7-15, assuming an ideal spectrum is captured by a spectrometer without any deterioration (an approximation which will be readdressed later in the section), the spectral interferogram can be written in terms of wavenumber k:

$$i_D(k) = \frac{1}{4}\alpha(k)G(k)c\left\{p_R^2 + \int_0^z p^2(z')dz'\right.$$

$$\left. + 2\text{Re}\left\{\int_0^z p(z')p_R \exp\left\{j4\pi k[n_s(k_0)z' - z_R]\right\}dz'\right\}\right\}$$

(7.17)

where $G(k)$ is the intensity spectrum of the light beam, $\alpha(k)$ represents a spectral coefficient for conversion from light intensity to any electronic entities (e.g., current, voltage, or charge density). Noise sources are ignored at this point. If $\alpha(k)$ is assumed to be the same as the responsiveness α in TD-OCT, applying the reverse Fourier transform to the recorded spectrum, we can get

$$F^{-1}\{i_D(k)\} = \frac{\alpha}{4}F^{-1}\{G(k)\}\otimes\left\{F^{-1}\left\{p_R^2 + \int_0^Z p^2(z')dz'\right\}\right.$$

$$\left. + p_R p\left(\frac{z'' + 2z_R}{2n_s(k_0)}\right) + p_R p\left(\frac{-z'' + 2z_R}{2n_s(k_0)}\right)\right\}$$

(7.18)

Both sides of Eq. 7.18 are the functions of a spatial variable z'' which is measured in free space. The right side is a convolution between a reverse Fourier transform of the spectrum of light source (wideband source or swept/tunable laser) $G(k)$ and a superposition of three elementary functions, which are included in brackets. Within the brackets, the Fourier transform of the first term is a delta function located at $z''=0$. The last two items are symmetrical around $z''=2z_R$. The diagnostic signal $p(z)$ is actually encoded and retrievable in either one of the last two items. Equation 7.18 also suggests choosing $|z_R|$ larger than the designated imaging depth,[7] otherwise the parasitic beams in the sample arm at $z''=0$ would overlap with the target signal. In other words, the pathlength in the reference arm is approximately a few hundred microns shorter than that in the sample arm. But as shown in Eq. 7.18, increasing $|z_R|$ could cause modulation depth distortion for a high frequency component in space k.[36] This balance represents a challenge in practically implementing the technology beyond the noise limitations to which we alluded.

Multiple schemes could accomplish the goal of spectral interferogram detection for SD-OCT. Probably the scanning spectrometer is the most intuitive choice as it is easy to implement with a dispersion component, slit, a single detector, and a set of scanning mechanics. But with this single detector approach, the bandwidth of the detecting electronics of scanning spectrometry has to be the same as that in TD-OCT for the same A-scan rate. Additionally the benefit of classical noise suppression in TD-OCT because of the bandpass performance would indicate its superior performance over single detector SD-OCT since signal integration is not utilized.

A popular spectral interferogram detection approach is using a spectrometer with a detector array. It is also called channeled spectrometry. A dispersion component is used to separate the different spectral components. The detector array captures a discrete spectrogram described as

$$i_D'(k) = i_D(k)\left\{ rect\left(\frac{k}{\Delta_k}\right)\left[rect\left(\frac{k}{\gamma\delta_k}\right) \otimes comb\left(\frac{k}{\delta_k}\right)\right]\right\} \tag{7.19}$$

where $comb\left(\frac{k}{\delta_k}\right) = \sum_{n=-\infty}^{\infty} \delta(k - n\delta_k) n = \ldots, -2, -1, 0, 1, 2, \ldots$

The items in brackets {} represent the discrete sampling process by the detector array in a limited spectral range, for example, the FWHM width in the k space Δ_k. The *comb* function represents the infinite impulse sampling series with the same spectral resolution δ_k. The number of detectors in the array, M, will be as Δ_k/δ_k. The coefficient γ in the rectangular function is the fill factor of an individual detector which conducts spectral average over δ_k. So the retrievable signal is as

$$F^{-1}[i_D'(k)] = F^{-1}[i_D(k)] \otimes \text{Sinc}(\Delta_k z'') \otimes [\text{Sinc}(\gamma\delta_k z'')comb(\delta_k z'')] \tag{7.20}$$

where $comb(\delta_k z) = \sum_{n=-\infty}^{\infty} \delta(z'' - \frac{n}{\delta_k}) n = \ldots, -2, -1, 0, 1, 2, \ldots$

Compared to Eq. 7.18, Eq. 7.20 indicates possible signal distortion in SD-OCT by the finite discrete spectrum detection, in addition to noise consideration. However, the spectrographic detection approach does eliminate the moving parts for the A-scan compared with TD-OCT.

For most FD-OCT, the detector array used is a CCD imager or a photodiode array that has a photo-generated signal integration function in addition to the photo conversion function. Such a signal integration process is called on-focal-plane signal process in electro-optic (E-O) imaging field. How the signal integration process affects the SNR and the sensitivity and dynamic range of an FD-OCT is important in comparing the technologies. The general principles are in Section 7.6.4 and now we look specifically at its influence on FD-OCT.

7.6.3 Detector Array Signal Integration

Before addressing the SNR issue, it is important to clarify how a noise $n(z)$ transfers to k space. According to the Parseval theorem, the SNR can be expressed as

$$SNR = \frac{\sqrt{\langle s(z)s^*(z)\rangle}}{\sqrt{\langle n(z)n^*(z)\rangle}} = \frac{\sqrt{\int_{-\infty}^{\infty} G_S(k)dk}}{\sqrt{\int_{-\infty}^{\infty} G_N(k)dk}} \tag{7.21}$$

where $G_S(k)$ and $G_N(k)$ represent the power spectrum of the signal $s(z)$ and the noise $n(z)$. Equation 7.21 indicates no SNR change by Fourier transform. From Eq. 7.18, the captured spectrum for single scatter at depth z_0 is expressed as

$$F(k) \propto G(k)[p_R^2 + p^2 + 2pp_R \cos(4\pi(z_0 - z_R)k)] + N(k) \tag{7.22}$$

The signal spectrum will be $\sqrt{2}pp_R G(k)$.

First we consider a simple example of using a scanning spectrometer that has a *single detector*. The captured spectral interferogram is time dependent with noise as

$$F(t) \propto G(V_{kt})[p_R^2 + p^2 + 2pp_R \cos(4\pi z_0 V_{kt})] + N(V_{kt}). \quad (7.23)$$

where V_k is the spectrum scan speed. Assuming this detector has the same noise intensity as the one in the TD-OCT (i.e., ignoring noise sources previously described), the SNR of such a FD-OCT can be expressed as

$$SNR = \frac{\sqrt{2I_R I_S}}{\sigma_n} \quad (7.24)$$

This equation tells us that the SNR over the quantum limit using scanning spectrum capturing must be very similar to TD-OCT. But again, an important difference is that the spectrum detection here is essentially a low pass band filter.

Commonly a detector array is used in an FD-OCT. For better comparison, we assume each detector element is the same as the one in TD-OCT. If a multiplexer is used for spectrum readout, the SNR will be the same as that of an FD-OCT using a scanning spectrometer. Obviously, there is no extraordinary benefit for using such FD-OCT setups. The signal integration function in the array is the real contributor to possible performance improvement. A typical detector array is a CCD imager. In each element, the photo-generated electrons are accumulated and stored as a packet. Even though it is different with all setups discussed above, the interferogram detection is essentially energy detection. Hence the spectral distribution of signal $Q_S(k)$ is proportional to $\sqrt{2}pp_R G(k)\Delta T_{FD}$, with ΔT_{FD} representing the integrating time. The incoherently integrated noise in each element Q_n will be proportional to $\sigma_n\sqrt{\Delta T_{FD}}$. It is important to realize that the random noise in k space has the same statistics as in a single element. Thus the SNR of an FD-OCT is determined by

$$SNR = \frac{Q_s}{Q_n} = \frac{\sqrt{2I_R I_S \Delta T_{FD}}}{\sigma_n} \quad (7.25)$$

Adapting the SNR expression of the TD-OCT into the version of energy detection, the SNR of a TD-OCT can be described as

$$SNR_{TD} = \frac{\sqrt{2I_R I_S \Delta T_{TD}}}{\sigma_n} \quad (7.26)$$

where ΔT_{TD} is the equivalent integrating time of a TD-OCT, which is the reverse of the bandwidth of the system as $\Delta T_{TD} = 1/(2\Delta f_{TD})$. We will assume both TD-OCT and FD-OCT have the same scanning depth. If the spectral resolution in FD-OCT is δk, the equivalent scan depth Δz_{max} in TD-OCT will be as $1/(2\delta_k)$. If the integrating time is part of the time needed for an a-scan in TD-OCT, associated with a duty coefficient $\gamma (0 \leq \gamma \leq 1)$, the equivalent scanning speed V_R is $\gamma \Delta f_{FD}/\delta_k$, and Δf_{FD} is the equivalent bandwidth of the detecting electronics in an FD-OCT equal to $\Delta f_{TD} = 1/(2\Delta T_{TD})$. The equivalent system bandwidth in TD-OCT and FD-OCT will be related as $\Delta f_{TD}/\Delta f_{FD} = \gamma \Delta_k/\delta_k = \gamma M$. Substituting them into Eqs. 7.25 and 7.26, the SNR difference between FD-OCT and TD-OCT is approximated to $\sqrt{\gamma M}$. Consequently the sensitivity of FD-OCT will be improved $\sqrt{\gamma M}$ times than TD-OCT.

An important disadvantage of using signal integration with OCT is that any vibrations in the system during the integration period will cause distortion in the image, limiting both the use of moving tissue and the length of the integration time.

Another important issue as discussed with FD-OCT is the dynamic range, in addition to losses associated with A-D conversion.

7.6.4 SS-OCT

An alternative way to obtain a spectrogram is to use a frequency-swept laser or tunable laser with just a single detector and without dispersion components, which is referred to as SS-OCT. From Eq. 7.17, the detected spectrogram can be described as

$$
i_D'(k) = \frac{1}{4}\alpha(k)\left\{ G(k)\left[\exp\left(\frac{-k^2}{2\sigma_k^2} \right) \otimes comb\left(\frac{k}{\delta_k'} \right) \right] \right\}
$$
$$
\left\{ p_R^2 + \int_0^z p^2(z')dz' + 2\mathrm{Re}\left\{ \int_0^z p(z')p_R \exp\left\{ j4\pi k[n_s(k_0)z' - z_R] \right\} dz' \right\} \right\}
$$

$$(7.27)$$

The spectrum of the light source is described as a discrete spectral intensity distribution with a line width σ_k (standard deviation) and tuning mode-hop δ_k'. So the retrieved signal can be expressed as

$$
F^{-1}\{i_D(k)\} = \frac{\alpha}{4}F^{-1}[G(k)] \otimes [\exp(-2\sigma_k^2 z''^2)]comb(\delta_k' z'')]\otimes
$$
$$
\left\{ F^{-1}\left\{ p_R^2 + \int_0^z p_s^2(z')dz' \right\} + p\left(\frac{z'' + 2z_R}{2n_s(k_0)} \right) + p\left(\frac{-z'' + 2z_R}{2n_s(k_0)} \right) \right\}
$$

$$(7.28)$$

From Eq. 7.28, it is seen that the mode-hop will limit the A-scan range and the laser line width will degrade the a-scan profile.

In SS-OCT, as in FD-OCT, no moving parts are required for axial scan (ignoring the tuning mechanism in the laser source). Possible sensitivity improvement is obtained through higher spectral intensity of the laser source but not by the signal integration process. Intrinsically, TD-OCT is a bandpass signal detecting system whereas SS-OCT is a low pass system. Although SS-OCT use as single detector as TD-OCT does, the detection electronics have a critical difference with TD-OCT which will cause possible degradation in sensitivity. In a swept or tunable laser source, the total intensity over the entire spectrum can be easily kept much higher than a wideband light source. Thus possible sensitivity improvement can be gained in SS-OCT. It is proposed to use a bandpass filter technique, by shifting the area of interest (ROI) of the sample fairly far off the plane O_S. Thus all the high frequency spectral oscillation introduced by the interested scatters will be kept while much of the low frequency noise is filtered out. However, if the frequency spectrum is shifted too high, it can lead to modulation depth distortion.

Since the low frequency noise presents significantly in the range through $0 \sim$ tens of kHz, the tuning speed of the swept source should also be high enough to shift the spectrum of the time-sequentially recorded spectral interferogram higher out of that range.

In SD-OCT, the spectral interferogram of a single reflector located at z_0 off the O_S can be expressed according to Eq. 7.17 as

$$i(k) = \frac{\alpha}{4} G(k)\{p_R^2 + p^2 + 2pp_R \cos[4\pi(z_0 - z_R)k]\} \qquad (7.29)$$

z_R is the mirror position off the O_R.

The theory behind the advantages of high tuning speed is as follows. For SS-OCT, such a spectral interferogram is detected at the output of a single photodiode by uniformly sweeping the wavelength of the laser. Assuming the source sweep over a range Δ_k in a period Δt from a starting wavenumber k_0,[14] the spectral interferogram is captured as a time-sequential signal

$$i(t) = \frac{\alpha}{4} G\left(k_0 + \frac{\Delta_k}{\Delta t}t\right)\left\{p_R^2 + p^2 + 2pp_R \cos[4\pi(z_0 - z_R)\frac{\Delta_k}{\Delta t}t + 4\pi(z_0 - z_R)k_0]\right\} \qquad (7.30)$$

This is an AM signal with DC offset. The central frequency is

$$f_c = \frac{2\Delta_k(z_R - z_0)}{\Delta t} \qquad (7.31)$$

It is proportional to the sweeping speed $\Delta_k/\Delta t$ and $(z_0 - z_R)$. So higher sweeping speed and larger z_R can all shift the working band to higher frequency. A bandpass filter can thus be employed for $1/f$ noise suppression. The lower cutoff frequency f_1 and upper cutoff frequency f_2 are defined by the border of the ROI.

$$f_1 = \frac{2\Delta_k(z_R - z_2)}{\Delta t} \qquad (7.32a)$$

$$f_2 = \frac{2\Delta_k(z_R - z_1)}{\Delta t} \qquad (7.32b)$$

z_1 and z_2 are defined as the nearest and far border of ROI, respectively, and follow as

$$0 < z_1 < z_2 < z_R \qquad (7.33)$$

It should be noted that, like the case with TD-OCT, nonlinearities at fast sources sweeps now need to be considered analogous to a moving mirror in the reference arm.

7.7 CONCLUSION

The summary of the theoretical comparison of the three technologies is listed in Table 7.1. Several conclusions can be made after evaluating noise sources,

Table 7.1 Summary

	TD-OCT	FD-OCT	SS-OCT
Classical noise	a. $1/f$ noise, dark current noise, Johnson noise, preamplifier noise; b. photon excess noise; c. DC signal; d. micro-vibration and nonlinear motion.	a. $1/f$ noise, dark current noise, Johnson noise, preamplifier noise; b. photon excess noise; c. DC signal; d. noise unique to CCD.	a. $1/f$ noise, dark current noise, Johnson noise, preamplifier noise; b. DC signal; c. nonlinear frequency tuning.
Mechanism to compensate for classical noise	a. band-pass filter; b. dual balanced detection; c. algorithms if micro-vibration is significant.	a. poorly compensate for as the system is a low-pass filter; b. time integration gives some improvement but limited by motion artifacts; c. z_0 offset can give some improvement but leads to modulation depth distortion; d. E-O phase modulation proposed but leads to issues analogous to mechanical translation; e. photon excess noise not easily compensated for.	a. low-pass filter but can be partially compensated for by z_0 offset. high source intensity across the bandwidth, and fast sweep rate.
CCD		a. read noise and reset noise; b. increased classical noise; c. time integration restricted by sample motion artifacts.	
A-D converter's dynamic range	a. $\log_{10}$ demodulation allows dynamic range approaching 80 dB.	a. the use of digital Fourier transform prevents $\log_{10}$ demodulation. dynamic range of 40dB.	a. same as FD-OCT.
Quantum noise	a. photon shot noise, radiation pressure vacuum fluctuations, and current shot noise. The first three have same magnitude to FD-OCT but less than SS-OCT.	a. same as TD-OCT.	a. photo shot noise, radiation pressure, and vacuum fluctuation greater than TD-OCT and FD-OCT.
Conclusion	a. through band-pass filter, dual balanced detection, and control of motion nonlinearity, SNR approaches quantum noise limit; b. dynamic range approaching 80dB.	a. low-pass filter allows high classical noise; b. dual balanced detector not possible to remove excess noise; c. CCD associated with unique noise sources; d. dynamic range approaches 40 dB; e. z_0 offset, time integration, and use of E-O phase modulation unlikely to overcome these limitations.	a. low-pass filter allows high classical noise; b. low excess noise; c. now limited to fast sweep rate; d. increased quantum noise; e. z_0 offset, high source intensity across the bandwidth, and fast sweep rate may lead to SNR near TD-OCT; f. dynamic range limited to 40 dB.

optoelectronics, and A-D conversion losses. For TD-OCT, through the use of low pass filtration, dual balanced detection, and control of mirror nonlinearities, SNR approaching the quantum noise limit can be achieved. In addition, the ability to use $\log_{10}$ demodulation allows dynamic ranges to approach 80 dB. For SD-OCT, its low pass filtering properties and use of CCD result in difficulties in eliminating classical noise. While z_0 offset, time integration, and/or the use of E-O phase modulation may improve performance, it is not likely to overcome the large amount of classical noise. In addition, due to the nature of the Fourier transform process, the dynamic range will only be on the order of 40 dB. With SS-OCT, the system is also low pass but several aspects of its embodiments may improve its performance. These include z_0 offset, high intensity across the bandwidth, and fast sweep rates, in addition to the low excess noise. However, due to the nature of the Fourier transform process, the dynamic range will only be on the order of 40 dB and quantum noise is in excess of that of the other two embodiments. Low SNR and dynamic range can be tolerated in transparent tissue such as the eye, but not in highly scattering tissue. Future work is required to assess if SS-OCT can generate SNR in the range of those of TD-OCT.

REFERENCES

1. Brezinski, M. E., Tearney, G. J., Bouma, B. E., Izatt, J. A., Hee, M. R., Swanson, E. A., Southern, J. F., and Fujimoto, F. G. (1996). Optical coherence tomography for optical biopsy — Properties and demonstration of vascular pathology. *Circulation* 93(6), 1206–1213.
2. Tearney, G. J., Brezinski, M. E., Bouma, B. E., Boppart, S. A., Pitris, C., Southern, J. F., and Fujimoto, J. G. (1997). In vivo endoscopic optical biopsy with optical coherence tomography. *Science* 276, 2037–2039.
3. Brezinski, M. E. and Fujimoto, J. G. (1999). Optical coherence tomography: High-resolution imaging in nontransparent tissue. *IEEE J. Selected To. Quant. Electron.* 5(4), 1185–1192.
4. Yabushita, H., Bouma, B. E., Houser, S. L., Aretz, H. T., Jang, I.-K., Schlendorf, K. H., Kauffman, C. R., Shishkov, M., Kang, D.-H., Halpern, E. F., and Tearney, G. J. (2002). Characterization of human atherosclerosis by optical coherence tomography. *Circulation* 106(13), 1640–1645.
5. Martin, S. D., Patel, N. A., Adams, S. B., Roberts, M. J., Plummer, S., Stamper, D. L., Brezinski, M. E., and Fujimoto, J. G. (2003). New technology for assessing microstructural components of tendons and ligaments. *Int. Orthopaed.* 27(3), 184–189.
6. Huang, D., Swanson, E. A., Lin, C. P., Schuman, J. S., Stinson, W. G., Chang, W., Hee, M. R., Flotte, T., Gregory, K., Puliafito, C. A., and Fujimoto, J. G. (1991). Optical coherence tomography. *Science* 254, 1178–1181.
7. Fercher, A. F., Hitzenberger, C. K., Kamp, G., and Elzaiat, S. Y. (1995). Measurement of intraocular distances by backscattering spectral interferometry. *Opt. Commun.* 117(1–2), 43–48.
8. Wojtkowski, M., Leitgeb, R., Kowalczyk, A., Bajraszewski, T., and Fercher, A. F. (2002). In vivo human retinal imaging by Fourier domain optical coherence tomography. *J. Biomed. Opt.* 7(3), 457–463.

9. Wax, A. Yang, C. H., and Izatt, J. A. (2003). Fourier-domain low-coherence interferometry for light-scattering spectroscopy. *Opt. Lett.* 28(14), 1230–1232.

10. Yun, S. H., Tearney, G. J., Bouma, B. E., Park, B. H., and deBoer, J. F. (2003). High-speed spectral-domain optical coherence tomography at 1.3 μm wavelength. *Opt. Express* 11(26), 3598–3604.

11. Leitgeb, R. A., Drexler, W., Unterhuber, A., Hermann, B., Bajraszewski, T., Le, T., Stingl, A., and Fercher, A. F. (2004). Ultrahigh resolution Fourier domain optical coherence tomography, *Opt. Express* 12(10), 2156–2165.

12. Nassif, N. A., Cense, B., Park, B. H., Pierce, M. C., Yun, S. H., Bouma, B. E., Tearney, G. J., Chen, T. C., and deBoer, J. F. (2004). In vivo high-resolution video-rate spectral-domain optical coherence tomography of the human retina and optic nerve. *Opt. Express* 12(3), 367–376.

13. Chinn, S. R., Swanson, E. A., and Fujimoto, J. G. (1997). Optical coherence tomography using a frequency-tunable optical source. *Opt. Lett.* 22(5), 340–342.

14. Choma, M. A., Sarunic, M. V., Yang, C. H., and Izatt, J. A. (2003). Sensitivity advantage of swept source and Fourier domain optical coherence tomography. *Opt. Express* 11(18), 2183–2189.

15. Zhang, J., Nelson, J. S., and Chen, Z. P. (2005). Removal of a mirror image and enhancement of the signal-to-noise ratio in Fourier-domain optical coherence tomography using an electro-optic phase modulator. *Opt. Lett.* 30(2), 147–149.

16. Peter, A., Knauer, M., Kiesewetter, F., and Gerd, H. (2000). Optical coherence tomography by spectral radar: Improvement of signal-to-noise ratio. In *Coherence Domain Optical Methods in Biomedical Science and Clinical Applications IV.* V. T. Valery, A. I. Joseph and G. F. James, (eds.), Vol. 3915. SPIE Press, Bellingham, WA, pp. 55–59.

17. Leitgeb, R., Hitzenberger, C. K., and Fercher, A. F. (2003). Performance of Fourier domain vs. time domain optical coherence tomography. *Opt. Express* 11(8), 889–894.

18. Peter, A., Michael, W. L., Juergen, M. H., Schultz, A., Konzog, M., Kiesewetter, F., and Gerd, H. (1999). Optical coherence tomography by spectral radar: Dynamic range estimation and in-vivo measurements of skin. In *Optical and Imaging Techniques for Biomonitoring IV.* F. Marco Dal, F. Hans-Jochen, K. Neville, M. Renato, and P. Halina, (eds.), Vol. 3567. SPIE Press, Bellingham, WA, pp. 78–87.

19. Mitsui, T. (1999). Dynamic range of optical reflectometry with spectral interferometry. *Jpn. J. Appl. Phys. Part 1-Regular Papers Short Notes & Review Papers* 38(10), 6133–6137.

20. Henry C. H. and Kazarinov, R. F. (1996). Quantum noise in photonics. *Rev. Mod. Phys.* 68(3), 801–853.

21. Hodara, H. (1965). Statistics of thermal and laser radiation. *Proc. IEEE* 53, 696–704.

22. Takada, K., Himeno, A., and Yukimatsu, K. (1991). Phase-noise and shot-noise limited operations of low coherence optical-time domain reflectometry. *Appl. Phys. Lett.* 59(20), 2483–2485.

23. Hanbury-Brown, R. and Twiss, R. Q. (1956). Correlation between photons in two coherent beams of light. *Nature* 177, 27–32.

24. Rosa, C. C. and Podoleanu, A. G. (2004). Limitation of the achievable signal-to-noise ratio in optical coherence tomography due to mismatch of the balanced receiver. *Appl. Opt.* 43(25), 4802–4815.

25. Caves, C. M. (1981). Quantum-mechanical noise in an interferometer. *Phys. Rev. D* 23(8), 1693–1708.

26. Haus, H. A. (2000). *Electromagnetic Noise and Quantum Optical Measurements.* Springer Press, New York.

27. Takada, K. (1998). Noise in optical low-coherence reflectometry. *IEEE J. Quantum Electron.* 34(7), 1098–1108.

28. Meers, B. J. and Strain, K. A. (1991). Modulation, signal, and quantum noise in interferometers. *Phys. Rev. A* 44(7), 4693–4703.

29. Horowitz, P. and Hill, W. (1989). *The Art of Electronics.* Cambridge University Press, New York.

30. Yariv, A. (1997). *Optical Electronics in Modern Communications.* Oxford University Press, New York.

31. Gordon, J. P., Zeiger, H. J., and Townes, C. H. (1955). The Maser — New type of microwave amplifier, frequency standard, and spectrometer. *Phys. Rev.* 99, 1264.

32. Schrieffer, J. R. (1964). *Theory of Superconductivity.* W. A. Benjamin, New York.

33. Bachor H.-A. and Fisk, P. T. H. (1989). Quantum noise a limit in photodetection. *Appl. Phys. B* 49, 291.

34. Theuwissen, A. J. P. (1995). *Solid-State Imaging with Charge-Coupled Devices.* Kluwer Academic Publishers, Boston, MA.

35. Gerd, H. and Michael Walter, L. (1998). "Coherence Radar" and "Spectral Radar" — New tools for dermatological diagnosis. *J. Biomed. Opt.* 3(1), 21–31.

36. Wojtkowski, M., Kowalczyk, A., Leitgeb, R., and Fercher, A. F. (2002). Full range complex spectral optical coherence tomography technique in eye imaging. *Opt. Lett.* 27(16), 1415–1417.

37. Schmitt, J. and Olszak, A. (2002). High-precision shape measurement by white-light interferometry with real-time scanner error correction. *Appl. Opt.* 41, 5943.

38. Yun, S. H., Tearney, G. J., deBoer, J. F., and Bouma, B. E. (2004). Pulsed-source and swept-source spectral domain optical coherence tomography with reduced motion artifacts. *Opt. Express* 12, 5614–5624.

39. Boppart, S. A. (1999). Optical Coherence Tomography: Doctoral Thesis. MIT, Cambridge, MA.

40. Liu, B. and Brezinski, M. E. (2006). Theoretical and practical considerations on detection of time domain, Fourier domain, and swept source optical coherence tomography, (submitted).

8   POLARIZATION AND POLARIZATION SENSITIVE OCT

Chapter Contents

8.1 GENERAL

With OCT, imaging of most tissue is polarization insensitive, meaning that it is independent of the incident light polarization state. This is due to the fact that the polarization states are randomized by the relatively minimally organized structure within the tissue (i.e., depolarizes). However, when the tissue contains highly organized structures, particularly parallel linear structures, it can be sensitive to the light polarization state used for imaging. Put another way, when light is passed through the tissue, the sample is capable of generating an altered state of backreflected polarized light. Detection of these altered states is the basis of polarization sensitive optical coherence tomography (PS-OCT). This chapter will examine both single and dual detector PS-OCT approaches. Assessment of polarization through the process of second harmonic generation and phase sensitive imaging will be discussed in Chapter 9. The emphasis will be on how PS-OCT approaches yield potentially clinically relevant information. In particular, it will emphasize that for the majority of applications, imaging needs to be performed through fiber optics, data are obtained in real time, and measured changes correlate with clinically relevant markers of pathology such as appropriately stained histopathology.

8.2 POLARIZATION PROPERTIES OF TISSUE

Although the distinction is not critical, PS-OCT is often described as a contrast-enhancing technique, but it also may be looked at as a spectroscopic technique. PS-OCT helps discern biochemical composition (or the state of biomolecules) rather than more sharply defining structure within the image, more like a spectroscopic technique, and the banding patterns often present produces artificial "structure" within the image (contrast is discussed in Chapter 11). Therefore, in this text, it will be described as a spectroscopic technique.

Highly organized biomolecules in tissue that alter incident light polarization include collagen, cholesterol crystals, actin-myosin complexes, nerve fibers (myelin), and calcium hydroxyapatite (enamel and dentin). As collagen is one of the most important and most studied molecules altering polarization in tissue, its biochemical properties will be discussed in considerable detail in the next section.

Biological substances can effect the polarization of incident light through at least three separate mechanisms: birefringence, dichroism, and optical rotation. Birefringence refers to tissue whose light propagation velocities (refractive index) are dependent on the spatial orientation of the sample. Generally speaking, tissue dichroism is the filtering of one polarization state and not its orthogonal state. Finally, certain molecules due to their complex structure, such as sugars, are able to rotate the state of polarization. Birefringence appears to be the most important of the polarization-altering mechanisms and will be the most extensively discussed in this chapter.

Among the factors that can affect birefringence include molecular concentration, chemical composition of the molecule(s) (such as collagen type), form and intrinsic birefringence, molecular angle, alterations by the local environment, molecular organization, and the presence of multiple birefringent materials. Studies have been and are being performed to assess if PS-OCT can identify these factors that may be useful in identifying pathological states.

Changes in molecular type, concentration, and organization can be signs of a variety of pathologic conditions, including coronary plaques likely to progress to myocardial infarction and cartilage developing early osteoarthritis.[10–13, 38] Therefore, PS-OCT may potentially be a powerful clinical technology.

Intrinsic birefringence is a characteristic of the molecule and results from different refractive indices for light traveling parallel and perpendicular to the fiber axis. Intrinsic birefringence is due to the chemical nature of the parallel molecules. Form birefringence, on the other hand, arises because the medium around the fiber binds with a high degree of organization, which results in a second refractive index profile parallel to the fiber.[13]

The local environment can alter tissue polarization properties through mechanisms other than form birefringence. We will see that different local environments, such as lipid versus water, can actually alter the intrinsic birefringence of molecules by changing their microstructure. For example, short chain hydroxylated lipids change the bundling of collagen from quasi-pentagonal to quasi-hexagonal, which then changes intrinsic birefringence.[2,3] We will see these changes detected by PS-OCT.

8.3 COLLAGEN

Because collagen may be the most important birefringent molecule, it will be discussed here in detail to better understand how its birefringence relates to pathophysiology. Collagen is a fibrous protein that has a rod-like shape and is insoluble in water. It is the most abundant protein in the body and is excreted into the extracellular matrix to provide the structural support for many tissues.[4] Its fundamental unit is tropocollagen, a molecule with three peptide chains, in a helix, of about 1000 residues each (Figure 8.1). These collagen molecules associate and cross link to form the larger fibers of tissue (Figure 8.2).

There are a variety of collagen forms that have different chemical compositions (at least 10). Examples are type I collagen, the most abundant collagen in the body, which consists of low carbohydrate concentrations, less than 10 hydroxylysines per chain, and two types of polypeptide chains.[4] In contrast, type III collagen contains low carbohydrate concentrations, high hydroxyproline, and glycine. As an example of why different collagen types have different optical properties, fibers with high proline-rich chains are more widely open with respect to the helix configuration, which alters birefringence.

Type I collagen is structurally very strong and is found in bone, tendon, dentine, and atherosclerotic arteries. The most important location of type II collagen is hyaline cartilage, which is present in synovial joints such as the knees, hips, and shoulders. Type II is also found in parts of the eye. Type III collagen is similar in

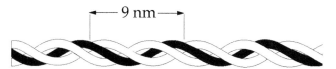

Figure 8.1 The fundamental unit of collagen. Tropocollagen is the fundamental unit of collagen consisting of three peptide chains in a helix.

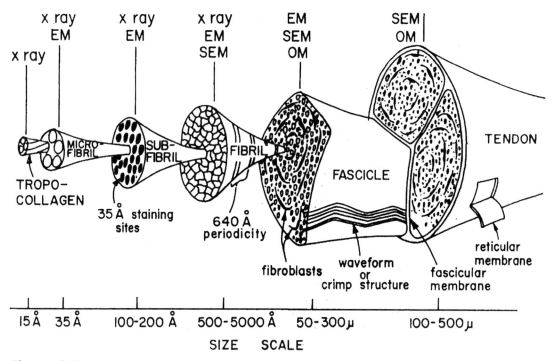

Figure 8.2 Organization of a tendon. A tendon consists primarily of collagen, but at multiple levels of organization. Courtesy of Kastelic, J. A., et al. (1978). *Connective Tissue Res. 6*, 11-23.

distribution to type I, except it has only a minimal presence in bone and tendon. It is also present in wound healing, such as in the skin or in the formation of plaque. Collagen type III is weakly organized so that when it replaces collagen type I in the reparative process, the tissue becomes structurally weak. Type IV collagen is found in the basement membrane of a wide range of tissue types. The ability to identify collagen types may be of considerable diagnostic value. For example, in a lipid-filled plaque, the replacement of the structurally strong collagen type I with the weaker type III, along with decreased overall concentrations, may be an indicator of plaque instability.

The orientation of collagen can also vary dramatically. In tendons, which consist predominately of collagen type I, the collagen bundles are arranged longitudinally.

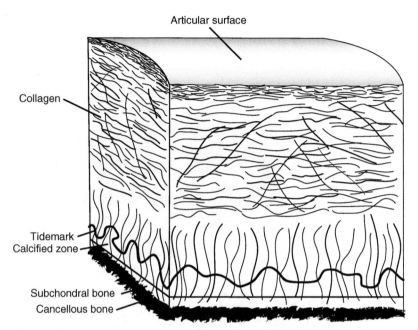

Articular surface

Collagen

Tidemark
Calcified zone

Subchondral bone
Cancellous bone

Figure 8.3 Knee cartilage organization. This slide illustrates the organization of knee articular cartilage, particularly with respect to collagen organization. It can be seen that strands near the surface run parallel and likely represent the source of its polarization sensitive. This is diagnostically useful as changes in surface collagen usually represent the sign of early osteoarthritis. Drawn by Karen Aneshansley.

Articular cartilage of the knee consists almost exclusively of type II collagen. The trilaminar orientation of the cartilage is shown in Figure 8.3.[5] The most superficial zone, the lamina splendens, accounts for 5–10% of the total thickness of the cartilage. Here the collagen fibrils are oriented tangentially to the articular surface. The transitional zone contains tangential fibers which blend with radial fibers that extend deep into the cartilage, near the bone interface. Another type of cartilage is meniscal cartilage, which is present in a ring shape in the knee joint (Figure 8.4).[6] Menisci consist predominately of type I collagen with a lesser amount of collagen types III, V, and VI present. Collagen fibers located in different areas of the meniscus will have a different orientation. The fibers located in the periphery are circumferentially oriented while those in the central region extend radially. We will see that PS-OCT detected changes in collagen orientation and concentration are signs of early osteoarthritis.

To validate the ability of PS-OCT to assess birefringent molecules, well-recognized gold standards are essential for correlation with OCT data. This generally means well-registered, histological or biochemical assessments. Unfortunately, too often simple hematoxylin and eosin (H&E) or Masson's trichrome blue stains are used to "quantify collagen," for which neither stain is appropriate. With collagen, appropriate validation techniques include picrosirius staining with polarization

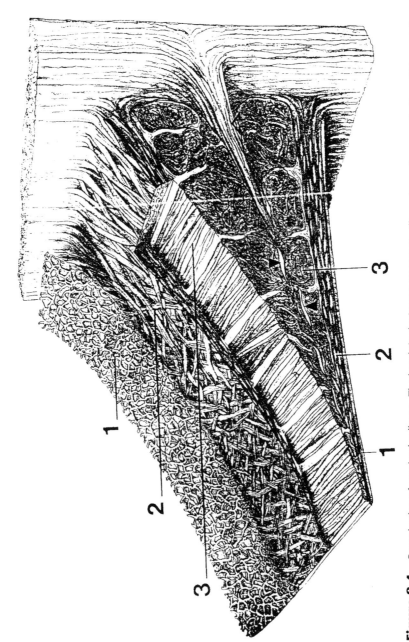

Figure 8.4 Organization of meniscal collagen. The bands in the image represent collagen in varying organizations. This needs to be taken into account when assessing this particular structure. Courtesy of Petersen, W. and Tillmann, B. (1998). *Anat. Embryol.* 197, 317-324.

microscopy, scanning electron microscopy, high-performance liquid chromatography, thin layer chromatography, immunohistochemistry, or the use of collagenase with a variety of imaging techniques.

Picrosirius staining is particularly straightforward to quantify and since it will be used in figures throughout the text, a brief description is provided. The stain binds collagen, producing increased linear birefringence (by a factor of 6×), but linear dichroism may also contribute if the appropriate suspension medium is not used. Picrosirius red staining has been shown to be superior to conventional polarization microscopy for the identification of organized collagen.[7,8] Organized collagen appears as a bright region via picrosirius staining as seen in Figure 8.5 in this human coronary artery, when compared with relatively diffuse staining by trichrome blue. Its color will be greener for thin fibers (type III collagen, 0.8 μm) and yellowish orange for thick fibers (type I collagen, 1.6–2.4 μm).[9] While picrosirius staining is an excellent technique for assessing collagen organization, those unfamiliar with the technique should solicit investigators with expertise because, for example, if the dye is allowed to become dehydrated it can be explosive.

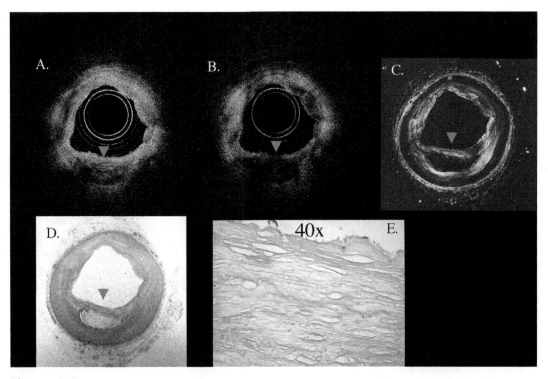

Figure 8.5 Picrosirus staining for identifying organized collagen. In this figure, polarization sensitivity of a human atherosclerotic coronary artery imaged with single detector PS-OCT is performed (A and B), as well as picrosirus staining (C) and trichrome blue (D and E). It can be seen that organized collagen is detected by OCT and picrosirus but not trichome blue. PS-OCT images were taken with the reference arm controllers in two different positions. Courtesy of Giattina, S.D. et al. (2006). *Intern. J. Cardiol.* 107, 400-409.

8.4   SINGLE DETECTOR PS-OCT

With the single detector approach, the lone detector, which is not sensitive to the polarization state of the incident light, is used (for now ignoring the use of dual balanced detection for noise reduction). Current single detector systems are attractive because they are simple in design, perform polarization assessments in real time from the image, and assessments can be performed through fiber optics with real-time eliminating of bending or compression artifacts (asymmetric stress). Furthermore, results are directly measured from the images without the need for complex data analysis. With SDPS-OCT, when birefringence is high, as will be seen, it can be measured directly from the banding pattern. When birefringence is low, it is assessed by changing the polarization in the reference arm with controllers.

Why SDPS-OCT minimizes artifacts when catheter bending/compression occurs is illustrated in Figures 8.6, 8.7, and 8.8. In Figure 8.6, the top images are

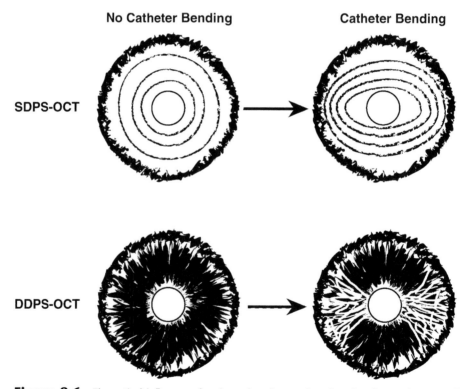

Figure 8.6   Theoretical influence of catheter bending on imaging. Bending catheters and endoscopes changes the polarization state of the light on the sample and returning from the sample arm. Since single detector approaches measure changes in birefringence as a function of depth (banding) rather than absolute values, banding shifts but spacing remains relatively constant. However, the dual detector approach measures absolute values of backscattering so that significant artifacts can result.

No Catheter Bending

Catheter Bending

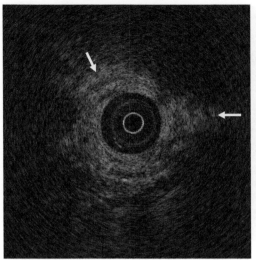
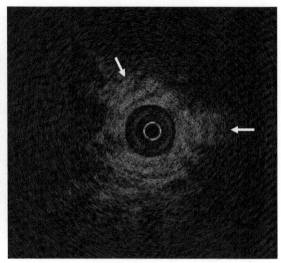

Figure 8.7 Single detector PS-OCT with and without catheter bending. Imaging is performed with a catheter placed in a tendon with the single detector approach. Little change is noted in the spacing of the banding pattern (unpublished).

simulations with a single detector system under the two conditions. In the single detector images, catheter bending results in a shift in the banding pattern, but the width of the bands remains relatively constant. This is due to the fact that relative polarization states are measured and therefore the birefringence measurements remain relatively constant. With the dual detector approach, when the catheter is bent, the incident light on the tissue changes its polarization depending on which portion of the artery is examined. This results in erroneous interpretation of birefringence as absolute values are being measured. In Figure 8.7, SDPS-OCT images of an ACL are shown with and without catheter bending. The areas are brightest when maximum differentiation between the detectors is seen and lowest when the intensities equalize. Catheter bending results in shifting of the banding pattern, but distances between the bands remain the same (arrow). With DDPS-OCT, Figure 8.8, catheter bending results in distinct backreflection patterns from baseline as absolute values vary around the catheter.

With dual detector PS-OCT (DDPS-OCT), the two orthogonal polarization states of the exiting light field from the interferometer are directed on to separate detectors to reconstruct the intensity, phase retardation, and/or optic axis rotation mapping. However, DDP-OCT systems may be associated with significant limitations. First, most systems are free space, which is generally not applicable to clinical medicine, and recent fiber-based techniques require complicated approaches to prevent artifacts from fiber bending or compression. Second, while the Jones and Stokes vectors are typically measured, it is actually the matrices of the tissue that need to be known. Third, acquiring polarization assessments is currently not done in real time but often

No Catheter Bending   **Catheter Bending**

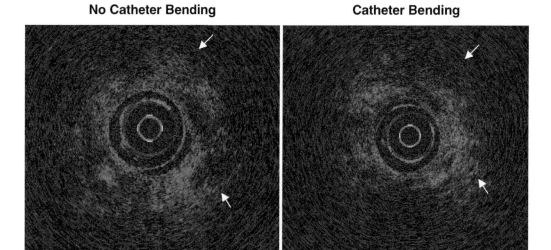

Figure 8.8 Dual detector PS-OCT with and without catheter bending. In this figure, imaging is now performed with a dual detector approach (measuring absolute phase). Substantial distortion of the image is noted with catheter bending (unpublished).

requires complex analysis of significant numbers of images. Fourth, the systems contain extra beam splitters and filters that reduce power. Fifth, retarders used in the systems such as a quarter wave plate (QWP) are designed for nearly monochromatic light, so that artifacts may be introduced with broadband light. Finally, and probably most important, while systems have been demonstrated to determine vectors, matrices, and the optical axis in phantoms and some tissue using relatively complex analysis, it is unclear how this information can be utilized to practically diagnose pathophysiology, particularly when the tissue has molecules of different polarization properties and the optical axis changes with depth. Therefore, SDPS-OCT is the focus of our group and will be discussed in this section. DDPS-OCT will be discussed in Section 8.5.

With SDPS-OCT, a standard TD-OCT Michelson interferometer is used. In Figure 8.9 two OCT images of a biceps tendon along with the corresponding picrosirius stained section demonstrate SDPS-OCT.[10] In this image of the birefringent tissue, a banding pattern is seen while the bright area in the picrosirius corresponds to organized collagen. This banding pattern occurs because the backreflected light from within the sample at different depths (within one band cycle) has different polarization states. The light is changing its polarization state as a function of depth due to the tissue birefringence. Bright bands result when the backreflected light is in a similar polarization state as the reference arm, whereas dark bands occur when the polarization state in each arm is orthogonal. With this embodiment, the polarization state in the reference arm, and therefore the pattern imaged in the sample, is changed with polarization controllers, which may include manual paddles, galvanometer

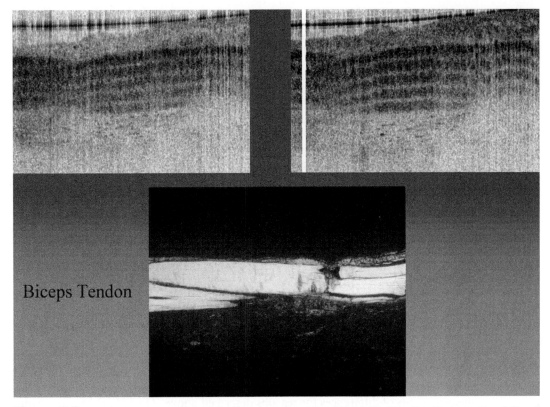

Figure 8.9 Birefringence assessment of a biceps tendon using SDPS-OCT. This slide illustrates the ability of SDPS-OCT to identify organized collagen in biceps tendon. It can be seen that the high collagen concentration is represented by the banding pattern that is noted at the two different controller positions in the reference arm. Courtesy of Martin, S.D., et al. (2003). *Int. Orthoped. 27*, 184-189.

driven paddles, or fiber squeezers. This changes the polarization pattern observed within the sample. Picrosirius confirms the presence of organized collagen (bright areas).

8.4.1 SDPS-OCT Theoretical Model

In this section, we examine theoretical modeling associated with SDPS-OCT imaging.[11] Most studies model a single slab of tissue with a constant birefringence profile. Here, modeling is achieved primarily through the use of theoretical modeling of a serially layered system of alternating birefringence. These data illustrate the complexities of interpreting polarization data of biological tissue but offer the potential to use these complexities to increase our understanding of tissue composition.

A mathematical description of an SD-OCT system is provided, by which the optical signal evolution can be evaluated. It should be noted that what is called SD-PS-OCT can use, and likely should use, polarization insensitive, dual balanced detection to remove photon excess noise. While it uses two detectors, unlike DDPD-OCT, the detectors are not measuring orthogonal states. In general, because of excess noise and the sources typically used, SDPS-OCT should use two detectors while DDPS-OCT should use four to remove excess noise (Chapter 7). Improved terminology for the embodiments is likely necessary.

A 2×2 nonpolarization-sensitive coupler is used with a Michelson interferometer for simplicity. In the reference arm, a linear moving mirror is used to change the optical group delay but essentially the same derivation would hold if we used, for example, with the grating based delay line. A polarization controller adjusts the polarization state in the reference arm as described above. As only A-scans are used in the modeling, the lateral scan in the sample arm is ignored. A temporally low coherent source around the central angular frequency ω_0 with full width half maximum (FWHM) bandwidth Δ_λ is employed. Light from the source can be either polarized, partially polarized, or unpolarized. For simplification of the theoretical analysis, the output of light source is assumed to be linearly polarized for which any polarization mode dispersion (PMD) in the fibers is compensated. In the theoretical formalization of the single detector system, changes induced by the single mode optical fibers can be treated as if they were in free space, since we are looking at birefringence *changes* induced by altering the reference arm polarization in a controlled manner and are not measuring absolute polarization values. Therefore, fiber-induced changes can be ignored due to the measurements of differences for reasons described above. The light field perturbation at any spot in the system can be fully characterized by two orthogonal linear oscillated fields with certain retardations in terms of a Jones vector (if we were including partially polarized or unpolarized light, we would prefer to use the Stokes vector). But for an instantaneous interval, the Jones vector fully characterizes the amplitude and phase and therefore polarization. The Jones vector can be written as

$$\vec{E}_0(t) = \begin{bmatrix} E_{x0} & (t) \\ E_{y0} & (t) \end{bmatrix} = \begin{bmatrix} A_{x0}(t) \exp\{ j[\phi_{x0}(t) - 2\pi\omega_0 t] \} \\ A_{y0}(t) \exp\{ j[\phi_{y0}(t) - 2\pi\omega_0 t] \} \end{bmatrix} \tag{8.1}$$

The two elements in the above array, $E_{x0}(t)$ and $E_{y0}(t)$, represent the field perturbation of incident light in the two orthogonal directions, whose amplitudes $A_{x0}(t)$ and $A_{y0}(t)$, and phases $\phi_{x0}(t)$ and $\phi_{y0}(t)$, are real and time dependent. Again, the light beam could be either totally polarized, partially polarized, or unpolarized in time average, but totally polarized instantaneously. This is dealt with in more detail in Appendix 8-1. Since OCT is based on white light interferometry, theoretically OCT can track the changes of the polarization state if the measurement is faster than the variation. We will see this with the coherence matrix in Appendix 8-1.

Assuming the linear polarized light beam from the light source is oriented 45 degrees to the x axis, the incident beam on the beam splitter can be characterized as (Chapter 5):

$$E_{x0}(t) = \frac{\sqrt{2}}{2} \int_0^\infty a(\omega) \exp[\phi_0(\omega) - j2\pi\omega t] d\omega \tag{8.2}$$

$$E_{y0}(t) = \frac{\sqrt{2}}{2} \int_0^\infty a(\omega) \exp[\phi_0(\omega) - j2\pi\omega t] d\omega \tag{8.3}$$

Both components have the same amplitude spectrum $a(\omega)$, initial phase $\phi_0(\omega)$, and intensity:

$$I_0 = \left\langle \vec{E}_0(t) \cdot \vec{E}_0^*(t) \right\rangle = I_{x0} + I_{y0} = 2 \int_{-\infty}^\infty G(\omega) d\omega \tag{8.4}$$

The intensity is the superposition of the intensity of the two orthogonal polarization states, E_{x0} and E_{y0}, with the total power spectrum $G(\omega)$. In Eq. 8.4, the angle brackets represent the mean value in unit time or time average.

In the reference arm, the light beam passes through a polarization controller with adjustable retardation and optical axis, which can be adjusted to all polarization states. With the reasonable assumption that there is not attenuation in the mirror, the return beam from the reference arm at the beam splitter is

$$\overline{E}_R(t) = \begin{bmatrix} E_{x\_R}(t) \\ E_{y\_R}(t) \end{bmatrix} =$$

$$\int_0^\infty \frac{\sqrt{2}}{2} \begin{bmatrix} \cos\theta_R & \sin\theta_R \\ -\sin\theta_R & \cos\theta_R \end{bmatrix} \begin{bmatrix} 2v_{x\_R}(\omega) \exp(-j2\pi\omega t) \\ 2v_{y\_R}(\omega) \exp(-j2\pi\omega t) \end{bmatrix} d\omega \tag{8.5}$$

$$v_{x\_R}(\omega) = \frac{1}{4} a(\omega) \exp\{j[\phi_0(\omega) + \phi_R(\omega, z_R)] + \delta\}, \ \omega \geq 0 \tag{8.6}$$

$$v_{y\_R}(\omega) = \frac{1}{4} a(\omega) \exp\{j[\phi_0(\omega) + \phi_R(\omega, z_R)] + \delta\}, \ \omega \geq 0 \tag{8.7}$$

The complex functions $v_{x\_R}(\omega)$ and $v_{y\_R}(\omega)$ are defined as the analytical signals representing the amplitude and phase (Appendix 5-1). The intensity of the returned reference beam approaching the detector can also be determined as a time average:

$$I_R = \left\langle E_R(t) E_R^*(t) \right\rangle = \frac{1}{2} \int_{-\infty}^\infty G(\omega) d\omega = \frac{I_0}{4} \tag{8.8}$$

As in previous chapters, in the sample arm, the origin of coordinates O_S is set up at the tissue surface. The origin of coordinates in the reference arm O_R is set up at a point, from which the optical path to the coupler is matching with the optical path between O_S and the coupler in the sample arm. The mirror position is off the O_R with z_R. $\phi_R(\omega, z_R)$ representing the phase delay induced in reference arm. δ is the adjustable retardation between both of the orthogonal states induced by the polarization controller.

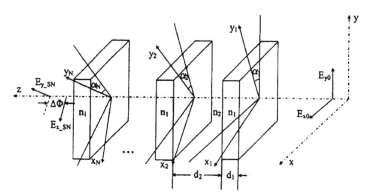

Figure 8.10   Schematic of simulation used to model SDPS-OCT. In the multilayer structure, the incident light beam propagates along z direction and its two orthogonal polarization components E_{x0} and E_{y0} are coincident with x and y axis respectively. The three blocks, which are parallel to the x-y plane, represent the 1st, 2nd and Nth layer of material 1 with d_1 thickness and average group index n_1. Material 2 is assumed to fill the gap between the each couple of blocks with d_2 thickness and average group index n_2. x_1 and y_1 represent the fast and slow axis of the 1st layer, which rotate α_1 angle respect to the incident axes. The axes of 2nd layer, x_2 and y_2, rotate α_2 angle respect to the axes of 1st layer. Analogously, the axes of Nth layer, rotate α_N respect to its former layer. The orthogonal components of output beam, E_{y_SN} and E_{y_SN}, exit along the axes of the last layer with retardation $\Delta\phi$. The denotation of the first layer is also applicable for the single slab model. Courtesy of Liu, B. (2005). *J. Opt. Soc. Am. A.* 22, 262-271.

As stated, most previous models examining polarization effects of OCT treat the sample as a single slab.[8] Here it is modeled as a multilayered structure with varying optic axes and effective indices for both orthogonal states (Figure 8.10). The incident light beam propagates along the z direction and its two orthogonal polarization components E_{x0} and E_{y0} are coincident with x and y axes, respectively. The three blocks, which are parallel to the x–y plane, represent the first, second, and Nth layer of material 1 with d_1 thickness and average group index n_1. Material 2 is assumed to fill the gap between each birefringent layer with d_2 thickness and average group index n_2. x_1 and y_1 represent the fast and slow axis of the first layer, which rotate the α_1 angle with respect to the incident axes. The axes of the second layer, x_2 and y_2, rotate the α_2 angle with respect to the axes of the first layer. Analogously, the axes of the Nth layer rotate α_N with respect to its former layer. The orthogonal components of the output beam, E_{y_SN} and E_{y_SN}, exit along the axes of the last layer with retardation $\Delta\phi$.

The layered material is assumed as birefringent with average group index n_1 and intra-layer material is assumed as non-birefringent material with a group index of n_2. The rotation angle of the optic axis in the first layer is α_1 (referring to the incident axis). Thus the incident light beam on the first layer can be described as two orthogonal modes that propagate along its optic axes x_1 and y_1:

$$\vec{E}_{S1}^i(t) = \begin{bmatrix} E_{x\_S1}^i(t) \\ E_{y\_S1}^i(t) \end{bmatrix} = \frac{1}{\sqrt{2}} \begin{bmatrix} \cos\alpha_1 & \sin\alpha_1 \\ -\sin\alpha_1 & \cos\alpha_1 \end{bmatrix} \begin{bmatrix} E_{x0}(t) \\ E_{y0}(t) \end{bmatrix} \qquad (8.9)$$

Within the layer, the orthogonal propagating modes experience different phase delays, $\phi_{x\_S}(\omega, z)$ and $\phi_{y\_S}(\omega, z)$. The exiting light from first layer is

$$
\vec{E}^o_{S1}(t) = \begin{bmatrix} E^0_{x\_S1}(t) \\ E^o_{y\_S1}(t) \end{bmatrix} = \begin{bmatrix} E^i_{x\_S1}(t, \phi_0 + \phi_{x\_S}) \\ E^i_{y\_S1}(t, \phi_0 + \phi_{y\_S}) \end{bmatrix}
$$

$$
= \frac{1}{\sqrt{2}} \begin{bmatrix} 2(\cos\alpha_1 + \sin\alpha_1) \int_0^\infty v_{x\_S}(\omega) \exp(-j2\pi\omega t) d\omega \\ 2(\cos\alpha_1 - \sin\alpha_1) \int_0^\infty v_{y\_S}(\omega) \exp(-j2\pi\omega t) d\omega \end{bmatrix} \tag{8.10}
$$

$$
v_{x\_S}(\omega) = \frac{1}{4} p_S(z) a(\omega) \exp\{j[\phi_0(\omega) + \phi_{x\_S}(\omega, z)]\}, \ \omega \geq 0 \tag{8.11}
$$

$$
v_{y\_S}(\omega) = \frac{1}{4} p_S(z) a(\omega) \exp\{j[\phi_0(\omega) + \phi_{y\_S}(\omega, z)]\}, \ \omega \geq 0 \tag{8.12}
$$

$p_S(z)$ is the amplitude backscattering coefficient distributions within the sample. z is measured from O_S. The complex functions $v_{x\_S}(\omega)$ and $v_{v\_S}(\omega)$ are defined as the analytical signals.

Analogous to Eq. 8.8 and 8.9, backscattering light from layer 2 through the Nth layer can be described as

$$
\vec{E}^i_{s2}(t) = \begin{bmatrix} E^i_{x\_S2}(t) \\ E^i_{y\_S2}(t) \end{bmatrix} = \begin{bmatrix} \cos\alpha_2 & \sin\alpha_2 \\ -\sin\alpha_2 & \cos\alpha_2 \end{bmatrix} \vec{E}^o_{s1}(t) \tag{8.13}
$$

$$
\vec{E}^i_{s2}(t) = \begin{bmatrix} E^o_{x\_S2}(t) \\ E^o_{y\_S2}(t) \end{bmatrix} = \begin{bmatrix} E^i_{x\_S2}(t, \phi_0 + \phi_{x\_S}) \\ E^i_{y\_S2}(t, \phi_0 + \phi_{y\_S}) \end{bmatrix} \tag{8.14}
$$

$$
\vec{E}^i_{SN}(t) = \begin{bmatrix} E^i_{x\_SN}(t) \\ E^i_{y\_SN}(t) \end{bmatrix} = \begin{bmatrix} \cos\alpha_N & \sin\alpha_N \\ -\sin\alpha_N & \cos\alpha_N \end{bmatrix} \vec{E}^o_{S(N-1)}(t) \tag{8.15}
$$

$$
\vec{E}^o_{SN}(t) = \begin{bmatrix} E^o_{x\_SN}(t) \\ E^o_{y\_SN}(t) \end{bmatrix} = \begin{bmatrix} E^i_{x\_SN}(t, \phi_0 + \phi_{x\_S}) \\ E^i_{y\_SN}(t, \phi_0 + \phi_{y\_S}) \end{bmatrix} \tag{8.16}
$$

Eventually, the backscattering light from a certain spot z within the sample can be described as two orthogonal polarization states oscillating along the optical axis of the last layer (first layer for double pass) with varying relative phases. Defining the axes of the last layer oriented as θ_S to x axis, at the beam splitter, the return beam from the sample arm would be

$$\vec{E}_S(t) = \begin{bmatrix} E_{x\_S}(t) \\ E_{y\_S}(t) \end{bmatrix} = \frac{\sqrt{2}}{2} \begin{bmatrix} \cos\theta_S & \sin\theta_S \\ -\sin\theta_S & \cos\theta_S \end{bmatrix} \vec{E}_{SN}^o(t)$$

$$= \frac{\sqrt{2}}{2} \begin{bmatrix} \cos\theta_s & \sin\theta_s \\ -\sin\theta_s & \cos\theta_s \end{bmatrix} \begin{bmatrix} 2\int_0^\infty v_{x\_S}(\omega)\exp(-j2\pi\omega t)d\omega \\ 2\int_0^\infty v_{y\_S}(\omega)\exp(-j2\pi\omega t)d\omega \end{bmatrix} \tag{8.17}$$

The amplitude backscattering coefficient distributions $p_S(z)$ and the depth-integrated optic axis rotation are encoded in Eq. 8.15 as complicated modulation functions $M_{x\_S}(z)$ and $M_{y\_S}(z)$. Equations 8.11 and 8.12 turn out to be

$$v_{x\_S}(\omega) = \frac{1}{4}M_{x\_S}(z)a(\omega)\exp\{j[\phi_0(\omega) + \phi_{x\_S}(\omega,z)]\}, \ \omega \geq 0$$

$$v_{y\_S}(\omega) = \frac{1}{4}M_{y\_S}(z)a(\omega)\exp\{j[\phi_0(\omega) + \phi_{y\_S}(\omega,z)]\}, \ \omega \geq 0 \tag{8.18}$$

The perturbation of light field at the beam splitter, contributed by the return sample beam, should be superimposed from all spots of a longitudinal line within the sample as

$$\vec{E}_s(t) = \int_0^z \begin{bmatrix} E_{x\_S}(t,z') \\ E_{y\_S}(t,z') \end{bmatrix} dz' \tag{8.19}$$

The total intensity of the backscattering light is

$$I_S = \frac{I_0}{8} \int_0^z \left[M_{x\_S}^2(z') + M_{y\_S}^2(z') \right] dz' \tag{8.20}$$

Combined with the light field from the reference arm, the total perturbation at the beam splitter is

$$\vec{E}_D(t) = \begin{bmatrix} E_{x\_D}(t) \\ E_{y\_D}(t) \end{bmatrix} = \vec{E}_S(t) + \vec{E}_R(t)$$

$$= \int_0^z \begin{bmatrix} E_{x\_S}(t) + E_{x\_R}(t) \\ E_{y\_S}(t) + E_{y\_R}(t) \end{bmatrix} dz' \tag{8.21}$$

The intensity of the light field, to which the photocurrent of the detector is proportional, is as

$$I_D = \left\langle \vec{E}_D(t) \cdot \vec{E}_D^*(t) \right\rangle$$

$$= \left\langle E_{x\_D}(t)E_{x\_D}^*(t) \right\rangle + \left\langle E_{y\_D}(t)E_{y\_D}^*(t) \right\rangle \tag{8.22}$$

Substituting Eqs. 8.4 to 8.13 and Eq. 8.14, the further derivation is

$$I_D = I_R + I_S + \int_0^z \psi dz' \tag{8.23}$$

$$
\begin{aligned}
\psi = {} & \frac{1}{2}\cos(\theta_S - \theta_R)M_{x\_S}(z)\text{Re}\left\{\int_0^\infty G(\omega)\exp[j(\phi_{x\_S} - \phi_{x\_R})]d\omega\right\} \\
& + \frac{1}{2}\cos(\theta_S - \theta_R)M_{y\_S}(z)\text{Re}\left\{\int_0^\infty G(\omega)\exp[j(\phi_{y\_S} - \phi_{y\_R})]d\omega\right\} \\
& - \frac{1}{2}\sin(\theta_S - \theta_R)M_{x\_S}(z)\text{Re}\left\{\int_0^\infty G(\omega)\exp[j(\phi_{x\_S} - \phi_{y\_R})]d\omega\right\} \\
& + \frac{1}{2}\sin(\theta_S - \theta_R)M_{y\_S}(z)\text{Re}\left\{\int_0^\infty G(\omega)\exp[j(\phi_{y\_S} - \phi_{x\_R})]d\omega\right\} \tag{8.24}
\end{aligned}
$$

The first two items on the right side of Eq. 8.23 are constant and correspond to the independent contribution of the reference and sample arms. The third item represents the interference signal, in which the diagnostic information $p_S(z)$ and optic axes rotation are encoded. Since light is propagating through the birefringent sample, different propagating states experience different phase delays, plus the rotation of the optical axis and different backscattering coefficients to the two orthogonal states. The interference signals now consist of a beat with polarization-induced effects. The detection scheme determines the measurement of the components in the signal. Since most sources used with OCT generate Gaussian spectral distributions, the power spectrum can be simplified to

$$G(\omega) = \frac{I_0}{4} \times \frac{1}{\sigma_\omega\sqrt{2\pi}}\left\{\exp\left[-\frac{(\omega - \omega_0)^2}{2\sigma_\omega^2}\right] + \exp\left[-\frac{(\omega + \omega_0)^2}{2\sigma_\omega^2}\right]\right\} \tag{8.25}$$

The derived power spectrum includes the negative frequency components and is an even function. σ_ω represents its standard deviation bandwidth. Assuming no dispersion exists in this OCT system, the refractive index in the sample becomes frequency independent as a function of $n_S(\omega_0)$. Equations 8.4–23 and 8.4–25 become:

$$
\begin{aligned}
\psi \approx {} & \frac{I_0}{8}\exp\left[-\frac{4\ln 2(n_s(\omega_0)z - z_R)^2}{\Delta_l^2}\right] \\
& \left\{\cos(\theta_S - \theta_R)M_{x\_S}(z)\cos\delta\cos\left[4\pi k_0(n_S(\omega_0)z - z_R + \frac{\Delta n_S(\omega_0)}{2}z)\right]\right. \\
& + \cos(\theta_S - \theta_R)M_{y\_S}(z)\cos\left[4\pi k_0(n_S(\omega_0)z - z_R - \frac{\Delta n_S(\omega_0)}{2}z)\right] \\
& - \sin(\theta_S - \theta_R)M_{x\_S}(z)\cos\left[4\pi k_0(n_S(\omega_0)z - z_R + \frac{\Delta n_S(\omega_0)}{2}z)\right] \\
& \left. + \sin(\theta_S - \theta_R)M_{y\_S}(z)\cos\delta\cos\left[4\pi k_0(n_S(\omega_0)z - z_R - \frac{\Delta n_S(\omega_0)}{2}z)\right]\right\} \tag{8.26}
\end{aligned}
$$

where k_0 is known as the central wave number in free space. Δ_l represents the FWHM coherence length of the light source (Chapter 5):

$$\Delta_l = \frac{2\ln 2\lambda_0^2}{\pi\Delta_\lambda} \tag{8.27}$$

To simplify the description of this state, two new z-dependent modulation functions $M_x(z)$ and $M_y(z)$ composed of all the diagnostic information are defined:

$$M_x(z) = \frac{1}{2}[\cos\delta\cos(\theta_S - \theta_R)M_{x\_S}(z) - \sin(\theta_S - \theta_R)M_{x\_S}(z)];$$

$$M_y(z) = \frac{1}{2}[\cos(\theta_S - \theta_R)M_{y\_S}(z) + \cos\delta\sin(\theta_S - \theta_R)M_{y\_S}(z)] \tag{8.28}$$

So Eq. 8.23 can be rewritten with a convolution expression, using wave number k instead of the angular frequency ω and Eq. 8.4–28.

$$I_D(z_R) = I_R + I_S$$

$$+ \frac{I_0}{4}\left\{2\cos[2\pi k_0\Delta n_S(\omega_0)z_R]M_y\left(\frac{z_R}{n_S(k_0)}\right) + \left[M_x\left(\frac{z_R}{n_S(k_0)}\right) - M_y\left(\frac{z_R}{n_S(k_0)}\right)\right]\right\}$$

$$\otimes\left\{\cos[4\pi k_0 z_R]\exp\left[-\frac{4\ln 2z_R^2}{\Delta_l^2}\right]\right\} \tag{8.29}$$

Equation 8.29 fully describes the OCT signal at the exit of the interferometer when imaging through a layered birefringent sample. It shows PS-OCT can be treated as a linear system with a system transfer function of (Chapter 5):

$$h(z_R) = \cos[4\pi k_0 z_R]\exp\left[-\frac{4\ln 2z_R^2}{\Delta_l^2}\right] \tag{8.30}$$

Usually Δ_l is determined as the axial resolution of the OCT system. It is important to note that the depth ranging is refractive index dependent in OCT, which is introduced. For TD-OCT, by scanning the mirror in the reference with constant velocity V_R, the designated PS-OCT signal can be read out time-sequentially:

$$i_D(t) = \alpha\left\{l_R + I_s + \frac{I_0}{4}\left\{2\cos[2\pi k_0\Delta n_S(\omega_0)V_Rt]M_y\left(\frac{V_Rt}{n_S(k_0)}\right)\right.\right.$$

$$\left.\left.+\left[M_x\left(\frac{V_Rt}{n_S(k_0)}\right) - M_y\left(\frac{V_Rt}{n_S(k_0)}\right)\right]\right\} \otimes\left\{\cos[4\pi k_0 V_Rt]\exp\left[-\frac{4\ln 2(V_Rt)^2}{\Delta_l^2}\right]\right\}\right\} \tag{8.31}$$

α represents a constant related to the performance of the detection electronics. From the detected OCT signal, a comprehensive mapping can be reconstructed that is representative of the backscattering amplitude and polarization state.

8.4.2 Simulations with Layered Birefringent Phantoms

Based on the theoretical model derived above, mathematical simulations on a one-dimensional phantom were carried out.[11] The program was coded with Matlab. Multiple factors or parameters were varied in the calculation to establish how the A-scan signals are affected. The simulated phantom is a layered structure composed by a birefringent and moderately scattering material 1 and a non-birefringent minimally scattering material 2. Material 1 has 9 layers with d_1 thickness. Its average group index, n_1, is varied in different simulations. The value of the differential group index for two orthogonal propagating modes is also varied. Material 2 has 8 layers with d_2 thickness, whose average group index n_2 is also defined. During the simulation, the scattering extinction coefficients of materials 1 and 2 are defined as 25 cm^{-1} and 0.625 cm^1, respectively. It is also designated that the surface and interface within the target has a higher backreflection. These studies are designed to better clarify how tissue birefringence effects OCT imaging.

8.4.3 Varying Differential Group Index (Layer Birefringence)

In this simulation, the differential group index is varied and is shown in Figure 8.11. In material 1, the average group index was chosen as 1.54. Scattering coefficients for both layers are described above. The value of the differential group index for the two orthogonal propagating modes was then varied in Figure 8.11, such that (a) is 0.001, (b) is 0.005, (c) is 0.01, (d) is 0.015, and (e) is 0.020. For this simulation, the widths of birefringent and non-birefringent materials were kept constant at 500 μm, along with the fixed extinction coefficients.

Each graph in Figure 8.11 includes two decaying A-scan profiles as the depth increases. The dashed curve represents a simulated A-scan signal assuming there is no birefringence in the phantom and purely exponential decay. It is provided for comparison with solid lines that show the simulated A-scans within the birefringent phantom. A minimum signal shows up within the layers of material 2 because it is assumed that minimum backreflection or scattering exists. Banding patterns within backscattering regions, due to the retardation of polarization states, are superimposed over the exponential decay. The most striking feature is how oscillations of intensity within layers, which vary with tissue birefringence, appear to produce multiple layers (false structure) where none actually exists in the model. Troughs and peaks are noted that purely relate to the mismatch and not other structural features of the phantom. It is also observed that the contrast of the banding pattern varies with the mismatch. As will be discussed, this modeling demonstrates the complications associated with interpreting PS-OCT data.

8.4.4 Different Intra-Layer Medium

In this simulation, the effect of changing the refractive index mismatch between the birefringent layer and the surrounding layer is examined, maintaining a constant birefringence within layer 1. The differential group index or birefringence is 0.001. It is assumed that the structure of the birefringent layer is not changed, an

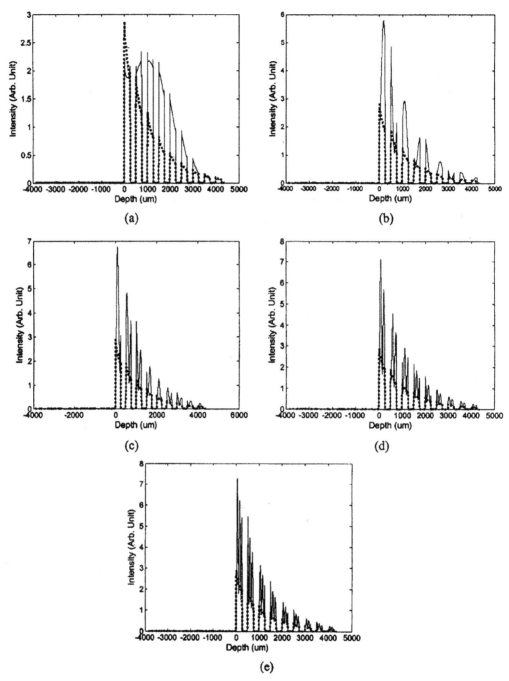

Figure 8.11 Simulated A-scan profiles of an alternating layered phantom, with varying differential group index. Courtesy of Liu, B. (2005). *J. Opt. Soc. Am. A.* 22, 262–271.

assumption which will be discussed below. Each layer is of equal width. The refractive index of material 2 is chosen as 1.34, 1.42, and 1.47, which are in the range of biological materials, associated to graphs a–b in Figure 8.12. The refractive index of material 1 was constant as 1.54. Less reflection from the interface will be seen in the OCT signal with less refractive index mismatch. From graphs a–c, the signal decays quicker when the refractive index mismatching decreases and a higher refractive index results in the least backreflection A-scan profile. Interestingly, the fast decay in this simulation makes the layers discernible. There are minimal gross oscillations of the bands as compared to Figure 8.10 due to the group refractive index mismatch. It is assumed that the different mediums do not alter the collagen.

8.4.5 Varying Layer Concentration

In this simulation, the relative width of material 1 to material 2 will be varied. The differential refractive index of the birefringent layer is 0.001. The relative concentration of material 1 was varied from 10, 20, 30, 40, and 50% of the total depth. Again, the simulated exponential decay is shown in the graphs of Figure 8.13. Oscillations are noted that affect interpretation of the PS-OCT data, although in this case the period is relatively long. Not surprisingly, the oscillatory rate increases with increasing concentration of the birefringent layer.

8.4.6 Periodical Rotation of Polarization State in Reference Arm

In Figure 8.14a and b, the polarization state is manipulated in the reference arm at intervals of $\pi/8$. Rotating the reference arm is critical to SDPS-OCT. Materials 1 and 2 are kept at equal width. The differential refractive index of the birefringence layer is 0.005. From graphs a–i, the banding pattern is changing with the rotation of polarization state in the reference arm. It should be noted that no birefringence effects are seen because the polarization states in both arms are in the same polarization state.

8.4.7 Use of the Fast Fourier Transform

To provide a simple method for quantifying oscillatory rates, and potentially identifying multiple birefringent components, the spatial frequency (frequencies) of the induced birefringence is measured with the fast Fourier transform (FFT).[12]

Figure 8.15 looks at birefringent models with different layer parameters and their corresponding FFT. In Figure 8.15A and B, the birefringence layers have a differential group refractive index of 0, but in Figure 8.15B exponential decay (mimicking tissue) occurs that is not present in 8.15A. It can be seen that the FFT of both A and B are similar. In Figure 8.15C, the differential refractive index is 0.02. It can now be seen that in the FFT, when compared to Figure 8.15A and B, in addition to the regular pattern of increasing spatial frequency from the true layers of the phantom, other peaks from the high mismatch are noted (arrows). Therefore, two phenomena are contributing to the FFT, the regular layer structure and the high refractive index mismatch. This figure illustrates an additional example of

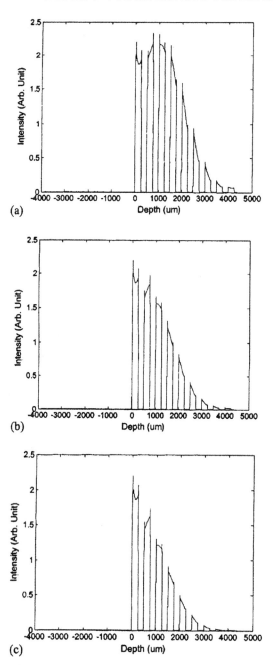

Figure 8.12 Simulated A-scan profiles of an alternating layered phantom, with different intra-layer medium. Courtesy of Liu, B. (2005). *J. Opt. Soc. Am. A. 22*, 262–271.

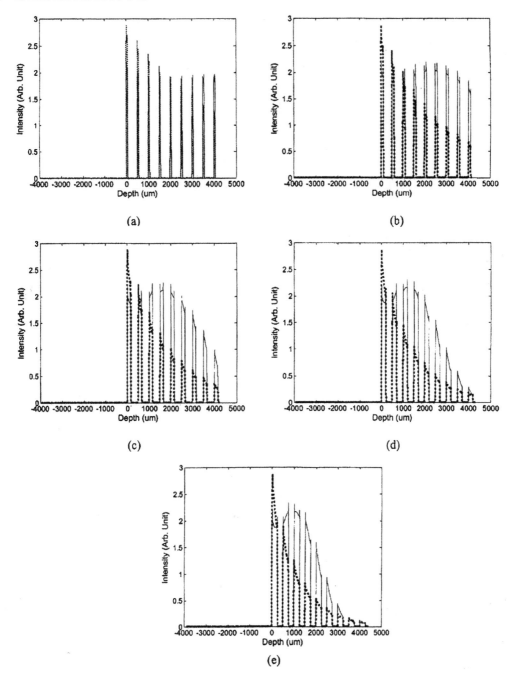

Figure 8.13 Simulated A-scan profiles of an alternating layered phantom, with relatively varying volume concentration of birefringent material 1. Courtesy of Liu, B. (2005). *J. Opt. Soc. Am. A.* 22, 262–271.

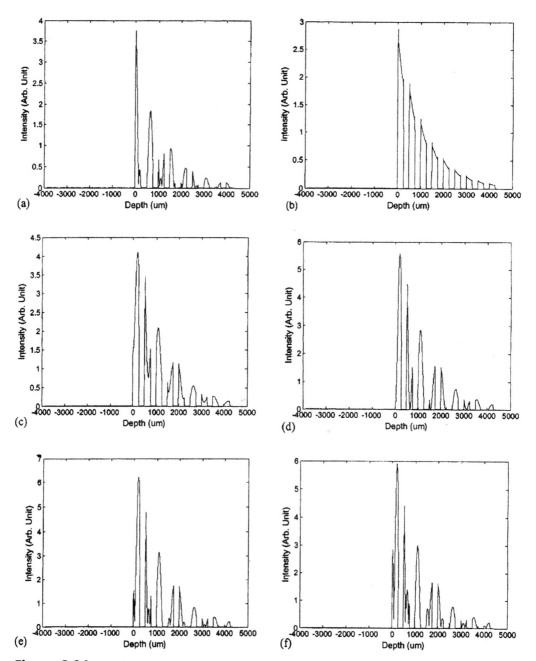

Figure 8.14 Simulated A-scan profiles of an alternating layered phantom, with relative retardations of polarization states in reference arm. From (a) through (i), the retardation increases with step p/8. Courtesy of Liu, B. (2005). *J. Opt. Soc. Am. A.* 22, 262–271.

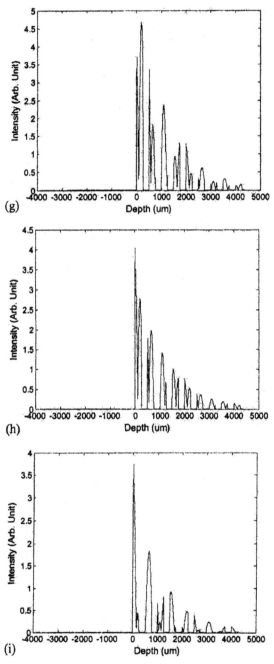

(g)

(h)

(i)

Figure 8.14 (Continued)

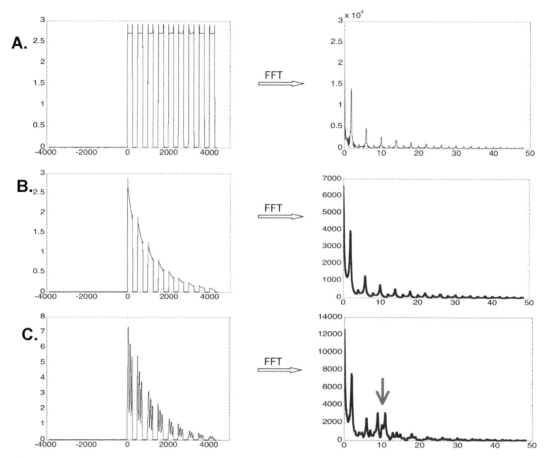

Figure 8.15   In images A and B, the birefringence layers have a differential group refractive index of 0, but in B exponential decay (mimicking tissue) occurs which is not present in A. It can be seen that the FFT of both are similar. In figure C, the differential refractive index is greater at 0.02. It can now be seen that in the FFT, when compared to figures A and B, in addition to the regular pattern of increasing wave number from the layers of the phantoms, additional peaks from the high mismatch are noted (arrows). Therefore, two phenomena are contributing to the FFT, the regular layer structure and the high refractive index mismatch. This figure illustrates an additional example of information gained from the FFT. In figure D, in the A-scan, the relative width of the birefringent layers is reduced to 10% of the total volume. In addition, the mismatch is set at 0.001. This results in a slow oscillation in the amplitude that obscures the exponential decay. The FFT shows two components. Because the layers are thin and square functions, a sinc function is superimposed over the spatial frequencies from the presence of multiple layers. The sinc function is prominent because of the relative small size of the layers. This ability of thin structures to generate sinc functions could be of diagnostic value when assessing tissue microstructure. The extremely slow oscillation due to the small differential refractive index that obscures the exponential decay in the A-scan is not obvious in the FFT because of its low spatial frequency. Courtesy Liu, B., et al. (2006). *Applied Optics*. 45, (in press June).

D.

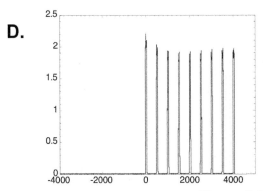
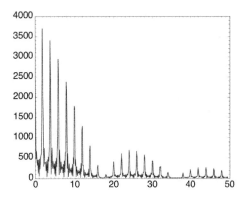

Figure 8.15 (Continued)

information gained from the FFT. In Figure 8.15D, in the A-scan, the relative width of the birefringent layers is reduced to 10% of the total volume. In addition, the mismatch is set at 0.001. This results in a slow oscillation in the amplitude that obscures the exponential decay. The FFT shows two components. Because the layers are thin and square functions, a sinc function is superimposed over the spatial frequencies from the presence of multiple layers. The sinc function is prominent because of the layer size. For larger layer thickness, the secondary peaks of the sinc function are too close to the main peak to be seen. The extremely slow oscillation due to the small differential refractive index that obscures the exponential decay in the A-scan is not obvious in the FFT because of its low spatial frequency.

In Figure 8.16, the differential group refractive index is kept constant at 0.005. However, the paddle positions in the reference arm are varied and two important observations are noted. In Figure 8.16A and C there are multiple frequencies noted in the FFT. However, in Figure 8.16B, the reference arm is exactly parallel to one of the phantom polarization axes and perpendicular to the other. This results in an FFT pattern similar to that seen in Figure 8.15B and serves as a potential method for identifying the two major optical axes of tissue. By eliminating the mismatch of a birefringent component, it could be possible to isolate it from a different component with a different optical axis. In Figure 8.16C, the small surface deflection on the A-scan, which is due to the presence of a partial surface band and results in artifact, peaks in the FFT.

In Figure 8.17, modeling is performed of a single slab with increasing differential group refractive index mismatch (A–C). With increasing mismatch, increasing oscillations are noted in the A-scan. In the FFT, a single peak is noted which increases in spatial frequencies with increasing mismatch.

8.4.8 SDPS-OCT of Human Tissue

Figure 8.18 is images of normal meniscus with their corresponding A-scan and FFT. In all images, the banding pattern appears relatively uniform in the B-scans.

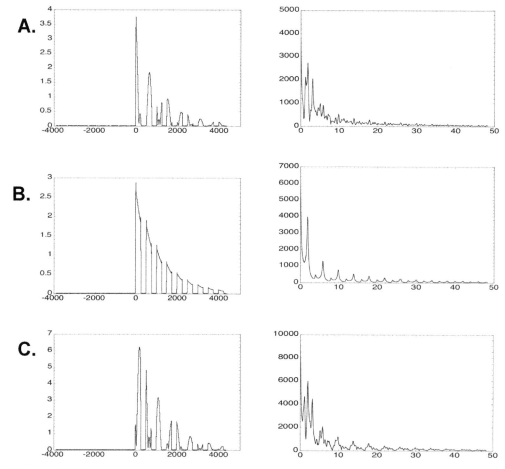

Figure 8.16 The differential group refractive index for this multilayered phantom is kept constant at 0.01. However, the paddle positions (polarization controllers) in the reference arm are varied and two important observations are noted. In both the figure A and C, there are higher spatial frequency components noted in the FFT. However, in figure B, the reference arm is exactly parallel to one the polarization axises and perpendicular to the other. This results in an FFT pattern similar to that seen in figure B and represents a method for identifying the optical axes of the phantom. Courtesy Liu, B., et al. (2006). *Applied Optics.* 45, (in press June).

These banding patterns can be used to measure regional birefringence of the tissue. However, the A-scans of Figure 8.18 demonstrate substantial amplitude fluctuations within the bands. These phenomena are seen most prominently in the FFTs where higher frequency components are noted (red arrows). This suggests, as seen in the modeling experiments, that multiple birefringent components are present within the image. Whether these multiple components are the result of different collagen configurations, different collagen types, or the alterations of form birefringence remains the source of future investigation.

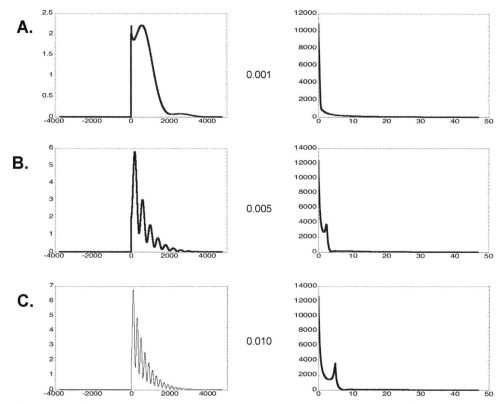

Figure 8.17 In this figure, modeling is performed of a single slab with increasing differential group refractive index mismatch (A-C). With increasing mismatch, increasing oscillations are noted in the A-scan. In the FFT, a single peak is noted which increases in spatial frequencies with increasing mismatch. In figures D and E, the mismatch is kept constant at 0.005. However, the paddle position in the reference arm is changed so that the peak spatial frequency is unchanged, but the presence of higher spatial frequency is noted due to the artifact caused by a partial peak at the surface. For this reason, again, the surface has been excluded from the FFT data of the meniscus to prevent this artifact. Courtesy Liu, B., et al. (2006). *Applied Optics*. 45, (in press June).

In Figure 8.19, OCT imaging was performed of a meniscus before and after treatment with collagenase. It can be seen in the B-scans that the banding pattern begins to deteriorate after 48 hours of treatment with collagenase. In the FFT, the loss in banding pattern can be seen as a loss of a spatial frequency peak (arrow). This is consistent with collagen being the source of birefringence. These collagenase results were also confirmed in a study with Helistat (Integra LifeSciences Corporation, Plainsboro, NJ 08536), a collagen type I phantom.[12] In the same experiment, the collagen was expanded by saline, ethanol, and lipid. Lipid resulted in a 75% increase in birefringence compared to saline, which must result from a change in the form birefringence of the collagen .[2,3]

We have examined polarization sensitivity of human cartilage both *in vitro* and *in vivo* using a SDPS-OCT[13,14,39] y. In Figure 8.20, images a, b, and c, taken at

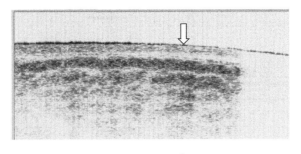

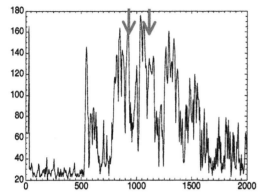
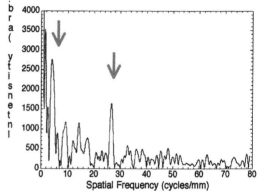

Figure 8.18   This figure demonstrates the B-scan, A-scan, and the corresponding FFT of a human meniscus. In the B-scan, the banding pattern due to tissue birefringence is noted. The A-scan (insert B) peaks are noted which correspond to the bands. The FFT shows two discrete peaks (arrows) which represent different birefringent material which can be distinguished in the A-scan or B-scan.

three different paddle positions in the reference arm, demonstrate the polarization sensitive change seen in normal *in vitro* cartilage. Smooth banding pattern changes with changing polarization state of incident light are seen due to organized collagen. It represents a rotation of the polarization state of the backreflected light as a function of depth (birefringence). Histologically, the cartilage appears normal (Figure 8.20d). In Figure 8.20e, a section stained with picrosirius red demonstrates the presence of organized collagen. Positive birefringence appears as brightly stained areas.

In Figure 8.21a, b, and c, polarization sensitivity is lost in the cartilage. The cartilage is thick and shows no gross fibrillations or erosions in the H&E (d). In both the picrosirius staining or OCT images, the banding pattern and birefringence is lost. There was a significant correlation between OCT polarization sensitivity and birefringence as determined by picrosirius staining.

In vivo human imaging of knee articular cartilage was performed via a hand-held probe described in Chapter 6 at 8 frames per second.[38] Imaging was performed in patients prior to knee replacement, which allowed histological correlation. In Figure 8.22, A and B demonstrate the abrupt loss of polarization sensitivity and backscattering intensity from left to right due to osteoarthritis. Denser alternating

FFT in the Presence or Absence of Collagenase

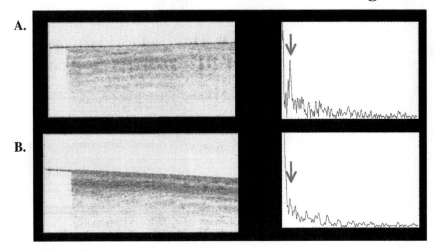

Figure 8.19 In this figure, OCT imaging was performed of a meniscus before and after treatment with collagenase. It can be seen in the B-scans that the banding pattern begins to deteriorate after 48 hrs of treatment with collagenase. In the FFT, the loss in banding pattern can be seen as a loss of a spatial frequency peak (arrow). Courtesy Liu, B., et al. (2006). *Applied Optics.* 45, (in press June).

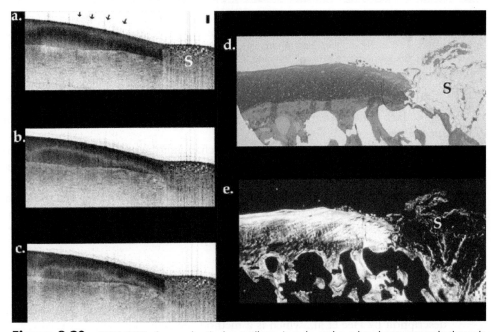

Figure 8.20 SDPS-OCT of normal articular cartilage. In a, b, and c, a band moves evenly through the cartilage as the polarization paddles are rotated in the reference arm. In the picrosirus stained section, the cartilage is bright identifying organized collagen. Courtesy of the Drexler, W., et al. (2001). *J. Rheum.* 28, 1311-1318.

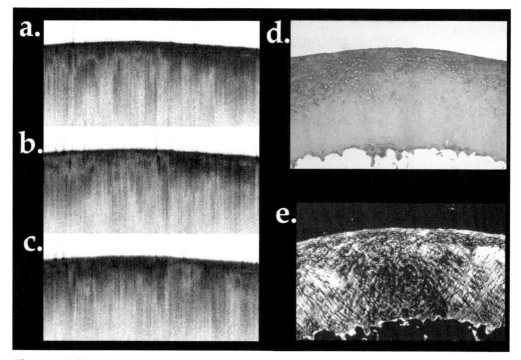

Figure 8.21 SDPS-OCT of osteoarthritic cartilage of relatively normal thickness. In a, b, and c, no significant banding patterns are noted. In e, the picrosirus stained section, no evidence of organized collagen is noted. In d however, the cartilage appears relatively normal. PS-OCT has therefore detected early osteoarthritic changes prior to cartilage thinning. Courtesy of the Drexler, W., et al. (2001). *J. Rheum.* 28, 1311-1318.

bright/dark band appearance on OCT image indicates stronger birefringence of the specimen.

When birefringence is relatively weak, changing the polarization of the reference arm with controllers is utilized. In Figure 8.23, a plaque with a wall thickness at points less than 60 μm in diameter is imaged.[13] In parts A and B, the white box demonstrates an area of thin intima with almost no organized collagen. This is confirmed in the picrosirius stained section. The picrosirius image (C) confirms the presence of variable organized collagen. The trichrome images are relatively uniform.

8.5 DUAL DETECTOR PS-OCT

DDPS-OCT has a variety of embodiments, so the initial focus will be on the original description by Hee et al., which is a method for detecting the Jones vector.[15] It has some similarities to a design for optical low coherence reflectometry (OLCR).[16] This technique has been applied to bovine tendon, myocardium, and skin.[17–19] The more complex approaches that measure the Stokes vector of the light and Mueller matrices

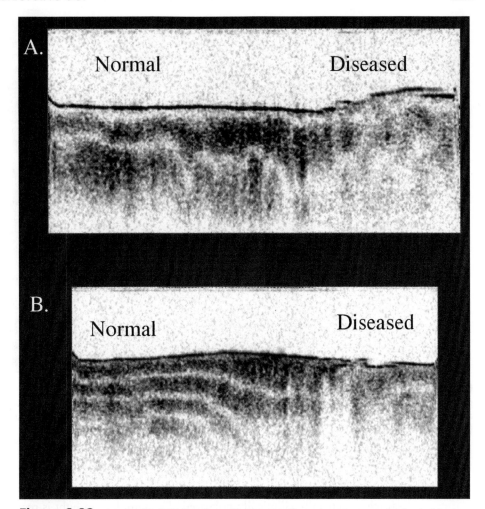

Figure 8.22 In vivo PS-OCT imaging of human cartilage. Images were obtained of human knee cartilage with a hand-held probe during open surgery. The left side of the knee shows relatively normal cartilage while the right demonstrates severe disease. This is demonstrated by a loss of banding in the diseased section. Courtesy Li, X. et al. (2005). *Arthritis Research and Therapy.* 7, R318-323.

of the tissue will then be examined. Jones and Mueller calculus were discussed in Chapter 1.

A schematic of the system is shown in Figure 8.24. Light from the source is linearly polarized parallel to the y axis. The light is then divided evenly by a nonpolarizing beam splitter. In the reference arm, a quarter wave plate (QWP) is placed with the slow axis 22.5 degrees from the x axis. It is then reflected off a mirror in the reference arm moved with a galvanometer (a grating based delay line can also be used). The light returning to the beam splitter is linearly polarized at 45 degrees to the x axis. In the sample arm, the QWP is 45 degrees from the x axis. This results in

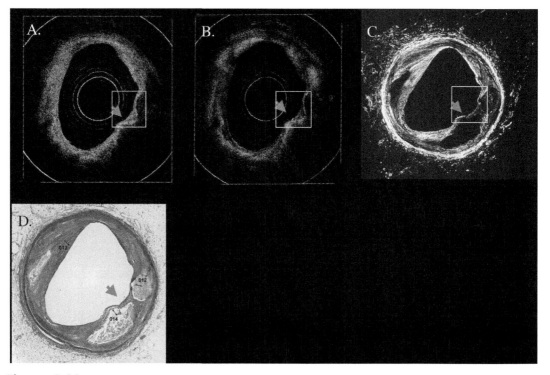

Figure 8.23   Polarization sensitive imaging of human atherosclerotic coronary arteries. This is a thin walled plaque with minimal organized collagen. In parts A and B, at two different controller positions, the white box shows an area of thin walled plaque. This is confirmed in the picrosirus stained section C, where brightness is minimal. Courtesy of Giattina, S.D. et al. (2006). *Intern. J. Cardiol.* 107, 400–409.

circularly polarized light directed at the sample. The light reflected from within the sample is generally elliptically polarized. The sample arm light is recombined at the beam splitter. In the detector arm, a polarizing beam splitter separates the light into orthogonal polarization states that are directed at separate balanced detectors.

Describing it in a more analytic fashion, the horizontal linearly polarized light is described by:

$$\mathbf{E(z)} = E(z)\begin{pmatrix}1\\0\end{pmatrix} \tag{8.32}$$

Phase changes associated with the beam splitter and mirrors described in Chapter 5 are not reintroduced here since changes in light from both the sample and reference arm are identical once they reach the detector. Light entering the reference arm after the beam splitter is given by:

$$\mathbf{E}_r(\mathbf{z}) = (1/2)^{0.5} E(z)\begin{pmatrix}1\\0\end{pmatrix} \tag{8.33}$$

The light then passes through a QWP whose slow axis is 2.5 degrees from the horizontal. The slow axis obviously represents the one with the highest RI.

The matrix representation of the QWP is:

$$\begin{bmatrix} e^{i\pi/4} & 0 \\ 0 & e^{-i\pi/4} \end{bmatrix} = e^{i\pi/4} \begin{bmatrix} 1 & 0 \\ 0 & -i \end{bmatrix}$$

The new matrix representation of the QWP is derived from:

$$T' = R(\theta)TR(-\theta) \text{ (see next page insert)} \tag{8.34}$$

where

$$R(\theta)R(-\theta) = \text{unit matrix} = \begin{bmatrix} 1 & 0 \\ 0 & 1 \end{bmatrix} \tag{8.35}$$

and:

$$R(\theta) = \begin{bmatrix} \cos(\theta) & -\sin(\theta) \\ \mathrm{Sin}(\theta) & \cos(\theta) \end{bmatrix} \tag{8.36}$$

$$R(-\theta) = \begin{bmatrix} \cos(\theta) & \sin(\theta) \\ -\mathrm{Sin}(\theta) & \cos(\theta) \end{bmatrix} \tag{8.37}$$

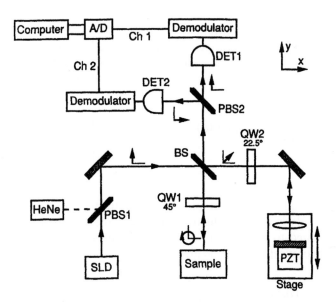

Figure 8.24 A schematic of a dual detector PS-OCT system. The system utilizes multiple beam splitters, wave plates, and detectors to isolate orthogonal polarization states. Courtesy of Hee, M., et al. (1992). *J. Opt. Soc. Am. B.* **9**, 903–908.

Insert

$$\psi_2 = T\psi_1$$

$$\psi_{2'} = R(\theta)J_2 = R(\theta)T\psi_1$$

$$\psi_1 = R(-\theta)\psi_{1'}$$

$$\psi_{2'} = R(\theta)TR(-\theta)\psi_{1'}$$

$$\psi_{2'} = T'\psi_{1'}$$

The light from the reference arm at the exit is now given by:

$$\mathbf{E_R}(z) = (1/2)^{0.5}E_Z \begin{pmatrix} 1 \\ 0 \end{pmatrix} \begin{bmatrix} \cos(22.5)^0 & -\sin(22.5)^0 \\ \sin(22.5) & \cos(22.5) \end{bmatrix} \begin{bmatrix} 1 & 0 \\ 0 & -i \end{bmatrix}^2$$

$$\begin{bmatrix} \cos(22.5)^0 & \sin(22.5)^0 \\ -\sin(22.5)^0 & \cos(22.5)^0 \end{bmatrix} = 1/2\,\mathbf{E(z)} \begin{pmatrix} 1 \\ 1 \end{pmatrix} \qquad (8.38)$$

The light in the exit from the reference arm is therefore linearly polarized at 45 degrees.

In the sample arm, the medium is generally assumed to be a linear homogeneous retarder with a constant orientation of the optical axis. The retardation is dependent on the wavelength, the amount of medium propagated through (z_R), and the difference in refractive index between the slow and fast axis (Δn). The round trip phase shift between orthogonal states is therefore given by $\delta = 2k_z\,\Delta n$. The Jones matrix for the sample is then given by:

$$S(z, \Delta n) = e^{-ikz\bar{n}} \begin{bmatrix} \cos(\alpha) & -\sin(\alpha) \\ \sin(\alpha) & \cos(\alpha) \end{bmatrix} \begin{bmatrix} \cos(22.5)^0 & \sin(22.5)^0 \\ -\sin(22.5)^0 & \cos(22.5)^0 \end{bmatrix}$$

$$(8.39)$$

Again, light in the sample arm passes through a QWP oriented at 45 degrees so that circularly polarized light is directed at the sample. Assuming the reflectivity at z_R is related to R_z, the round trip light passing through the sample arm (including the theoretical medium) is given by[37]:

$$\mathbf{E_R}(z_R) = (1/2)^{0.5}E_Z \begin{pmatrix} 1 \\ 0 \end{pmatrix} QWP(45)S(z, \Delta n)(R_z)^{0.5}S(z, \Delta n)QWP(45)\alpha R_z$$

$$(8.40)$$

$$\int_0^\infty a(k)\exp\{-2ik(z + z_r n_m)\}.$$

As previous:

$$\langle I \rangle = \langle I_R \rangle + \langle I_S \rangle + 2\mathrm{Re}\langle \psi_R^*(z)\psi_s(z)\rangle \tag{8.41}$$

We will assume that the source is Gaussian, there is no dispersion or chromatic dependence of the QWP, and the resolution is:

$$\Delta_l = \frac{2\ln 2\lambda_0^2}{\pi\Delta_\lambda}$$

The AC term of Eq. 8.41 can be expressed as:

$$\mathrm{Re}\langle \psi_R^*(z)\psi_s(z)\rangle = A_H + A_V \tag{8.42}$$

$A_H \alpha \ [\exp-(\Delta z/\Delta_l)^2] \cos(2k_0\Delta z + 2\phi) \sin(k_0 z_R \Delta n)$
$A_V \alpha \ [\exp-(\Delta z/\Delta_l)^2] \cos(2k_0\Delta z) \cos(k_0 z_R \Delta n)$

where Δz is the difference between optical pathlengths in the sample and reference arm, ϕ is the angle of the optical axis of the sample with respect to the horizontal, and Δn is the refractive index difference between the slow and fast axis. The term θ represents the offset angle of the QWP in the reference arm. The two detectors, measuring the orthogonal states, are assumed to be balanced. Signal processing needs to include phase-sensitive demodulation with respect to θ. The intensity after signal processing is given by:

$$I_H(z)\alpha \sin^2(k_0 z_R \Delta n)$$
$$I_v(z)\alpha \cos^2(k_0 z_R \Delta n) \tag{8.43}$$

Assuming equal reflectivity for both polarization states, the total phase retardation is given by:

$$\delta = \arctan (I_v(z)/I_H(z))^{0.5} = k_0 z_R \Delta n \tag{8.44}$$

An alternate approach to the above, using a very similar physical setup to that described by Hee et al., uses a Hilbert transform previously used in Doppler OCT and differential phase contrast OCT.[15,20] The amplitude and phase are found from a single A-scan through the complex function $A'_k(z)$:

$$A'_k(z) = I_k(z) + i\,\mathrm{H}\{I_k(z)\} = A_k(z)\exp[i\phi(z)] \tag{8.45}$$

where H is the Hilbert transform, and $A_k(z)$ is the amplitude of the oscillatory interference term while $\varphi(z)$ is the phase.

Example images of porcine myocardial tissue, which has large amounts of collagen, as well as myosin using this approach are shown in Figure 8.25.[21] The first image plots the phase retardation δ while the second sums the two detectors (no birefringence measurements). Using dual detector approaches in highly birefringent tissue such as tendon and cartilage which exhibit strong banding patterns is of less clear diagnostic value, but in tissue with low birefringence plots like Figure 8.25 it may be of use. With a single detector, low birefringence is measured by changing

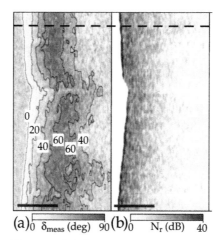

Figure 8.25 Dual detector PS-OCT image of porcine myocardial tissue. The image on the left represents the measured phase retardation in degrees while the right is a polarization insensitive image. Courtesy of Everett, M.J., et al. (1998). *Opt. Lett.* 23, 228–230.

the reference arm polarization state. Myocardium is likely less birefringent then other muscle tissue because the fibers bifurcate, so they are not as parallel.

8.5.1 Measurement of the Stokes Vector, Mueller Matrix, and Poincare's Sphere

The Stokes vector and Mueller matrix are discussed in Chapter 2. The Stokes vector and Mueller matrix provide a more complete description of the polarization state, including the optical axis and the degree of polarization. With early OCT systems, the determination of the Mueller matrix required 16 separate PS-OCT images. Due to the slow time for data acquisition, these techniques were impractical for clinical use.[22,23] These early PS-OCT systems also were free space with bulk optical components. Since most clinical OCT imaging is performed through endoscopes and catheters, fiber-based designs are essential. Furthermore, fiber-based systems allow for ease of handling and alignment.

Recently, several groups have introduced fiber-based systems with partial dynamic compensation of polarization distortion from fiber optics in the sample arm.[24–26] Most require tracking polarization changes at the tissue surface, which is difficult to do with *in vivo* clinical imaging as the surface and overlying medium may be changing rapidly. Furthermore, the acquisition rates at this time are impractically slow. In addition, the data for Stokes vector/Mueller matrices are complex to interpret and it is unclear how they can be used by the clinician. An example set of a Mueller data set is shown in Figure 8.26.[27] Probably the best way to display the

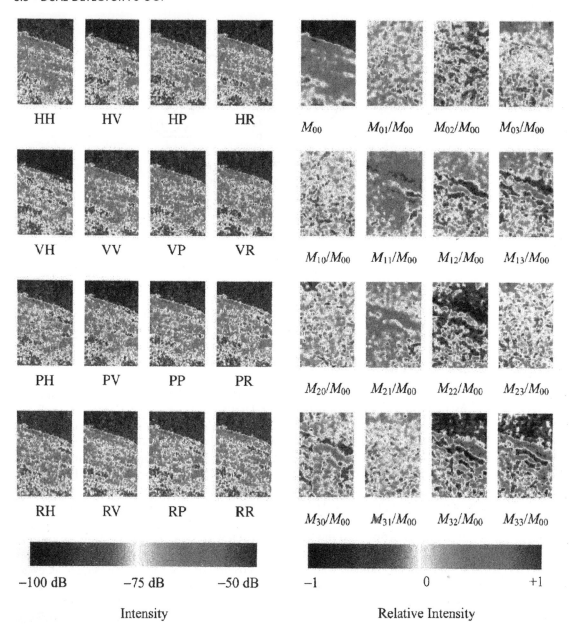

Figure 8.26   Mueller matrix of a tissue sample. The left are the raw OCT images; the right are the corresponding Mueller matrix parameters. Details of each image are not discussed here, but can be found in the original publication of this. The image is provided as an example of the current difficulty of interpreting data from a single Mueller's matrix. Courtesy of Jiao, S., et al. (2000). *Applied Optics*. 39, 6318-6325.

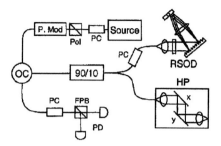

Figure 8.27 More recent embodiment of DDPS-OCT. The specifics of the embodiment are discussed in the text. Figure courtesy Pierce, M.C., et al. (2002). *Optics Letters.* 27, 1534. [32]

Stokes vector is in terms of the Poincare's sphere, which is explained in Appendix 8-1 and shown in Figure 8A.1.[28] When looking at these data, it should be noted that it is the Mueller matrix, or the change in the vector (Jones or Stokes) as a function of depth, which provides potential diagnostic information about the tissue. Although a large number of papers exist on DDPS-OCT, few actually achieve this through a fiber-based approach with appropriate correction for fiber distortion.

The merits and limitations of two of the most common DDPS-OCT approaches have recently been debated in *Optics Letters.*[29,30] Rather than summarizing these arguments, readers are encouraged to read both and draw their own conclusions.

As there are many embodiments for DDPS-OCT, one newer approach will be discussed here as shown in Figure 8.27.[31,32] References are provided for other recent embodiments.[33–35] This includes a publication of polarization sensitive SS-OCT.[36]

With the technique in Figure 8.27, only two polarization states of light are examined; one during the forward scan and one during the reverse scan. Light from the source is directed toward an electro-optic polarization modulator at 45 degrees to the modulator optic axis. The modulator is then driven by a two-step voltage function. The light is then coupled into a polarization-independent optical circulator followed by direction into the sample and reference arm at a 90:10 ratio. A grating based delay line is placed in the reference arm. Horizontal and vertical detection is performed in the sample with sine and cosine components of the signal at each detector. The sine and cosine terms are obtained by multiplying them by a sine and cosine term at the carrier frequency ω_0 and averaging over a few cycles within the coherence length. The additional signal processing is described in the original reference, but the Stokes parameters are obtained from:

$$
\begin{aligned}
I &= \sin_H{}^2 + \cos_H{}^2 + \sin_V{}^2 + \cos_V{}^2, \\
Q &= \sin_H{}^2 + \cos_H{}^2 - \sin_V{}^2 - \cos_V{}^2, \\
U &= 2\sin_H \times \sin_V + 2\cos_H \times \cos_V, \\
V &= 2\sin_H \times \cos_V - 2\cos_H \times \sin_V
\end{aligned}
\tag{8.46}
$$

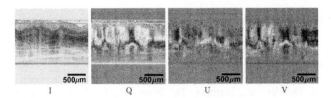

Figure 8.28 Stokes parameters using the embodiment in Figure 8.25 of human skin. Courtesy Saxes, C.E. et al. (2000). *Optics Letters.* 25, 1356.

In Figure 8.27, I, Q, U, and V images are shown using this technique. While these images are intriguing, how they can be reduced to diagnostic information is the source of future work.

Noise considerations are the same as those outlined in Chapter 5. Of note should be the fact that with DDPS-OCT, when one detector receives substantially more signal then the other, the sensitivity of PS-OCT decreases because of the low power on the opposite detector. This may occur with low birefringence.

REFERENCES

1. Pimentel, E. R. (1981). Form birefringence of collagen bundles. *Acta Histochem. Cytochem.* 14, 35–41.
2. Folkhard, W., Knorzer, E., Mosler, E., and Nemitschek, T. (1984). Packing of collagen molecules modified with 2-propanol. *J. Mol. Biol.* 177, 841–844.
3. Fraser, R. D., MacRae, T. P., Miller, A., and Suzuki, E. (1983). Molecular conformation and packing in collagen fibers. *J. Mol. Biol.* 167, 497–521.
4. Goldberg, B. and Rabinovitch, M. (1983). Connective tissue. *Histology: In Cell and Tissue Biology*, 5th edition. L. Weiss (ed.). Elsevier Biomedical, New York, 144–154.
5. Neuman, A. P. (1998). Articular cartilage repair. *Am. J. Sports Med.* 26, 309–324.
6. Petersen, W. and Tillmann, B. (1998). Collagenous fibril texture of the human knee joint menisci. *Anat. Embryol.* 197, 317–324.
7. Junqueira, L. C., Figueiredo, M., Torloni, H., and Montes, G. S. (1986). A study on human osteosarcoma collagen by the histochemical picrosirius-polarization method. *J. Pathol.* 148, 189–196.
8. Junqueira, L. C., Bignolas G., and Brentani, R. R. (1979). Picrosirius staining plus polarization microscopy, a specific method for collagen detection. *Histochem. J.* 11, 447–455.
9. Dayan, D., Hiss, Y., Hirshberg, A. et al. (1989). Are the polarization colors of picrosirius red stained collagen determined only by the diameter of the fibers? *Histochemistry* 93, 27–29.
10. Martin, S. D., Patel, N. A., Adams, S., Roberts, M. J., Plummer, S., Stamper, D. L., Fujimoto, J., and Brezinski, M. B. (2003). New technology for assessing the microstructural components of tendons and ligaments. *Int. Orthoped.* 27(3), 184–189.
11. Liu, B., Harman, M., and Brezinski, M. E. (2005). Variables affecting polarization sensitive optical coherence tomography imaging examined through the modeling of birefringent phantoms. *J. Opt. Soc. Am. A* 22, 262–271.

12. Liu, B., Harman, M. et al. (2006) Characterizing of tissue microstructure with single detector polarization sensitive optical coherence tomography. *Applied Optics* 45(18) June.

13. Li, X., Martin, S. D., Pitris, C., et al. (2005). High resolution optical coherence tomography during open knee surgery. *Anth. Res. Ther.* 7: R318–R323.

14. Drexler, W., Stamper, D., Jesser, C., Li, X. D., Pitris, C., Saunders, K. et al. (2001). Correlation of collagen organization with polarization sensitive imaging in cartilage: Implications for osteoarthritis. *J. Rheumatol.* 28, 1311–1318.

15. Hee, M., Huang, D., Swanson, E. A., and Fujimoto, J. G. (1992). Polarization sensitive low coherence reflectometer for birefringence characterization and ranging. *J. Opt. Soc. Am. B* 9, 903–908.

16. Kobayashi, M., Hanafusa, H., Takada, K., and Noda, J. (1991). Polarization independent interferometric optical time domain reflectometer. *J. Lightwave Technol.* 9, 623–628.

17. de Boer, J. F., Milner, T. E., van Gemert, M. J. et al. (1997). Two dimensional birefringence imaging in biological tissue by PS-OCT. *Opt. Lett.* 22, 934–936.

18. Everett, M. J., Schoenenberger, K., Colston, B. W. et al. (1998). Birefringence characterization of biological tissue by use of optical coherence tomography. *Opt. Lett.* 23, 228–230.

19. de Boer, J. F., Srinivas, S. M., Malekafzali, A. et al. (1998). Imaging thermally damaged tissue by polarization sensitive optical coherence tomography. *Opt. Express* 3, 212–218.

20. Hitzenberger, C. K., Gotzinger, E., Sticker, M. et al. (2001). Measurement and imaging of birefringence and optic axis orientation by phase resolved polarization sensitive optical coherence tomography. *Opt. Express* 9, 780–790.

21. Everett, M. J., Schoenenberger, K., Colston, B. W., and DaSilva, L. B. (1998). Birefringence characterization of biological tissue by use of optical coherence tomography. *Opt. Lett.* 23, 228–230.

22. Yao, G. and Wang, L. W. (1999). Two dimensional depth resolved Mueller matrix characterization of biological tissue by optical coherence tomography. *Opt. Lett.* 24, 537–539.

23. Ducros, M. G., de Boer, J. F., Huang, H. E. et al. (1999). Polarization sensitive optical coherence tomography of the rabbit eye. *IEEE J. Selected Top. Quantum Electron.* 5, 1159–1167.

24. Saxer, C. E., de Boer, J. F., Park, B. H. et al. (2000). High speed fiber based polarization sensitive optical coherence tomography of in vivo human skin. *Opt. Lett.* 25, 1355–1357.

25. Jiao, S., Yu, W., Stoica, G., and Wang, L. V. (2003). Optical fiber based Mueller optical coherence tomography. *Opt. Lett.* 28, 1206–1208.

26. Guo, S., Zhang, J., Wang, L. et al. (2004). Depth resolved birefringence and differential optical axis orientation measurements with fiber based polarization sensitive optical coherence tomography. *Opt. Lett.* 29, 2025–2027.

27. Park, H. E., Pierce, M. C., Cense, B., and de Boer, J. F. (2004). Jones matrix analysis for a polarization sensitive optical coherence tomography system using fiber optic components. *Opt. Lett.* 29, 2512–2514.

28. Haurd, S. (1997). *Polarization of Light*. John Wiley & Sons, New York.

29. Jiao, S. and Wang, L. (2004). Reply to comment on "optical fiber based Mueller optical coherence tomography." *Opt. Lett.* 29, 2875–2877.

30. Park, B. H., Pierce, M. C., and de Boer, J. F. (2004). Comment on "optical fiber based Mueller optical coherence tomography." *Opt. Lett.* 28, 2873–2874.

31. Park, H. E., Pierce, M. C., Cense, B., and de Boer, J. F. (2004). Jones matrix analysis for a polarization sensitive optical coherence tomography system using fiber optic components. *Opt. Lett.* 29, 2512–2514.

32. Pierce, M. C., Hyle Park, B., Cense, B., and Boer, J. F. (2002). Simultaneous intensity, birefringence, and flow measurements with high speed fiber based optical coherence tomography. *Opt. Lett.* 27, 1534–1536.

33. Todorovic, M., Jiao, S., Wang, L. V., and Stoica, G. (2004). Determination of local polarization properties of biological samples in the presence of diattenuation by use of Mueller optical coherence tomography. *Opt. Lett.* 29, 2402–2404.

34. Guo, S., Zhang, J., Wang, L. et al. (2004). Depth resolved birefringence and differential optical axis orientation measurements with fiber based polarization sensitive optical coherence tomography. *Opt. Lett.* 29, 2025–2027.

35. Hitzenberger, C. K., Gotzinger, E. et al. (2001). Measurement and imaging of birefringence and optic axis orientation by phase resolved polarization sensitive optical coherence tomography. *Opt. Express* 9, 780–790.

36. Zhang, J., Jung, W., Nelson, J., and Chen, Z. (2004). Full range polarization sensitive Fourier domain optical coherence tomography. *Opt. Express* 12, 6033–6039.

37. de Boer, J. F. and Milner, T. E. (2002). Review of polarization sensitive OCT and Stokes vector determination. *J. Biomed. Opt.* 7, 359–371.

38. Gia Hina, S. D., Courtney, B. K., Herz, P. R. et al. (2006). Assessment of coronary plague collagen with polarization sensitive optical coherence tomography. *Int. J. of Cardial.* 107(3), 400–409.

APPENDIX 8-1

POINCARE'S SPHERE AND REPRESENTING PARTIALLY POLARIZED STATES

Poincare's Sphere

The Poincare's sphere is a way to geometrically represent the polarization state described by the Stokes vector, as shown in figure 8A.1 (1). In this representation, the

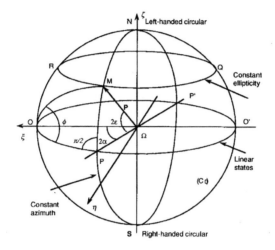

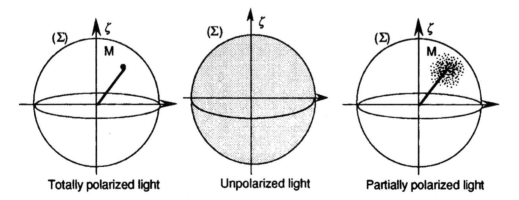

Figure 8A.1   Poincare's sphere. (A) Demonstration of the value of different points on the sphere. (B) Representation of light of different coherence properties on the sphere. Drawing based on concepts from Haurd, S. (1997). *Polarization of Light*. [1]

equatorial plane represents linearly polarized states, with x and y linearly polarized light along the x or y axis. The north pole is a state of left handed circularly polarized light while the south pole represents right handed circularly polarized light. Between the poles and equator are elliptically polarized states. The upper hemisphere corresponds to the left handed elliptically polarized light where as the lower hemisphere is right handed elliptically polarized light. The lower images display how light of different polarization states is displayed.

Stokes Parameters and Partially Coherent Light

The Stokes vector is given by:

$$S = \begin{bmatrix} P_0 \\ P_1 \\ P_2 \\ P_3 \end{bmatrix} = \begin{bmatrix} \langle I_x + I_y \rangle \\ \langle I_x - I_y \rangle \\ \langle I_{+45°} - I_{-45°} \rangle \\ \langle I_L - I_R \rangle \end{bmatrix}$$

Here, I_x and I_Y are time average linearly polarized states, I_L and I_R are time averaged circularly polarized states, and I_{+45} and I_{45} are time averaged linearly polarized states at 45 and –45 degree polarization. These parameters are measurable quantities with a single detector and filters. The P_0, P_1, P_2, and P_3 are sometimes written as I, Q, U, and V. Stokes vector can also be written as:

$$S = \begin{bmatrix} P_0 \\ P_1 \\ P_2 \\ P_3 \end{bmatrix} = \begin{bmatrix} A_x^2 + A_y^2 \\ A_x^2 - A_y^2 \\ 2A_x A_y \cos \phi \\ 2A_x A_y \sin \phi \end{bmatrix}$$

Here A_X and A_y are the time average field at the x and y detector in the dual detector scheme. The angle ϕ is the phase difference between the two orthogonal fields. In this form, the Stokes vector can be directly measured from the DDPS-OCT.

Partially polarized light can be represented mathematically in several forms. The first is in the Stokes vector form which is the sum of polarized and unpolarized light:

$$S = \begin{bmatrix} P_0 \\ P_1 \\ P_2 \\ P_3 \end{bmatrix} = \begin{bmatrix} pP_0 \\ P_1 \\ P_2 \\ P_3 \end{bmatrix} + \begin{bmatrix} (1-p)P_0 \\ 0 \\ 0 \\ 0 \end{bmatrix} = S_P + S_{UP}$$

Here p is a measure of the degree of coherence and is given by $(P_1^2 + P_2^2 + P_3^2)/P_0$. The first matrix ($S_P$) is the totally polarized light while the second (S_{UP}) is unpolarized light. This decomposition is useful as both vectors can be transformed through the optical system separately and added at the conclusion.

Partially polarized and unpolarized light cannot always be quantified with a Jones vector as it can with the Stokes vector. However, when an interferometer (although not limited to) is used, the degree of polarization can be determined by the polarization or coherence matrix. In other word, the coherence can be used to determine the degree of polarization with the Jones vector. The coherence matrix is discussed elsewhere in detail, but some general properties are discussed here (2,3). The coherence matrix is given by:

$$\mathbf{J} \equiv \begin{bmatrix} \langle E_x^*(t)E_x(t) \rangle & \langle E_x^*(t)E_y(t) \rangle \\ \langle E_y^*(t)E_x(t) \rangle & \langle E_y^*(t)E_y(t) \rangle \end{bmatrix}$$

where $E_x(t)$ and $E_y(t)$ are the complex electric fields in the x and y directions respectively. The trace of the matrix gives the total intensity:

$$\begin{aligned} \mathrm{tr}\mathbf{J} &\equiv J_{xx} + J_{yy} \\ &= \langle E_x^*(t)E_x(t) \rangle + \langle E_y^*(t)E_y(t) \rangle \end{aligned}$$

The degree of correlation between the two orthogonal field components given by:

$$j_{xy} \equiv |j_{xy}|e^{i\beta_{xy}} \equiv \frac{J_{xy}}{(J_{xx})^{1/2}(J_{yy})^{1/2}}.$$

where the absolute value of j_{xy} varies between 0 and 1. The degree of polarization is then given by:

$$p = (1 - \{4\det\mathbf{J}/\mathrm{Tr}(\mathbf{J})^2)$$

The determinant of $\mathbf{J}$ is given by:

$$\det\mathbf{J} \equiv J_{xx}J_{yy} - J_{xy}J_{yx} \geqslant 0.$$

The coherence matrix has been applied to interferometric ellipsometry and recently with OCT (4,5).

REFERENCES

1. Haurd, S. (1997). *Polarization of Light*. John Wiley & Sons, New York.
2. Mandel, L., and Wolf, E. (1995). *Optical Coherence and Quantum Optics*. Cambridge Press, New York, New York, 342–355.
3. Barakat, R. (1963). Theory of the Coherency Matrix for Light of Arbitrary Spectral Bandwidth. *J. Opt. Soc. Am.* 53: 317–323.
4. Hazebroek, H.F., and Holscher, A.A. (1973). Interferometric Ellipsometry. *J. Phys. E. Sci. Instrum.* 6:822–826.
5. de Boer, J.F., and Milner, T.E. (2002). Review of Polarization Sensitive OCT and Stokes Vector Determination. *J. of Biomedical Optics.* 7:359–371.

9 ADJUVANT TECHNIQUES: ABSORPTION SPECTROSCOPY, CONTRAST PROBES, PHASE CONTRAST, ELASTOGRAPHY, AND ENTANGLED PHOTONS

Chapter Contents

9.1   GENERAL

In this chapter methods are discussed for improving the information gained from OCT images, not discussed in previous sections, above those of pure structural imaging. These techniques are still in the early stage of development and include absorption spectroscopy/contrast agents, elastography, phase contrast, and the use of entangled photons.

9.2   ABSORPTION SPECTROSCOPY

9.2.1   Absorption

The interaction of light with matter provides an opportunity to evaluate the biochemical and microstructural properties of tissue. Measuring light absorption, in particular, has attracted a great deal of attention. However, with OCT, direct measurements of tissue biomolecule absorption can be considerably difficult for several reasons. These include broad overlapping absorption peaks for most molecules, the inability to quantify scattering independent of absorption, the large number of different molecular species absorbing, shallow imaging depth, and the generally low concentration of the particular absorbing molecule of interest. In addition, the local environment can have a substantial effect on the absorption profile of molecules as it does for fluorescence. However, several approaches can be utilized to improve the diagnostic potential of OCT absorption-based techniques. These can be divided into those techniques that utilize the intrinsic absorption and those that use dyes/probes. Some of the dyes/probes will be scattering rather then absorbing agents, but are added here because they add contrast in an analogous manner. It should be noted that while OCT imaging in nontransparent tissue is performed near 1300 nm, most of these absorption-based techniques are done with imaging below 1000 nm. Furthermore, absorption-based approaches in general (not just with OCT), with the clear exception of pulse oximetry, have not found a major role in clinical medicine over the years.

9.2.2 Intrinsic Absorption

Absorption OCT techniques that work by intrinsic absorption include the use of a single broadband source (spectroscopic OCT), dual source same frequency (differential absorption OCT), second harmonic generation (SHG), and third order nonlinear effects.

9.2.3 Dual Source

This is a technique similar to lidar measurements used to assess gases in the atmosphere, as well as other areas of the physical sciences.[1,2] Sources at two distinct median wavelengths are introduced into the interferometer and OCT imaging is performed of the sample at the separate wavelengths.[3,4] In one embodiment, a wavelength division multiplexer is used to separate the spectrum by the median wavelengths and independent interferograms are obtained with separate detection and filtering schemes.

Broadly, the theory assumes that the tissue scatters light approximately equally (or at a constant ratio) at both median wavelengths. Then, if a compound of interest (e.g., water) absorbs much more strongly at one wavelength relative to the other, and no other compounds present have large differential absorption, then the concentration of the compound of interest can be determined. Most studies are done at 1300 and 1500 nm, where water absorption is measured and is significantly higher at the latter. A major limitation is that at these wavelengths scattering is on the same order of magnitude as absorption, which makes measurements susceptible to significant error. This can partially be compensated for by imaging at 1800 nm rather than 1500 nm, where water absorption is increased by a factor of three. An additional difficulty is that scattering can be highly variable from tissue to tissue, further complicating this situation. To date, this technique has not been of wide diagnostic value.

9.2.4 Single Source Same Frequency (Spectroscopic OCT)

This technique uses a single ultrabroad bandwidth source for absorption measurement. The spectral profile of the light is backreflected from within the sample and it is correlated with the original source signal (reference arm) in terms of wavelength distribution.[5,6,8,9] Changes in the OCT interferogram are used as a measure of absorption. The single ultrabroadband source typically used is a mode-locked femtosecond titanium sapphire laser. Unlike most OCT systems, the entire spectral profile of the single backreflected signal is extracted using digital signal processing. In one publication the digital processing technique used was a Morlet wavelet transform, which reduces windowing artifacts compared with many other sampling methods. The spectrum at each point in the A- or B-scan is represented by either the center of mass or a single frequency. Data are multidimensional, containing both backreflection intensity and spectral profile, so they cannot be represented by simple gray scale look-up tables. Instead, saturation and luminance data display is used. This technique is associated with several significant problems. First, as previously discussed, scattering is wavelength dependent. Therefore, a spectral shift occurs as a function of depth even if absorption is negligible, so the technique appears limited to weakly scattering medium. Second, absorptive is accumulative, with assessments at a particular depth being the sum of absorption at that point and which has occurred in the layers above. Third, as stated, the absorption spectrum of the molecules is typically broad, particularly in the visible region. With bandwidths below 1000 nm, there appears to be limited potential endogenous molecules to observe, with the exception of oxy- and deoxyhemoglobin, which have slightly different attenuations as a function of wavelength.[8] Fourth, if absorption is too large, the spectrum will no longer be Gaussian and the interferogram will be non-Gaussian, reducing ranging ability. Finally, the technique in its original form suffers from the trade-offs between time resolution and frequency resolution (time frequency uncertainty principle). Therefore, techniques other than short time Fourier transforms (STFT) and continuous wavelet transforms have recently been examined.[9] Each technology has its advantages and disadvantages. The Cohen class TFD can provide the most compact time frequency (TF) analysis compared with STFT, but suffers from the fact that multicomponent signals generate artifacts. Therefore, when endogenous tissue absorption or low concentrations of dyes are measured, long pathlengths can be necessary and therefore the STFT may be superior.

9.2.5   Second Harmonic Generation

Several nonlinear phenomena are important for endogenous spectroscopic OCT imaging. These are used in coherent detection techniques, which include second and third harmonic generation. Here the sample signals, at distinct frequencies from the incident light, are mixed with a reference generator at the appropriate frequency. Second harmonic generation occurs when the intensity on the sample is sufficient so that the relationship between the polarizability and the applied electric field is no longer linear, resulting in backreflected light with a different frequency then the incident light. Classical treatment of light assumes a linear relationship between the field and medium, such as refraction, reflection, and superposition. This follows from the oscillating dipole model discussed in Chapter 3. However, with sufficient intensities, such as a focused beam of substantial intensity, the "spring" is overdriven. In general, the phenomenon is particular, under the dipole approximation, to molecules that are not centro-symmetric. Molecules that demonstrate inversion symmetry will not exhibit second harmonic generation.

For a low intensity (minimally focused) scalar light wave, transversing a linear isotropic medium, the polarizability is given by:

$$P = \varepsilon_0 \chi E \tag{9.1}$$

where χ is of course the dimensionless constant known as the electric susceptibility. At high field strengths, P can be considered saturated and no longer increases linearly with E. For a medium exposed to a high field, the polarizability is given by:

$$P = \varepsilon_0 \chi E_0 \sin \omega t + \varepsilon_0 \chi_2 E_0^2 \sin^2 \omega t + \varepsilon_0 \chi_3 E_0^3 \sin^3 \omega t \ldots \qquad (9.2)$$

where $\chi \gg \chi_2$. Limiting our attention to the first two terms at the right side of the equal sign and using an identity from trigonometry, the equation becomes:

$$P \alpha \varepsilon_0 \chi E_0 \sin \omega t + (\varepsilon_0 \chi_2 E_0^2)(1 - \cos 2\omega t)/2 \qquad (9.3)$$

The second term then contains two important components. The first is the E_0^2 term that is the DC or bias term often referred to as optical rectification. In other words, the second harmonic radiation is offset by the value of the DC term. The term $\cos 2\omega t$ is twice the fundamental (incident) frequency. The second harmonic therefore consists of twice the frequency of the fundamental term offset by the DC term. An example of the incident wave and the second harmonic is shown in Figure 9.1. Second harmonic generation is proportional to the square of the input laser intensity. Again, for an isotropic medium that exhibits inversion symmetry, such as water or glass, $P(E)$ is an odd function and even terms in Eq. 9.2 vanish. Therefore, no second harmonic generation occurs.

Second harmonic generation assessment with OCT has not yet reached a point where it can be applied for human diagnostics, but several features suggest it may ultimately have clinical utility.[10–12] In particular, second harmonic generation has received increased attention for identifying collagen.[13,14] In addition, since it is a nonresonant phenomenon, it is not subject to photobleaching- and photodamage-like processes such as fluorescence.

To date, three studies have examined second harmonic generation with OCT. Although these studies used free space devices, high focal intensities, and were not performed near real time, they demonstrate potential for future work.[10–12]

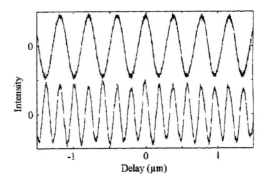

Figure 9.1 Second harmonic generation. This slide illustrates an incident OCT beam (top) and the corresponding second harmonic generation. Courtesy Jiang, Y., et al. (2004). *Optics Letters.* 29, 1091.

In one of the studies, which also examined birefringence, a femtosecond Nd:glass laser was used. Total power on the sample was not noted but peak laser power was 170 mW. Second harmonic light was generated in the reference arm using a KDP crystal and linearly polarized at 45 degrees with a quarter wave plate. In the detector arm, a dichroic mirror was used to separate first and second harmonic light. The second harmonic light was further separated into orthogonal polarization states. Second harmonic images of both phantoms and fish scales were generated. Although no quantitative technique, such as picrosirius staining or SEM was used, to verify organized collagen, the ability to generate images with second harmonic light was significant. Although the technique shows promise, in addition to the limitations described, separate dispersion compensation for the two frequencies remains a challenge as dispersion is frequency dependent.

Third harmonic OCT imaging has recently been introduced, although the technique is currently based on transillumination.[15,16] This is either the coherent anti-Stokes Raman scattering (CARS) or the phase sensitive approach, nonlinear interferometric vibrational imaging (NIVI). Future work is needed to ultimately establish the value of this approach.

9.2.6   OCT Dyes/Probes

In tissue where contrast is inadequate, OCT contrast agents can be used to improve imaging. These agents are typically based either on absorption or scattering.

9.2.6.1. Absorptive Agents

The absorptive agents can be divided into two groups, those where the absorbing properties of the agent are modified (active absorption) and those in which the absorptive properties are passively observed.

The active absorption group includes pump probe and pump suppression techniques. These techniques are performed in three steps. First, a baseline OCT image is acquired. Second, the contrast agent in the tissue is optically altered. Finally, another OCT image is acquired after the contrast agent is optically altered. The pump probe technique was first reported with methylene blue.[17] An OCT image is obtained as a baseline at 850 nm in the presence of the methylene blue. A high energy pulse laser is then used as the pump light, and for methylene blue, is at a power of 16 mW/cm^2 and a wavelength of approximately 650 nm. The pump light is used to cycle the molecule between the ground state and a higher singlet state, in addition to a relatively long-lived triplet state (approximately 2 μsec). A schematic is shown in Figure 9.2.[65] A probability exists of the triplet transitioning to another triplet state by a probe light at a different wavelength. Here, it is at approximately 850 nm, so this is the median wavelength at which the OCT image is performed. Therefore, the OCT backreflected signal is absorbed when the triplet transition occurs and the image is distinct from that of the baseline.

Several problems exist with this technique. First, it requires knowledge of the probe transition state in the local environment. Second, the 2-μsec lifetime of

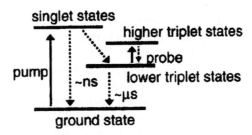

Figure 9.2 Pump probe technique. Principles are discussed in the text. Courtesy Yang, C. (2005). *Photochem. Photobiol.* 81, 215-237.

the triplet state may be long in terms of molecular transitions, but it is relatively rapid when compared to the imaging time of OCT system, so only a limited amount of data can be acquired. Third, relatively high intensities are required which may not be possible for clinical use.

The pump suppression technique overcomes the short transition time and involves a conformational change in the molecule of interest.[18] The absorption peak of the molecule changes when the molecule undergoes the structural change. With the pump suppression technique, the molecule is kept in one state by OCT imaging at a given wavelength. When a probing light is then used to illuminate the sample, the molecule transitions into a second state which absorbs at the OCT imaging wavelength, changing the image from baseline.

Two molecules have shown potential for pump suppression. They are bacteriorhodopsin (bR) and phytochrome A (Phy A). Phys A has been used with OCT. It has a strong absorption peak in the near IR region, which shifts with a conformational change of the molecule. The molecule has two possible states, which we will call A (Pfr, absorption 740 nm) and B (Pr, absorption, 670 nm). OCT imaging is performed at 750 nm, where the molecule is the Pr state. Absorption of the OCT signal is minimal. The tissue is then illuminated at 660 nm (suppression pump), which changes the absorption peak to 740 nm. After about 500 ms, OCT imaging is performed where absorption by the probe has now become significant. Because the relaxation time is at least six orders of magnitude larger, the intensities used can be substantially reduced.

9.2.6.2. Scattering Contrast Agents

While scattering probes do not absorb, they are analogous to absorbing probes in terms of their ability to generate contrast and therefore are discussed here. The difficulty with using a scattering probe to increase contrast is that the sample reflectivity needs to be increased substantially. In addition, these agents must target something relatively specific. Two types of scattering agents will be examined here. These are "engineered" scatterers and magnetically modified contrast agents.

One engineered scatterer used with OCT is the approximately 50 nm thick protein shell.[19] The average diameter is between 0.5 and 15 μm. Various scattering materials can be incorporated within the shell. The particles can be designed, for example, to be incorporated by macrophages or as antibodies that can be attached to bind to specific antigens on a certain cancer cell.

With magnetically modified contrast agents, typically ferromagnetic particles are used.[20] The particles are imaged with OCT in the presence of an oscillating magnetic field and the contrast signal can be obtained by finding the frequency component of the interference signal oscillating at the magnetic field frequency. By locking into the image signal change that is occurring at the oscillation frequency of the magnetic field, the distribution in the sample can be found. The molecules are either identified by moving in and out of the beam volume or by the molecule rotating in place.

Work on contrast agents is ongoing and future studies evaluating their *in vivo* use are critical.

9.3 ELASTOGRAPHY

9.3.1 Basic Concepts

Ultrasound and OCT elastography measure the mechanical properties of tissue. The basic concept is that images are obtained of individual tissue elements as they are compressed, stressed, or vibrated. Briefly reviewing tissue elasticity, we will assume that if an applied force is small enough, the relative displacement of points in the tissue will be proportional to the force. The force per unit area on an object is referred to as the stress.[21] The stretch (change of length) Δl per unit length l is called the strain: Strain = $\Delta l/l$. The stress and strain are related to each other through the Young's modulus (Y): Stress = $Y X$ Strain. When an object is stretched along its length, generally the width contracts. The contraction in width ($\Delta \omega$) is proportional to the width and strain:

$$\Delta w/w = -\sigma \Delta l/l. \tag{9.4}$$

The proportionality constant σ is the Poisson's ratio. Y and σ specify completely the elastic properties for a homogeneous, isotropic medium (that is not crystalline).

It is of interest to examine not only length or width but also volume. The volume strain ($\Delta V/V$) is given by $P = -K \Delta V/V$ where P is the pressure and K is known as the bulk modulus. The bulk modulus is given by $K = Y/3(1 - 2\sigma)$.

9.3.2 Ultrasound Elastography

The term ultrasound elastography was introduced for the first time by Ophir and colleagues in 1991.[22] The elastography technique is based on deformation of tissue typically assuming it is linear and isotropic elastic material. The tissue under inspection is deformed in a controlled fashion and the strain between pairs of ultrasound image is determined.[23–25] Its most significant applications currently are with assessing breast tissue for malignancy and atherosclerotic arteries for vulnerability to myocardial infarction. The main differences among the ultrasound elastography approaches are in the means of excitation (vibration or quasi-static) and the detection scheme (based on radio frequency or envelope analysis).[26] More specifically, there are basically seven techniques for assessing strain with ultrasound: low frequency amplitude imaging, high frequency amplitude, original elastography, envelope-based elastography, spectral tissue strain, phase-sensitive speckle tracking, and broadband rf-based elastography. While ultrasound elastography has shown some promise, it is limited by a resolution between 80 and 100 μm at high frequency, which is likely insufficient for vascular assessments.

9.3.3 OCT Elastography

Recently, elastography with OCT has been demonstrated using speckle tracking.[27–31] A primary strength of OCT-based elastography is the potential to evaluate the mechanics of intact tissue on a scale that has been out of the reach of ultrasound.

OCT imaging is performed at two different compressions. One technique for measuring the displacement is to first compute the normalized cross correlation of the pre- and post-compression medium within a defined window. The displacement maximum of the resultant cross-correlation function is then assessed. Specifically, the image data are processed pixel by pixel and the total displacement at a given pixel is computed using a cross-correlation coefficient between original and post-compressed images within a predefined area of $m_1 \times m_2$ pixels (this area is also known as kernel or window). The cross-correlation coefficient R between pixels in area $m_1 \times m_2$ in original image X and a potential target region in image Y, which will be moved by n pixels axially and k pixels laterally, is defined by:

$$R_{l,k} = \frac{\sum_{i=1}^{m_1} \sum_{j=1}^{m_2} (X_{i,j} - \bar{X})(Y_{i+l,j+k} - \bar{Y})}{\sqrt{\sum_{i=1}^{m_1} \sum_{j=1}^{m_2} (X_{i,j} - \bar{X})^2 (Y_{i+l,j+k} - \bar{Y})^2}} \qquad (9.5)$$

where $\bar{X}$ and $\bar{Y}$ are the mean values in $m_1 \times m_2$ areas on images X and Y, respectively. For identical regions, the cross-correlation coefficient $R = 1$, for uncorrelated regions, $R = 0$, and the values of R between 0 and 1 indicate the degree of correlation between image regions. Over the search region, an array of correlation coefficients will be calculated. This array is the cross-correlation function and its peak value identifies the target destination. For a given pixel, the axial and lateral displacements will be defined as values for which the cross-correlation function obtained maximum. The axial and lateral displacement values can be displayed as two-dimensional maps or combined together as displacement vector maps.[30,31] In addition, maximum cross correlation can also be displayed as a two-dimensional map. The size of the cross-correlation window is a very important parameter of the cross-correlation method, and it must be at least twice as large as the maximum pixel displacement.[30] If the window is too small, the displacement results will be incorrect.

On the other side, large windows tend to average out the differences in displacements of small features in images and decrease the ability of speckle tracking to make microstructural assessments.

In the original Schmitt work, an OCT system was used with a wide numerical aperture and 4-channel angular compounding to reduce interference noise. Dynamic focusing was also employed. Gelatin models, pork, and human skin were compressed while OCT imaging was performed. Strain was calculated as a function of depth. However, the data were not compared with any standard measure of tensile strength.

In another work by Rogowska et al.,[31] the importance of kernel size was demonstrated in gelatin phantoms which contained charcoal for scattering and aorta. In this study, phantom images were processed using cross correlation with several kernel sizes varying from 21×21 to 61×61 pixels. Larger kernel sizes were not able to track the small charcoal particles, and were not used in this study. The corresponding displacement vectors obtained using cross correlation with kernels of (A) 21×21, (B) 31×31, (C) 41×41, (D) 51×51, and (E) 61×61 pixels are shown on Figure 9.3. In terms of a percent error between calculated and measured displacements, the best results for phantoms were obtained with a 41×41 kernel (percent error 1.88%). For both phantom and aorta images, it was found that with the increasing size of the cross-correlation kernel, the axial and lateral displacement maps are less noisy, and the displacement vectors are more clearly defined. However, the large kernels tend to average out the differences in displacements of small particles in phantoms and decrease the ability of speckle tracking to make micro-structural assessments. Therefore, it was concluded that selection of the kernel size should be done carefully, based on the image features.

In another study,[33] the results of OCTE elasticity measurements were validated and quantitated with nonimaging mechanical measurements of the same phantoms. Stress-strain results on 1, 2, and 3% phantoms using caliper measurements are shown in Figure 9.4, while using OCTE measurements are shown in Figure 9.5. For each phantom, the linear regression equations are displayed, where slope is the estimated Young's modulus. Figure 9.6 demonstrates a comparison between the two. The average caliper and OCTE elastic modulus values of 1% phantoms were 115.7 $\pm$ 16.99 kPa and 125.82 $\pm$ 12.42 kPa, respectively, with an average error of 8.7%. The average caliper and OCTE elastic modulus values of 2% phantoms were 188.88 $\pm$ 24.25 kPa and 160.63 $\pm$ 6.26 kPa, with an average error of 14.9%. The average caliper and OCTE elastic modulus values of 3% phantoms were 218.28 $\pm$ 26.82 kPa and 200.46 $\pm$ 22.87 kPa, with an average error of 8.2%.

The displacement vectors for atherosclerotic aorta are shown in Figure 9.7. The average estimated Young's modulus for aorta sample was equal to 6.11 kPa.

9.3.4 Limitations of the Elastography Techniques

Significant obstacles exist in implementing OCT elastography *in vivo*. First, difficulties arise in applying controlled pressure *in vivo*. When the transmission of pressure is not uniform, lateral (or buckling), rather than axial, strain results. This can be minimized by keeping the displacement small, but as stated, the

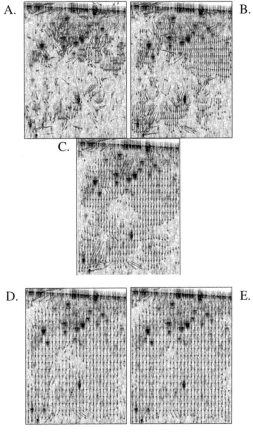

Figure 9.3 Influence of kernel size on elastography assessment. Kernels are increased from 21 × 21 (A) to 61 × 61 (E). The best results were obtained with kernels of 41 × 41 (C). Courtesy Rogowska, J. et al. (2004). *Heart.* 90, 556–562.

cross-correlation function works best with large displacements. Pressure changes will likely need to be provided *in vivo* for intra-arterial imaging via an occlusion balloon or with saline flushes. A second artifact arises because stress is typically not constant, where it is higher near the boundaries of the compressor. With ultrasound, several groups have addressed this through various processing techniques.[35–37] One group used the framework of inverse problem solutions to overcome some of the limitations. This technique does remove some of the artifacts associated with elastography, but its application to *in vivo* situations, such as imaging the coronary artery, is questionable. In another work, a weighted interpolation method has been shown with some potential. Another group found that the sum absolute differences (SAD) algorithm was superior to the cross-correlation function. This technique

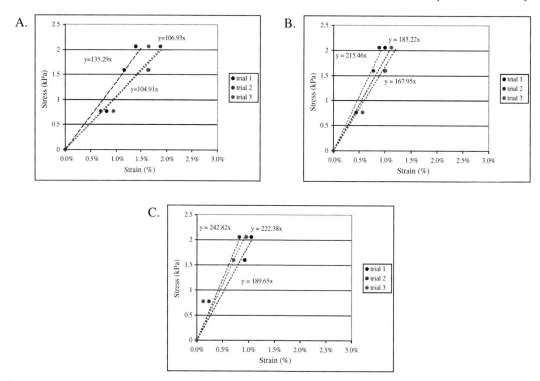

Figure 9.4 Strain-stress curves for (A) 1%, (B) 2%, and (C) 3% phantoms using three caliper measurements (trials 1-3). Young's modulus estimators are displayed as slope values of linear regression equations. Courtesy Rogowska, J. et al. (2002). *OSA Technical Digest*, PD20.1–20.3. [32,33]

subtracts two signals and computes the sum of the absolute differences for all the samples. Third, most models of elastography, both with OCT and ultrasound, assume that the tissue is nearly incompressible (the Poisson's ratio is about 0.5). In lipid-filled plaques, the validity of this assumption needs to be examined. Fourth, the cross-correlation function data are obtained from two different compressions. Data would likely be superior if analysis could be performed over a continuum.

9.4   DIFFERENTIAL PHASE MEASUREMENTS

9.4.1   Embodiment

Most OCT system designs measure the amplitude or the envelope of the interference signal but not the phase. In these traditional single channel designs, accurate phase measurements are difficult due to phase noise in the system, including air currents, temperature fluctuations, moving parts, and vibrations. Systems designs have been used for phase-sensitive and Doppler imaging, which incorporate decorrelation

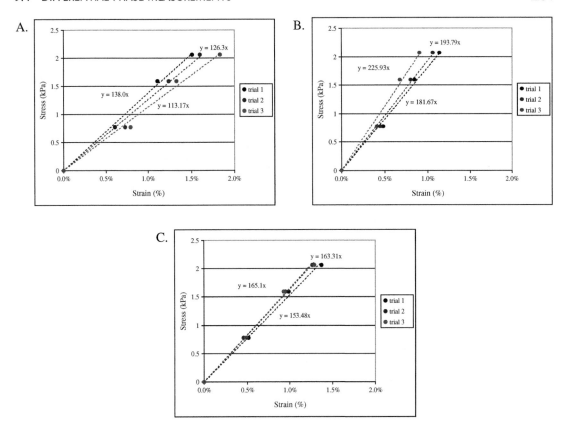

Figure 9.5 Strain-stress relationship for (A) 1%, (B) 2%, and (C) 3% phantoms using three OCT elastography measurements (trials 1-3). Young's modulus estimators are displayed as slope values of linear regression equations. Courtesy Rogowska, J. et al. (2002). *OSA Technical Digest*, PD20.1–20.3. [32]

of orthogonal polarization states, frequency modulation, and the analysis of phase with a Hilbert transform.[38–43]

The schematic of one system is shown in Figure 9.8 and described in detail below. The system uses a fiber-based dual channel Michelson interferometer where the two channels correspond to orthogonal and decorrelated polarization modes in the birefringence fiber. Polarization maintaining (PM) fibers are used throughout the system. Light from a partially polarized source (AFC Technologies) is completely depolarized by a Lyot depolarizer.[38,39] This is achieved by splicing two segments of birefringent fibers so that the orthogonal axes of the two fibers are at 45 degrees with respect to each other. Then the light passes through fiber loops, which changes their relative orientation in space and time. This ensures equal amplitudes of the two linearly polarized states and that the states are decorrelated after a short propagation distance. At the 2×2 PM coupler, the two independent and linearly decorrelated modes are split down each arm and form the two signal channels designated as 1 and 2.

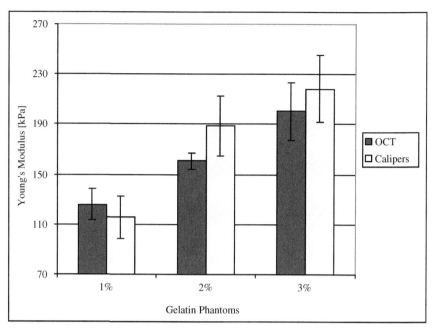

Figure 9.6   Young's modulus for 1%, 2%, and 3% phantoms estimated using two techniques: calipers' measurements and OCT elastography. The bars indicate standard deviation. Courtesy Rogowska, J., et al. (2006). *Br. J. Rad.* (in press).

In the reference arm, in the first embodiment, a dual channel electro-optic modulator (EOM)[38,39] is placed prior to the grating-based delay line (GBD) to allow precision control of the phase velocity in the reference arm. Because the EOM accepts light from only one linearly polarized state, use of a dual channel EOM (for each orthogonal state) is necessary. This is shown in Figure 9.9, but not in Figure 9.8. This allows each orthogonal state to be modulated at different frequencies. Phase modulation can be performed independently between 0 and 500 MHz. The EOM is driven with a ramp waveform with voltage amplitudes that gives sinusoidal fringe signals to reduce noise. The phase modulation frequency is set around 30 kHz.

The sample arm configuration, which is enlarged in Figure 9.10, is slightly more complex than conventional OCT designs. In the sample arm, a Wollaston prism is used.[40–42] The prism is used to separate polarization states. A lens is used to focus light on the Wollaston prism. The phase shift induced by the prism is determined by the equation:

$$\Delta\phi = \frac{4\pi\Delta n}{\lambda}\left[1 + \frac{1}{\cos(\theta/2)}\right]\Delta x \tan\alpha, \tag{9.6}$$

where $\Delta\phi$ is the delay, Δn is the birefringence of the prism, and $\theta/2$ is the divergence half angle of light transmitted through the prism. The two orthogonal states are

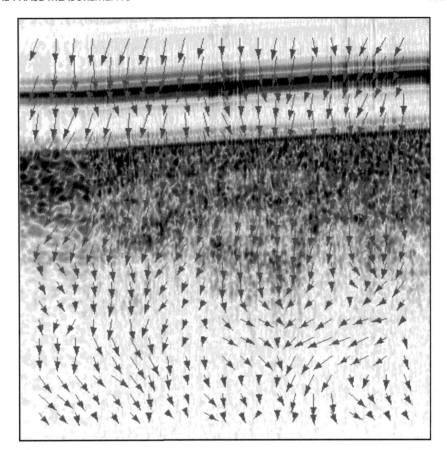

Figure 9.7 Displacement vectors using elastography on in vitro human aorta. Courtesy Rogowska, J., et al. (2006). *Br. J. Rad.* (in press).

separated by the prism and then focused by a separate lens into the tissue. This allows backscattering from both orthogonal states to be sampled from regions in close approximation. This is done to avoid errors associated with variable scattering/ refractive index and an uneven surface that may introduce unwanted phase delays. After the backreflected light passes through the lens and prism system on the return trip, it recombines with light from the reference arm. The two plates of the Wollaston prism are moved relative to one another to change the phase delay using a motorized stage.

Half the irradiance from the sample and reference arm is directed into the detector arm. In the detector arm, the recombined light is split with a fiber-based polarization sensitive beam splitter. This divides the orthogonal light components into different channels fiber coupled with separate detectors.

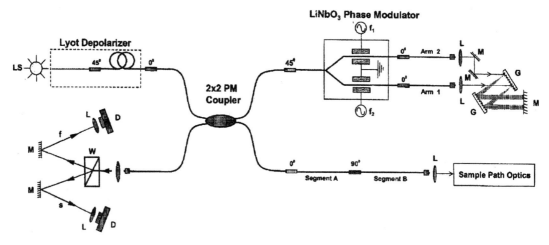

Figure 9.8 Schematic of a phase sensitive OCT system. Courtesy Dave, D.P., et al. (2001). *Optics Communication*, 193, 40.

Dual LiNbO₃ Phase Modulator

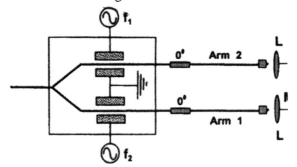

Figure 9.9 Schematic of the E-O modulator based on Figure 9.8. Concept courtesy Dave, D.P., et al. (2001). *Optics Communication*, 193, 40.

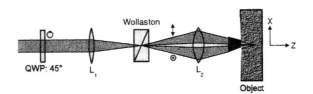

Figure 9.10 Schemtic of the sample arm based on Figure 9.8. Concept courtesy Dave, D.P., et al. (2001). *Optics Communication*, 193, 40.

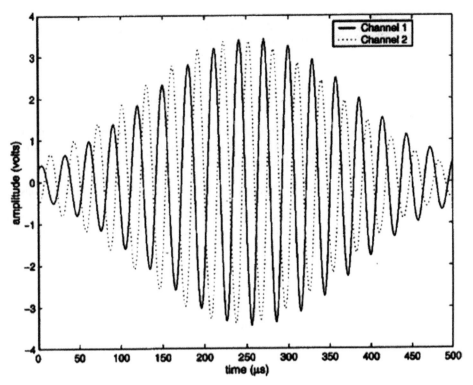

Figure 9.11 Detection through two detectors of phase sensitive OCT. Courtesy Dave, D., et al. (2000). *Optics Letters* 25(4), 227–229.

After storing the data, a signal processing algorithm is applied to extract the phase difference between the two fringe signals (Figure 9.11). The phase can be found through its relationship with the amplitude by using well-defined principles of optics. An algorithm that has previously been reported in optics (Appendix 5-1), including Doppler and differential phase OCT, is used to determine the amplitude ($A_k(z)$) and the phase ($\phi_k(z)$) of the oscillating term of each interference signal at any depths. These quantities can be extracted from the complex function:

$$\tilde{A}_k(z) = I_k(z) + i \cdot H\{I_k(z)\} = A_k(z) \cdot \exp[i \cdot \Phi_k(z)] \tag{9.7}$$

Knowing that for a standard Michelson interferometer the irradiance is described by:

$$I_k(z) = I_{r,k} + I_{s,k} + 2\sqrt{I_{r,k}I_{s,k}} \cdot \left|\gamma(z - z_0)\right| \cdot \cos(\Phi_k) \tag{9.8}$$

where the normalized modulus of the complex degree of interference is or the autocorrelation function. Here, $\phi_k = 2\pi (z - z_0)/\lambda_0$ and $I_{r,k}$ and $I_{s,k}$ are the reference and sample mirror intensities (at the detector). The reflectivity $(R(z))$ and phase retardation $(\delta(z))$ are then given by:

$$R(z) \sim A_1(z)^2 + A_2(z)^2$$
$$\delta(z) = \arctan\left(\frac{A_2(z)}{A_1(z)}\right) \tag{9.9}$$

Now, assuming the source spectral density is Gaussian with a spectral width $\Delta\omega$, the different frequencies, and system noise, the interference term (complex degree of interference) above is given by[39]:

$$I_m(\tau) \, \alpha \, e^{-(\tau/4a^2)} \cos[(f_1\tau + \phi_{1,m} + \phi_N) + \cos(f_2\tau + \phi_{2,m} + \phi_N)] \tag{9.10}$$

Here, $a = (\sqrt{\ln 2})/\Delta\omega$, and f_1 and f_2 are the modulation frequencies. The term ϕ_N is the noise in the interferometer due to environmental perturbation, and ϕ is the phase factor that carries the physically relevant information. The frequency multiplexed signals are separated by filtering and then the phase of each frequency channel is calculated by computing the arctangent of the filtered signal divided by its Hilbert transform.

In one filtering approach, this algorithm is implemented by a bandpass type II Chebyshev filter, which is monotonic in the passband and equiripple in the stop band and is designed to de-noise fringe data. To ensure a zero-phase response, a forward and reverse filtering technique is applied to both channels.

9.4.2 Limitations

This technology has several limitations. First, in the sample arm, the focusing after the Wollaston prism may require scanning of the focus in a manner analogous to confocal microscopy to maintain the beam waist at all depth levels. Second, due to the fact that imaging is performed in a highly scattering medium, phase shifts will occur as the light passes through the tissue due to multiple scattering. Third, motion artifacts, which induce phase shifts independent from changes in the refractive index, need to be addressed. Corrections can potentially be done by both using dual frequencies, as described above, and measuring phase shifts from surfaces inside the catheter or off the tissue surface through a guide beam.

9.5 ENTANGLED PHOTONS

Quantum mechanical phenomena have not been examined extensively in OCT research, although they are likely to play a more significant role in future investigations. One exception is the use of entangled photons to reduce or eliminate dispersion. While this technique is not yet viable for clinical imaging, it lays the groundwork for potential future approaches. This section will first examine the physics of two-particle entangled states (nonlocality), then it will move on to combining it with OCT.

9.5.1 General

The two-particle entangled state was initially described by Schrodinger and represents a state that does not factor into a product of the individual states.[45] With entanglement, a system of more than one particle must be treated as a single unit. The system of multiple particles is treated with one state vector that is a space function of all the particles with a single time function. Expressed another way, the amplitudes and uncertainties of entangled particles are correlated, as opposed to separable particles where the uncertainties are additive as they represent separate physical realities.

One of the most striking outcomes of entanglement is non-locality, where prior to collapse of the wavefunction of the entangled particles, the value of an observable such as momentum for each subsystem is not determined. However, once a measurement has been made on one subsystem (particle), irrespective of the distance between its entangled counterpart, the observable momentum of all other subsystems is exactly defined (with certainty).

This was brought to the forefront by the famous Einstein, Podolsky, and Rosen (EPR) experiments.[46] The EPR wavefunction is given by:

$$|\Psi\rangle = \sum_{a,b} \delta(a + b - c_0)|a\rangle|b\rangle \tag{9.11}$$

where a and b represent the position or momentum of particle 1 and 2, with c_0 being a constant. The delta function ensures that wavefunctions of the subsystems are not separable. More applicable to the following discussion are entangled photons, referred to as Bell states (or EPR-Bohm-Bell states). Ignoring for now the wavepackets or space-time aspect of the photons, the Bell states are given by four orthogonal states that form complete orthogonal basis and are given by[48]

$$\left|\Phi_{12}^{(\pm)}\right\rangle = \frac{1}{\sqrt{2}}\{|0_1 0_2\rangle \pm |1_1 1_2\rangle\}$$

$$\left|\Psi_{12}^{(\pm)}\right\rangle = \frac{1}{\sqrt{2}}\{|0_1 1_2\rangle \pm |1_1 0_2\rangle\} \tag{9.12}$$

Here, the kets of 0 and 1 represent the two orthogonal polarization states.

9.5.2 Entangled States Generation

Entangled photons can be produced by spontaneous parametric down conversion (SPDC).[49] With SPDC, light from a cw (or in some cases pulsed) laser is passed through a nonlinear crystal such as lithium niobate, potassium diphosphate (KDP), or barium borate, where a minority of the photons "break" into two photons of lower frequencies produced by nonlinear susceptibility. They are paired both with respect to direction and polarization. The photons are referred to as signal (*s*) and idler (*i*), a terminology that is a holdover from when early experiments produced a detected visible light photon and one in the IR region (not detected or idler).

The photons must obey the relationships:

$$\omega_p = \omega_s + \omega_i$$
$$k_p = k_s + k_i$$

(9.13)

representing the respective angular frequencies and wave vector. The SPDC process is not monochromatic so it leads to a range of frequency and momentum components. The SPDC processes can be divided into type, degenerate/nondegenerate, and direction. Type I has both s and i photons with the same polarization state while with type II they are orthogonal. The state is degenerate if angular frequencies are equal. The system is said to be collinear if the photons propagate in the same direction and noncollinear if they propagate in different directions.

The state function integral form can be obtained with first order perturbation theory and a nonlinear interaction Hamiltonian. Assuming a cw monochromatic source exciting the crystal with infinite interaction time, the integral form of the entangled photons is given by[47]:

$$|\Psi\rangle = A \int d^3k_s d^3k_i \delta(\omega_p - \omega_s(k_s) - \omega_i(k_i))$$
$$\times \delta(k_p - k_s - k_i)a_s^\dagger(k_s)\, a_i^\dagger(k_i)|0\rangle$$

(9.14)

where the 0 ket represents the initial vacuum state, $a_s^\dagger$ and $a_i^\dagger$ are the respective creation operators, and A is a normalization constant. The wave vector is used as a measure of momentum, which is easier for notation, as it differs only by a multiple of Planck's constant.

The concept of a two-photon wavepacket or biphoton will be critical for the fourth order interference experiments to be discussed and their application to OCT. "According to the quantum theory of optical coherence, the electromagnetic field can be expanded in modes having a single variety of polychromatic photons excited."[50] Given a monochromatic field in one dimension and an infinite cavity length, the wavepacket operator is given by:

$$\hat{A}^\dagger(\epsilon) \equiv \int_0^\infty \epsilon(\omega)\hat{a}^\dagger(\omega)\, d\omega$$

(9.15)

where the first term in the integral is the complex amplitude and the second is the continuum raise operator at ω. By allowing this operator to act on the vacuum, the single-photon wavepacket state is obtained:

$$|1;\epsilon\rangle \equiv \hat{A}^\dagger(\epsilon)|0\rangle = \int_0^\infty \epsilon(\omega)|1_\omega\rangle\, d\omega$$

(9.16)

where $|1_\omega\rangle \equiv \hat{a}^\dagger(\omega)|0\rangle$ are the monochromatic single photon states. The wavepacket state is associated with a coherence length analogous to conventional OCT.

9.5.3   Interference Experiments

Several points need to be addressed before interferometry of entangled photons is discussed. First, the fundamental reason why second order interference, such as is measured with Young's experiments or classic OCT, does not typically measure with entangled photons is that there is no definite phase relationship between two photons (although this is not absolute). Therefore, fourth order interference with entangled photons (second order correlation) is measured using two detectors similar in principle to those in a Brown-Twiss interferometer. Second, as modified from Dirac, a pair of photons can only interfer with themselves; two different photons (photon pairs) never interfere.[50,51] This is linked to the concept of indistinguishability, which is required for interference to occur.

The discussion will begin with a correlation interferometer that utilizes a beam splitter, which is shown in Figure 9.12.[47] In a configuration without a beam splitter, shown in Figure 9.13, dispersion can be canceled by having the two arms contain dispersive elements with dispersion constants that are equal in sign magnitude but opposite in magnitude. This is purely a quantum mechanical phenomenon.

In the beam splitter embodiment, when no dispersive elements are present, as the pathlength is scanned, a dip in correlation occurs at the position of equal arm pathlength with the width due to the coherence length of the light. This is due to destructive interference between the two Feynman paths (indistinguishable outcomes). When a dispersive element is present in one arm (Figure 9.13), dispersion

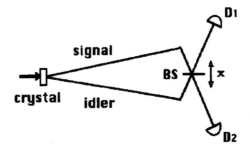

Figure 9.12 Correlation interferometry that uses a beam splitter. Moving the beam splitters in the x-direction alters the amount of correlation as will be seen. Courtesy Shin, Y. (2003). *IEEE of Selected Topics in Quantum Electronics*, 9(6), 1458.

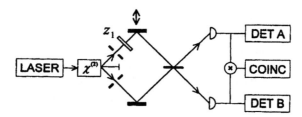

Figure 9.13 Use of a dispersive element in a correlation interferometer with beam splitter. A dispersive element in one arm is one method for correction/altering dispersion. Courtesy Larchuk, T.S., et al. (1995). *Phys. Rev. A.* 52(5), 4145.

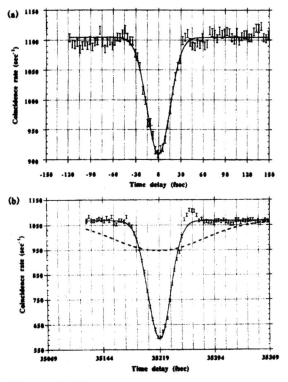

Figure 9.14   Plot of correlations with beam splitter movement. In (a), the beam splitter is moved and the maximum correlation occurs at zero splitter displacement. In (b), a dispersive element is added to the system which shift the point of correlation off zero. Courtesy Steinberg, A.M., et al. (1992). *Phys. Rev. Lett.* 68(16), 2424.

cancellation of the lowest order dispersive effects (and other odd orders) can still be performed. Here, with the dispersive medium placed in one arm, the dip remains the same in intensity but the relative location of the peak is changed as shown in Figure 9.14.[53] This dispersion cancellation occurs when the coherence gate is slow. The time resolution decreases with a fast detector system, as dispersion terms become significant in the approximation of a long interval. In other words, the time window needs to be longer than the dispersion-induced broadening.

The detector counting rate, R_C, of two detectors over the time interval T is given by the formula of recent Nobel Laureate R. Glauber[54]:

$$R_c \propto \frac{1}{T} \int_0^T \int_0^T dt_1 \, dt_2 \left| \langle 0 | E_2^{(+)}(t_2, r_2) E_1^{(+)}(t_1, r_1) | \Psi \rangle \right|^2 \tag{9.17}$$

where $E_{a,b}{}^{(\pm)}(t_1, r_1)$ and $E_{a,b}{}^{(\pm)}(t_2, r_2)$ are the fields at detector 1 and 2.

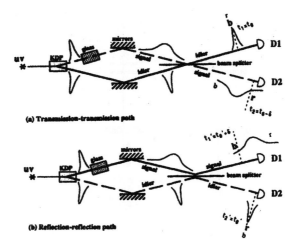

Figure 9.15 Feynman representation of the process occurring in Figure 9.14. Courtesy Steinberg, A.M., et al. (1992). *Phys. Rev. Lett.* 68(16), 2423.

While the dip and its shifting can be interpreted in terms of photon bunching and antibunching, Feynman diagrams will be employed here for explaining the phenomena, where it is due to frequency anticorrelation between the two conjugate photons. The two Feynman paths must lead to the same pair of detections.

The reason why degradation of the image does not occur in the presence of a dispersive element is as follows. Figure 9.15 demonstrates the two equivalent Feynman paths, one where both photons are transmitted and one where both are reflected.[53] The conjugate photons are depicted as minimum uncertainty Gaussian wavepackets. A dispersive element is present in the upper arm. To within the relatively long coherence gate window, the correlations depend only on the centers of the wavepacket meeting simultaneously at the beam splitter. The interference occurs as a result of the long coherence gate, and because the outcomes of the two paths do not differ in any observable way and therefore are fundamentally indistinguishable.

9.5.4 Ghost Imaging

Ghost imaging relates to entanglement and may have implications for OCT. With ghost imaging, an aperture is placed in one arm (signal) while the other arm (idler) is scanned in a transverse plane with a mobile detector (Figure 9.16).[55,56] The result is a sharp, magnified image of the aperture found in coincident counting (Figure 9.17) where the aperture here is the initials of the University of Maryland Baltimore County, even though detector single counting rates remain constant. Although aspects of the ghost image can be reproduced through classical approaches, several features of the effect are purely quantum mechanical. This includes the magnification properties, potential for 100% visibility, and resolution which may surpass the diffraction limit.

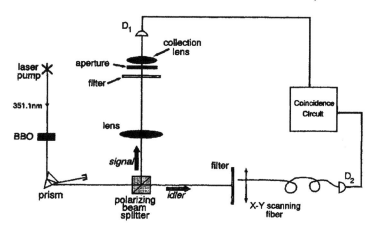

Figure 9.16   Schematic of a ghost imaging system. With this ghost imaging approach, an aperture is placed in one arm while the other arm is scanned in a transverse plane with a mobile detector. Figure 9.17 is an example image. Courtesy Pittman, T.B., et al. (1995). *Phys. Review A.* 52(5), 3430.

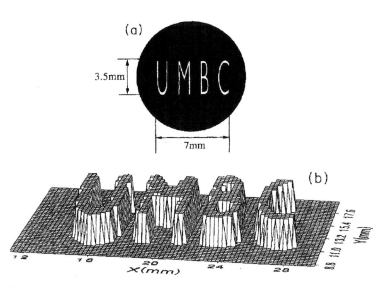

Figure 9.17   Reproduction of an aperature with the University of Maryland Baltimore county initials. Courtesy Pittman, T.B., et al. (1995). *Phys. Review A.* 52(5), 3430.

Magnification of the aperture is not due to different pathlength distance from the two paths, but to a quantum mechanical phenomenon. The relative magnification is due to the difference in the distance between the lens and the aperture (S) and *the distance from the lens back to the source forward through the beam splitter to the imaging plane(S').* This can be seen in Figure 9.18 where the system is unfolded with

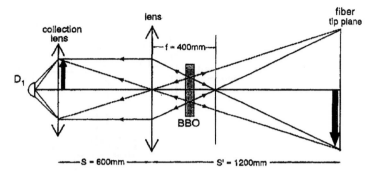

Figure 9.18 Schematic illustrating the non-classical magnification of the system in Figure 9.16 (unfolded). The relative magnification is due to the difference in distance between the lens and the aperature (S) and the distance from the lens back to the source forward through the beam splitter to the imaging plane (S'). Courtesy Pittman, T.B., et al. (1995). *Phys. Review A.* 52(5), 3431.

the source as the hinge. The relationship between the distances and the focal length of the lens is given by:

$$1/S + 1/S' = 1/f$$

This equation is referred to as the "quantum Gaussian thin lens equation."

Entangled photons exhibit: (1) coherent superposition of the two photon amplitudes and (2) correlation of the position and momentum variables. The second results in the violation of the inequality (but not the uncertainty) principle:

$$\Delta(p_1 + p_2))\Delta(x_1 - x_2) \geq \hbar \tag{9.18}$$

Because of these properties, the potential visibility is 100% (no background noise) and lateral resolution is only limited by the numerical aperture. In addition, it has been suggested that with the appropriate setup, resolution beyond the diffraction limit may be possible.[58]

A classical source, such as a thermal one, can simulate some aspects of ghost imaging.[58,59] A thermal source can generate an image that can be understood as the addition of two photon probability amplitudes. However, there is no correlation between the position and momentum variables, which has led the process to be referred to as incoherent two photon imaging. This lack of correlation has two important consequences. First, the visibility is limited to about 33% by a constant background noise, although some detection schemes may improve upon this. Second, for the same reason, it can not surpass the classical Rayleigh diffraction limit. Nevertheless, the use of incoherent two-photon states may have a role in OCT, particularly in view of the difficulties in generating photons through SPDC.

9.5.5 Entanglement and OCT

The use of entangled photons in OCT imaging has been studied largely through the efforts of a group at Boston University.[61,62] These publications refer to the technique

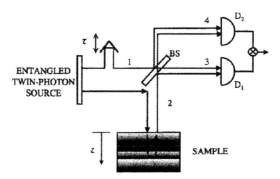

Figure 9.19 OCT system utilizing entangled photons. Courtesy Abouraddy, et al. (2002). *Phys. Rev. A. 65*, 053817-2.

as quantum OCT (Q-OCT), but I will refrain from using this terminology as it is likely that other quantum phenomena will be applied to OCT, such as squeezing, quantum nondemolition, and the Aharonov-Bohm effect. Therefore, I will refer to the technique as the use of entangled photons with OCT.

Figure 9.19 is a schematic of a system utilizing entangled photons. It is similar to the Hong, Oh, Mandel interferometer, except light from one arm is directed at the sample.[51] This setup operates under the theory described above. The coincidence rate $C(\tau)$ is given by:

$$C(\tau) \propto \Lambda_0 - \text{Re}\{\Lambda(2\tau)\} \qquad (9.19)$$

where τ is the delay and the self-interference term is given by:

$$\Lambda_0 = \int d\Omega |H(\omega_0 + \Omega)|^2 S(\Omega) \qquad (9.20)$$

and the cross-interference term is given by:

$$\Lambda(\tau) = \int d\Omega H(\omega_0 + \Omega) H^*(\omega_0 - \Omega) S(\Omega) e^{-i\Omega\tau}$$
$$= h_q(\tau)^* s(\tau) \qquad (9.21)$$

where H is the transfer function and S is the source spectrum. Ω is the frequency shift from the two-photon process. The cross-correlation function in Eq. 9.19 has the delay multiplied by a factor of two, which corresponds up to doubling of the axial resolution for a given source bandwidth and median wavelength.

When taking into account dispersion, the cross-correlation function can be written as the sum of two terms, which have been derived elsewhere[60]:

$$\Lambda(\tau) = \sum_j |r_j|^2 s\left(\tau - 4\frac{z_j}{v_0}\right)$$
$$+ \sum_{j \neq k} r_j r_k^* s_d^{(jk)}\left(\tau - 2\frac{z_j + z_k}{v_0}\right) e^{i2\beta_0(z_j - z_k)} \qquad (9.22)$$

Here, r represents reflectivity at a given layer, z is the distance to a layer, s is the correlation function of the source, and β is the first order dispersion. The s_d is the Fresnel transform given by the equation:

$$s_d^{(jk)}(\tau) = \int d\Omega S(\Omega) e^{i2\beta''\Omega^2(z_j - z_k)} e^{-i\Omega r} \tag{9.23}$$

The second term arises from the cross interference between reflection amplitudes associated with surfaces at different depths and is sensitive to dispersion. Therefore, it is proposed that the first term can be used to assess backreflection as a function of depth while the second term can be used to determine the refractive index between layers.[62] Separating the two terms can be done either by using a technique which involves varying the pump frequency or through a Wigner distribution.[63] Our group is attempting to achieve this with conventional OCT by varying the dispersion in the reference arm.[64]

Comparative OCT and entangled photon images of the two surfaces of a piece of silica after passing through a dispersive medium (ZeSe) is shown in Figure 9.20. In the entangled photon image the two surfaces of the silica are sharply defined at the predicted resolution (18 μm). In addition, a peak is found in between the two surfaces due to interference between the two surfaces. This is a quantum mechanical phenomenon that is altered by the refractive index of the medium. This peak represents the second term of Eq. 9.22. In the OCT image, the two surfaces are poorly defined due to dispersion.

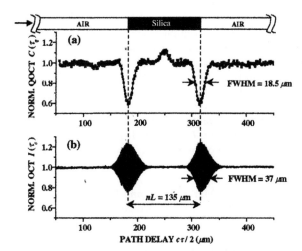

Figure 9.20 Comparative images of entangled photon (a) and conventional OCT images (b). Images were performed of two air-silica interfaces. A higher resolution is noted with the entangled photons. The center peak is likely due to a quantum interference effect between the two peaks. Courtesy of Nasr, M.B., 2003. *Phys. Rev. Lett.* 91(8), 083601-3.

The technique suffers from several significant limitations. First, the SDPC sources do not have sufficient intensity for performing biomedical imaging. This may potentially be compensated for with thermal or pseudothermal sources, although the lack of correlation of momentum and position uncertainty as described above reduce the attractiveness of the technique. Second, a slow time window is required and those used in published works are on the order of nanoseconds. Third, assessing the refractive index of medium between two peaks due to the interference between two surfaces will be difficult in scattering medium as the peak is likely to be obscured or altered. Nevertheless, the use of entangled photons, and quantum phenomena in general, with OCT does represent an exciting area of future research.

9.6 CONCLUSION

The techniques of absorption spectroscopy, contrast probes, phase contrast, elastography, and entangled photons have not yet reached the stage of clinical utility. However, these approaches demonstrate considerable promise and may ultimately be powerful tools for assessing pathology in humans.

REFERENCES

1. Sasano, Y. (1988). Simultaneous determination of aerosol and gas distribution by DIAL measurements. *Appl. Opt.* 27, 2640–2641.
2. Papayannis, A., Ancellet, G., Pelon, J., and Meggie, G. (1990). Multiwavelength lidar for ozone measurement in the troposphere and lower stratosphere. *Appl. Opt.* 29, 467–476.
3. Schmitt, J. M., Xiang, S. H., and Yung, K. M. (1998). Differential absorption imaging with optical coherence tomography. *J. Opt. Soc. Am. A* 15, 2288–2296.
4. Sathyam, U. S., Colston, B. W., Da Silva, L. W., and Evert, M. J. (1999). Evaluation of optical coherence quantitation of analytes in turbid media by use of two wavelengths. *Appl. Opt.* 38, 2097–2104.
5. Morgner, U., Drexler, W., Kartner, F. X. et al. (2000). Spectroscopic optical coherence tomography. *Opt. Lett.* 25, 111–113.
6. Leitgeb, R., Wojkowski, M., Kowalczyk, A. et al. (2000). Spectral measurement of absorption by spectroscopic frequency domain optical coherence tomography. *Opt. Lett.* 25, 820–822.
7. Xu, C., Ye, J., Marks, D. L., and Boppart, S. A. (2004). Near infrared dyes as contrast enhancing agents for spectroscopic optical coherence tomography. *Opt. Lett.* 29, 1647–1650.
8. Faber, D. J., Mik, E. G., Aalders, M. C., and van Leeuwen, T. G. (2005). Toward assessment of blood oxygen saturation by spectroscopic optical coherence tomography. *Opt. Lett.* 30, 1015–1017.
9. Xu, C., Kamalabadi, F., and Boppart, S. A. (2005). Comparative performance analysis of time frequency distributions (TFD) for spectroscopic optical coherence tomography. *Appl. Opt.* 44, 1813–1822.
10. Applegate, B. E., Yang, C., Rollins, A. H., and Izatt, J. A. (2004). Polarization resolved second harmonic generation optical coherence tomography. *Opt. Lett.* 29, 2252–2254.

11. Jiang, Y., Tomov, I., Wang, Y., and Chen, Z. (2004). Second harmonic generation optical coherence tomography. *Opt. Lett.* 29, 1089–1092.

12. Yazdanfar, S., Laiho, L. H., and Co, P. T. (2004). Interferometric second harmonic generation microscopy. *Opt. Express* 12, 2739–2742.

13. Roth, S. and Freund, I. (1981). Optical second harmonic scattering in rat tail tendon particularly distinguishing collagen type I from type III. *Biopolymers* 20, 1271–1290.

14. Cox, G., Kable, E., Jones, A. et al. (2003). 3-Dimensional imaging of collagen using second harmonic generation. *J. Struct.l Biol.* 141, 53–62.

15. Marks, D. L. and Boppart, S. A. (2004). Nonlinear interferometric vibrational imaging. *Phys. Rev. Lett.* 92, 123905-1–123905-5.

16. Vinegoni, C., Bredfeldt, J. S., Marks, D., and Boppart, S. A. (2004). Nonlinear optical contrast enhancement for optical coherence tomography. *Opt. Express* 12, 331–341.

17. Rao, K., Choma, M. A., Yandanfar, S. et al. (2003). Molecular contrast in optical coherence tomography by use of pump. *Opt. Lett.* 28, 340–342.

18. Yang, C., Choma, M. A., Lamb, L. E. et al. (2004). Protein based molecular contrast OCT with phytochrome as the contrast agent. *Opt. Lett.* 29, 1396–1402.

19. Lee, T. M., Oldenburg, A. L., Sitafalwall, S. et al. (2003). Engineered microsphere contrast agents for optical coherence tomography. *Opt. Lett.* 28, 1546–1548.

20. Oldenburg, A. L., Gunther, J. R., and Boppart, S. A. (2005). Imaging magnetically labeled cells magnetomotive optical coherence tomography. *Opt. Lett.* 30, 747–749.

21. Feynman, R. P., Leighton, R. B., and Sands, M. (1989). *The Feyman Lectures on Physics. Volume II.* Addison-Wesley, Redwood City, CA.

22. Ophir, J., Cespedes, I., Ponnekanti, H., Yanzi, Y., and Li, X. (1991). Elastography: A quantitative method for imaging the elasticity of biological tissues. *Ultrason. Imaging* 13, 111–134.

23. de Korte, C. L., Cespedes, I., van der Steen, A. F., and Lancee, C. T. (1997). Intravascular elasticity imaging using ultrasound. *Ultrasound Med. Biol.* 23, 725–746.

24. de Korte, C. L. and van der Steen, A. F. (2002). Intravascular ultrasound elastography: An overview. *Ultrasonics* 40(1–8), 859–865.

25. Erkamp, R. Q., Wiggins, P., Skovoroda, A. R., Emelianov, S. Y., and O'Donnell, M. (1998). Measuring the elastic modulus of small tissue samples. *Ultrason. Imaging* 20, 17–28.

26. van der Steen, A. F. W., de Korte, C. L., and Cespedes, E. I. (1998). Intravascular elastography. *Ultraschall. Med.* 19, 196–201.

27. Schmitt, J. M. (1998). OCT elastography: Imaging microscopic deformation and strain in tissue. *Opt. Express* 3(6), 199–211.

28. Chan, R., Chau, A., Karl, W. C. et al. (2004). Vascular optical coherence elastography: Assessment of conventional velocimetry applied to OCT. *Opt. Express* 12(19), 4558–4572.

29. Hansen, K. A., Weiss, J. A., and Barton, J. K. (2002). Recruitment of tendon crimp with applied tensile strain. *J. Biomech. Eng.* 124(1), 72–77.

30. Rogowska, J., Patel, N. A., Fujimoto, J. G., and Brezinski, M. E. (2002). OCT elastography of the vascular tissue — Importance of cross-correlation kernel size. OSA Biomedical Topical Meetings, *OSA Technical Digest.* Optical Society of America, Washington, D. C., pp. PD20.1–20.3.

31. Rogowska, J., Patel, N. A., Fujimoto, J. G., and Brezinski, M. E. (2004). Quantitative OCT elastography technique for measuring deformation and strain of the atherosclerotic tissues. *Heart* 90(5), 556–562.

32. Rogowska, J., Patel, N. A., Fujimoto, J. G., and Brezinski, M. E. (2002). OCT elastography of the vascular tissue — Importance of cross-correlation kernel size. OSA Biomedical Topical Meetings, *OSA Technical Digest.* Optical Society of America, Washington, D. C., pp. PD20.1–20.3.

33. Rogowska, J., Patel, N. A., Plummer, S., and Brezinski, M. E. (2006). Quantitative OCT elastography: Method for assessing arterial mechanical properties. *Br. J. Radial.* (in press).

34. Kallel, F. and Bertrand, M. (1996). Tissue elasticity reconstruction using linear perturbation method. *IEEE Trans. Med. Imaging*, 15, 299–313.

35. Konofagou, E. and Ophir, J. (1998). A new elastographic method for estimation and imaging of lateral displacements, lateral strains, and poisson's ratios in tissues. *Ultrasound Med. Biol.* 24, 1183–1199.

36. Janssen, C. R., de Korte, C. L., and van der Heiden, M. S. (2000). Angle matching in intravascular elastography. *Ultrasonics* 38, 417–423.

37. Dave, D. and Milner, T. (2000). Optical low coherence reflectometer for differential phase measurement. *Opt. Lett.* 25, 227–229.

38. Dave, D., Akkin, T., Milner, T. et al. (2001). Phase sensitive frequency multiplexed optical low coherence reflectometry. *Opt. Commun.* 193, 39–43.

39. Zhao, Y., Chen, Z., Saxer, C. et al. (2000). Phase resolved OCT and ODT for imaging blood flow in human skin with fast scanning speed and high velocity sensitivity. *Opt. Lett.* 25, 114–116.

40. Hitzenberger, C. K., Gotzinger, E., Sticker, M. et al. (2001). Imaging of polarization properties of transparent and scattering structure by phase resolved polarization sensitive optical coherence tomography. *Opt. Technol. Biomed. Med. III* 120–125.

41. Hitzenberger, C. K. and Fercher, A. F. (1999). Differential phase contrast optical coherence tomography. *Opt. Lett* 24, 622–624.

42. Picher, M., Hitzenberger, C. K., Sticker, M. et al. (2002). Absorption and dispersion measurements of water, D_2O and acetone by phase resolved PCI and OCT in the mid-infrared range 1.3 μm to 2.0 μm. *Coherence Domain Opt. Methods Biomed. Sci. Clin. Appl. VI* 4619, 82–89.

43. Yariv, A. (1997). *Optical Electronics in Modern Communications*, Fifth Edition. Oxford University Press, New York.

44. Wilson, J. and Hawkes, J. (1998). *Optoelecrtronics*, third edition. Prentice Hall, London, England.

45. Schrodinger, E. (1935). Die gegenwartige situation in der quantenamechanik. *Naturwissenschaften* 23, 823–844.

46. Einstein, A., Podolsky, B., and Rosen, N. (1935). Can quantum mechanical description of reality be considered complete? *Phys. Rev.* 47, 777.

47. Shih, Y. (2003). Entangled photons. *IEEE J. Selected Top. Quantum Electron.* 9, 1455–1467.

48. Harris, S. E., Oshman, M. K., and Byer, R. L. (1967). Observation of tunable optical parametric fluorescence. *Phys. Rev. Lett.* 18, 732–734.

49. Campos, R. A., Saleh, B. E., Teich, M. C. (1990). Fourth order interference of joint single photon wave packets in lossless optical systems. *Phys. Rev. A* 42, 4127–4137.

50. Dirac, P. A. M. *The Principles of Quantum Mechanics*, third edition. Clarendon Press, Oxford.

51. Hong, C. K., Ou, Z. Y., and Mandel, L. (1987). Measurement of subpicosecond time intervals between two photons by interference. *Phys. Rev. Lett.* 59, 2044–2046.

52. Franson, J. D. (1992). Nonlocal cancellation of dispersion. *Phys. Rev. A* 45, 3126–3132.

53. Steinberg, A. M., Kwiat, P. G., and Chiao, R. Y. (1992). Dispersion cancellation in a measurement of the single photon propagation velocity in glass. *Phys. Rev. Lett.* 68, 2421–2424.

54. Glauber, R. J. (1963). The quantum theory of optical coherence. *Phys. Rev.* 130, 2529–2539; Glauber, R. J. (1963). Photon correlations. *Phys. Rev. Lett.* 10, 84–86.

55. Pittman, T. B., Shih, Y. H., Strekalov, D. V., and Sergienko, A. V. (1995). Optical imaging by means of two photon quantum entanglement. *Phys. Rev. A* 52, R3429–R3432.

56. Angelo, M. D., Valencia, A., Rubin, M. H., and Shih, Y. (2005). Resolution of quantum and classical ghost imaging. *Phys. Rev. A* 72, 013810-1–013810-19.

57. Boto, A. N., Kok, P., Abrams, D. S. et al. (2000). Quantum interferometric optical lithography: Exploiting entanglement to beat the diffraction limit. *Phys. Rev. Lett.* 13, 2733–2736.

58. Valencia, A., Scarcelli, G., D'Angelo, M., and Shih, Y. (2005). Two photon imaging with thermal light. *Phys Rev. Lett.* 94, 063601-1–063601-4.

59. Shapiro, J. H. and Sun, K. (1994). Semiclassical versus quantum behavior in fourth-order interference. *J. Opt. Soc. Am. B* 11, 1130–1141.

60. Abouraddy, A. F., Nasir, M. B., Saleh, B. A. et al. (2002). Quantum optical coherence tomography with dispersion cancellation. *Phys. Rev. A* 65, 053817-1–053817-6.

61. Nasr, M. B., Saleh, B. E., Sergienko, A.V., and Teich, M. C. et al. (2003). Demonstration of dispersion cancellation quantum optical tomography. *Phys. Rev. Lett.* 91, 083601-1–083601-4.

62. Nasr, M. B., Saleh, B., Serienko, A. V., and Teich, M. C. (2004). Dispersion cancelled and dispersion sensitive quantum optical coherence tomography. *Opt. Express* 12, 1353–1362.

63. Ben-Aryeh, Y., Shih, Y. H., and Rubin, M. H. (1999). Optical Wigner functions for two photon entangled wavepackets. *J. Opt. Soc. Am. B* 16, 1730–1736.

64. Liu, B., MacDonald, E. A., Stamper, D. L., and Brezinski, M. E. (2004). Group velocity dispersion effects with water and lipid in 1.3 μm OCT System. *Phys. Med. Biol.* 49, 1–8.

65. Yang, C. (2005). Molecular contrast optical coherence tomography. *A. Rev. Photochem. Photo.* 81, 215–237.

10 DOPPLER OPTICAL COHERENCE TOMOGRAPHY

Bin Liu, PhD

Brigham and Women's Hospital, Harvard Medical School

Chapter Contents

Since the emergence of optical coherence tomography (OCT) technology,[15] more attention has been paid to Doppler OCT. The research on this topic has been very active and fruitful.[11,12,16,18,42,47,61] In previous chapters, we have discussed and demonstrated that OCT has a high potential in imaging nontransparent biological tissues with near pathological resolution and multidimensional microstructure imaging capability.[7,8,38] For many OCT applications, the fluid-flow dynamics or moving scatters in a sample is also of interest in addition to morphological

277

information. For instance, the blood flow velocity distribution in a biological tissue can provide more diagnostic information. Doppler OCT can provide this kind of functional extension of OCT by measuring the Doppler frequency shift in the OCT signal. Thus local flow velocity can be measured and mapped with few optical changes to the conventional OCT image. Compared with Doppler ultrasound, another Doppler imaging technique in medicine, Doppler OCT has an intrinsic advantage in velocity measurement resolution since the wavelength of lightwave is much shorter than ultrasound. As a prevalent and developed medical imaging technology, Doppler ultrasound might prove an aid to Doppler OCT signal processing. For instance, relevant signal processing algorithms in Doppler ultrasound have been applicable for Doppler OCT[50,55] in autocorrelation velocity estimation. Doppler OCT can also be benefited by sophisticated electronic techniques for signal processing.

In this chapter, laser Doppler velocimetry and Doppler OCT will be discussed through the aspect of working principle to the aspect of signal processing. Our primary interest will focus on the physical mechanisms and mathematical modeling, while applications will be discussed in Section III.

10.1   THE PRINCIPLE OF DOPPLER OCT

Doppler OCT measures and maps flow velocity by combining OCT technique with laser Doppler velocimetry, a modality developed in the 1960s.[57] Laser Doppler velocimetry has had a variety of applications such as blood flow measurement[5,37] since the early 1970s. Conventional laser Doppler velocimetry is typically a nonimaging technique, whereas Doppler OCT can provide local velocity mapping with high spatial resolution. As a result, Doppler OCT has found distinct applications in the field of laser Doppler velocimetry. Thus we begin with the concept of laser Doppler velocimetry.

10.1.1   Doppler Shift and Laser Doppler Velocimetry

Suppose we consider a radiation detection experiment such as that depicted in Figure 10.1A. A monochromatic light beam with frequency f_0 is projected to a detector. If the light source and the detector have a relative motion, the frequency of the detected light field will vary f_D or $-f_D$ off f_0. This is called Doppler shift, which occurs in all forms of wave phenomena. Lasers are the most popular light source used for studying the Doppler shift in the optical frequency range. We will use the specified term laser Doppler shift throughout this chapter. For the simplification of mathematical formulism, the light field from the source is assumed to be a plane wave propagating in Cartesian coordinates x, y, and z:

$$E = E_0 \exp[-i(2\pi f_0 t - \mathbf{K}_i \cdot \mathbf{r})] \qquad (10.1)$$

where E_0 represents the amplitude of the light field; the vector $\mathbf{r}$ points to a certain position O in the coordinates x, y, and z, which could be the observation point

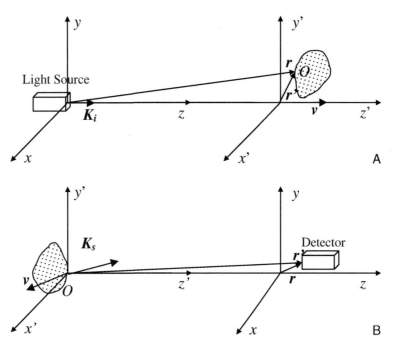

Figure 10.1 Schematic of laser Doppler shift measurement. (a) The observation point is moving relative to the light source. (b) The scattering object is moving while the light source and the detector is fixed.

or the light scattering point; and the vector K_i denotes the propagation coefficient of the incident light. If spot O is moving with a velocity v in the fixed reference frame x, y, and z, the frequency of the detected light field at spot O will differ from the incident light.

Alternatively spot O can be described as a fixed point with a position vector r' in new Cartesian coordinates x', y', and z', which translates in velocity v. Thus, the light field observed in this moving frame can be described as

$$E = E_0 \exp[-i(2\pi f_0 t - K_i \cdot (r' + vt)]$$ (10.2)

Usually the origin of the Cartesian coordinates x', y', and z' is chosen at spot O. In other words, r' equals zero. The frequency of the detected light field will be $f_0 - 1/(2\pi)K_i \cdot v$. The dot product of the vectors K_i and v explicitly indicates that v can be determined by measuring the frequency shift once the angle between the incident light direction and moving direction are known. If the light source and the detector are approaching one another, the frequency of the detected light field will be greater than the incident light ($f_D > 0$). If they are moving away from one another, the frequency will be smaller ($f_D < 0$). This represents the basic concept underlying laser Doppler velocimetry.

In practical scenarios, the positions of the light source and the detector are fixed but the detected objects are moving. In Figure 10.1B, the assumption is made that the incident light encounters a scattering object at position O, the origin of the moving coordinates. In this moving reference frame, the scattered light field will keep the same frequency as the incident light known as $f_0 - 1/(2\pi)\boldsymbol{K}_i \cdot \boldsymbol{v}$. Subsequently the scattered light field can be described as

$$E_s = \gamma E_0 \exp\{-i[(2\pi f_0 t - \boldsymbol{K}_i \cdot vt) - \boldsymbol{K}_s \cdot \boldsymbol{r'}]\} \tag{10.3}$$

where γ represents the scattering coefficient, $\boldsymbol{K}_s$ is the propagation coefficient of scattering light, and $\boldsymbol{r'}$ here denotes the position vector of a certain observation point in the coordinates x', y', and z'. By a very straightforward linear coordinates transformation $\boldsymbol{r'} = \boldsymbol{r} - \boldsymbol{v}t$, the detected scattering light field at any spot $\boldsymbol{r}$ in the fixed reference frame x, y, and z can be described as

$$E_s = \gamma E_0 \exp\{-i[(2\pi f_0 t - \boldsymbol{K}_i \cdot vt) - \boldsymbol{K}_s \cdot (\boldsymbol{r} - \boldsymbol{v}t)]\} \tag{10.4}$$

If the detector is fixed at the origin of this reference frame, Eq. 10.4 can be simplified as

$$E_s = \gamma E_0 \exp\{-i[2\pi\omega_0 t - (\boldsymbol{K}_i - \boldsymbol{K}_s) \cdot \boldsymbol{v}t]\} \tag{10.5}$$

The frequency of the scattering light field by a moving object will be $\omega_0 - 1/(2\pi)(\boldsymbol{K}_i - \boldsymbol{K}_s) \cdot \boldsymbol{v}$. The Doppler shift is therefore produced.

In a typical Doppler OCT setup, the light illumination and the scattering light collection usually share a same optical path in the sample arm, as shown in Figure 10.2. Thus the vector subtraction in Eq. 10.5 is equal to twice the propagation coefficient of the incident light $\boldsymbol{K}_i$. The frequency of the detected scattering light is

$$f_s = f_0 - \frac{1}{\pi}\boldsymbol{K}_i \cdot v = f_0 - \frac{2V_D n(\omega_o)}{\lambda_0}\cos\theta \tag{10.6}$$

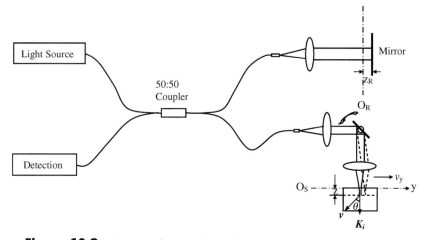

Figure 10.2   Diagram of essential optical setup for a fiberoptic Doppler OCT.

where θ denotes the angle between the direction of the incident beam and the moving direction of the object. The absolute value of the velocity is V_D. $n(f_0)$ is the refractive index of the medium around the object. λ_0 is the wavelength in a vacuum. Compared to optical frequency f_0, the Doppler shift introduced by many objects in the world is extremely small. Thus the scattering light field is always detected by heterodyne techniques which only retrieve the Doppler shift

$$f_D = \left| \frac{2V_D n(\omega_0)}{\lambda_0} \cos\theta \right| \tag{10.7}$$

Thus the value of the velocity can be determined but not the motion direction. If the light is modulated up to a radio frequency f_m, the Doppler shift would be

$$f_D = f_m \pm \left| \frac{2V_D n(\omega_0)}{\lambda_0} \cos\theta \right| \tag{10.8}$$

where the sign determines the direction.

10.1.2 Optical Signal Evolution in Doppler OCT

A general optical setup of a Doppler OCT system is depicted in Figure 10.2, which has been introduced in previous chapters to demonstrate a conventional OCT system. New denotations are included to represent the velocity v and the motion direction angled with the incident beam θ. Analogous to the analysis of a conventional OCT, scalar wave analysis is used. The perturbation of the incident light field at the coupler is described as a Fourier integral.[63]

$$E_0(t) = \int_0^\infty a_0(f) \exp\{i[\phi_0(f) - 2\pi ft]\} df \tag{10.9}$$

$a_0(f)$ and $\phi_0(f)$ represent the amplitude spectrum and phase spectrum of the incident light, respectively. The bandwidth of the incident light is usually finite around a certain frequency f_0, with certain intensity I_0 and spectral density/power spectrum $G(f)$. Here $G(f)$ is defined as the spectrum of an analytical signal whose real part is $E_0(t)$. So $G(f)$ is a real and even function including the negative frequency components. Its positive frequency portion is identical to the real power spectrum of the light source. Apart from a constant, the total intensity I_0 is related to $G(f)$ as

$$I_0 = 2\langle E_0^2(t) \rangle = 2 \int_{-\infty}^\infty G(f) df \tag{10.10}$$

where the value in the angle brackets represents the mean value in a unit time and the factor of 2 indicates that the intensity of the imaginary part of the analytical signal is accounted for.

Like a typical Michelson interferometer, the incident light beam in this Doppler OCT is split evenly into the sample and the reference beam by the coupler. In the reference arm, there is assumed no insertion loss. The light field, which is reflected back toward the exit of the interferometer by the mirror, can be expressed as

$$E_R(t)\frac{1}{2}\int_{-\infty}^{\infty} a_0(f)\exp\left\{i\left[\phi_0(f)+\delta+4\pi\frac{f}{c}z_R-2\pi ft\right]\right\}df \qquad (10.11)$$

In the sample arm, the origin of the coordinate O_S is set up at the sample surface. The origin of another coordinate in the reference arm O_R is set up at a point, where the optical path to the coupler is matched with the optical path between O_S and the coupler in the sample arm. δ represents this fixed phase delay. The mirror position is z_R off the O_R. c denotes the light speed in a vacuum. The intensity of the returning reference beam will be as

$$I_R = 2\langle E_R^2(t)\rangle = \frac{1}{2}\int_{\infty}^{\infty} G(f)df = \frac{I_0}{4} \qquad (10.12)$$

In the sample arm, the light beam encounters scatters in the sample. Define $p(z)$ as the amplitude backscattering coefficient distribution in the tissue, and z is measured from O_S. The distribution $p(z)$ carries the backscattering information. Considering the movement of the scatters, the backscattered light from one moving object can be written as

$$E_S(t, z) = \frac{1}{2}\int_{-\infty}^{\infty} a_0(f)p(z)\exp\left\{i\left[\phi_0(f)+\delta+4\pi\frac{n(f)f}{c}z-2\pi f_s t\right]\right\}df \qquad (10.13)$$

where $n(f)$ represents the refractive index of the sample. f_s is the frequency of the backscattering light from moving scatters, which can be directly cited from Eq. 10.6 considering that the incident beam is polychromatic.

$$f_S = f - \frac{1}{\pi}K_i \cdot v = f - \frac{2V_D f n(f)}{c}\cos\theta \qquad (10.14)$$

The total intensity of the returned scattering light is contributed incoherently by all the scatters within the tissue. This can be expressed as

$$I_s = 2\int_0^z \langle E_s^2(t, z')\rangle dz' = \frac{I_0}{4}\int_0^z p^2(z')dz' \qquad (10.15)$$

The function $p^2(z)$ in Eq. 10.15 represents the scattering coefficient in terms of intensity. Substituting Eq. 10.14 into Eq. 10.13, we could find that the velocity information is encoded in and can be extracted from the light field of the sample arm.

At the exit of the interferometer, the two return beams from the reference arm and the sample arm recombine coherently, described as

$$E_D(t, z) = \int_0^z [E_S(t, z')+E_R(t)]dz' \qquad (10.16)$$

The total intensity of the light field at the exit will be

$$I_D = 2\langle E_D^2(t, z')\rangle = 2\langle E_R^2(t)\rangle + 2\left\langle \int_0^z E_S^2(t, z')dz'\right\rangle + 4\int_0^z \langle E_S(t, z')E_R(t)\rangle dz' \quad (10.17)$$

Equation (10.17) can be further derived as

$$I_D = I_R + I_S + \int_0^z \text{Re}\left\{\int_{-\infty}^{\infty} G(f)p(z')\exp\left\{i4\pi\frac{f}{c}[n(f)(z' - V_D t \cos\theta) - z_R]\right\}df\right\}dz'$$

$$(10.18)$$

Equation (10.18) describes the intensity perturbation of the light field at the exit of the interferometer. The first two terms in the right of this equation are constant and correspond to the independent contribution from the reference arm and the sample arm. The third term represents the interferometric signal, where the backscattering coefficient distribution $p(z)$ and velocity information is encoded.

Usually in the theoretical analysis, it is assumed that the power spectrum $G(f)$ has a Gaussian distribution as

$$G(f) = \frac{I_0}{4} \times \frac{1}{\sigma_f\sqrt{2\pi}}\left\{\exp\left[-\frac{(f - f_0)^2}{2\sigma_f^2}\right] + \exp\left[-\frac{(f + f_0)^2}{2\sigma_f^2}\right]\right\}$$

$$(10.19)$$

Here $G(f)$ is defined according to Eq. 10.10 as the power spectrum includes the negative frequency components and is an even function. σ_f represents the standard deviation of the bandwidth. Assuming there is no dispersion exiting in this OCT system, the refractive index becomes frequency independent as $n(f_0)$. Substituted by Eq. 10.19, Eq. 10.18 becomes

$$I_D = I_R + I_S + \frac{I_0}{2}\int_0^z p(z')\cos\left\{2\pi f_0\left[\frac{n(f_0)(2z' - 2V_D t \cos\theta)}{c} - \frac{2z_R}{c}\right]\right\}$$

$$\times \exp\left\{-2\pi^2\sigma_f^2\left[\frac{n(f_0)(2z' - 2V_D t \cos\theta)}{c} - \frac{2z_R}{c}\right]^2\right\}dz'$$

$$(10.20)$$

Equation 10.20 can also be rewritten with a convolution expression, using wave number k instead of frequency f.

$$I_D(z_R, t) = I_R + I_S + \frac{I_0}{2}p\left(\frac{z_R + V_D t \cos\theta n(k_0)}{n(k_0)}\right)$$

$$\otimes \left\{\cos[4\pi k_0(z_R + V_D t \cos\theta n(k_0))]\exp\left[-\frac{4\ln 2(z_R + V_D t \cos\theta n(k_0))^2}{\Delta_l^2}\right]\right\}$$

$$(10.21a)$$

$$\Delta_l = \frac{2\ln 2\lambda_0^2}{\pi\Delta_\lambda}$$

$$(10.21b)$$

where Δ_l is the full width half maximum (FWHM) coherence length and Δ_λ is the FWHM spectral bandwidth of the light source, defined in previous chapters. k_0 denotes the central wavenumber in a vacuum. Equation (10.21a) indicates that the light intensity at the exit alternates with the mirror position z_R and time/velocity of object movement. The equation is a full description of a single axial Doppler OCT signal in terms of the A-scan. It is important to be aware that the convolution in Eq. 10.21a is the operation between the arguments of two functions. The backscattering $p(z)$ and velocity distribution is encoded in the AC part of the AM-FM signal. Using a temporal-spatial function $h(z_R, t)$ to represent the portion of Eq. 10.21a, Eq. 10.21a can be rewritten as

$$I_D(z_R, t) = I_R + I_S + \frac{I_0}{2} p\left(\frac{z_R + V_D t \cos\theta n(k_0)}{n(k_0)}\right) \otimes h(z_R, t) \qquad (10.22)$$

If there is no any movement of the sample, Eq. 10.22 is only the function of mirror position z_R as

$$I_D(z_R) = I_R + I_S + \frac{I_0}{2} p\left(\frac{z_R}{n(k_0)}\right) \otimes h(z_R) \qquad (10.23)$$

This one-dimensional modeling of an OCT signal is enough for the majority of performance characteristics of the conventional OCT such as what we have done in previous chapters. But for Doppler OCT, two-dimensional modeling might be necessary because a lot of velocity mapping methods and algorithms use correlations of multiple adjacent A-scans. For two-dimensional OCT imaging, the sample beam is steering along the lateral direction by a galvanometer-driven mirror as shown in Figure 10.2. The scanning beam focuses on the area of interest (ROI) through an objective lens. If the pivot of the mirror is located at the focal point of the lens, the incident beam will scan in parallel. So the tome-sequentially captured A-scans can be used to reconstruct a two-dimensional map of $p(y, z)$ and velocity, in terms of the B-scan. Here the y axis represents the lateral scan direction. So the cross-sectional imaging of the sample is along the y-z plane. A detailed discussion is presented in Chapter 5 about the relevance of the incident light beam to a Gaussian beam. Like other incoherent imaging modalities, the properties of this Gaussian beam determine the transverse resolution of OCT. Based on this decoupling nature of OCT in both lateral and axial directions, we define a Gaussian-type function $h_y(y)$ as a transfer function of the system in the y direction. Assuming the lateral scanning speed of beam is v_y, we can describe the AC part of a B-scan signal as

$$I_D(z_R, t) \propto \int_{-\infty}^{\infty} \left\{ p\left[y, \frac{z_R + V_D t \cos\theta n(k_0)}{n(k_0)}\right] \otimes h(z_R, t) \right\} h_y(y - v_y t)\, dy \qquad (10.24)$$

As usual we do not use a symbolic format to express the convolution in the y direction. However, Eq. 10.24 clearly reflects the spatial average in the y direction by the Gaussian beam whose beam waist determines the transverse resolution of OCT. The expression in Eq. 10.24 is also explicit for the description of a time-sequential scan and more flexible for scanning protocols whatever axial scan or lateral scan is in priority.

10.1.3 Interferogram Detection in Doppler OCT

As indicated above, the microstructure and flow velocity information are all encoded in the interferogram at the exit of the interferometer. Proper interferogram detection and analysis/processing can fully extract the target information. Two major interferogram detection methods — time domain and spectral domain — will be discussed here regarding the specialty for velocimetry.

10.1.3.1 Time domain doppler OCT

Like any standard time domain OCT, the mirror in the reference arm of a Doppler OCT is scanning with constant velocity V_R. Assuming axial scan in priority of the lateral scan, the interferogram can be read out time-sequentially by a single detector placed at the exit of the interferometer. The readout signal can be described as a function of a single variable t:

$$i_D(t) \propto \sum_{m=0}^{M-1} \begin{aligned} & \alpha p\left\{ mv_y T, \frac{[V_R + V_D n(k_0)\cos\theta](t - mT)}{n(k_0)} \right\} \\ & \otimes \{\cos\{4\pi k_0 [V_R + V_D n(k_0)\cos\theta](t - mT)\}\} \\ & \exp\left\{ -\frac{4\ln 2[V_R + V_D n(k_0)\cos\theta]^2(t - mT)^2}{\Delta_l^2} \right\} \otimes h_y(mv_y T) \end{aligned} \tag{10.25}$$

where α represents a uniform spectral responsiveness of the detector and T is the time period that a single A-scan takes. m is an integer from 0 to $M-1$, representing the mth A-scan. Therefore M denotes the total number of A-scans in a B-scan. The time delay mT and the linear summation reflect the nature of time-sequential scanning. The lateral scanning scheme practically guarantees that $v_y T \geq W$, ensuring that the adjacent A-scan signal will not overlap in a given time sequence. Here W denotes the lateral scan range.

As a result, the AM-FM signal has instantaneous carrier frequency f_c and FWHM bandwidth Δf_D as

$$f_c = 2k_0[V_R + V_D n(k_0)\cos\theta] \tag{10.26a}$$

$$\Delta f_D = 2V_R \Delta_k = \frac{2[V_R + V_D n(k_0)\cos\theta]\Delta_\lambda}{\lambda_0^2} \tag{10.26b}$$

Figure 10.3 gives a numerical simulation of a Doppler OCT A-scan signal that has a parabolic velocity profile. The envelope detection can easily extract the entire microstructure profile with high spatial resolution. However, the variant flow velocity deteriorates the ranging accuracy since the time baseline is no longer linear. The carrier frequency f_c varies with time instantaneously. The localized frequency shift or velocity can be detected by specific schemes or algorithms, which are discussed in section 10.2. Equation 10.26a also indicates that the scanning of a mirror not only provides the longitudinal scan but also modulates the OCT signal up to a higher frequency. As previously stated in the discussion regarding Doppler

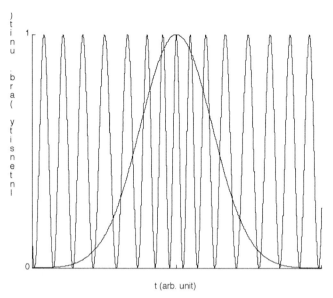

t (arb. unit)

Figure 10.3   Demonstration of a local Doppler OCT A-scan signal with a Gaussian type velocity profile.

velocimetry, this type of modulation is useful for distinguishing the flow direction. If the flow velocity keeps constant while the scanning mirror velocity is known, Fourier transform over the entire A-scan signal can determine the Doppler shift directly. However, the process is more difficult in practical applications because the flow velocity varies with the depth/time. Multiple schemes or algorithms relating to the velocity estimation will be introduced in the part of Doppler OCT signal processed. In Eq. 10.26, there is an underlying assumption that the sample has no movement in the y direction. In practice, the B-scan plane is usually perpendicular to the flow plane to simplify the measurement.

10.1.3.2 Spectral domain doppler OCT

Recently, an alternative OCT modality in terms of spectral domain OCT (SD-OCT) emerged and attracted more interest in this field.[14] As stated in Chapter 5, SD-OCT detects the spectral interferogram in which each A-scan is entirely encoded. By Fourier transform, the A-scan and associated velocity distribution can be consequently extracted without any group delay varying in the reference arm. As a result, no mechanical movement is necessary and a high axial scanning rate may be achievable, which potentially alleviates the major complexity of a TD-OCT system. SD-OCT is generally divided into two categories of swept source OCT (SS-OCT) and Fourier domain OCT (FD-OCT) or spectral radar. The former utilizes a swept/tunable laser source whereas the latter uses a wideband source. Most recently, researchers have begun to extend the SD-OCT technique to have a Doppler velocimetry function.[18–20,25–27,39,44] Since SD-OCT has been introduced in detail previously, the discussion here will focus on the concept of Doppler velocimetry with SD-OCT and performance evaluation.

In the system depicted in Figure 10.2, replacing the single detector with an optical spectrometer, or using a swept/tunable laser source instead of a wideband light source, a spectral interferogram can be obtained at the exit of the interferometer. Multiple spectral detection approaches may be used, but the full description of the spectral interferogram of the moving scattering object is the same and can be derived from Eq. 10.18. Assuming an ideal spectrometer is used that does not induce any distortions, the spectral interferogram can be written in terms of wavenumber k:

$$
I_D(k, t) = \frac{1}{4} G(k) \left\{ 1 + \int_0^z p^2(z') dz' \right.
$$
$$
\left. + 2\mathrm{Re}\left\{ \int_0^z p(z') \exp\{i4\pi k[n(k)(z' - V_D t \cos\theta) - z_R]\} dz' \right\} \right\}
\tag{10.27}
$$

where $G(k)$ denotes the spectral density/power spectrum of the light source defined as early. The specialty in Doppler SD-OCT is that the movement of the sample causes temporally varying phase shift of the spectrum. In other words, the spectral interferogram is time-variant. We cannot expect this in conventional SD-OCT. Issues will be brought out toward Doppler SD-OCT since we know that all spectrums are obtained on a time-average base. For example, in SD-OCT, signal integration is a common way to maintain high signal to noise ratio (SNR). If there is no motion in the sample, the spectral interferogram will be time-invariable. That means the spectrum can be captured without distortion in any time period. The targeted signal $p(z)$ can be extracted by a standard process described in earlier chapters, and all the performance characteristics including the axial resolution will be the same as standard SD-OCT. But the instantaneous phase shift of spectral interferogram will cause problems of which we should be aware.

Ideally, if the spectrum capturing and the Fourier transform are all instantaneous, the retrieved A-scan signal will be a time-variant function that can be presented as

$$
I_D'(z'', t) = F^{-1}[I_D(k)] = \frac{1}{4} F^{-1}\{G(k)\} \otimes \left\{ F^{-1}\left\{ 1 + \int_0^z p^2(z') dz' \right\} \right.
$$
$$
\left. + p\left(\frac{z'' + 2z_R + 2n(k_0)V_D t \cos\theta}{2n(k_0)} \right) + p\left(\frac{-z'' + 2z_R + 2n(k_0)V_D t \cos\theta}{2n(k_0)} \right) \right\}
\tag{10.28}
$$

As we derived in Chapter 5, the Fourier transform of $G(k)$ is as follows:

$$
F^{-1}\{G(k)\} = I_0 \cos(2\pi k_0 z') \exp\left[-\frac{4\ln 2(z'')^2}{(2\Delta_l)^2} \right]
\tag{10.29}
$$

Substituting Eq. 10.29 into 10.28 while only considering the AC portion of the signal, the transformed A-scan signal is simply represented as

$$I_D'(z'',t) \propto \left\{ \cos(2\pi k_0 z'') \exp\left[-\frac{4\ln 2(z'')^2}{(2\Delta_l)^2} \right] \right\}$$
$$\otimes \left\{ p\left(\frac{z'' + 2z_R + 2n(k_0)V_D t \cos\theta}{2n(k_0)} \right) + p\left(\frac{-z'' + 2z_R + 2n(k_0)V_D t \cos\theta}{2n(k_0)} \right) \right\}$$

$$(10.30)$$

It is known that convolution satisfies the properties of shift as

$$f_1(t - \tau) \otimes f_2(t) = f_1(t - \tau) \otimes f_2(t - \tau) \tag{10.31}$$

Equation 10.31 can be rewritten as

$$I_D'(z'', t) \propto \left\{ \cos\{2\pi k_0[z'' + 2z_R + 2n(k_0)V_D t \cos\theta]\} \right.$$
$$\left. \exp\left[-\frac{4\ln 2[z'' + 2z_R + 2n(k_0)V_D t \cos\theta]^2}{(2\Delta_l)^2} \right] \right\}$$
$$\otimes \left\{ p\left(\frac{z'' + 2z_R + 2n(k_0)V_D t \cos\theta}{2n(k_0)} \right) + p\left(\frac{-z'' + 2z_R + 2n(k_0)V_D t \cos\theta}{2n(k_0)} \right) \right\}$$

$$(10.32)$$

Compared to Eq. 10.21a, Eq. 10.32 clearly indicates that both TD-OCT and SD-OCT have the same Doppler motion effects. The axial structure profile is encoded in the envelope of the A-scan and the velocity distribution is encoded in the instantaneous frequency modulation of the carrier.

Considering the two-dimensional imaging process, the time-sequential B-scan signal can be derived from Eq. 10.32 as follows:

$$I_D'(z, t) \propto \sum_{m=0}^{M-1} \left\{ \begin{array}{l} \cos\{2\pi k_0[z'' + 2z_R + 2n(k_0)V_D mT \cos\theta]\} \\ \times \exp\left[-\dfrac{4\ln 2[z'' + 2z_R + 2n(k_0)V_D mT \cos\theta]^2}{(2\Delta_l)^2} \right] \end{array} \right\}$$
$$\otimes \left\{ \begin{array}{l} p\left(mv_y T, \dfrac{z'' + 2z_R + 2n(k_0)V_D mT \cos\theta}{2n(k_0)} \right) \\ + p\left(mv_y T, \dfrac{-z + 2z_R + 2n(k_0)V_D mT \cos\theta}{2n(k_0)} \right) \end{array} \right\} \otimes h_y(mv_y T)$$

$$(10.33)$$

The corresponding structure profile and velocity profile can subsequently be extracted according to either Eq. 10.28 or Eq. 10.33. Relevant methods will be discussed in the section on Doppler OCT signal processing.

An important issue associated with spectral domain Doppler OCT is that a spectral interferogram is often obtained by time integration such as in Fourier domain OCT. (Signal integration has been discussed in previous chapters.) The real spectral interferogram is actually a time average and can be described as

$$I_D(k) \propto \frac{1}{T} \int_{-\frac{T}{2}}^{\frac{T}{2}} G(k) \times \text{Re}\left\{ \int_0^z p(z')\{\exp\{j4\pi k[n(k)(z' - V_D t \cos\theta) - z_R]\}dz'\} \right\} dt \tag{10.34}$$

Using convolution and rectangular function,[64] we can simplify Eq. 10.34:

$$I_D(k) \propto \frac{G(k)}{T} \text{Re}\left\{ \int_{-\infty}^{\infty} rect\left(\frac{t}{T}\right) \exp[-i4\pi kn(k)V_D t \cos\theta]dt \right.$$

$$\left. \int_0^z p(z')\{\exp\{i4\pi k[n(k)z' - z_R]\}dz'\} \right\} \tag{10.35}$$

where T is the integral time and is usually the same as an A-scan period. Rectangular function $rect(t/T)$ represents the time average of the spectrum. This kind of time integration in the spectrum domain inevitably causes spectrum blur if the phase of the spectrum varies during integration. The blurred spectrum can be represented as follows:

$$I_D(k) \propto \frac{G(k)}{T} \sin c\{[2Tn(k)V_D \cos\theta]k\} \text{Re}\left\{ \int_0^z p(z') \exp\{i4\pi k[n(k)z' - z_R]\}dz' \right\} \tag{10.36}$$

The sinc function in Eq. 10.36 reflects the reality that spectrum filtering is induced by time average. The bandwidth of the sinc function is usually defined as

$$B \approx \frac{1}{2Tn(k)V_D \cos\theta} \tag{10.37}$$

Figure 10.4 demonstrates the effects of spectrum integration. The solid curve illustrates the spectral interferogram of a single scatter at z_0 without movement. The dot curve is simply a sinc function whose bandwidth is determined by the product of integration period T and velocity V_D. The dash curve clearly demonstrates that a longer integration period or faster flow velocity will cause more spectrum deterioration. That is not expected in Doppler SD-OCT.

10.2   SIGNAL PROCESSING IN DOPPLER OCT

As mentioned earlier, the diagnostic information including microstructure distribution and localized velocity are all encoded in the detected Doppler OCT signal.

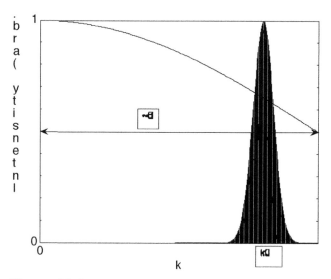

Figure 10.4 Graphic demonstration of the time integration effects on spectral interferogram in a Doppler SD-OCT.

Up to this point, we have successfully modeled the signal evolution and detection in a Doppler OCT. Now we will discuss multiple signal processing methods or algorithms that extract the information based on the signal model developed earlier.

10.2.1 Signal Processing in Time Domain Doppler OCT

As indicated by either Eq. 10.21a or 10.25, a time domain Doppler OCT signal is intrinsically an AM-FM signal with time-variant instantaneous frequency (IF). The IF of a signal indicates that the signal frequency peak varies with time. The IF signal is referred to as "nonstationary."[3,4] The IF of a Doppler OCT signal is proportional to the Doppler shift/flow velocity. The envelope of this AM-FM signal is proportional to the corresponding axial scattering coefficient distribution. Simple envelope detection or power detection can directly retrieve structure profiles but lose the phase information. Specific schemes or algorithms have been designed for IF/Doppler velocity estimation.

10.2.1.1 Short time fourier transform

An IF signal is usually analyzed by time-frequency analysis which operates in temporal-frequency domain. Short time Fourier transform (STFT) is one of the methods of linear time-frequency analysis that can provide localized spectrum in time domain by applying Fourier transform in a localized time window. The STFT method has been developed as a standard method since the early stage of Doppler OCT technique.[9,10,16,42]

The instantaneous Doppler shift is approximated by the peak of the STFT spectra of signal. The STFT of an individual A-scan Doppler OCT signal can be expressed as

$$STFT_i(t, f) = \int_{-\infty}^{\infty} [i_D(t')w^*(t' - t)] \exp(-j2\pi f t')dt' \qquad (10.38)$$

where $w(t)$ is the window function in which Fourier transform is conducted. i denotes the ith time point for STFT. $STFT_i(t, f)$ is thus a two-dimensional distribution associated with time axis and frequency axis. The fact of matter is that the window function $w^*(t)$ is critical to the STFT result and the performance of the velocity estimation. For Doppler OCT operation, STFT is usually operated in discrete time domain with digital signal processing. However, there is no difference with the concept elucidated here.

As mentioned earlier, Eq. 10.24 can be denoted as a convolution between axial profile and the transfer function of the Doppler system

$$i_D(t) = p(t) \otimes h(t) \qquad (10.39)$$

The $STFT_i(t, f)$ can also be denoted as the reverse Fourier transform of the signal spectra $P(f)H(f)$ and spectra of window function $W^*(f)$, which might help us understand the physical meaning of this kind of signal processing.

$$STFT_i(t, f) = \exp(-j2\pi tf) \int_{-\infty}^{\infty} P(f')H(f')W^*(f - f') \exp(j2\pi tf')df' \qquad (10.40)$$

Equation 10.40 indicates that the Doppler OCT signal passes through an analysis bandpass filter whose transfer function is $W^*(f - f')$. f is the analysis frequency. The relationship between frequency resolution δf and time resolution δt of STFT follows Heisenberg uncertainty:

$$\delta t \delta f \geq \frac{1}{4\pi} \qquad (10.41)$$

Higher time resolution, in other words narrower Fourier transform window, will result in lower frequency resolution. For Doppler OCT, higher temporal resolution means better axial resolution in spatial mapping of velocity but worse velocity resolution because of the lower frequency resolution. So, with the system using STFT, there will inevitably be a trade-off between time/depth resolution and frequency/velocity resolution.

The velocity mapping is constructed by calculating the centroid of $STFT_i(t, f)$ at certain time t.

$$\overline{f_c}(t) = \frac{\int_{\Delta f} f STFT_i(t, f)df}{\int_{\Delta f} STFT_i(t, f)df} \qquad (10.42)$$

Multiple centroid estimation algorithms were reported for accuracy modification,[28] but they will not be reviewed here. Substituting Eq. 10.42 into Eq. 10.26a, the time/depth dependent velocity profile is shown as

$$V_D(t) = \left[\frac{\overline{f}_c(t)}{2k_0} - V_R \right] / \cos\theta n(k_0) = \frac{\lambda_0 [\overline{f}_c(t) - f_0]}{2\cos\theta n(\lambda_0)} \qquad (10.43)$$

The velocity resolution and spatial resolution of the two A-scan profiles will be discussed in the following section.

As shown in Eq. 10.23, the time-dependent backscattering profile $p(t)$ is filtered by a bandpass filter that has instantaneous central frequency f_c. If this instantaneous central frequency is predictable, theoretically the structure profile $p(z)$ can be extracted with a spatial resolution Δ_l, as stated above. For instance, if the flow velocity is fixed, a bandpass filter with certain central frequency and the envelope detection can easily establish the time-dependent structure $p(t)$ as many conventional OCTs have done. But this is not applicable in practical circumstances because the flow velocity is unpredictable. STFT gives an estimation within a certain time window; however, the spatial resolution of velocity will be affected by this window. In addition to the coherence length of light source Δ_l, the window function chosen for STFT is another limitation for the spatial resolution of Doppler velocity extracted by the STFT algorithm. A wider window means lower axial resolution in Doppler velocity images.

From Eqs. 10.30 and 10.31, the velocity resolution in Doppler velocity mapping is determined by the measurement accuracy of the Doppler shift. According to the reconstruction method of Doppler velocity mapping, the centroid average on all STFT spectra is still a linear operation. So the Doppler spectrum resolution determines the accuracy of the Doppler shift, hence the Doppler velocity resolution is represented by

$$\delta V_D = \frac{\lambda_0}{2n(\lambda_0)\cos\theta}\delta f \qquad (10.44)$$

As stated above, the frequency resolution of the Doppler velocimetry is determined by the width of the window function δt, which follows the Heisenberg uncertainty principle. Wider window width can provide better frequency resolution, but there are some limitations. First, a wider window deteriorates spatial resolution of the Doppler velocity. Thus the time window has an upper limit for a certain spatial resolution δl. The Doppler velocity resolution turns out to be

$$\delta V_D = \frac{\lambda_0}{2n(\lambda_0)\cos\theta\delta t} = \frac{\lambda_0 V_R}{2n(\lambda_0)\cos\theta\delta l} \qquad (10.45)$$

Equation 10.45 shows that the reduced mirror velocity in the reference arm might improve the velocity resolution as well as keep a certain spatial resolution. But lower mirror velocity means the slower Doppler image captures speed in terms of frame rate. This might not be tolerated for many applications. Low frame rate would also introduce trouble for flow direction judgment if the flow velocity is relatively high.

Another important issue related to the velocity resolution is the light source. So far all of the discussion is based on the assumption that the light source is a cw source. Short pulse laser is another kind of light source for OCT application. The occupation ratio of the short pulse γ ($0 \le \gamma \le 1$) becomes one of the critical parameters of velocity resolution. For a given time window width δt, the velocity resolution of the Doppler OCT using short pulse laser is

$$\delta V_D = \frac{\lambda_0}{2\gamma n(\lambda_0)\cos\theta\delta t} = \frac{\lambda_0 V_R}{2\gamma n(\lambda_0)\cos\theta\delta l} \tag{10.46}$$

Since the short pulse occupation ratio γ is always smaller than one, the velocity resolution of a short pulse Doppler OCT is always worse than a cw Doppler OCT if compared with the same spatial resolution and frame rate.

10.2.1.1 Phase resolved velocity estimation

STFT-based Doppler velocity estimation cannot simultaneously maintain high spatial resolution and velocity resolution. It is commonly referred to that Doppler velocity resolution and spatial resolution are coupled. Based upon the restrictive definition, STFT lends itself very well to a pure nonstationary signal. But as shown in Eq. 10.25, the total B-scan signal of a Doppler OCT is actually a time-sequential cascade of nonstationary signal slices in terms of A-scans, whereas the adjacent A-scans have a fixed phase shift:

$$\Delta\phi = 4\pi k_0 V_D n(k_0) T \cos\theta \tag{10.47}$$

where T is the A-scan period as defined earlier. That means the phase shift between consecutive A-scans is stationary. Thus, the Doppler velocity can be retrieved by measuring the phase variation between consecutive A-scans as

$$V_D = \frac{\Delta\phi}{T 4\pi k_0 n(k_0)\cos\theta} \tag{10.48}$$

Consequently the velocity resolution will be

$$\delta V_D = \frac{\lambda_0}{2n(\lambda_0)\cos\theta T} \tag{10.49}$$

Comparing Eq. 10.49 with Eq. 10.47, it is clear that the demonstrated method will have higher velocity resolution than the STFT method, since T is always larger than the STFT window width δt. Meanwhile, the high spatial resolution of the velocity measurement is maintained. In other words, they are decoupled. This method is referred to as the phase-resolved velocity estimation.[34,58,61]

Usually the phase shift inter A-scans is not directly picked up from the captured real signals $i_D(t)$. However, each A-scan signal can be converted into its corresponding analytical signal by Hilbert transform (Appendix 5-1).

$$\tilde{i}_D(t) = i_D(t) + \frac{i}{\pi} P \int_{-\infty}^{\infty} \frac{i_D(\tau)}{\tau - t} dt = A(t)\exp[i\phi(t)] \tag{10.50}$$

where P denotes the Cauchy principle value. $A(t)$ and $\phi(t)$ represents the mode and phase of the analytical signal, respectively. From the calculated analytical signal, the phase shift between consecutive A-scans $\phi(t)$ can be recovered as a function of time/depth. Using Eq. 10.48, the axial velocity profile can be consequently calculated. The spatial resolution will only be determined by the inter A-scan phase shift uncertainty, which might be caused by the turbid flow velocity and noise in measurement.

A variety of sources, including velocity gradient, turbulence, Brownian motion, speckle, and probing optics, cause the fluctuation of inter A-scan phase shift in terms of the broadening of the Doppler spectrum.[33] When flow velocity is low, Brownian motion dominates the broadening, whereas probing optics dominate when velocity is high. In practical application of phase-resolved velocity estimation, the mean phase shift of several consecutive A-scans is measured instead of two adjacent A-scans and the standard deviation is also calculated.[60] This modification is triggered by the concern that the velocity measurement is sensitive to the angle θ as shown in Eq. 10.48. For example, when the probing beam is almost perpendicular to the flow, the inter A-scan phase shift will be insensitive to the velocity change. But the standard deviation will still keep the high sensitivity and is linearly proportional to the velocity if above a threshold determined by Brownian motion. So the Doppler bandwidth measurement permits the lateral velocity accuracy of flow velocity without the need for precise determination of flow direction when the Doppler flow angle is within ± 15 degrees perpendicular to the probing beam.[33] It is important to notice that the transverse spatial resolution will degrade with this method because multiple A-scans are grouped for average.

An issue with this method is that the phase shift measurement has 2Π ambiguity, so there is an upper limit for the velocity measurement. Another issue is that the numerical Hilbert transform is very time-consuming, which will degrade the capability of real-time imaging for Doppler OCT. A modified method using hardware in-phase and quadrature (I&Q) demodulator with software implementation of the Kasai autocorrelation velocity estimation algorithm has been reported.[50,51] An alternative way using optical Hilbert transform was also reported.[59] All these hardware implementations will significantly alleviate the heavy duty of computation for real-time imaging.

10.2.1.3 Doppler velocity estimation based on correlation detection

Autocorrelation-based signal processing has been used for a long time in radar, sonar, ultrasound imaging, and other fields. The underlying mechanism is to calculate the autocorrelation function of a signal. Based on the properties of autocorrelation function, correlation detection of a signal can phenomenally suppress the random noise. Another potential advantage of this method is that the algorithm is flexible for either analogue or digital realization. Thus the autocorrelation-detection-based Doppler velocity estimation was utilized soon after the emergence of the Doppler OCT technique.[35] Figure 10.5 is the hardware-based signal processing diagram.

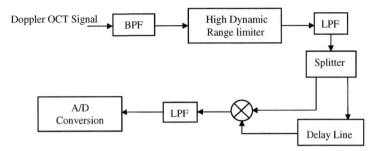

Figure 10.5   Flowchart of the autocorrelation processing on Doppler OCT signal based on hardware implementation.

The Doppler OCT signal is first fed into a bandpass filter (BPF) for noise suppression. Then the amplitude of the signal is clipped by a high dynamic range limiter, keeping the phase information only. The signal is further conditioned by a low pass filter (LPF) and fed into a splitter. The two outputs of the splitter have a relative time delay when these two signals are multiplexed and passed through an LPF. The output will be the autocorrelation of the input signal. The process of the correlation can be expressed as

$$R(\tau) = \langle g(t)g^*(t-\tau) \rangle = \lim_{\Delta T \to \infty} \frac{1}{\Delta T} \int_0^{\Delta T} g(t)g^*(t-\tau)dt \qquad (10.51)$$

where τ denotes the delay time and ΔT is the integration time which is inversely proportional to the bandwidth of the LPF. The angular brackets denote the time average on the analytical signal $g(t)$ and its delayed conjugate.

From Eq. 10.25, an amplitude-limited A-scan signal can be expressed as

$$i_D(t) \propto A \cos\{4\pi k_0[V_R - V_D n(k_0)\cos\theta]t\} \qquad (10.52)$$

where amplitude A is constant. Substituting Eq. 10.52 into Eq. 10.51, the autocorrelation function will be

$$\begin{aligned} R_D(\tau)e &\propto \cos\{4\pi k_0[V_R - V_D n(k_0)\cos\theta]\tau\} \\ &= \cos[4\pi k_0 V_R \tau - 4\pi k_0 V_D n(k_0)\cos\theta\tau] \end{aligned} \qquad (10.53)$$

Here we assume the amplitude of the autocorrelation function is one. Carefully selecting the fixed Doppler velocity V_R in the reference arm and fixed time delay τ_0 of the delay line gives the first phase item in Eq. 10.53 a value of $\pi/2$. The autocorrelation becomes

$$R_D(V_D) \propto \sin[4\pi k_0 V_D n(k_0)\cos\theta\tau_0] \qquad (10.54)$$

Equation 10.54 clearly demonstrates that the autocorrelation is a function of the variable of velocity V_D, so the Doppler velocity can be extracted from the autocorrelation function. Based on the same mechanism, the Doppler OCT using digital autocorrelation processing was reported.[13,43] A similar scheme using phase-lock-loop (PLL) has also been reported by Zvyagin et al.[62]

Intrinsically, the Doppler velocity has a nonlinear relationship with the signal of autocorrelation as shown in Eq. 10.54. Around $\pm \pi$ range, the phase shift is less sensitive to velocity, so a linear approximation is often chosen practically as indicated by the dash line in Figure 10.6. The slope of the dashed approximation line is proportional to delay time τ_0. Thus the velocity sensitivity is determined as

$$\gamma_D = \frac{\lambda_0}{2n(\lambda_0)\cos\theta\tau_0} \tag{10.55}$$

Longer delay times will provide higher velocity sensitivity. It is noted that the output of the autocorrelator will sinusoidally "wrap around" for higher velocity. So there is a theoretical limitation of velocity estimation range because of these $\pm \pi$ ambiguities. The range will be

$$-\frac{\lambda_0}{8n(\lambda_0)\cos\theta\tau_0} \leq V_D \leq \frac{\lambda_0}{8n(\lambda_0)\cos\theta\tau_0} \tag{10.56}$$

Otherwise periodical banding patterns will inevitably occur in the velocity profile, which is unexpected.

The autocorrelation detection method confronts the same issue as the STFT method, that of the spatial resolution of velocity measurement. Equation 10.51 indicates that the integration time ΔT must be longer than at least one period of the resolvable frequency component, which is equivalent to τ_0. Unfortunately, longer integration time will definitely reduce the spatial resolution. To overcome this problem, an inter A-scan correlation was suggested in terms of cross-correlation

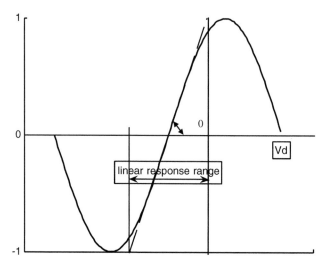

Figure 10.6   Illustration of the linear velocity measurement range and the dependence of velocity sensitivity on delay time.

technique by calculating the correlation of successive analytical A-scan signals at the same time/depth point.[28] With this method, the integration time could be multiplied by the A-scan period T while keeping the high spatial resolution. In practical scenarios, whatever frequency shift or phase shift in Doppler OCT is measured there is the presence of multiple error sources including speckles, turbid flow dynamics and noises. A fiber-optic differential phase contrast Doppler OCT system was reported to eliminate such interferences.[54] This scheme employs a Wollaston prism to separate the sample beam into two orthogonal polarization states (Chapter 9). The interferogram of each polarization state is recorded independently in terms of channel signals. The inter A-scan correlation is conducted to calculate the correlation function. Thus differential Doppler frequency is observed by calculating the differential phase of the autocorrelations from each channel. The phase fluctuation will be canceled in terms of self-referenced autocorrelation detection.

10.2.2 Signal Processing in Spectral Domain Doppler OCT

As indicated in Eq. 10.33, in a spectral domain Doppler OCT, the microstructure profile is the envelope of a calculated A-scan by applying a Fourier transform on a spectral interferogram, whereas the velocity information is encoded in the phase of the interferogram. Analogous to the method of phase resolved velocity estimation in the time domain Doppler OCT, the Doppler velocity can be extracted by measuring the phase shift between two successive calculated A-scans in spectral domain Doppler OCT. The formulas for velocity and accuracy are exactly the same as Eqs. 10.48 and 10.49. That means spectral domain Doppler OCT does not need to compromise between velocity resolution and spatial resolution. As mentioned in section 10.2.1.1, the phase shift between two successive A-scans is actually measured within the corresponding analytical signals by a Hilbert transform in time domain OCT. This is not necessary for the spectral domain OCT since the calculated A-scans are all complex functions terms of

$$\tilde{i}_D(z'') = \text{Re}[\tilde{i}_D(z'')] + i\text{Im}[\tilde{i}_D(z'')] = \left|\tilde{i}_D(z'')\right| \exp[i\phi(z'')] \qquad (10.57)$$

Therefore the phase of this complex function is exactly the phase information of each A-scan described as

$$\phi(z'') = \arctan\left\{\frac{\text{Im}[\tilde{i}_D(z'')]}{\text{Re}[\tilde{i}_D(z'')]}\right\} \qquad (10.58)$$

The Doppler velocity could be obtained by measuring the phase shift $\Delta\phi(z'')$. Meanwhile the structure profile is easily found from the amplitude of the complex function shown by Eq. 10.57. It had to be clarified that the general description of a spectral domain Doppler signal in Eq. 10.33 is a real function. It is a real function because for simplicity of derivation, the spectral interferogram includes the negative components that are symmetrical to the positive part. In practical circumstances the Fourier transform is always performed upon the spectral interferogram with wavenumber above zero, so there will have to be an imaginary part in the result.

10.3   APPLICATIONS OF DOPPLER OCT

This technique has found a wide range of potential applications, such as in ophthalmology,[19,44,45,55,56] dermatology,[2,30,40] gastroenterology,[6,53] blood cardiovascular medicine,[10,16,23,52] rheumatology,[24] tissue engineering,[21] photodynamic therapy,[49]

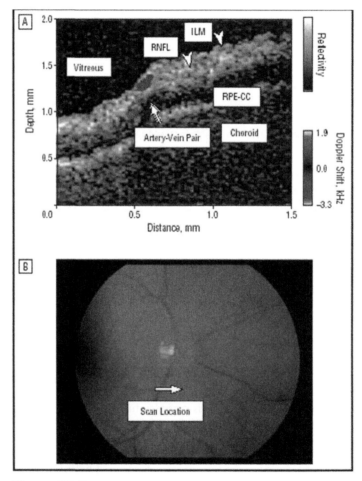

Figure 10.7   (A) *In vivo* color Doppler OCT (CD-OCT) image (2048 axial × 100 lateral pixels) of bidirectional flow in the human retina acquired in 10 seconds. The axial dimensions indicate optical path. CD-OCT is able to distinguish various layers of tissue and to quantify blood flow magnitude and direction in sub-100-μm diameter retinal blood vessels. (B) Fundus photograph marked to illustrate the position of the linear scan inferior to the optic nerve head. ILM, inner limiting membrane; RNFL, retinal nerve fiber layer; and RPE-CC, retinal pigment epithelium-choriocapillaris complex. (Courtesy of Dr. J. A. Izatt, *Arch Ophthamol.* 2003, 121, 235–239.)

oncology,[22] flow dynamics,[31,41] and microfluidic devices.[47] Several Doppler OCT images with different applications are included (Figures 10.7 through 10.10). Hopefully these figures and the detailed figure legends can provide the readers with a direct sense of Doppler OCT imaging.

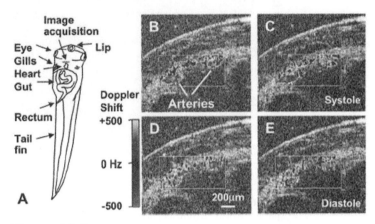

Figure 10.8 Demonstration of *in vivo* real-time Doppler flow imaging. (A) Schematic ventral view of a 17-day-old tadpole, *Xenopus laevis*. The imaging plane is slightly superior to the heart, transecting two branches of the ductus arteriosus. (B–E) Subsequent frames acquired at 8 frames per second. The blood pulses first through the left artery (B–C) and then, slightly delayed, through the right artery (D). (Courtesy of Dr. J. A. Izatt, *Opt. Lett.* 2002, 27, 34–36.)

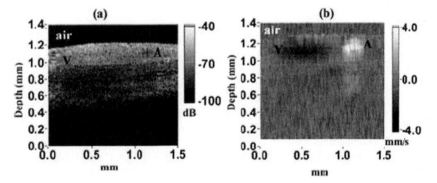

Figure 10.9 (a) Amplitude and (b) Doppler images of native skin in the hamster dorsal skin flap window preparation taken from the subdermal side of the native skin. The flow velocity in the arteriole (A) is 4 mm/sec, and in the venule (V) it is 3.5 mm/sec. (Image axes: *x* axis, lateral width; *y* axis, depth; *z* axis, amplitude in [a] and flow in [b].) (Courtesy of Dr. G. Vargas, *Photochem. Photobiol.* 2003, 77, 541–549.)

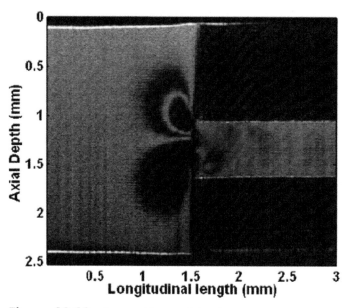

Figure 10.10   Structural image of the phantom and image of one specific velocity (8 mm/sec) of the flow with complex geometry ($\alpha = 90°$, every second A-scan is shown). (Courtesy of Dr. R. K. Wang, *Phys. Med. Biol.* 2004, 49, 1265–1276.)

REFERENCES

1. Barton, J. K. and Stromski, S. (2005). Flow measurement without phase information in optical coherence tomography images. *Opt. Express* 13, 5234–5239.
2. Barton, J. K., Welch, A. J., and Izatt, J. A. (1998). Investigating pulsed dye laser-blood vessel interaction with color Doppler optical coherence tomography. *Opt. Express* 3, 251–256.
3. Boashash, B. (1992a). Estimating and interpreting the instantaneous frequency of a signal, Part 1: Fundamentals. *Proc. IEEE* 80, 520.
4. Boashash, B. (1992b). Estimating and interpreting the instantaneous frequency of a signal, Part 2: Algorithms and applications. *Proc. IEEE* 80, 540.
5. Bonner, R. and Nossal, R. (1981). Model for laser Doppler measurements of blood-flow in tissue. *Appl. Opt.* 20, 2097–2107.
6. Brand, S., Poneros, J. M., Bouma, B. E., Tearney, G. J., Compton, C. C., and Nishioka, N. S. (2000). Optical coherence tomography in the gastrointestinal tract. *Endoscopy* 32, 796–803.
7. Brezinski, M. E. and Fujimoto, J. G. (1999). Optical coherence tomography: High-resolution imaging in nontransparent tissue. *IEEE J. Selected Top. Quantum Electron.* 5, 1185–1192.
8. Brezinski, M. E., Tearney, G. J., Bouma, B. E., Izatt, J. A., Hee, M. R., Swanson, E. A., Southern, J. F., and Fujimoto, J. G. (1996). Optical coherence tomography for optical biopsy — Properties and demonstration of vascular pathology. *Circulation* 93, 1206–1213.

9. Chen, Z. P., Milner, T. E., Dave, D., and Nelson, J. S. (1997a). Optical Doppler tomographic imaging of fluid flow velocity in highly scattering media. *Opt. Lett.* 22, 64.

10. Chen, Z. P., Milner, T. E., Srinivas, S., Wang, X. J., Malekafzali, A., Vangemert, M. J. C., and Nelson, J. S. (1997b). Noninvasive imaging of in vivo blood flow velocity using optical Doppler tomography. *Opt. Lett.* 22, 1119–1121.

11. Chen, Z. P., Zhao, Y. H., Srinivas, S. M., Nelson, J. S., Prakash, N., and Frostig, R. D. (1999). Optical Doppler tomography. *IEEE J, Selected Top. Quantum Electron.* 5, 1134–1142.

12. de Boer, J. F., Cense, B., Nassif, N., White, B. R., Park, B. H., Pierce, M. C., Tearney, G. J., Bouma, B. E., and Chen, T. C. (2005). Ultra-high speed and ultra-high resolution optical coherence tomography and optical Doppler tomography. *Invest. Ophthalmol. Vis. Sci.* 46, 116 (Suppl).

13. Ding, Z. H., Zhao, Y. H., Ren, H. W., Nelson, J. S., and Chen, Z. P. (2002). Real-time phase-resolved optical coherence tomography and optical Doppler tomography. *Opt. Express* 10, 236–245.

14. Fercher, A. F., Hitzenberger, C. K., Kamp, G., and Elzaiat, S. Y. (1995). Measurement of intraocular distances by backscattering spectral interferometry. *Opt. Commun.* 117, 43–48.

15. Huang, D., Swanson, E. A., Lin, C. P., Schuman, J. S., Stinson, W. G., Chang, W., Hee, M. R., Flotte, T., Gregory, K., Puliafito, C. A., and Fujimoto, J. G. (1991). Optical coherence tomography. *Science* 254, 1178–1181.

16. Izatt, J. A., Kulkami, M. D., Yazdanfar, S., Barton, J. K., and Welch, A. J. (1997). In vivo bidirectional color Doppler flow imaging of picoliter blood volumes using optical coherence tomography. *Opt. Lett.* 22, 1439–1441.

17. Kariya, R., Mathine, D. L., and Barton, J. K. (2004). Analog CMOS circuit design and characterization for optical coherence tomography signal processing. *IEEE Trans. Biomed. Eng.* 51, 2160–2163.

18. Leitgeb, R. A., Drexler, W., Schmetterer, L., Bajraszewski, T., and Fercher, A. F. (2004a). High speed, ultra high resolution morphologic and real time Doppler flow imaging of the human retina by Fourier domain optical coherence tomography. *Invest. Ophthalmol. Vis. Sci.* 45, U796–U796.

19. Leitgeb, R. A., Schmetterer, L., Drexler, W., Fercher, A. F., Zawadzki, R. J., and Bajraszewski, T. (2003). Real-time assessment of retinal blood flow with ultrafast acquisition by color Doppler Fourier domain optical coherence tomography. *Opt. Express* 11, 3116–3121.

20. Leitgeb, R. A., Schmetterer, L., Hitzenberger, C. K., Fercher, A. F., Berisha, F., Wojtkowski, M., and Bajraszewski, T. (2004b). Real-time measurement of in vitro flow by Fourier-domain color Doppler optical coherence tomography. *Opt. Lett.* 29, 171–173.

21. Mason, C., Markusen, J. F., Town, M. A., Dunnill, P., and Wang, R. K. (2004). Doppler optical coherence tomography for measuring flow in engineered tissue. *Biosens. Bioelectron.* 20, 414–423.

22. Matheny, E. S., Hanna, N. M., Jung, W. G., Chen, Z. P., Wilder-Smith, P., Mina-Araghi, R., and Brenner, M. (2004). Optical coherence tomography of malignancy in hamster cheek pouches. *J. Biomed. Opt.* 9, 978–981.

23. Moger, J., Matcher, S. J., Winlove, C. P., and Shore, A. (2004). Measuring red blood cell flow dynamics in a glass capillary using Doppler optical coherence tomography and Doppler amplitude optical coherence tomography. *J. Biomed. Opt.* 9, 982–994.

24. Murray, A. K., Herrick, A. L., and King, T. A. (2004). Laser Doppler imaging: A developing technique for application in the rheumatic diseases. *Rheumatology* 43, 1210–1218.

25. Park, B. H., Pierce, M. C., Cense, B., and de Boer, J. F. (2003). Real-time multi-functional optical coherence tomography. *Opt. Express* 11, 782–793.

26. Park, B. H., Pierce, M. C., Cense, B., Yun, S. H., Mujat, M., Tearney, G. J., Bouma, B. E., and de Boer, J. F. (2005). Real-time fiber-based multi-functional spectral-domain optical coherence tomography at 1.3 μm. *Opt. Express* 13, 3931–3944.

27. Pedersen, C. J., Yazdanfar, S., Westphal, V., and Rollins, A. M. (2005). Phase-referenced Doppler optical coherence tomography in scattering media. *Opt. Lett.* 30, 2125–2127.

28. Piao, D. Q., Otis, L. L., Dutta, N. K., and Zhu, Q. (2002). Quantitative assessment of flow velocity-estimation algorithms for optical Doppler tomography imaging. *Appl. Opt.* 41, 6118.

29. Piao, D. Q. and Zhu, Q. (2003). Quantifying Doppler angle and mapping flow velocity by a combination of Doppler-shift and Doppler-bandwidth measurements in optical Doppler tomography. *Appl. Opt.* 42, 5158.

30. Pierce, M. C., Strasswimmer, J., Park, B. H., Cense, B., and de Boer, J. F. (2004). Advances in optical coherence tomography imaging for dermatology. *J. Invest. Dermatol.* 123, 458–463.

31. Proskurin, S. G., He, Y. H., and Wang, R. K. (2004). Doppler optical coherence imaging of converging flow. *Phys. Med. Biol.* 49, 1265–1276.

32. Proskurin, S. G., He, Y. H., and Wang, R. K. K. (2003). Determination of flow velocity vector based on Doppler shift and spectrum broadening with optical coherence tomography. *Opt. Lett.* 28, 1227–1229.

33. Ren, H. W., Brecke, K. M., Ding, Z. H., Zhao, Y. H., Nelson, J. S., and Chen, Z. P. (2002a). Imaging and quantifying transverse flow velocity with the Doppler bandwidth in a phase-resolved functional optical coherence tomography. *Opt. Lett.* 27, 409–411.

34. Ren, H. W., Ding, Z. H., Zhao, Y. H., Miao, J. J., Nelson, J. S., and Chen, Z. P. (2002b). Phase-resolved functional optical coherence tomography: Simultaneous imaging of in situ tissue structure, blood flow velocity, standard deviation, birefringence, and Stokes vectors in human skin. *Opt. Lett.* 27, 1702–1704.

35. Rollins, A. M., Yazdanfar, S., Barton, J. K., and Izatt, J. A. (2002). Real-time in vivo colors Doppler optical coherence tomography. *J. Biomed. Opt.* 7, 123–129.

36. Schaefer, A. W., Reynolds, J. J., Marks, D. L., and Boppart, S. A. (2004). Real-time digital signal processing-based optical coherence tomography and Doppler optical coherence tomography. *IEEE Trans. Biomed. Eng.* 51, 186–190.

37. Tanaka, T., Riva, C., and Bensira, I. (1974). Blood velocity-measurements in human retinal-vessels. *Science* 186, 830–831.

38. Tearney, G. J., Brezinski, M. E., Bouma, B. E., Boppart, S. A., Pitris, C., Southern, J. F., and Fujimoto, J. G. (1997). In vivo endoscopic optical biopsy with optical coherence tomography. *Science* 276, 2037–2039.

39. Vakoc, B. J., Yun, S. H., de Boer, J. F., Tearney, G. J., and Bouma, B. E. (2005). Phase-resolved optical frequency domain imaging. *Opt. Express* 13, 5483–5493.

40. Vargas, G., Readinger, A., Dozier, S. S., and Welch, A. J. (2003). Morphological changes in blood vessels produced by hyperosmotic agents and measured by optical coherence tomography. *Photochem. Photobiol.* 77, 541–549.

41. Wang, R. K. K. (2004). High-resolution visualization of fluid dynamics with Doppler optical coherence tomography. *Measurement Sci. Technol.* 15, 725–733.

42. Wang, X. J., Milner, T. E., and Nelson, J. S. (1995). Characterization of fluid-flow velocity by optical Doppler tomography. *Opt. Lett.* 20, 1337–1339.

43. Westphal, V., Yazdanfar, S., Rollins, A. M., and Izatt, J. A. (2002). Real-time, high velocity-resolution color Doppler optical coherence tomography. *Opt. Lett.* 27, 34–36.

44. White, B. R., Pierce, M. C., Nassif, N., Cense, B., Park, B. H., Tearney, G. J., and Bouma, B. E. (2003). In vivo dynamic human retinal blood flow imaging using ultra-high-speed spectral domain optical Doppler tomography. *Opt. Express* 11, 3490.

45. Wu, F. I. and Glucksberg, M. R. (2005). Choroidal perfusion measurements made with optical coherence tomography. *Appl. Opt.* 44, 1426–1433.

46. Wu, L. (2004). Simultaneous measurement of flow velocity and Doppler angle by the use of Doppler optical coherence tomography. *Opt. Lasers Eng.* 42, 303–313.

47. Xi, C. W., Marks, D. L., Parikh, D. S., Raskin, L., and Boppart, S. A. (2004). Structural and functional imaging of 3D microfluidic mixers using optical coherence tomography. *Proc. Natl. Acad. Sci. USA* 101, 7516–7521.

48. Yan, S. K., Piao, D. Q., Chen, Y. L., and Zhu, Q. (2004). Digital signal processor-based real-time optical Doppler tomography system. *J. Biomed. Opt.* 9, 454–463.

49. Yang, V., Song, L., Buttar, N., Wang, K., Gordon, M., Yue, E. S., Bisland, S., Marcon, N., Wilson, B., and Vitkin, A. (2003a). Monitoring of photodynamic therapy-induced vascular response using Doppler optical coherence tomography. *Gastrointest. Endosc.* 57, AB175–AB177.

50. Yang, V. X. D., Gordon, M. L., Mok, A., Zhao, Y. H., Chen, Z. P., Cobbold, R. S. C., Wilson, B. C., and Vitkin, I. A. (2002). Improved phase-resolved optical Doppler tomography using the Kasai velocity estimator and histogram segmentation. *Opt. Commun.* 208, 209–214.

51. Yang, V. X. D., Gordon, M. L., Qi, B., Pekar, J., Lo, S., Seng-Yue, E., Mok, A., Wilson, B. C., and Vitkin, I. A. (2003b). High speed, wide velocity dynamic range Doppler optical coherence tomography (Part I): System design, signal processing, and performance. *Opt. Express* 11, 794–809.

52. Yang, V. X. D., Gordon, M. L., Seng-Yue, E., Lo, S., Qi, B., Pekar, J., Mok, A., Wilson, B. C., and Vitkin, I. A. (2003c). High speed, wide velocity dynamic range Doppler optical coherence tomography (Part II): Imaging in vivo cardiac dynamics of *Xenopus laevis*. *Opt. Express* 11, 1650–1658.

53. Yang, V. X. D., Gordon, M. L., Tang, S. J., Marcon, N. E., Gardiner, G., Qi, B., Bisland, S., Seng-Yue, E., Lo, S., Pekar, J., Wilson, B. C., and Vitkin, I. A. (2003d). High speed, wide velocity dynamic range Doppler optical coherence tomography (Part III): In vivo endoscopic imaging of blood flow in the rat and human gastrointestinal tracts. *Opt. Express* 11, 2416–2424.

54. Yazdanfar, S. and Izatt, J. A. (2002). Self-referenced Doppler optical coherence tomography. *Opt. Lett.* 27, 2085–2087.

55. Yazdanfar, S., Rollins, A. M., and Izatt, J. A. (2000). Imaging and velocimetry of the human retinal circulation with color Doppler optical coherence tomography. *Opt. Lett.* 25, 1448–1450.

56. Yazdanfar, S., Rollins, A. M., and Izatt, J. A. (2003). In vivo imaging of human retinal flow dynamics by color Doppler optical coherence tomography. *Arch. Ophthalmol.* 121, 235–239.

57. Yeh, Y. and Cummins, H. Z. (1964). Localized fluid flow measurement with an He-Ne laser spectrometer. *Appl. Phys. Lett.* 4, 176.

58. Zhao, Y. H., Brecke, K. M., Ren, H. W., Ding, Z. H., Nelson, J. S., and Chen, Z. P. (2001). Three-dimensional reconstruction of in vivo blood vessels in human skin using

phase-resolved optical Doppler tomography. *IEEE J. Selected Top. Quantum Electron.* 7, 931–935.

59. Zhao, Y. H., Chen, Z. P., Ding, Z. H., Ren, H. W., and Nelson, J. S. (2002). Real-time phase-resolved functional optical coherence tomography by use of optical Hilbert transformation. *Opt. Lett.* 27, 98–100.

60. Zhao, Y. H., Chen, Z. P., Saxer, C., Shen, Q. M., Xiang, S. H., de Boer, J. F., and Nelson, J. S. (2000a). Doppler standard deviation imaging for clinical monitoring of *in vivo* human skin blood flow. *Opt. Lett.* 25, 1358.

61. Zhao, Y. H., Chen, Z. P., Saxer, C., Xiang, S. H., de Boer, J. F., and Nelson, J. S. (2000b). Phase-resolved optical coherence tomography and optical Doppler tomography for imaging blood flow in human skin with fast scanning speed and high velocity sensitivity. *Opt. Lett.* 25, 114–116.

62. Zvyagin, A. V., Fitzgerald, J. B., Silva, K., and Sampson, D. D. (2000). Real-time detection technique for Doppler optical coherence tomography. *Opt. Lett.* 25, 1645–1647.

63. Born, M. and Wolf, E. (1999). *Principles of Optics*. Cambridge University Press, Cambridge, UK.

64. Goodman, J. W. (1996). *Introduction to Fourier Optics*. McGraw-Hill, New York.

11 DIGITAL IMAGE PROCESSING TECHNIQUES FOR SPECKLE REDUCTION, ENHANCEMENT, AND SEGMENTATION OF OPTICAL COHERENCE TOMOGRAPHY (OCT) IMAGES

Jadwiga Rogowska, PhD

Assistant Professor, Harvard Medical School

Chapter Contents

11.1 INTRODUCTION

Optical coherence tomography (OCT) is a new imaging technique that allows noninvasive, high resolution, cross-sectional imaging of transparent and nontransparent structures. The greatest advantage of OCT is its resolution. Standard-resolution OCT can achieve axial resolution of 10 to 15 μm.[8,23] A recently developed high resolution OCT increases the resolution to the subcellular level of 1–2 μm.[17,23,42] Advances in OCT technology have made it possible to apply OCT in a wide variety of applications, including diagnosing and monitoring retinal diseases; imaging atherosclerotic plaque; tumor detection in gastrointestinal, urinary, and respiratory tracts; and early osteoarthritic changes in cartilage.[7,11,20,23,71,84,85]

Like other coherent imaging techniques, OCT suffers from speckle noise, which makes detection of the boundaries and image segmentation problematic and unreliable. Therefore, speckle removal and feature enhancement is a critical preprocessing step, which may improve human interpretation and the accuracy of computer-assisted procedures. To date, very limited research has been done in the area of image enhancement and speckle reduction in OCT imaging.[36,58,63,68,69,87,88] The OCT speckle reduction techniques include spatial averaging,[3,67] frequency compounding,[53,75] polarization averaging,[15] and image processing techniques.[57]

Image segmentation is another important step for feature extraction, image measurements, and image display.[57] In some applications it may be useful to classify image pixels into anatomical regions, such as blood vessels or skin layers, while in others into pathological regions, such as cystoids and subretinal fluid spaces.[19] In some studies the goal is to divide the entire image into subregions (for example, bone and cartilage,[63] or to extract one specific structure (for example, retinal borders).[35] Only a few researchers evaluated the edge detectors and image segmentation techniques on OCT images. In the area of edge detection, a Marr-Hildreth (Laplacian of a Gaussian) algorithm has been used in OCT for detecting edges of retina.[35] In the area of image segmentation, manual boundary tracing and thresholding[24,71] are commonly used techniques. More recently, an automatic technique using a Markov boundary model was developed for detecting retinal boundaries on OCT images.[35] Rogowska et al.[63] applied edge detection followed by edge linking using graph searching to the images of cartilage.

In this chapter we will review digital image processing techniques for speckle removal, image enhancement, and segmentation of OCT images. The techniques will be illustrated on several OCT images of different anatomical structures. Because the ultrasound images are similar to OCT images in appearance (speckle), we will also review some of the techniques that can be adapted from the ultrasound applications.

11.2 SPECKLE REDUCTION TECHNIQUES

Figures 11.1 and 11.2 show OCT images of rabbit cartilage and human coronary plaque. Most of the granular pattern on those images does not correspond to the real tissue microstructure. Speckle degrades the quality of the images and makes it difficult to differentiate anatomical structures. The OCT speckle is similar to the speckle in ultrasound or radar imaging. It is a complex phenomenon, but in general it

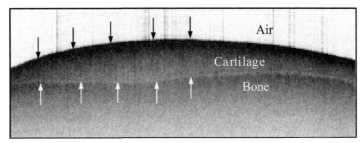

Figure 11.1 Original OCT image of the cartilage. The black arrows indicate the cartilage-air boundary, and the white arrows point to cartilage-bone boundary. Reprinted with permission from J. Rogowska and M. E. Brezinski, *Phys Med Biol*, 47:641–655, 2002.

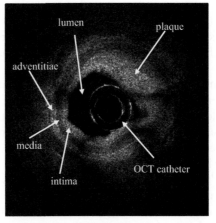

Figure 11.2 Original OCT image of the human coronary plaque.

occurs when light from a coherent source illuminates scatters separated by distances near that of the coherence length of the source. Speckle is not truly a noise in the typical engineering sense. It is signal-dependent and its texture carries information about the tissue being imaged.[69,70]

Experienced OCT users can easily interpret different textures of the speckle as features. However, texture makes computer detection of boundaries difficult. Therefore, it is important to suppress signal-degrading speckle and accentuate signal-carrying speckle.[12]

It is well known that speckle averaging can be used for speckle reduction.[3,33,53] However, applying this technique requires imaging with special instrumental setups. The filtering methods discussed here are mathematical manipulations of image gray levels.

The most popular image processing methods for speckle reduction are spatial filters, including mean, median, and hybrid median.[9,25,54,64,65] Other techniques include rotational kernel transformation (RKT),[13,39,58,59] Wiener filtering, multiresolution wavelet analysis,[12,28] and adaptive smoothing and anisotropic diffusion.[4,52,83,89]

11.2.1   Mean, Median, and Hybrid Median Filters

One of the noise removal techniques is image smoothing (low pass filtering) by using mean (average), median, and Gaussian filters.[31,64,69] In the mean filter, the value of each pixel is replaced by the average of all pixel values in a local neighborhood (usually an $N \times N$ window, where $N = 3, 5, 7$, etc.). In the median filter, the value of each pixel is replaced by a median value calculated in a local neighborhood. Median smoothing, unlike the mean filter, while smoothing out small textural variations, does not blur the edges of regions larger than the window used. The conventional median filter does not shift boundaries, but it can round off corners. The hybrid median overcomes this limitation by ranking the pixels separately as a function of direction and then combining the results. Another popular smoothing method is Gaussian filtering, where the kernel coefficients have Gaussian distribution. Chauhan et al.[11] used median and Gaussian smoothing to the OCT images of retina.

The amount of smoothing in filtering techniques depends on the kernel size. As the kernel size increases, the speckle in the image is blurred and the edges are less visible (Figure 11.3). Figure 11.4 shows the results of speckle removal techniques using mean, median, hybrid median, and Gaussian filters with a kernel of 7×7 pixels.

Many investigators have shown that mean filtering suppresses high frequency noise.[54,64,65] However, low pass filtering also smears the edges, which is not desirable in border detection. A conventional mean low pass filter is, therefore, not very useful in OCT imaging (Figures 11.3A and 11.4A). The other filters, such as median and hybrid median, are more promising. Mean and median filters have been applied to OCT images of the retina[35] and atherosclerotic plaque.[77]

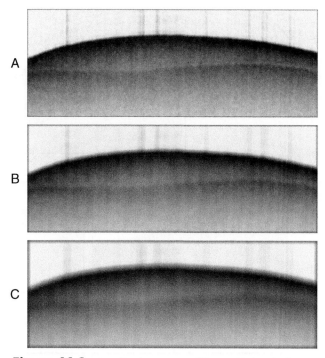

Figure 11.3 Results of an average filter applied to the image on Figure 11.1 with a kernel of (A) 5×5 pixels, (B) 9×9 pixels, and (C) 15×15 pixels. Reprinted with permission from J. Rogowska and M. E. Brezinski, *Phys Med Biol.*, 47:641–655, 2002.

11.2.2 Adaptive Filtering

There are other, more complicated filtering methods that are used for specific types of images. They are called adaptive filtering techniques, where the algorithm is modified locally based on the pixel's neighborhood.[30,41] If, for example, the neighborhood is constant, we can assume that we are within an object with constant features and thus we can apply an isotropic smoothing operation to this pixel to reduce the noise level. If an edge has been detected in the neighborhood, we could still apply some smoothing, but only along the edge. Adaptive filtering combines an efficient noise reduction and an ability to preserve and even enhance the edges of image structures. Local adaptive filters have been applied successfully in ultrasound.[32]

One of the adaptive filtering techniques successfully applied to the OCT images is RKT, or "sticks" technique.[13,39,58,60] The RKT technique operates by filtering an image with a set of templates (kernels), and retaining the largest filter output at each pixel. This method enhances the image features by emphasizing thin edges while suppressing a noisy background.

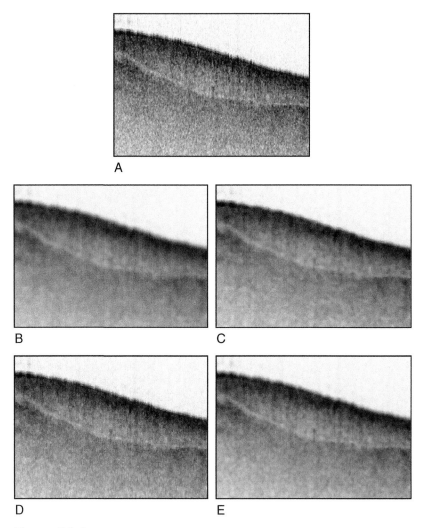

Figure 11.4 The results of different smoothing filters applied to the tibial cartilage image; (A) the original image, (B) average filter, (C) median filter, (D) hybrid median filter, and (E) Gaussian filter. The smoothing kernel was 7 × 7 pixels.

In this technique the input image is convolved with a kernel that is rotated discretely in small steps through 360 degrees. The convolution can be written as:

$$S_\theta(x,y) = I(x,y) \cdot K_\theta(x,y) \tag{11.1}$$

where I is the input image, and K_θ is the kernel oriented at rotation angle θ. The maximum values are calculated over all rotated kernels, and the output image is defined as:

$$O(x, y) = \text{maximum}\{S_\theta(x,y): 0 \leq \theta < 360°\} \tag{11.2}$$

This RKT algorithm is related to the class of rotating kernel min-max transformation (RKMT), which was investigated by Lee and Rhodes.[39] The difference is that the output image in the RKMT technique is defined as the arithmetic difference between the maximum and the minimum of S_θ while in the RKT algorithm, it is the maximum.

In the RKT technique, with increasing kernel size, the speckle in the image is blurred and the edges are more clearly emphasized. Figure 11.5 presents the results of the RKT filter with the selected kernels: 5×5, 11×11, 21×21, and 31×31. Since the cartilage edge is long, it is possible to use large kernel sizes for edge enhancement. However, large kernels cause excessive blurring. Please note the artifacts introduced at the borders of RKT-processed images. These are effects of the implementation of the RKT filter via convolution, which affects the pixels at the border (equal to half the kernel size) of the image.

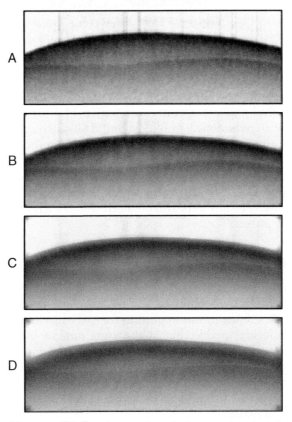

Figure 11.5 The results of the rotating kernel transformation filter with kernels: (A) 5×5, (B) 11×11, (C) 21×21, and (D) 31×31 pixels, applied to the original image shown on Fig. 1. Reprinted with permission from J. Rogowska and M. E. Brezinski, *Phys Med Biol*, 47:641–655, 2002.

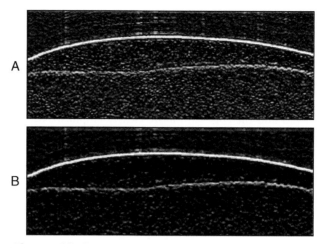

Figure 11.6 Results of the vertical gradient applied to the (A) original image from Figure 11.1 and (B) hybrid median image from Figure 11.3C. Reprinted with permission from J. Rogowska and M. E. Brezinski, *Phys Med Biol*, 47:641–655, 2002.

To test which filter is the best in detecting the cartilage-bone border, we applied a vertical gradient operator (see section 11.3 for description of the gradient) to the original, hybrid-median filtered and RKT-filtered images. The results are presented in Figures 11.6 and 11.7. The cartilage-bone border in Figure 11.6 is shown as a thick, broken line, which would be very difficult to detect automatically. Similar results are presented in Figure 11.7A, B. However, when the RKT kernel size is larger (Figure 11.7C, D), the cartilage-bone border is shown as a thin, solid line.[61] Very promising results were also obtained with the RKT technique applied to the coronary OCT images.[58]

11.2.3 Other Techniques

A review of a few other techniques for OCT speckle reduction can be found in a paper by Schmitt et al.[69] A technique called (ZAP), which operates in the complex domain, was applied to the OCT images by Yung et al. It showed promise in reducing speckle, but at the same time, it blurred the boundaries of features of interest. Other complex-domain processing methods applied to OCT are iterative point deconvolution, called CLEAN,[68] and constrained iterative deconvolution.[36] The deconvolution techniques require *a priori* information about the point spread function of the imaging optics, as well as optical properties of the imaged tissue.

Speckle reduction using a wavelet filter[18,25] was also investigated in ultrasound[28] and OCT.[1][Xiang 97] The technique was first introduced by Guo et al.[27] After transforming an image into the wavelet domain, the components in some scales are discarded, and the de-noised image is obtained by an inverse wavelet transform.

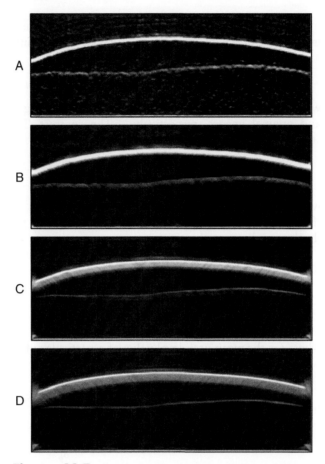

Figure 11.7 Results of the vertical gradient applied to the rotating kernel transformation filter with kernels: (A) 5×5, (B) 11×11, (C) 21×21, and (D) 31×31 pixels from Figure 11.5. Reprinted with permission from J. Rogowska and M. E. Brezinski, *Phys Med Biol*, 47:641–655, 2002.

In the last few years a new class of image processing operators, called diffusion filters, have been introduced.[4,52,83,89] There are two classes of diffusion filters: inhomogeneous diffusion (smoothing stops completely and in all directions at edges, leaving the edges noisy) and anisotropic diffusion (edges are blurred by diffusion perpendicular to them while diffusion parallel to edges stabilizes them). The anisotropic diffusion filter was applied in the procedure for measuring retinal thickness.[19]

11.2.4 Evaluation of Image Enhancement Results

Improvement of image quality after speckle reduction and image enhancement is often very difficult to measure. A processed image is enhanced over the original image if it allows the observer to better perceive the desirable information in the image. One of the quantitative measures of contrast improvement can be defined by a Contrast Improvement Index (CII):

$$CII = C_{processed}/C_{original} \tag{11.3}$$

where $C_{processed}$ and $C_{original}$ are the contrasts for a region of interest in the processed and original images, respectively.[38] The contrast C can be calculated using the standard formula:

$$C = \frac{f - b}{f + b} \tag{11.4}$$

where f is the mean (average) gray-level value of a particular object in an image (foreground) and b is the mean gray-level value of the surrounding region media (background).

Another measure of contrast enhancement is the contrast to noise ratio (CNR).[58,74,88] It is defined as:

$$CNR = \frac{f - b}{\sqrt{\delta_f^2 + \delta_b^2}} \tag{11.5}$$

where f is the mean gray-level value of a particular object in an image (foreground), b is the mean gray-level value of the surrounding region (background), and δ_f and δ_b are the corresponding standard deviations.

Two variations of the CNR can also be used as evaluation measures: ASNR and PSNR.[40,76] They are defined by:

$$ASNR = \frac{f - b}{\delta_b} \tag{11.6}$$

$$PSNR = \frac{p - b}{\delta_b} \tag{11.7}$$

where p is the maximum gray-level value of a foreground.

Other evaluation techniques can be found in Lu (1991), Fernandez (2005), and Singh (2005).[19,43,72]

11.3 IMAGE SEGMENTATION TECHNIQUES

The principal goal of the segmentation process is to partition an image into regions (also called regions, classes, or subsets) that are homogeneous with respect to one or more characteristics or features.[9,22,25,64,73]

A wide variety of segmentation techniques have been proposed in the digital image processing literature (see surveys in Freixenet, 2002; Fu, 1981; Haralick, 1985; Pal, 1993; and Weszka, 1978).[21,22,29,50,86] However, there is no one standard segmentation technique that can produce satisfactory results for all imaging applications. The definition of the goal of segmentation varies according to the goal of the study and the type of the image data.

The most common segmentation techniques can be divided into two broad categories: (1) region segmentation techniques that look for the regions satisfying a given homogeneity criterion and (2) edge segmentation techniques that look for edges between regions with different characteristics.[9,25,64,73]

Thresholding is a common region extraction method.[14,50,66,73,86] In this technique, a threshold is selected and an image is divided into groups of pixels having values less than the threshold and groups of pixels with values greater than or equal to the threshold. There are several thresholding methods: global methods based on gray-level histograms, global methods based on local properties, local threshold selection, and dynamic thresholding. Clustering is the name of another class of algorithms for image segmentation.[31] Clustering segments the image in terms of sets or clusters of pixels with the strong similarity in the feature space. The basic operation is to examine each pixel and assign it to the cluster that best represents the value of its characteristic vector of features of interest. Both thresholding and clustering operate at the pixel level. Other methods, such as region growing, operate on groups of pixels. Region growing is a process by which two adjacent regions are assigned to the same segment, if their image values are close enough, according to some pre-selected criterion of closeness.[54]

The strategy of edge-based segmentation algorithms is to find object boundaries and segment regions enclosed by the boundaries[9,25,29,44,64] These algorithms usually operate on edge magnitude and/or phase images produced by an edge operator suited to the expected characteristics of the image. For example, most gradient operators such as the Prewitt, Kirsch, or Roberts operator are based on the existence of an ideal step edge. Other edge-based segmentation techniques are graph searching and contour following.[2,73]

The following sections will present some of the segmentation techniques that are commonly used in medical imaging.

11.3.1 Thresholding

There are several thresholding techniques.[9,14,25,29,30,64–66,73,86] Some of them are based on image histograms; others are based on local properties, such as local mean value and standard deviation or the local gradient. The most intuitive approach is global thresholding. When only one threshold is selected for the entire image (based on the image histogram), thresholding is called global. If the threshold depends on local properties of some image regions (e.g., the local average gray value), the thresholding is called local. If the local thresholds are selected independently for each pixel (or groups of pixels), the thresholding is called dynamic or adaptive.

Global thresholding is based on the assumption that an image has a bimodal histogram and, therefore, the object can be extracted from the background by a simple operation that compares image values with a threshold value.[14,73]

If an image contains more than two types of regions, it may still be possible to segment it by applying several individual thresholds,[64] or by using a multi-thresholding technique.[55] With the increasing number of regions, the histogram modes are more difficult to distinguish, and threshold selection becomes more difficult.

Global thresholding is computationally simple and fast. It works well on images that contain objects with uniform intensity values on a contrasting background. However, it fails if there is a low contrast between the object and the background, or if the image is noisy. As an example, detection of the cartilage-bone borders by thresholding technique is shown in Figure 11.8. Thresholds were selected based on the image histograms. Figure 11.8A presents the results of thresholding performed on the original image. The image in Figure 11.8B demonstrates the border obtained by thresholding the vertical gradient of the hybrid median-processed image (from Figure 11.6B), while Figure 11.8C demonstrates the vertical gradient of the RKT image (from Figure 11.7D). Based on Figure 11.8, we conclude that the thresholding technique failed to detect the cartilage-bone border on the original and vertical gradient of hybrid median-processed images, and it detected a solid continuous edge on vertical gradient of the RKT-processed image (31×31 kernel).[61]

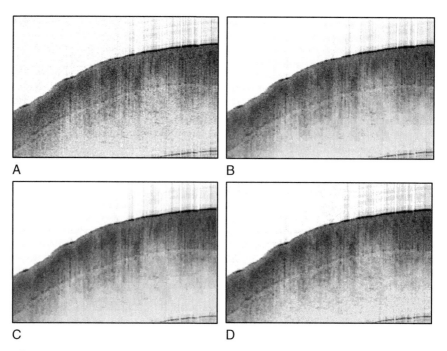

A   B

C   D

Figure 11.8   (A) The original image of cartilage. Results of speckle reduction using (B) wavelet processing, (C) the wavelet filter followed by the adaptive 7×7 median filter, (D) the 7×7 hybrid median filter.

11.3.1.1 Local (adaptive) thresholding

In many applications, a global threshold cannot be found from a histogram, or a single threshold cannot give good segmentation results over an entire image. For example, when the background is not constant and the contrast of objects varies across the image, thresholding may work well in one part of the image, but may produce unsatisfactory results in other areas. If the background variations can be described by some known function of position in the image, the background can be subtracted and a single threshold may work for the entire image. Another solution is to apply local (adaptive) thresholding.[2,5,14,29,37,48,86]

Local thresholds can be determined by (1) splitting an image into subimages and calculating thresholds for each subimage, or by (2) examining the image intensities in the neighborhood of each pixel. In the former method, an image is first divided into rectangular overlapping subimages and the histograms are calculated for each subimage. The subimages used should be large enough to include both object and background pixels. If a subimage has a bimodal histogram, then the minimum between the histogram peaks should determine a local threshold. If a histogram is unimodal, the threshold can be assigned by interpolation from the local thresholds found for nearby subimages. In the final step, a second interpolation is necessary to find the correct thresholds at each pixel.

In the latter method, a threshold can be selected using a mean value of the local intensity distribution. Sometimes other statistics can be used, such as mean plus standard deviation, mean of the maximum and minimum values,[9,14] or a statistic based on the local intensity gradient magnitude.[14,34]

Modifications of the above two methods can be found in Rosenfeld, 1982; Fu, 1981; Haralick, 1985; and Nakagawa, 1979.[22,29,48,64] In general, local thresholding is computationally more expensive than global thresholding. It is very useful for segmenting objects from a varying background, and also for extraction of regions that are very small and sparse.

11.3.2 Region Growing

While thresholding focuses on the difference of pixel intensities, the region growing method looks for groups of pixels with similar intensities. Region growing (also called region merging) starts with a pixel or a group of pixels (called seeds) that belong to the structure of interest. Seeds can be chosen by an operator, or provided by an automatic seed finding procedure. In the next step, if they are sufficiently similar based on a uniformity test (also called a homogeneity criterion), neighboring pixels are examined one at a time and added to the growing region. The procedure continues until no more pixels can be added. The object is then represented by all pixels that have been accepted during the growing procedure.[2,25,54,64,73,79,80]

One example of the uniformity test is comparing the difference between the pixel intensity value and the mean intensity value over a region. If the difference is less than some predefined value (for example, less than two standard deviations of the mean over the region), the pixel is included in the region; otherwise, it is defined as an edge pixel. The results of region growing depend strongly on the selection of the homogeneity criterion. If it is not properly chosen, the regions leak out into adjoining areas or merge with regions that do not belong to the object of interest. Another problem of region growing is that different starting points may not grow into identical regions.

Instead of region merging, it is possible to start with some initial segmentation and subdivide the regions that do not satisfy a given uniformity test. This technique is called the splitting technique.[29,64,73] A combination of splitting and merging combines the advantages of both approaches.[2,51] Various approaches to region growing segmentation have been described by Zucker.[90] Excellent reviews of region growing techniques were done by Fu and Mui,[22] Haralick and Shapiro,[29] and Rosenfeld and Kak.[64]

11.3.3   Watershed Algorithm

Watershed segmentation is a region-based technique based on image morphology.[9,73] It requires identifying at least one marker (seed point) interior to each object of the image, including the background as a separate object. The markers are chosen by an operator or are provided by an automatic procedure that takes into account the application-specific knowledge of the objects. Once the objects are marked, they can be grown using a morphological watershed transformation.[6] A very intuitive description of watersheds can be found in Castleman.[9] To understand the watershed, one can think of the image as a surface where the bright pixels represent mountaintops and the dark pixels represent valleys. The surface is punctured in some of the valleys, and then slowly submerged into a water bath. The water will pour into each puncture and start to fill the valleys. However, the water from different punctures is not allowed to mix, and, therefore, dams need to be built at the points of first contact. These dams are the boundaries of the water basins and also the boundaries of image objects.

An application of watershed segmentation to extract lesion borders on OCT images is shown in Figure 11.9 for the original image and images preprocessed with the RKT technique. In this implementation, a 3×3 Sobel edge operator[25,64] is used in place of the morphological gradient to extract edge strength. The original OCT image is shown in Figure 11.8A. In the first step, the operator positions a cursor inside the lesion. All pixels within a radius of two pixels of the mark are used as seed points for the lesion. To mark the exterior of the lesion, the operator drags the cursor outside of the lesion to define a circular region, which completely encloses the lesion. All pixels outside this circle mark the background. In the next step, an edge image is created using the Sobel edge operator. The edge image has high values for the pixels with strong edges. With the seed point marking the lesion interior, the circle marking the background, and the edge image generated by the Sobel operator, the segmentation proceeds directly with the watershed operation. The watershed algorithm operates on an edge image to separate the lesion from the surrounding tissue. By using a technique called simulated immersion,[81] the

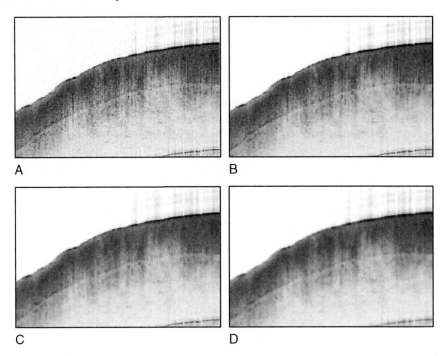

Figure 11.9 Results of speckle reduction using the anisotropic diffusion algorithm applied to the original image on Fig. 8A. The strength of diffusion varied with parameter K: (A) K = 5, (B) K = 10, (C) K = 20, (D) K = 40 [Perona 90].

watershed considers whether a drop of water at each point in the edge image will flow to the interior seed point or the exterior marker. Points that drain into the interior belong to the lesion, while points that drain to the exterior belong to the surrounding tissue. More formal discussions of morphological segmentation can be found in Meyer[46] and Vincent.[81,82]

The Sobel/watershed algorithm was repeated for images processed with an adaptive filter using RKT with two kernel sizes: 11×11 and 21×21 pixels. As shown on Figure 11.9B and C, the segmentation edges are smoother on filtered images.

11.3.4 Edge-Based Segmentation Techniques

An edge or boundary on an image is defined by the local pixel intensity gradient. A gradient is an approximation of the first order derivative of the image function. For a given image $f(x,y)$, we can calculate the magnitude of the gradient as:

$$|G| = \sqrt{\left[G_x^2 + G_y^2 \right]} = \sqrt{\left[\left(\frac{\partial f}{\partial x} \right)^2 + \left(\frac{\partial f}{\partial y} \right)^2 \right]} \qquad (11.8)$$

and the direction of the gradient as:

$$D = \tan^{-1}\left(\frac{G_y}{G_x}\right) \tag{11.9}$$

where G_x and G_y are gradients in directions x and y, respectively. Since the discrete nature of digital images does not allow the direct application of continuous differentiation, calculation of the gradient is done by differencing.[25]

Both magnitude and direction of the gradient can be displayed as images. The magnitude image will have gray levels that are proportional to the magnitude of the local intensity changes, while the direction image will have gray levels representing the direction of the maximum local gradient in the original image.

Most gradient operators in digital images involve calculation of convolutions, e.g., weighted summations of the pixel intensities in local neighborhoods. The weights can be listed as a numerical array in a form corresponding to the local image neighborhood (also known as a mask, window, or kernel). For example, in the case of a 3×3 Sobel edge operator, there are two 3×3 masks:

-1	-2	-1
0	0	0
1	2	1

-1	0	1
-2	0	2
-1	0	1

The first mask is used to compute G_x while the second computes G_y. The gradient magnitude image is generated by combining G_x and G_y using Eq. 11.8. The results of edge detection depend on the gradient mask. Some of the other edge operators are Roberts, Prewitt, Robinson, Kirsch, and Frei-Chen.[25,29,31,64,65]

Many edge detection methods use a gradient operator, followed by a threshold operation on the gradient to decide whether an edge has been found.[9,14,25,29,44,64,65,73,79] As a result, the output is a binary image indicating where the edges are. For example, Figure 11.8B and C shows the results of thresholding applied to the vertical gradient for detection of the cartilage-bone border. Please note that the selection of the appropriate threshold is a difficult task. Edges may include some background pixels around the border or may not enclose the object completely. To form closed boundaries surrounding regions, a post processing step of linking or grouping of the edges that correspond to a single boundary is required. The simplest approach to edge linking involves examining pixels in a small neighborhood of the edge pixel (3×3, 5×5, etc.) and linking pixels with similar edge magnitude and/or edge direction. In general, edge linking is computationally expensive and not very reliable. One solution is to make the edge linking semi-automatic and allow a user to draw the edge when the automatic tracing becomes ambiguous. A technique of graph searching for border detection has been used in many medical applications.[2,49,73] In this technique, each image pixel corresponds to a graph node and each path in a graph corresponds to a possible edge in an image. Each node has a cost associated with it, which is usually calculated using the local edge magnitude, edge direction, and *a priori* knowledge about the boundary shape or location. The cost of a path through the graph is the sum of costs of all nodes that are included in the path. By finding the optimal low-cost path in the graph, the optimal border can be defined. The graph-searching technique is very powerful, but it strongly depends on an application-specific cost function. A review of graph searching algorithms and cost function selection can be found in Sonka.[73]

Since the peaks in the first order derivative correspond to zeros in the second order derivative, the Laplacian operator (which approximates the second order derivative) can also be used to detect edges.[9,25,64]

The Laplacian operator ∇^2 of a function $f(x,y)$ is a second order derivative and is defined as:

$$\nabla^2 f(x,y) = \frac{\partial^2 f(x,y)}{\partial x^2} + \frac{\partial^2 f(x,y)}{\partial y^2} \tag{11.10}$$

The Laplacian is approximated in digital images by an N by N convolution mask.[25,73] Here are three examples of 3×3 Laplacian masks that represent different approximations of the Laplacian operator:

0	−1	0
−1	4	−1
0	−1	0

−1	−1	−1
−1	8	−1
−1	−1	−1

1	−2	1
−2	4	−2
1	−2	1

The image edges can be found by locating pixels where the Laplacian makes a transition through zero (zero crossings).

All edge detection methods that are based on a gradient or Laplacian are very sensitive to noise. In some applications, noise effects can be reduced by smoothing the image before applying an edge operation. Marr and Hildreth[44] proposed smoothing the image with a Gaussian filter before applying the Laplacian (this operation is called Laplacian of Gaussian, LoG). Figure 11.8B shows a result of a 7×7 Gaussian followed by a 7×7 Laplacian applied to the original image in Figure 11.7A. The zero crossings of the LoG operator are shown in Figure 11.8D. The advantage of a LoG operator compared to a Laplacian is that the edges of the blood vessels are smoother and better outlined. However, in Figure 11.8C and D, the nonsignificant edges are detected in regions of almost constant gray level. To solve this problem, the information about the edges obtained using first and second derivatives can be combined.[73]

Other edge-finding algorithms can be found in Davies, 1975; Rosenfeld, 1982; Gonzalez, 2002; and Fu, 1981.[14,22,25,64]

Rogowska et al.[62,63] used Sobel edge detection for delineating the cartilage boundaries and measurement of cartilage thickness in OCT. The procedure consists of the RKT technique for image enhancement and speckle reduction, 3×3 Sobel edge detection, and edge linking by graph searching. Figures 11.10 and 11.11 show the results of the Sobel edge detector and the corresponding cost images used in graph searching. The magnitude of edge images was used in the edge linking

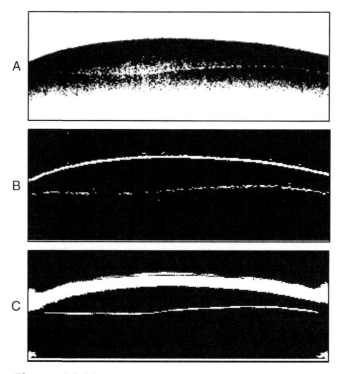

Figure 11.10   Results of thresholding for detection of the cartilage-bone border, performed on the (A) original image, (B) vertical gradient of the hybrid median-processed image (7×7 kernel), (C) vertical gradient of the RKT-processed image (31×31 kernel). Reprinted with permission from J. Rogowska and M. E. Brezinski, *Phys Med Biol*, 47:641–655, 2002.

procedure to calculate cost functions. Therefore, the cost images displayed in Figures 11.10 and 11.11 show inverted Sobel magnitude images. Figure 11.10 presents two cartilage borders, outer (green), and inner (red), which were found using our algorithm applied to the original image. The cartilage-bone border on Figure 11.11A, B is the result of the algorithm applied to the RKT-processed image with a 11×11 pixel kernel. Similarly, Figure 11.7C, D shows the result of the algorithm applied to the RKT-processed image with a 31×31 pixel kernel. The final result of cartilage edge detection is presented in Figure 11.12, where the outer and inner cartilage boundaries are superimposed on the original image. The outer boundary (in green) is from Figure 11.10A, while the inner (in red) boundary is from Figure 11.11C.

11.3.5   Other Segmentation Techniques

Other strategies, such as active contours models[45] and texture,[2,26,47] can also be used in image segmentation. In OCT active contours models (snakes) were used

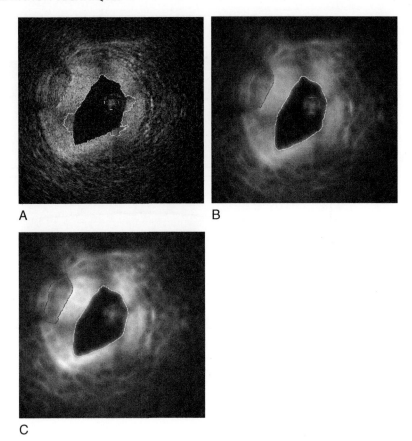

A   B

C

Figure 11.11 Results of segmentation using a Sobel/watershed algorithm applied to (A) the original image, (B) the image processed RKT technique (11 × 11 pixels kernel), and (C) the image processed with the RKT technique (21 × 21 pixels kernel). Red and yellow outlines represent the lipid lesion and lumen, respectively.

for outlining fluid-filled regions within retina.[19] Gossage et al.[26] evaluated the application of statistical and spectral texture analysis techniques for differentiating tissue types based on the speckle content in OCT images of mouse skin, fat, and lungs.

A new algorithm for detecting retinal boundaries was presented by Koozekanani et al.[35] In their technique, a median filter is used first to reduce the speckle noise and one-dimensional Laplacian of a Gaussian detects the edges. Then, the edges are linked together using the Markov boundary model. Once the boundary is detected, it is smoothed with the cubic B-spline. It was shown that the technique is robust and can determine average retinal thickness with high accuracy.

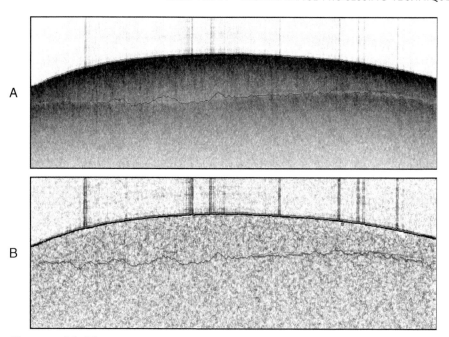

Figure 11.12 Results of the Sobel edge detection and graph searching algorithm applied to the original image; (A) two cartilage borders: outer (green), and inner (red), and (B) the corresponding cost image used in graph searching. Reprinted with permission from J. Rogowska, C.M. Bryant, and M. E. Brezinski, *J Opt Soc Am A*, 20(2):357–367, 2003.

11.4. SUMMARY

Speckle reduction, image enhancement, and segmentation are important steps in many applications involving measurements and visualization. This chapter is a brief introduction to the fundamental concepts of methods that are common in digital image processing in addition to methods for quantitating cartilage. These image-processing techniques can easily be adapted to many applications, such as detection of tumors, retinal layers, coronary plaques, and skin lesions on OCT images. Selection of the "correct" technique for a given application is a difficult task. In many cases, a combination of several techniques may be necessary to obtain the image enhancement or image segmentation goal.

As new and more sophisticated techniques are developed, the need for objective evaluation and quantitative testing procedures grows.[10,16,43,72] Evaluation of new algorithms using standardized protocols will be useful for selection of methods for a particular application.

REFERENCES

1. Adler, D. C., Ko, T. H., and Fujimoto, J. G. (2004). Speckle reduction in optical coherence tomography images by use of a spatially adaptive wavelet filter. *Opt. Lett.* 29, 2878–2880.

2. Ballard, D. G. and Brown, C. M. (1982). *Computer Vision.* Prentice Hall, Englewood Cliffs, NJ.

3. Bashkansky, M. and Reintjes, J. (2000). Statistics and reduction of speckle in optical coherence tomography. *Opt. Lett.* 25, 545–547.

4. Bayraktar, B. and Analoui, M. (2004). Image denoising via fundamental anisotropic diffusion and wavelet shrinkage: A comparative study. *Computational Imaging II.* C. A. Bouman, C. Miller, and E. L. Miller (eds.). Proceedings of the SPIE, Volume 5299, pp. 387–398.

5. Bernsen, J. (1986). Dynamic thresholding of gray-level images. Proceedings of the 8th International Conference on Pattern Recognition, Paris, France, pp. 1251–1255.

6. Beucher, S. (1990). Segmentation tools in mathematical morphology. *Image Algebra and Morphological Image Processing.* SPIE, Vol. 1350, pp. 70–84.

7. Brezinski, M. E., Tearney, G. J., Weissman, N. J., Boppart, S. A., Bouma, B. E., Hee, M. R., Weyman, A. E., Swanson, E. A., Southern, J. F., and Fujimoto, J. G. (1997). Assessing atherosclerotic plaque morphology: Comparison of optical coherence tomography and high frequency intravascular ultrasound. *Heart* 77(5), 397–403.

8. Brezinski, M. E., Tearney, G. J., Bouma, B., Boppart, S. A., Pitris, C., Southern, J. F., and Fujimoto, J. G. (1998). Optical biopsy with optical coherence tomography. *Ann. N. Y. Acad. Sci.* 838, 68–74.

9. Castleman, K. R. (1996). *Digital Image Processing.* Prentice Hall, Englewood Cliffs, NJ.

10. Chalana, V. and Kim, Y. (1997). A methodology for evaluation of boundary detection algorithms on medical images. *IEEE Med. Imaging* 16(5), 634–652.

11. Chauhan, D. S. and Marshall, J. (1999). The interpretation of optical coherence tomography images of the retina. *Invest. Ophthalmol. Vis. Sci.* 40(10), 2332–2342.

12. Choi, H., Milner, T. E., Rylander, II, H. G., and Bovik, A. C. (2003). Speckle noise reduction and segmentation on polarization sensitive optical coherence tomography images. 25th Annual International Conference Of the IEEE Engineering in Medicine and Biology Society, Vol. 2, Cancun, Mexico, 1062–1065.

13. Czerwinski, R. N., Jones, D. L., and O'Brien, W. D. (1999). Detection of lines and boundaries in speckle images — Application to medical ultrasound. *IEEE Med. Imaging* 18, 126–136.

14. Davies, E. R. (1997). *Machine Vision.* Academic Press, San Diego, CA.

15. de Boer, J. F., Drinivas, S. M., Park, B. H., Pham, T. H., Chen, Z. P., Milner, T. E., and Nelson, J. S. (1999). Polarization effects in optical coherence tomography of various biological tissues. *IEEE J. Selections Top. Quantum Electron.* 5, 1200–1204.

16. deGraaf, C., Koster, A., Vincken, K., Viergever, M. (1992). A methodology for the validation of image segmentation algorithms. *Proc. IEEE Symp. Computer-Based Med. Systems.* 17–24.

17. Drexler, W. (2004). Ultrahigh-resolution optical coherence tomography. *J. Biomed. Opt.* 9(1), 47–74.

18. Donoho, D. L. (1995). De-noising by soft thresholding. *IEEE Trans. Inform. Theory* 41, 613–627.

19. Fernandez, D. C. (2005). Delineating fluid-filled region boundaries in optical coherence tomography images of the retina. *IEEE Med. Imaging* 24(8), 929–945.

20. Fercher, A. F., Drexler, W., Hitzenberger, C. K., and Lasser, T. (2003). Optical coherence tomography — Principles and applications. *Rep. Prog. Phys.* 66, 239–303.

21. Freixenet, J., Munoz, X., Raba, D., Marti, J., and Cufi, X. (2002). Yet another survey on image segmentation: Region and boundary information integration source lecture notes in computer science. Proceedings of the 7th European Conference on Computer Vision — Part III, 2352, pp. 408–422.

22. Fu, K. S. and Mui, J. K. (1981). A survey on image segmentation. *Pattern Recognition* 13(1), 3–16.

23. Fujimoto, J. G. (2003). Optical coherence tomography for ultrahigh resolution in vivo imaging. *Nature Biotechnol.* 21(11), 1361–1367.

24. George, A., Dillenseger, J. L., Weber, M., and Pechereau, A. (2000). Optical coherence tomography image processing. *Invest. Ophthalmol. Vis. Sci.* 40, S173.

25. Gonzalez, R. C. and Woods, R. E. (2002). *Digital Image Processing*. Prentice Hall, Englewood Cliffs, NJ.

26. Gossage, K. W., Tkaczyk, T. S., Rodriguez, J. J., and Barton, J. K. (2003). Texture analysis of optical coherence tomography images: Feasibility for tissue classification. *J. Biomed. Opt.* 8(3), 570–575.

27. Guo, H., Odegard, J. E., Lang, M., Gopinath, R. A., Selesnick, I. W., and Burrus, C. S. (1994). Wavelet based speckle reduction with application to SAR based ATD/R. *Proc. Int. Conf. Image Processing.* 1, 75–79.

28. Hao, X., Gao, S., and Gao, X. (1999), A novel multiscale nonlinear thresholding method for ultrasonic speckle suppressing. *IEEE Med. Imaging* 18(9), 787–794.

29. Haralick, R. M. and Shapiro, L. G. (1985). Survey: Image segmentation techniques. *Comput. Vision Graph. Image Process.* 29, 100–132.

30. Jahne, B. (2004). *Practical Handbook on Image Processing for Scientific Applications*. CRC Press, Boca Raton, FL.

31. Jain, A. K. (1989). *Fundamentals of Digital Image Processing*. Prentice Hall, Englewood Cliffs, NJ.

32. Karaman, M., Kutay, A., and Bozdagi, G. (1995). An adaptive speckle suppression filter for medical ultrasonic imaging. *IEEE Trans. Med. Imaging* 14, 283–292.

33. Kholodnykh, A. I., Petrova, I. Y., Larin, K. V., Motamedi, M., and Esenaliev, R. O. (2003). Precision of measurement of tissue optical properties with optical coherence tomography. *Appl. Opt.* 42(16), 3027–3037.

34. Kittler, J., Illingworth, J., and Foglein, J. (1985). Threshold based on a simple image statistic. *Comput. Vision Graph. Image Process.* 30, 125–147.

35. Koozekanani, D., Boyer, K., and Roberts, C. (2001). Retinal thickness measurements from optical coherence tomography using a Markov boundary model. *IEEE Trans. Med. Imaging* 20(9), 900–914.

36. Kulkarni, M. D., Thomas, C. W., and Izatt, J. A. (1997). Image enhancement in optical coherence tomography using deconvolution. *Electron. Lett.* 33, 1365–1367.

37. Kundu, A. (1990). Local segmentation of biomedical images. *Comput. Med. Imaging Graph.* 14, 173–183.

38. Laine, A. F., Schuler, S., Fan, J., and Huda, W. (1994). Mammographic feature enhancement by multiscale analysis. *IEEE Trans. Med. Imaging* 13(4), 725–740.

39. Lee, Y.-K. and Rhodes, W. T. (1990). Nonlinear image processing by a rotating kernel transformation *Opt. Lett.* 15, 43–46.

40. Li, H., Liu, K. J. R., Lo, S.-C. B. (1997). Fractal modeling and segmentation for the enhancement of microcalcifications in digital mammograms. *IEEE Trans. Med. Imaging* 16(6), 785–798.

41. Lim, J. S. (1990). *Two-Dimensional Signal and Image Processing*. Prentice Hall, Englewood Cliffs, NJ.

42. Lim, H., Jiang, Y., Wang, Y., Huang, Y. C., Chen, Z., and Wise, F. W. (2005). Ultrahigh-resolution optical coherence tomography with a fiber laser source at 1 micron. *Opt. Lett.* 30(10), 1171–1173.

43. Lu, H.-Q. (1991). Quantitative evaluation of image enhancement algorithms. *Human Vision, Visual Processing, and Digital Display II*. B. E. Rogowitz; M. H. Brill, and J. P. Allebach (eds.). Proc. SPIE Vol. 1453, pp. 223–234.

44. Marr, D. and Hildreth, E. (1980). Theory of edge detection. *Proc. R. Soc. London* 27, 187–217.

45. McInerney, T. and Terzopoulos, D. (1996). Deformable models in medical image analysis: A survey. *Med. Image Anal.* 1(2), 91–108.

46. Meyer, F. and Beucher, S. (1990). Morphological segmentation. *J. Vis. Commun. Image Rep.* 1(1), 21–46.

47. Muzzolini, R., Yang, Y.-H., and Pierson, R. (1993). Multiresolution texture segmentation with application to diagnostic ultrasound images. *IEEE Trans. Med. Imaging* 12(1), 108–123.

48. Nakagawa, Y. and Rosenfeld, A. (1979). Some experiments on variable thresholding. *Pattern Recognition* 11, 191–204.

49. Nilsson, N. J. (1971). *Problem Solving Methods in Artificial Intelligence*. McGraw-Hill, New York.

50. Pal, N. R. and Pal, S. K. (1993). A review on image segmentation techniques. *Pattern Recognition* 26(9), 1227–1249.

51. Pavlidis, T. (1977). *Structural Pattern Recognition*. Springer-Verlag, Berlin.

52. Perona, P. and Malik, J. (1990). Scale-space and edge detection using anisotropic diffusion. *PAMI* 12(7), 629–639.

53. Pircher, M., Gotzinger, E., Leitgeb, R., Fercher, A. F., and Hitzenberger, C. K. (2003). Speckle reduction in optical coherence tomography by frequency compounding. *J. Biomed. Opt.* 8(3), 565–569.

54. Pratt, W. K. (1991). *Digital Image Processing*. J. Wiley & Sons, New York.

55. Preston, K. and Duff, M. J. (1984). *Modern Cellular Automata*. Plenum Press, New York.

56. Refregier, A. (2003). Shapelets: I. A method for image analysis. *Mon. Not. Roy. Astron. Soc.* 338, 35–47.

57. Rogowska, J. (2000). Overview of medical image segmentation techniques. *Handbook of Medical Image Processing*. I. Bankman (ed.). Academic Press, New York.

58. Rogowska, J. and Brezinski, M. E. (2000a) Evaluation of the adaptive speckle suppression filter for coronary optical coherence tomography imaging. *IEEE Trans. Med. Imaging* 19(12), 1261–1266.

59. Rogowska, J. and Brezinski, M. E. (2000b). Optical coherence tomography image enhancement using sticks technique. Conference on Lasers and Electro-Optics, CLEO 2000, pp. 418–419.

60. Rogowska, J. and Brezinski, M. E. (2000c). Evaluation of rotational kernel transformation technique for enhancement of coronary optical coherence tomography images. *Image Reconstruction from Incomplete Data*. M. Fiddy and R. P. Millane (eds.). Proceedings of the SPIE, Vol. 4123, pp. 285–294.

61. Rogowska, J. and Brezinski, M. E. (2002). Image processing techniques for noise removal, enhancement and segmentation of cartilage OCT images. *Phys. Med. Biol.* 47(4), 641–655.

62. Rogowska, J., Bryant, C. M., and Brezinski, M. E. (2002a). Boundary detection system for cartilage thickness measurement on OCT images. OSA Biomedical Topical Meetings, *OSA Technical Digest*, (Optical Society of America, Washington, D.C., 2002), pp. 78–80.

63. Rogowska, J., Bryant, C. M., and Brezinski, M. E. (2003). Cartilage thickness measurements from optical coherence tomography. *J. Opt. Soc. Am. A Opt. Image Sci. Vis.* 20(2), 357–367.

64. Rosenfeld, A. and Kak, A. C. (1982). *Digital Image Processing.* Academic Press, New York.

65. Russ, J. C. (2002). *The Image Processing Handbook* CRC Press, Boca Raton, FL.

66. Sahoo, P. K., Soltani, S., Wond, A. K., and Chen, Y. C. (1988). A survey of thresholding techniques. *Comput. Vision Graph. Image Process.* 41, 233–260.

67. Schmitt, J. M. (1997). Array detection for speckle reduction in optical coherence microscopy. *Phys. Med. Biol.* 42, 1427–1439.

68. Schmitt, J. M. (1998). Restoration of optical coherence images of living tissue using the CLEAN algorithm *J. Biomed. Opt.* 3, 66–75.

69. Schmitt, J. M., Xiang, S. H., and Yung, K. M. (1999). Speckle in optical coherence tomography. *J. Biomed. Opt.* 4, 95–105.

70. Schmitt, J. M., Xiang, S. H., and Yung, K. M. (2002). Speckle reduction techniques. Chapter 7 in *Handbook of Optical Coherence Tomography.* B. E. Bouma and G. J. Tearney, Eds., pp. 175–202, Marcel Dekker, New York.

71. Schuman, J. S., Hee, M. R., Puliafito, C. A., Wong, C., Pedut-Kloizman, T., Lin, P. C., Hertzmark, E., Izatt, J. A., Swanson, E., and Fujimoto, J. G. (1995). Quantification of nerve fiber layer thickness in normal and glaucomatous eyes using optical coherence tomography *Arch. Ophthalmol.* 113, 586–596.

72. Singh, S. and Bovis, S. K. (2005). An evaluation of contrast enhancement techniques for mammographic breast masses. *IEEE Trans. Inf. Technol. Biomed.* 9(1), 109–119.

73. Sonka, M., Hlavac, V., and Boyle, R. (1999). *Image Processing, Analysis, and Machine Vision.* PWS Publishing, Boston, MA.

74. Stetson, P. F., Sommer, F. G., and Macovski, A. (1997). Lesion contrast enhancement in medical ultrasound imaging. *IEEE Med. Imaging* 16(4), 416–425.

75. Støren, T., Røyset, A., Giskeødegård, N.-H., Pedersen, H. M., and Lindmo, T. (2004). Comparison of speckle reduction using polarization diversity and frequency compounding in optical coherence tomography. *SPIE Proc.* 5316, 196–204.

76. Tapiovaara, M. J. and Wagner, R. F. (1993). SNR and noise measurements for medical imaging: I. A practical approach based on statistical decision theory. *Phys. Med. Biol.* 38, 71–92.

77. Tearney, G. J., Yabushita, H., Houser, S. L., Aretz, H. T., Jang, I. K., Schlendorf, K. H., Kauffman, C. R., Shishkov, M., Halpern, E. F., and Bouma, B. E. (2003). Quantification of macrophage content in atherosclerotic plaques by optical coherence tomography. *Circulation* 7;107(1), 113–119.

78. Tomlins, P. H. and Wang, R. K. (2005). Theory, developments and applications of optical coherence tomography *J. Phys. D: Appl. Phys.* 38, 2519–2535.

79. Torre, V. and Poggio, T. A. (1986). On edge detection. *IEEE Trans. PAMI.* 8, 147–163.

80. Umbaugh, S. E. (1998). *Computer Vision and Image Processing: A Practical Approach Using CVIPtools*. Prentice Hall, Upper Saddle River, NJ.

81. Vincent, L. and Soille, P. (1991). Watersheds in digital spaces: An efficient algorithm based on immersion simulations. *IEEE Trans. PAMI* 13(6), 583–598.

82. Vincent, L. (1993). Morphological grayscale reconstruction in image analysis: Applications and efficient algorithms. *IEEE Trans. Image Process.* 2(2), 176–201.

83. Wang, R. K. (2005). Reduction of speckle noise for optical coherence tomography by the use of nonlinear anisotropic diffusion. *Coherence Domain Optical Methods and Optical Coherence Tomography in Biomedicine IX*, V. V. Tuchin, J. A. Izatt, and J. G. Fujimoto (eds.). Proceedings of the SPIE, Volume 5690, pp. 380–385.

84. Westphal, V., Rollins, A. M., Willis, J., Sivak, M.. V., and Izatt, J. A. (2005). Correlation of endoscopic optical coherence tomography with histology in the lower-GI tract. *Gastrointest. Endosc.* 61(4), 537–546.

85. Welzel, J. (2001). Optical coherence tomography in dermatology: A review. *Skin Res. Technol.* 7(1), 1–9.

86. Weszka, J. S. (1978). A survey of threshold selection techniques. *Comput. Graph. Image Process.* 7, 259–265.

87. Xiang, S. H., Zhou, L., and Schmitt, J. M. (1997). Speckle noise reduction for optical coherence tomography. *SPIE Proc.*, 3196, 79–88.

88. Yung, K. M., Lee, S. L., and Schmitt, J. M. (1999). Phase-domain processing of optical coherence tomography images. *J. Biomed. Opt.* 4, 125–136.

89. Yu, Y. and Acton, S. T. (2002). Speckle reducing anisotropic diffusion. *IEEE Trans. Image Process.* 11, 1260–1270.

90. Zucker, S. W. (1976). Region growing: Childhood and adolescence. *Comput. Graph. Image Process.* 5, 382–399.

SECTION III

12 APPLICATION OF OCT TO CLINICAL IMAGING: INTRODUCTION

Chapter Contents

12.1   INTRODUCTION

This introduction to clinical imaging covers several topics relevant to developing OCT for clinical imaging. These topics include areas where OCT shows promise as a clinical imaging device, factors that influence decisions on its clinical relevance, methods for validating the structure in OCT images, animal models, statistical analysis, and the role of human perception on image interpretation. The topics are somewhat loosely linked but all are relevant to subsequent chapters.

12.2   AREAS WHERE OCT SHOWS PROMISE AS A CLINICAL IMAGING DEVICE

The general areas where OCT is likely to be most effective have not changed significantly since our original publication in the *IEEE Journal of Selected Topics in Quantum Electronics*.[1] These are in situations where biopsy can not be performed, where sampling areas with conventional biopsies are likely, and which involve guiding surgical/microsurgical procedures. An area not discussed in the original article, which will likely be of value, is the three-dimensional reconstruction of *in vitro* pathology. Of the areas, the most promising ones to the author in the near future, for reasons discussed in the text, are cardiovascular disease and arthritis, followed by guiding microsurgery and screening of certain early malignancies.

12.2.1   When Biopsy Cannot Be Performed

OCT has often been referred to as an optical biopsy technique, imaging approximately over the distance of a biopsy but without the need for tissue removal. Probably the most important area where OCT can be applied is when biopsy is either hazardous or difficult to obtain. As stated previously, this book focuses on imaging in nontransparent tissue (at a median wavelength 1280 nm), so ophthalmology will not be discussed here.[2]

Therefore, ignoring the eye, OCT is most likely to have the greatest impact during biopsy of the cardiovascular and musculoskeletal systems (specifically joints). While OCT has demonstrated a feasibility for assessing dental pathology, cost barriers may prevent its application in the near future.

Cardiovascular disease, myocardial infarction, and stoke remain the leading cause of death world wide and each can be viewed at first approximation as a single disease process. Therefore, with the cardiovascular system the OCT technology developed is essentially going to have one design and approach. This is in contrast to cancer, which represents a diverse spectrum of diseases of widely varying characteristics. With early cancer detection a wide range of probe designs and approaches will likely be needed for different neoplasia. The use of OCT for both cardiovascular disease and oncology will be discussed in Chapters 15 and 17.

Osteoarthritis is analogous to cardiovascular disease, except that it is a leading cause of morbidity rather than mortality, and that at first approximation is one disease. It is primarily a disorder of articular cartilage in predominately synovial joints. OCT has shown considerable promise as a method for diagnosing osteoarthritis and this application is discussed in Chapter 16.

12.2.2 Where Sampling Errors with Conventional Biopsy Are Likely

A sampling error occurs when, screening for a disease process, the biopsy misses the region containing the cancer (for example), thereby producing a false negative. This can occur when screening for esophageal cancer in patients with Barrett's disease and patients at risk for recurrence of transitional cell carcinoma of the bladder. Again, this will be addressed in the oncology chapter.

12.2.3 Guiding Surgical and Microsurgical Procedures

A variety of surgical and microsurgical procedures could benefit from subsurface high resolution, real-time imaging. These include nerve reconstruction, prostate resection, tumor resection (particularly of the nervous system), caries removal, and the assessment of vascular integrity of tissue grafts. This will be discussed in the conclusion chapter.

12.2.4 Three-Dimensional Reconstruction of *In Vitro* Pathology

When *in vitro* pathology is typically assessed, histological sections are obtained at regular intervals. This has several problems associated with it. First, cost constraints limit the number of samples that can be taken. This can result in important clinical findings being missed due to sampling errors. Second, tissue reconstructed in three dimensions can give a dramatically different impression than a single cross section (Figure 12.1). To date, only a limited amount of work has been produced on the three-dimensional reconstruction of pathology.

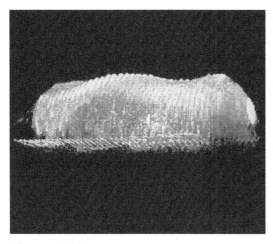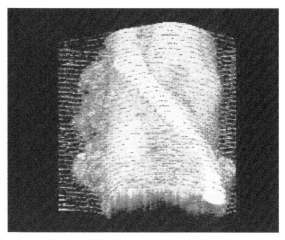

Figure 12.1   Three dimensional imaging of a myelinated nerve. Since OCT is a high speed (near video rate) imaging modality it allows many cross-sectional images to be obtained at high resolution in short time intervals. In this slide the winding tract of fascicles within the nerve can be seen. Courtesy of Boppart, S.A. et al., 1998. *Radiology*, 208, 81–86. [24]

12.3   FACTORS THAT INFLUENCE DECISIONS ON THE CLINICAL RELEVANCE OF OCT

Before moving on to how interpretation of OCT images are validated with histology or biochemical techniques, factors affecting whether an application will achieve widespread use and reimbursement should be discussed. Even if OCT is effective in identifying pathology, other factors could come into play that affect its value as a diagnostic entity. These include the prevalence of the pathology relative to the size of the screening population; whether a treatment exists; cost relative to the benefit, safety, effectiveness of identifying the pathology at a treatable stage; false positives (which can lead to considerable patient anxiety and often invasive follow-up tests); the presence of a superior diagnostic technique; and if an appropriate screening population can be identified.

Some examples of these are provided for illustrative purposes. First, the general use of OCT for cervical cancer screening demonstrates several of these points. Initial studies suggest a feasibility of OCT for identifying early cervical neoplasia. However, for several reasons, at least in the near future, OCT will not be applied for the general screening of cervical cancer. First, frequent screening with Pap smears is very effective, with only approximately 12,000 cervical cancers deaths occurring per year in the United States. Of these cases, approximately two-thirds of the deaths are in patients who have not had a Pap smear over the previous 5-year period. Second, to pick up these 4000 cases missed by the Pap smear screening you would have to screen a population in excess of 50 million, making it impractical from a cost-benefits perspective. Cost-effectiveness is generally expressed as the difference in cost between screening and not screening (or screening at a different interval) divided by the quality adjusted life years (QALYs). The QALYs are the numbers of years of life gained adjusted for the morbidity. Third, the cost-effectiveness of Pap smears performed at intervals of 1 year is a subject of debate and in Europe, they are often performed at 3–5 year intervals. This is gone through in more detail in Chapter 17. In view of these cost concerns, it is unclear how to add or substitute OCT into the screening procedure of cervical cancer. Therefore, size of the screening population relative to the total number of cases, cost-effectiveness concerns, and relatively effective competing screening technology represent significant barriers to its widespread use. However, it may have a limited role in patients with equivocal or positive Pap smears, and in guiding colposcopy.

OCT has demonstrated potential in the field of dentistry for diagnosing a range of disorders from early caries to early oral cancer. At first glance, this would appear to be an extremely promising market in view of the large number of dental offices, the effectiveness of OCT imaging, and the frequency of dental procedures. However, the upper limit of the price point for dental instruments is currently in the range of $20 to 30,000. In view of the current cost of OCT systems in excess of $100,000, it is unclear how OCT can extensively penetrate this market at its current cost. Therefore, instrument cost and not its effectiveness is likely to be the limiting factor to its implementation.

The use of OCT to the diagnosis of malignant melanoma is subject to significant hurtles even if the technique can identify pathology with a high degree of accuracy. Although it may ultimately find clinical utility in this field, it requires replacing conventional biopsy techniques in many circumstances. Malignant melanoma is an almost universally fatal disease unless completely excised, and histopathologic diagnosis has a high degree of accuracy. Therefore, it is unlike the situations where conventional biopsy cannot be performed in that imaging techniques must prove to be superior to an established technique for a potentially curable, otherwise lethal disease. However, OCT, along with dermoscopy and confocal laser microscopy, is currently being evaluated for the purpose of improved localization of early cancers, but the "burden of proof" of it becoming a widespread diagnostic technology is likely much higher than that for many other applications.[3,4]

12.4 OCT IMAGING OF HUMAN TISSUE

Several issues regarding obtaining and using human tissue, which investigators are sometimes unaware of, are discussed in this section. This is not meant to be a comprehensive review of obtaining and handling human tissue, and it is highly recommended that the reader consult healthcare professionals and institutional officials (as well as read the appropriate literature) on both government and institutional regulations/oversight. These include but are not exclusive to the topics of patient's rights, tissue handling, pathogens, and toxic substances (such as formalin).

Many factors need to be considered when imaging human tissue including patients rights, regulatory compliance, preventing the transmission of disease, and maintaining tissue integrity. This section will focus on the handling of *in vitro* tissue while issues surrounding clinical trials are discussed in Chapter 14.

Patients have rights and regulations have become stricter under the new HIPPA regulations. These include rights regarding excised tissue, even after the patient is deceased, and in some instances, the data obtained from their tissue. Therefore, investigators need to familiarize themselves with these regulations to avoid job termination, loss of funding, civil litigation, and/or criminal litigation.

Working with human tissue is associated with the potential of transmission of infectious organisms. Proper precautions should be taken and therefore investigators again need to seek the appropriate training. Significant risks can often be overlooked. Prions or slow viruses (such as mad cow disease) found in nervous tissue (including the retina) and the cornea are highly resilient to most disinfectants or fixation techniques, so special precautions are necessary. Another example is the risk of transmitting tuberculosis in lung samples, even those which have undergone short periods of fixation. Investigators should be aware of potential pathogens and how the risk of transmission can be reduced. Most research institutions offer courses on this topic.

Many of the substances used in tissue handling and preparation can be considered hazardous if used incorrectly. These include fixatives (such as formalin), dyes, and solvents used for tissue processing, high-performance liquid chromatography (HPLC), thin-layer chromatography (TLC), and electron microscopy. In addition some compounds can have unusual hazards, such as the solution for picrosirius staining, which if allowed to become dry is potentially explosive. Government and institutional regulations exist for the appropriate handling of materials used in the laboratory, which should be reviewed, as well as courses provided at most research institutions.

Preventing the breakdown of tissue samples prior to and during imaging can be challenging for OCT studies. Cartilage and teeth are uncommon tissue in that their breakdown can take a relatively long period of time. In contrast, other tissues are substantially more delicate, such as bowel. The surface of bowel is highly dependent on the vascular supply, so that once the tissue is removed, no contact can be made with the surface, even from the adjacent tissue surface. If the lumen is not opened immediately, the researcher runs the risk of epithelium sloughing off. Similarly, enzyme assays require special tissue handling. When trying to distinguish motor to sensory nerves with the cholinesterase assay, the tissue must be imaged either *in vivo* or immediately post-resection because of autolysis of the enzyme (described below). In addition, the enzyme cannot survive normal histological processing, so it is cut in a cryostat and the enzyme assay performed as quickly as possible.

Most tissue of interest falls in between these two extremes, but all will eventually degrade unless some type of stabilization is performed. The use of an agent like sodium azide may help to slow autolysis, but it is toxic (somewhat like cyanide) and may lead to tissue breakdown in high enough concentrations. A procedure we have found particularly useful is to fix the sample immediately after it is excised with formalin in buffered saline for 24 hours. The sample is then placed in saline for subsequent imaging. We have studied this technique with arterial and tendon samples and no contrast or polarization sensitivity difference exists at 0 hours, 24 hours, and 1 week. This process seems to minimize both autolysis- and formalin-induced changes in imaging. However, the technique cannot be used in all instances. It can take over a week for cartilage to be fixed, so the tissue is unlikely to be sufficiently treated at 24 hours. Also, the sample should be cut into relatively small pieces when fixing to allow adequate penetration of formalin. Furthermore, fixation changes the mechanical properties of tissue so this is not a viable approach when studying OCT elastography.

12.5 METHODS FOR VALIDATING THE STRUCTURE IN OCT IMAGES

12.5.1 General

Ultimately, the interpretation of structure within OCT images requires validation by comparison with a "gold standard." This generally means accurately registered histopathology (or in some cases biochemical techniques) that has been appropriately prepared and then stained with the optimal technique. Data are then analyzed for statistical significance. Impressions of images that are not compared to a gold standard, where the incorrect staining procedure is used, or that have not been validated by previous work demonstrating statistical significance often represent the opinion of the investigator and not valid science. Because using the appropriate histological or biochemical approach and performing it correctly is critical to interpreting OCT imaging of tissue, this topic is discussed in considerable detail. Validating OCT images with these techniques is difficult, requiring accurate registration between the images and where testing is performed, precision performance of these tests. For quality control, we have performed most, if not all of these procedures in our lab for over a decade.

12.5.2 Histopathology

Steps in traditional histological preparation are fixation, embedding, sectioning, removal of embedding medium, staining, mounting, and observation by light microscopy.[5,6] Important exceptions to this sequence include specimens prepared by cryostats, thin specimens embedded by resin rather than paraffin, and samples for electron microscopy. A more detailed discussion of techniques and hazards is found elsewhere.[6]

12.5.2.1 Fixation

Once tissue is obtained, it typically is fixed for several reasons. These include attempting to maintain the tissue architecture, preventing autolysis, making it more amenable for staining, and hardening the sample to allow microtoming into thin sections.

Most fixatives work by stabilizing proteins by promoting cross-linking. The most common fixation for light microscopy is formaldehyde, which typically is diluted with normal saline, and works by cross linking lysine groups on the exterior of protein.[7] For electron microscopy, glutaraldehyde and/or osmium tetraoxide are typically used. Blocks used for both light and electron microscopy can be fixed with formalin followed by glutaraldehyde.[8] As with most agents used in tissue processing, these are toxic and should be used with the appropriate precautions outlined by the EPA/OSHA, manufacturer, and environmental services of your institutions.

Limitations of fixation include tissue shrinkage, the formation of formalin pigment when acid is present, movement of large molecules of interest such as hemoglobin to other locations in the sample, and failure to fix certain relevant molecules such as lipid and mucopolysaccharides (which are subsequently lost in processing).

After fixation, some samples require decalcification including cartilage, dentin, enamel, and arterial plaque. The decalcification is needed to prevent tissue cracking and tearing during microtoming. Methods for decalcification include treatment with acid or EDTA. Acid decalcification is a rapid technique, but is not recommended when macromolecules such as collagen are measured or when immunohistochemistry is employed because it denatures many biomolecules. EDTA treatment is less destructive, but months of exposure may be necessary before processing can be performed.

12.5.2.2 Processing

Dehydration is the next step in processing. The sample ultimately needs to be embedded with paraffin or resin, both of which are hydrophobic (do not mix with water) for microtoming. The water then needs to be removed. This is typically done in a tissue processor but can be done manually. It is sequential dehydration with increasing concentrations of ethanol followed by treatment with xylene or toluene. Again, these solvents need considerable precautions when used. Tissue hardening and loss of lipids are undesirable effects of the dehydration process. When these effects are problematic to sample interpretation, such as evaluating lipid in atherosclerotic plaque, alternative methods such as those described below are available.

12.5.2.3 Embedding

With embedding, the xylene is replaced by molten wax resulting in complete infiltration of the tissue. The wax is allowed to harden on a frozen block. Most commonly paraffin is used, which allows sectioning down to around 2 μm. When thinner or stronger sections are required, such as with electron microscopy, plastic resin can be used rather than wax, and requires microtoming knives that are made of stronger material such as tungsten or diamond.

12.5.2.4 Microtoming

Microtoming is the cutting of tissue into thin sections to be put on a slide and stained. Basically, the tissue embedded in the paraffin is shaved by the blade of the microtome. The thin tissue segments are then guided onto the water bath. The floating pieces of tissue are then swept up on a glass slide. Microtoming requires considerable experience and practice to produce intact sections.

12.5.2.5 Artifacts

Generally, tissue processing for most tissue results in tissue shrinking. In an artery section, one study found a 19% reduction in cross section.[9] An exception is cartilage, which when cut and the collagen disrupted, water influx occurs (the tight collagen network normally limits water uptake), leading to expansion in the cross-sectional area.

Because tissue shrinkage occurs during processing, when comparing distance measurements between OCT images and histology other approaches are necessary. One approach is to use McDowell's solution followed by embedding in glycol methacrylate, which only has a 4% reduction of arterial cross section seen in the study mentioned above.[9] Our group uses an alternate approach, using frozen sectioning with a cryostat. Tissue can be hardened by freezing and cut in a cryostat, which microtomes the tissue at -20 to $-40°C$. The sections are then placed on a slide and stained. The cryostat can also be used to prevent loss of lipid, movement in tissue of macromolecules such as hemoglobin, and when enzyme activity is being measured, as described below.

12.2.5.6 Staining

Staining generally refers to the binding of dye to tissue to improve the contrast of specific structures in the sample. There are many types of chemical bonds formed between dye and tissue. For example, acidic dyes bind to cellular cytoplasm and collagen in connective tissue while basic dyes are more suitable for nucleic acids and phospholipids. The two most common general dyes used for staining are hematoxylin and eosin (H&E) and Masson's trichrome blue. H&E is a relatively nonspecific dye with nuclei stained blue and cytoplasm red. With Masson's trichrome blue, cytoplasm is pink, red, or greenish blue, erythrocytes are yellow, muscle cells red, and collagen and matrix are also green to bluish green. The diagnostic ability of these stains is very often overstated. For example, some authors have used these dyes to quantitate collagen content, but neither is appropriate for this function. Other techniques such as picrosirius staining, immunohistochemistry, scanning electron microscopy (SEM), TLC, and HPLC are more appropriate methods for quantitating and assessing collagen. This is illustrated below.

Important special stains include acid fuchsin for elastin, sudan red for lipid, picrosirius for collagen, and metal impregnation for electron microscopy. Other cell constitutes may be better quantified with immunohistochemical techniques (smooth muscle) or enzyme assays (motor nerves).

Most studies seek to quantitate data generated by staining techniques. One method of quantitation is to use blinded investigators grading the slide with a pre-determined scaling system. In a second approach, when looking at total intensity or for a specific color, quantitative data can be obtained by producing digital images of the sample and processing them in a program such as MATLAB or Photoshop. A variety of grading schemes are present in the literature, for example, the Wakitani scoring system for cartilage.[10]

Picrosirius staining for collagen will be described in more detail due to its relevance to OCT imaging. Collagen content and organization is important to assess in many tissues such as cartilage, arteries, tendons, and teeth. Collagen cannot be quantitated with H&E or Masson trichrome blue.[11] However, staining with picrosirius is specific for collagen. Picrosirius stain, an acid dye of picric acid with sirius red, has been shown to be more sensitive than polarization microscopy alone for assessing organized collagen.[12,13] Organized collagen appears as high intensity in a picrosirius stained section. Its color, generated in part through an interference phenomenon, correlates with collagen type. Large fibrils (greater than 60 nm) are typically yellow to red and are generally collagen type I, whereas fibrils less than 40 nm appear green and correlate with collagen type III.[14] To improve distinguishing fiber types, sections should be kept between 3 and 6 μm, and illumination with circularly polarized light, use of the optimal mounting agent, and correct alignment of the polarization filter are necessary. Example images generated of a coronary artery stained by picrosirius and trichrome blue are shown in Figure 12.2.

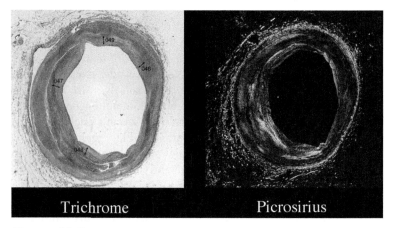

Trichrome   Picrosirius

Figure 12.2   Picrosirus staining. Using the appropriate staining techniques is critical for hypothesis testing. One the left is a trichrome stained section of coronary artery, a technique which is sometimes used by investigators to 'quantitate' collagen concentrations. On the right is a picrosirus stained section that demonstrates organized collagen by bright areas, which is clearly more informative than the trichrome image. Courtesy of Stamper, D., et al., 2006. Plaque characterization with optical coherence tomography. *J. Am. Coll. Cardiol.* 47(8), 69–79. [25]

12.5.3 Immunohistochemistry

A limitation of dye-based techniques is that staining intensity and color are variable and interpretation is often dependent on the expertise of the investigator. An alternate approach for characterizing tissue components is immunohisto-chemistry. This technique relies on antibodies binding to very specific sites on molecules. A label is attached to the primary or secondary antibody in the form of an enzyme, fluorescent dye, metal, or radioactive isotope. This discussion will focus on the indirect approach because the direct approach suffers from little signal amplitude.

An avidin-biotin complex (ABC) is probably the most favored technique. In the first step, a primary unlabeled antibody to the antigen of interest is added. In the second step, a secondary antibody labeled with biotin reacts with the primary antibody. In the third step, a preformed complex of avidin and peroxidase-labeled biotin is used. With this technique, binding sites are often not preserved during fixation. Therefore, the samples are carefully heated in the presence of heavy metals or salts, which open sites for poorly understood reasons. However, with each new immunohistochemical assay, the laboratory needs to establish the appropriate heating protocol as too much heat destroys the antigen of interest while too little fails to open the site. Many common antigens have commercially available immuno-histochemical kits.

An example immunohistoassay used by our group is for smooth muscles in arterial intima. Smooth muscle content in the intima is assessed by measuring α-actin within the smooth muscle (Sigma, St. Louis, MO). Actin is a 43,000-MW structural and contractile microfilament protein in smooth muscle. With this technique, background staining is virtually eliminated. Tissue is fixed in 10% neutral buffered formalin and processed as described. Then, endogenous peroxide in tissue sections is quenched with 2 drops of 3% hydrogen peroxide and allowed to incubate for 5 minutes. Sections are then incubated with normal goat serum (used as a blocking agent) for 10 minutes. Next, the primary antibody, mouse monoclonal anti-α-smooth muscle actin, is reacted with the sample for 60 minutes. The biotinylated secondary antibody is then added (goat anti-mouse immuno-globin) for 20 minutes, followed by peroxidase reagent added for an additional 20 minutes. Specific binding is detected with AEC Chromogen and 3% hydrogen peroxide. Mayer's hematoxylin is used as a counterstain. This protocol results in the cytoplasm of the smooth muscle cell appearing rose to brownish red with nuclei staining dark blue/purple. Image analysis is performed using a digital camera color analysis system.

12.5.4 Enzymatic Assay

In some cases, tissue needs to be analyzed through enzyme analysis. This is the case when attempting to distinguish motor from sensory nerves. The clinical relevance is addressed in the conclusion chapter. To analyze enzyme activity, the sample must be examined almost immediately after removal from the body and does not undergo fixation or processing (which would destroy activity). Instead, the sample is rapidly frozen and cut in a cryostat (under −10°C). The procedure we use to distinguish motor from sensory nerves is the acetylcholinesterase assay. This will be used as an example enzymatic assay.

After imaging, fresh samples are marked with tissue dye for registration, and flash frozen at −38°C for two minutes.[15] Each sample is then placed in a cryo-mold and embedded with Tissue-Tek OCT (optimal cutting temperature) compound. The molds are placed at −38°C until the embedding medium is completely frozen. Samples are removed from the mold and attached to chucks with Tissue-Tek OCT medium. The chuck is placed in the Leica CM 1900 cryostat set at −18°C and the fresh, frozen samples are cut in 10- to 20-μm thick sections. The samples are placed on the slide in the appropriate orientation and the imaged plane is etched, using a diamond tip pen, onto the slide to ensure proper correlation with OCT images.

In the acetylcholinesterase assay, the slide is incubated with a staining solution of citrate stock, $CuSO_4$ solution, ferricyanide, and OMPA at 45°C for 45 minutes. It is then dehydrated with sequential solutions of 85, 95, and 100% ethanol then xylene. Mounting is performed with Permount and counterstained with neutral red. The acetylcholinesterase assay results in motor fascicles stained brown to black and sensory nerve fascicles unstained or having small dark spots in the center. An example image of a fresh bovine nerve is seen in Figure 12.3.

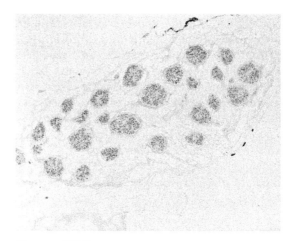

Figure 12.3 Enzyme assay to distinguish sensory and motor fascicles. In this slides, an acetylcholinesterase assay was used to separate motor from fascicles, which is critical in nerve repair. The acetylcholinesterase assay stains the motor fascicles black and leaves the sensory minimally or unstained.

12.5.5 High-Performance Liquid Chromatography

HPLC is a powerful technique for measuring the type and concentration of biomolecules in tissue with extremely high precision.[16] For instance, in the example below, HPLC will be used to measure lipids and their individual subtypes. HLPC involves first taking the tissue and extracting compounds of a similar class (such as lipid chains). The individual molecules are separated within an HPLC column. The HPLC system consists of an attached packing material that compounds introduced in the mobile phase bind to at different rates. The mobile phase consists of a combination of liquid solvents in addition to the compounds of interest. By controlling and varying the compositions of the solvents and electrolytes for a fixed immobile phase composition, different compounds of interest progress through the column at different rates. This occurs because the interaction with the immobile phase is different for each compound due to the fine control of the environment. A detector is at the end of the column. By comparing the transition times and amplitudes of the compounds studied with those of known standards, the biomolecules are identified and quantified.

An example HPLC protocol used by our group for lipid isolation in plaque is described. The media of the plaque are dissected away with a low numeric aperture microscope.[17,18] These samples are stored at $-80°C$ before extraction of lipids using redistilled chloroform/methanol (2:1 vol/vol containing 0.01% wt/vol butylated hydroxytoluene as an antioxidant), followed by a Folch wash to remove non-lipid contaminants. HPLC analysis is performed on a Beckman System Gold Programmable Solvent Module #125 with a Beckman System Gold Programmable Detector Module. A reverse phase C_{18} analytical column is used. The free cholesterol and cholesterol esters are eluted isocratically at 1.2 ml/min with acetonitrile/isopropanol (50:50, v/v) at 45°C, and detected by UV absorption at 210 nm. The mixture is vortexed and centrifuged, and the supernatant is injected into the injector of the HPLC system.

Standard curves for the various cholesteryl esters are obtained by comparing the peak height to that of the internal standard and plotting the height ratios as a function of the cholesteryl ester concentration. The standard mixture contains cholesterol, cholesterol arachidonate, cholesteryl linoleate, cholesteryl oleate, and cholesterol palmitate in chloroform.

12.5.6 Scanning Electron Microscopy

A scanning electron microscope is an electron microscope capable of high resolution, three-dimensional images of a surface of a sample. While it does not have the atom level resolution of transmission electron microscopy (TEM), it does achieve images in the range of 1 to 20 nm. It has additional advantages including the ability to image large areas and bulk materials. The samples are fixed as described above. With SEM, an electron beam is generated at the sample, with the spot size primarily determining the resolution. The electrons are inelastically scattered from the sample. They are then detected by a scintillator-photomultiplier device.

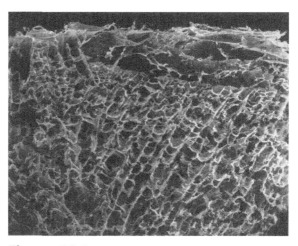

Figure 12.4 Scanning Electron Microscope (SEM) image. This is a SEM image of a collagen phantom of a collagen phantom, (Helistat® Integra LifeSciences Corporation, Plainsboro, NJ 08536 USA), showing the collagen bands. Courtesy Liu, B., et al. (2006). *Applied Optics*, 45, (in press June).

An example SEM protocol used in our laboratory is described here. Samples are fixed in 2% glutaraldehyde and 2% paraformaldehyde (0.1 M PBS, pH 7.2) for 1 hour at 4°C and with 1% OsO4 for 1 hour at 4°C and then treated with 2% tannic acid with 1% OsO4 for 1 hour at 4°C. The tissue is then dehydrated by increasing concentrations of alcohol, followed by critical point drying and Au sputter coated. The SEM used is an Amray 1000A scanning electron microscope operating at 20 KV. An SEM image of a collagen phantom (collagen type I) is shown in Figure 12.4.

12.6 ANIMAL MODELS

Animal models are important for evaluating OCT, but there are criteria that should be met so that animal welfare is appropriately protected. Some of these which are typically required by USDA Regulations (9 CFR Part 1-3) and the Animal Welfare Act, in addition to the given institutions, standing committee on animals as required by federal law are listed below. When animal studies are explored for evaluating OCT technology or its ability to assess pathology, some of the issues that need to be addressed are reviewed in the next sections.

12.6.1 Justification of Need for Animals

One of the most important questions to ask is whether the information needed can be obtained through other methods. Just as with human studies, has the technology been validated sufficiently with phantoms and *in vitro* studies before going to animal studies? Will the information gained significantly advance the field? For example, the function of an OCT arthroscope may be appropriately assessed in a postmortem human knee while the ability of OCT to longitudinally follow therapeutics for preventing arthritis (below) may be better suited in the appropriate animal model. Typically, a literature search and specific aims that address a clinically relevant hypothesis are required by oversight committees.

12.6.2 Justification for the Species

To test certain hypotheses, sometimes the choices of animal species are limited, but other times there are a variety of options that need to be weighted. While we have had the option of monitoring therapeutics for osteoarthritis in primates, dogs, and rabbits, we were able to develop a technique using rats instead. Generally, large animals, particularly primates, require substantially more justification than small animal models.

12.6.3 Justification of the Total Number of Animals

Estimation of the appropriate number of animals is important is designing an experiment. This is typically done through the statistical power. The power of a statistical test refers to its ability to find differences in distributions when they are really there. We do not want to construct a hypothesis driven experiment only to fail to find a significant result because the number of animals in experimental groups is too low. Similarly, the use of more animals than statistically required results in the sacrificing of animal's lives for no scientific reason.

For example, in the osteoarthritis experiment below the power was determined as follows. We start by assuming all treated animals will develop the disease and progress to at least 75% of the total arthritis score (achieved in an earlier study). With each group, at each time point we will use 4 animals and assume a standard deviation about the mean of 25%. Then a power of 0.88 is achieved with an alpha of 0.05.

12.6.4 Drugs and Their Dosages Used for Anesthesia and Euthanasia

Drugs and dosages used for anesthesia and euthanasia are scrutinized to ensure that animals are not subject to discomfort purely from the investigator's lack of knowledge in the area. The USDA requires that you list the names and dosages of analgesics or tranquilizing drugs to be used for each species of animals under study. Particular attention is paid to adequate use of tranquilizing agents when a paralyzing agent is also employed. In addition to a thorough literature search, communication with the veterinary staff and animal committee will likely decrease the possibility of protocol rejection.

12.6.5 Will the Animal Experience Discomfort and Is a Survival Model Utilized?

Any protocol that involves the animal experiencing discomfort or where the animal is followed for a period of time after a surgical procedure (survival study) will require a rigorous defense of the approach and evidence that no alternative is available.

12.6.6 Use of Experienced Personnel

Before contemplating an animal experiment, personnel experienced with animal studies should be involved through the development of the protocol to the training of the remainder of the team and finally the actual experiment.

12.6.7 Registration

OCT is a micron scale imaging technology. Therefore, whether performing *in vitro* studies, *in vivo* animal experiments, or *in vivo* human experiments, the most accurate method of tissue registration should be employed. In animal studies, particularly survival studies examining internal structures and requiring repeat imaging of the same site, this can be challenging. For intravascular imaging, this may involve placing an intravascular stent a known distance from the imaging site and monitoring it under fluoroscopy. In other models, the position of the imaging site from known landmarks may be employed. Registration techniques should be carefully thought through as they can mean the difference between a solid scientific work and personal speculation on the interpretation of OCT images.

12.6.8 Example Rat Protocol

An example rat osteoarthritis protocol used by our group is described. We initially evaluated the rabbit model performing imaging with an OCT arthroscope, but found that a rat model combined with a hand-held OCT probe was preferable based on cost, the use of a small animal, reduced quantities of therapeutics needed (when examining experimental therapeutics), and more rapid time to initiation of arthritis. Arthritis is ultimately induced with meniscal and anterior cruciate ligament (ACL) ligation. Anesthesia is chosen after multiple discussions with the veterinarian and IACUC staff. The prestudy power determinations are as described above.

All surgical procedures will follow aseptic guidelines and be approved by the IACUC of Brigham and Women's Hospital and Harvard Medical School. All rats receive buprenorphine one-half hour prior to induction of anesthesia with an intraperitoneal injection of Nembutal. Twenty of the rats had osteoarthritis (OA), induced in their left hind leg via ligation. The remaining animals will undergo a sham operation and serve as controls. The reason that the contralateral knee is not used as a control is that previous work has shown that by week four, signs of OA develop in the contralateral (untreated) knee as well.[19]

For treated and untreated rats, the operative leg is shaved preoperatively on day 0. Aseptic technique is maintained throughout the surgery and is begun with three alternating scrubs of Betadine followed by 70% alcohol. A medial parapatellar incision is made on the skin and the medial retinacular and the posterior segment of the posterior capsule opened. In the OA rats, the medial meniscus and ACL are cut through the full thickness to simulate a complete tear. The tibial and femoral condyles are the focus of imaging though other supportive tissue, particularly the ACL, and both menisci are examined. From experience, the greatest disease is found in the medial tibial condyle with this model. At each time point, for each condyle, imaging is performed along 6 adjacent sagittal planes spaced approximately 0.5 mm apart (proceeding from a medial to lateral direction). During imaging, the OCT hand-held probe will be clamped in place above the knee and accurately positioned with a micrometer stage. At each location both structural and PS-OCT is performed. Registration is achieved using a visible light-guiding beam oriented relative to landmarks. The position of the beam relative to the landmarks is digitally photographed to reestablish position at different time points. The transected meniscus is imaged to look at degenerative changes in the collagen during later imaging. The joint is irrigated with sterile saline and closed. The capsule is closed with a running suture supplemented with interrupted sutures of 4-0 nylon. Control animals have a sham operation on one knee. OCT imaging is completed similar to the control group. Following surgery, rats are maintained in cages and fed ad lib.

In vivo serial OCT imaging is performed on the experimental and control knees on weeks 0, 1, 2, 3, and 4 by reopening the suture site as described. Four OA and four control rats are sacrificed at each time point with pentobarbital for histological analysis of the imaged tissue using trichrome blue and picrosirius stains.

12.7 STATISTICAL ANALYSIS

In the early years of OCT imaging of nontransparent tissue, statistical analysis was not critical as most studies were either feasibility experiments or technology development. However, as the technology has now reached more widespread acceptance verification of data both with techniques discussed previously and through careful statistical analysis is required.

An example of the current level of scrutiny is a paper we recently had accepted where one of the reviewer's comments was "the statistics used were parametric and should have been non-parametric. In addition, a Bonferroni adjustment of the non-parametrics should be used." Statistical analysis typically beyond simple paired *t* tests is needed to effectively test the hypothesis to the satisfaction of the scientific community.

Unfortunately, the statistical background necessary represents a textbook by itself, so the topic will not be dealt with in detail here. A cursory review of the topic is addressed at the end of Chapter 14. However, those without experience in this area are strongly encouraged to familiarize themselves with it. Fortunately, a wide variety of textbooks are available on the topic which can be purchased at most book chains or major internet book sites.

12.8 ROLE OF HUMAN PERCEPTION ON IMAGE INTERPRETATION

A brief discussion of the limitations of imaging data interpretation by the medical personnel will be discussed as it is relevant to subsequent chapters.[20–22] In general, trained physicians and healthcare workers visualize an image from the "bottom up." They approach the image in a global manner and then formally decide to focus on specific areas. When I was originally trained to read EKGs or X-rays, I, like other trainees, approached the image in a step-by-step manner moving through sections of the data systematically. Training does ultimately improve the ability to detect abnormalities within images. As investigators become experienced they approach the image globally, immediately putting attention on abnormalities without a systematic approach, and then fixate on a few sites within the image that are considered abnormal. It is interesting that after a certain point, the longer a healthcare worker looks at an image, diagnostic accuracy does not improve and the likelihood of a false positive increases significantly. Ultimately, while healthcare workers are well trained, mistakes are still made in interpreting many images. The next few paragraphs will be spent on why these errors occur, because they ultimately affect the validity of virtually any imaging technology.

To evaluate an image, criteria that become important include the image/data quality, perception/recognition, and the interpretation of the findings, whether it is OCT or another imaging modality. Physical attributes of images that affect interpretation include edge gradients of the abnormalities and contrast within the image site of interest. Physicians usually do not take advantage of the ability to control aspects of the physical interpretation process, such as minimizing indirect light. Another ability that physicians do not take advantage of is that fine structures are best detected at shorter viewing distances while contrast is best appreciated at a few meters from the images.

Perception/recognition requires interpretation from the human nervous system. The brain is likely the source of most interpretation errors and will likely play a role in misinterpretation of OCT imaging. Biological factors contribute to these errors which include biological variation among patients, the visual acuity of the physician, and the mechanisms by which retinal image is integrated into the brain. However, other factors contributing to the misinterpretation include ambiguous results, physician state of mind, use of motion, and the "satisfaction search."

In ambiguous circumstances, differences can occur with up to one-third of qualified physicians. These differences in observation commonly occur even when a given individual is subject to blinded reinterpretation. The competent physician in a blinded study will disagree with his/her own interpretation 20% of the time.

About 80% of cases of mistakes by radiologists happen while reading positive cases (false negative).

It would be expected that the major source of error is lack of knowledge on the part of the physician, but this typically is not the case. This only accounts for less than 5% of clinical mistakes in interpreting imaging. Larger sources of mistake include prejudice on the part of the physician based often on previous experience and the state of mind of the physician (boredom/distraction).

Motion can also be an important tool for analyzing images. This should be remembered when discussing the technologies in the next chapter that rely on fixed numerical data. Aspects within an image are often better defined when the tissue is in motion than when it is static. The motion may give information about the tissue that may reveal unusual states within tissue sections that cannot be detected within a static image. This is particularly helpful when evaluating single detector PS-OCT images.

Satisfaction search is another source of error, even among experienced physicians, and occurs when more then one abnormality is present on the image. This results in the investigator focusing on a specific aspect of the image and missing other abnormalities within the image.

12.9 CONCLUSION

This chapter reviews topics relevant to performing research on OCT imaging of human tissue. Safety and patient rights issues were briefly discussed, but researchers are strongly encouraged to pursue more in-depth literature on these topics.

REFERENCES

1. Brezinski, M. E. and Fujimoto, J. G. (1999). Optical coherence tomography: High resolution imaging in nontransparent tissue. *IEEE J. Selected Top. Quantum Electron.* 5, 1185–1192.
2. Schuman, J. S., Puliafito, C. A., and Fujimoto, J. G. (2004). *Optical Coherence Tomography for Ocular Diseases.* Slack, New York.
3. Pellacani, G. A., Cesinaro, M., Longo, C. et al. (2005). In vivo description of pigment network in nevi and melanomas. *Arch. Dermatol.* 141, 147–154.
4. Braun, R. P., Rabinovitz, H. S., Oliviero, M. et al. (2005). Dermoscopy of pigmented skin lesions. *J. Am. Acad. Dermatol.* 52, 109–121.
5. Moore, J. and Zouridalis, G. (2004). *Biomedical Technology and Devices Handbook.* CRC Press, Boca Raton, FL, pp. 16-1 to 16-21.
6. Carson, F. L. (1997). *Histotechnology: A Self- Instructional Text*, 2nd edition. ASCP Press, Chicago.
7. Baker, J. R. (1958). *Principles of Biological Microtechnique*, 1st edition. Methuem & Co. Ltd., London.
8. Tandler, B. (1990). Improved slides of semithin sections. *J. Electron. Microsc. Tech.* 14, 285–286.

9. Dobin, P. B. (1996). Effect of histologic preparation on the cross sectional area of arterial rings. *J. Surg. Res.* 61, 631–633.

10. Wakitani, S., Goto, T., Pineda, S. J. et al. (1994). Mesenchymal cell-based repair of large, full thickness defects of articular cartilage. *J. Bone Joint Surg. Br.* 76, 579–592.

11. Whittaker, P., Kloner, R. A., Boughner, D. R., and Pickering, J. G. (1994). Quantitative assessment of myocardial collagen with picrosirius red staining and circularly polarized light. *Basic Res. Cardiol.* 89, 397–410.

12. Junqueira, L. C., Figueiredo, M., Torloni, H., and Montes, G. S. (1986). A study on human osteosarcoma collagen by the histochemical picrosirius-polarization method. *J. Pathol.* 148, 189–196.

13. Junqueira, L. C., Bignolas, G., and Brentani, R. R. (1979) Picrosirius staining plus polarization microscopy, a specific method for collagen detection. *Histochem. J.* 11, 447–455.

14. Wolman, M. and Kasten, F. H. (1986). Polarized light microscopy in the study of the molecular structure of collagen and reticulin. *Histochemistry* 85, 41–49.

15. Kanaya, F., Ogden, L., Breidenbach, W. C., Tsai, T. M., and Scheker, L. (1991). Sensory and motor fiber differentiation with Karnovsky staining. *J. Hand Surg. Am.* 16(5), 851–858.

16. Brezinski, M. E. and Serhan, C. N. (1990). Selective incorporation of 15 hydroxyeicosatetraenoic acid in phosphatidylinositol of human neutrophils. *Proc. Natl. Acad. Sci.* 87, 6248–6252.

17. Folch, J., Lees, M., and Sloane-Stanley, G. H. (1957). A simple method for the isolation and purification of total lipids from animal tissue. *J. Biol. Chem.* 226, 497–509.

18. Vercaemst, R., Union, R. A., and Rosseneu, M. (1989). Quantitation of plasma free cholesterol and cholesterol esters by high performance liquid chromatography. *Atherosclerosis* 78, 245–250.

19. Roberts, M. J., Adams, S. B., Patel, N. A., Stamper, D. L., Westmore, M. S., Martin, S. D., Fujimoto, J. G., and Brezinski, M. E. (2003). A new model for assessing early osteoarthritis in the rat. *Anal. Bioanal. Chem.* 377, 1003–1006.

20. Krupinski, E. A. (2000). Medical image perception: Evaluating the role of experience. *SPIE Proc.* 3959, 281–286.

21. Swensson, R. G. (1996). Unified measurement of observer performance in detecting and localizing target objects on images. *Med. Phys.* 23, 1709–1725.

22. Herman, P. G. and Hessel, S. J. (1975). Accuracy and its relationship to experience in the interpretation of chest radiographs. *Invest. Radiol.* 10, 62–67.

23. Krupinski, E. A. (2000). The importance of perception research in medical imaging. *Radiat. Med.* 18, 329–334.

24. Boppart, S. A., Bouma, B. E., Pitris, C., et al. (1998). Intraoperative assessment of microsurgery with three dimensional optical coherence tomography. *Radiology.* 208, 81–86.

25. Stamper, D., Weissman, N. J., and Brezinski, M. E. (2006). Plaque characterization with optical coherence tomography. *J. Am. Coll. Card.* 47(8):69–79.

13 OTHER TECHNOLOGIES

Chapter Contents

13.1 GENERAL

This chapter reviews some of the technologies which may be competitive with OCT in a given organ system. Others are introduced in subsequent chapters. They are discussed on an elementary level, but other texts are readily available that describe the physical basis of these technologies in greater detail. The technologies that will be discussed can be divided into either spectroscopic and/or structural imaging techniques. In general, spectroscopic techniques depend on measurements of the absorbing properties of tissue, which to a large degree are assessing biochemical parameters, while structural imaging depends predominately on the scattering properties of tissue. This distinction is somewhat of an oversimplification since there is some overlap between the technologies, but it is sufficient for the purposes of this chapter.

13.2   STRUCTURAL IMAGING

The optical technologies that characterize the properties of tissue through assessing tissue morphology or structure include two-photon laser microscopy and confocal microscopy. In addition, a brief mention of non-optical technologies, such as high frequency ultrasound, CT, and MRI, will be made since they may be considered a competitive technology in some organ systems.

13.2.1   Confocal Microscopy

Conventional microscopy typically illuminates a wide area with a condenser lens. This configuration allows light from outside the focus to enter the aperture, reducing the resolution. The smallest object resolvable by conventional light microscopy can be calculated using the Abbe equation[2]:

$$Z = 0.6\lambda/NA \qquad (13.1)$$

where Z is the resolution, λ is the wavelength, and NA is the numerical aperture of the objective lens. The NA is defined as $n\sin\theta$ which represents the ability of the optics to collect light and is a common optics parameter used to define the properties of the system.[1] In this equation, θ is the half angle of the light collected by the objective. For angles greater than θ light can no longer be accepted by the system (i.e., lens). Since we are generally dealing with tissue that has a refractive index greater than air, the NA must be corrected for the refractive index of the medium (n) viewed.

The limit of the lateral resolution for conventional microscopy, which is wavelength dependent, is on the order of 0.2 μm. The length of the focus in the axial direction is referred to as the depth of field. For a conventional microscope, the depth of field is on the order of a few microns. Therefore, there exists a significant difference in lateral and axial resolution. This difference in the lateral and axial resolution degrades the overall image viewed. In addition, thick specimens lead to out-of-focus scattered light from different planes detected which could lead to even further optical deterioration.

Due to the limitations of conventional microscopy, confocal microscopy has recently become popular. Although Marvin Minsky first patented the confocal microscope design in 1957, its general acceptance took decades and commercial systems were not available until the late 1980s.[3] Confocal microscopy eliminates much of the limitations of conventional microscopy by having a smaller field of depth and rejecting much of the out-of-focus information. It does this by requiring illumination and detection to occur at the same point.[4] To achieve this, the illumination optics and detection optics must be near perfectly aligned. This is usually done by using the same lens for illumination and detection. This coalignment is shown in Figure 13.1. The objective lens simultaneously focuses on a pinhole

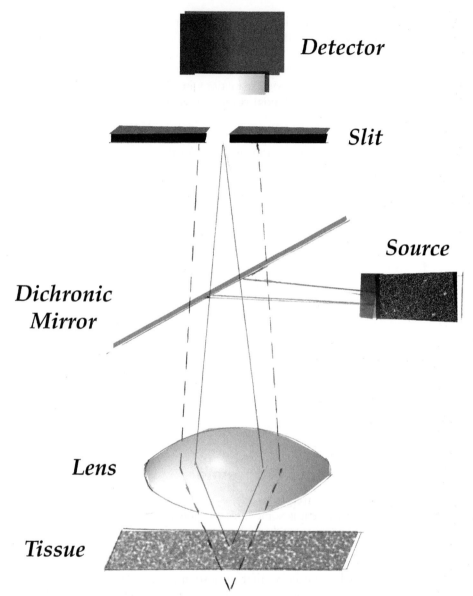

Figure 13.1 Schematic of a confocal microscope.

between the lens and the detector. Light returning from the focus passes through the pinhole and onto a detector. Out-of-plane light returning from above or below the focal point is focused behind or in front of the pinhole, reducing or eliminating its contribution and/or interference, therefore improving resolution. A lateral scan generator is also present to allow information to be obtained from an x-y plane rather than a single point.

The axial resolution with confocal microscopy is determined by the size of the pinhole relative to the magnification and NA. The smaller the pinhole at a given lens parameter, the higher the resolution. When the pinhole size approaches the diffraction limited spot size of the focused light from the lens, the resolution can no longer be improved with reductions in the pinhole size. If the system is diffraction limited, the maximal axial resolution is given by[2]:

$$Z = 2n\lambda/(NA)^2 \qquad (13.2)$$

Maximal lateral resolutions with confocal microscopy can be in the range of 0.5 μm with depth resolutions of 0.5 μm. Unfortunately, confocal microscopy does have several disadvantages for diagnostic imaging. First, in highly scattering tissue, which represents most biological tissue, its penetration is only in the range of 100 μm. Second, imaging is typically performed in the x-y plane, which makes it more difficult to interpret than the usual x-z data obtained with technologies such as ultrasound and OCT. Finally, confocal microscopy typically requires direct contact with the tissue, making its use impractical for routine use in many clinical scenarios. However, confocal microscopy remains a powerful tool in experimental biology and some endoscopic applications have shown promise, particularly in the bladder.[5]

A recent modification of endoscopic confocal microscopy is spectrally encoded confocal microscopy.[6] This approach uses a quasimonochromatic light source and a transmission diffraction grating to detect the reflectivity at multiple points along a transverse line within the sample. This method does not require fast scanning within the probe since this function is taken over by the grating. Therefore, the equipment may ultimately be miniaturized and incorporated into a catheter or endoscope. Therefore, it represents another "optical biopsy" technique, although only over a few 100 microns.

13.2.2   High Frequency Ultrasound

Although high frequency ultrasound is not an optical technology, it deserves discussion because it is sometimes considered a competitive technology to OCT. Sound generally propagates through a medium via mechanical compression and expansion, although some transverse component can occur under the right circumstances. Ultrasound (or echocardiography when imaging the heart) is a technology commonly performed in medicine that uses sound waves for diagnostic purposes. When the term ultrasound is used, it generally refers to sound waves with a frequency greater than that which can be detected by the ear (>20 kHz). Ultrasound is used in a manner analogous to that described for OCT. Acoustical waves are generated at the sample or tissue. The time for the sound to be reflected back, or echo delay time, is used to measure distances. The echo delay time is measured electronically.

In the standard clinical device, the ultrasound waves are generally produced by vibrating disks or plates (transducer). These disks serve both as the source of ultrasound waves and as the detector. Transducers are typically made of crystals such as lead zirconate titanate (PZT) or polyvinylidine difluoride (PDVF).[7] If voltage is placed across the transducer (crystal), it will expand and if it is reversed, the crystal contracts. The voltage across the crystal controls the relative size and rate of change of the transducer. If an alternating current is placed across the transducer, the transducer will change in a sinusoidal manner. The changing shape of the transducer results in a sound wave that compresses or expands the adjacent medium. The deformation propagates through the tissue. As stated, the crystal serves both as a sound generator or detector so that, if the detector receives backreflected acoustical waves, voltage is generated by the transducer that can be detected electronically.

For distance measurements, pulsed ultrasound is used for reasons analogous to that described for OCT. Also, similar to OCT, the shorter the pulse duration (or larger the bandwidth), the greater the resolution.[8] The principles as to why backreflection of sound occurs in tissue are very similar to that of light, except that the mismatch of acoustical impedance (density/velocity of sound) rather than the refractive index is the source of scattering. Scattering theory with respect to size and shape of the object is also similar to that of light, and is generally covered under what is known as Mie's theory.[9]

Sound is a wave, so it is not surprising that it has many physical properties similar to light. For instance, sound exhibits diffraction and interference when coherent just like light. An important difference between light and sound is that the attenuation of light in air is minimal (as in OCT imaging) but much more significant than sound waves. Due to these limitations, ultrasound imaging, for all practical purposes, cannot be performed in air and requires a transducing medium.

The ultrasound transducer can be either single or an array.[8] For single detectors, focusing, for the purpose of imaging, is performed either by having a curved piezoelectric array or through the addition of an acoustic lens. If the outer ring of the array is energized first, then the rest of the ring is activated sequentially inward, the different waves coming from the various transducers interfere in such a way as to form a focus. Since the different rings can be controlled electronically, the focus is adjustable with system electronics. The lateral resolution of ultrasound is therefore dependent on the properties of a lens-like system, similar to an optical lens as was previously described for OCT. Similar to light, the smaller the spot size, the shorter the distance over which the focus is maintained or in other words, the focus falls off rapidly.

Ultrasound's axial resolution is increased with increasing frequency and decreasing bandwidth. However, there is a trade-off between frequency and penetration. The higher the frequency, the less the penetration. The penetration is approximated by the formula:

$$\text{penetration (cm)} = 40/\text{frequency (MHz)} \tag{13.3}$$

Currently, the ultrasound system used clinically with the highest resolution is the 40 MHz intravascular imaging catheter which has an axial resolution of approximately 80–90 μm.[10] The penetration is on the order of a few millimeters, slightly better than OCT.

While the resolution of high frequency ultrasound is high, although far less than that of OCT, it has significant disadvantages for most applications. First, since the transducer is present within the catheter/endoscope, its size must be relatively large and the device expensive. The catheters/endoscopes are generally greater than 1 mm in diameter. Second, ultrasound cannot be performed effectively through air. Therefore a transducing medium or fluid (such as blood) is required, which is difficult to implement routinely in most organs. This is currently attained by the use of a fluid-filled balloon that must occlude the entire airway. Third, the crystal used must be the emitter or detector at any given time. Therefore only a finite time period exists for the transition between emitting and detecting, which leads to artifacts.

13.2.3   Magnetic Response Imaging

Magnetic resonance imaging (MRI) is an important medical imaging technology that uses radio waves and magnetic fields to interact with the proton of the hydrogen atom.[12,13] The proton of the hydrogen atom can be viewed classically as spinning about an axis. Since the proton is charged, it can be viewed as having a magnetic or dipole moment. When not in the presence of a magnetic field, the dipole moments are randomly distributed.

With MRI, a patient is placed in a magnetic field and then is exposed to radio waves. The radio waves are then turned off and secondary radio waves are then readmitted from the patient. The readmitted radio waves are then reconstructed to form an image. The absorbed and measured readmitted waves typically come from nuclei of hydrogen atoms.

Patients are generally placed in a magnetic field set at a value somewhere between 1.0 and 3.0 Tesla. The magnetic field causes the magnetic dipoles to become more organized, pointing either toward or away from the magnetic field axis. These states will be referred to as spin up and spin down. For thermodynamic reasons, one state is in higher proportion to the other. MRI works by determining the difference between the two states.

For reasons described by quantum mechanics, the proton proceeds about the axis of the magnetic field and is not exactly aligned with it. The stronger the magnetic field, the higher the frequency of procession. The procession is also dependent on a factor known as the gyromagnetic ratio, analogous to the stiffness of a string, which is different for different nuclei. RF pulses are generated from coils around the patient. If the RF frequency is close to the resonance frequency of the procession, the energy difference between spin up and down protons, protons are rotated from spin up to spin down. With MRI, series of RF pulses are generated at the patient at intervals known as the repetition time. Once the pulse is over, the proton remains in this state for a short period of time. This rotating charge generates a small alternating voltage in the original RF coil. After the signal is amplified, the amplitude at the resonance frequency is measured. The actual signal intensity measured is determined by proton density, gyromagnetic ratio, and the static field. Only mobile atoms are measured, with water and fat the most important.

The radiated RF signal is greatest immediately after the pulse. After that, the dipoles, at different times, begin returning to their original position. This is due to energy loss via the phenomena known as spin lattice and spin-spin relaxation. The dipoles exchange energy with the rest of the molecule or near molecules, which is called spin-lattice or T_1 relaxation. The T_1 value is the time to recover 63% of the original maximum. Bound water and fat have short T_1 relaxation. Free water and cerebrospinal fluid (CSF) have relatively long T_1 relaxation. Fixed macromolecules have very long T_1.

After the protons have partially recovered, a second pulse converts the available up protons into down protons. However, as the protons flip, dephasing occurs as other neighboring protons begin interfering with the field of the flipped protons. This is spin-spin or T_2 relaxation. The T_2 value is defined as the intensity when the MR signal falls to 37% of its maximum value. Since the local variation in solids such as bone and tendon is large, they dephase quickly. The least effect is in free water, CSF, and urine, where the T_2 time is long.

There are inherent errors associated with MRI imaging in this manner such as inhomogeneity in the magnetic field and engineering imperfections. To overcome these and other sources of noise, a technique of spin-echo is used. With this technique, the initial pulse is followed by a second pulse with twice the energy. The reason why this technique eliminates noise is discussed by Farr, R.F. et al.[13]

Generally, images are either T_1 weighted, T_2 weighted, or proton density. To generate each type of image, the time between pulses and the measurements are different for each type. The times are referred to as the time to echo (TE) and the time to repeat (TR). Fat is typically bright in T_1 weighted images while water is dark. In T_2 weighted images, water is bright while fat is dark. With different MR techniques, different pulse durations, intensities, or intervals are used to maximize contrast. There are a variety of other techniques that can be applied. An example is echo-planar techniques for imaging moving blood.

Most nuclei other than hydrogen cannot be used for MRI since the protons are paired and therefore cancel each other out. However, some atoms that can be used include ^{16}O, ^{14}N, ^{12}C, and ^{31}P. This becomes relevant with magnetic resonance spectroscopy (MRS), which uses chemical shifts to analyze tissue. For example, phosphorus nuclei have different resonant frequencies when bound to different molecules. By using a broad enough RF source, all the different phosphorus nuclei can be made to resonate. The frequency intensities are then analyzed. This allows, for example, the relative concentration of the various metabolites of ATP (which obviously contains phosphorous) to be measured.

In addition to using protons, electrons can be used in MRI. Contrast agents can be used for this purpose. An example is gadolinium. This is used as a contrast agent because it has unpaired electrons. MRI and MRS will be discussed with various pathophysiologies in later chapters.

13.2.4 Computer Tomography

The basis of computer tomography (CT) is that beams of X-rays pass through the body and are partially absorbed or scattered by structures present.[12,13] The absorption depends on the physical density and biochemical composition of the structure, as well as the energy of the X-rays. The beam is attenuated as it passes through the body and the beam on the detector is a superposition of structures along the line between the source and detector. To distinguish different structures along the pass, with CT the X-ray path and detector is changed to different angles. The beam is collimated and in the early embodiment, the detector and X-ray emitter are moved in parallel. There are several schemes that serve as the basis of CT devices with parallel beam as one example. Image reconstruction at the various angles is performed through various algorithms. The absorption intensity is calculated for each pixel in the image at the different angles. There are a variety of methods for reconstructing the angular data, with direct analytical methods such as convolution or filtered back-projection the most common. First in the convolution step the projection data with a function is selected to negate the blurring process and then in the second step (back-projection step), the projections are filtered. Theoretically, these are integrals. By defining the conditions appropriately, the integrals can be replaced by sums.

CT is a powerful diagnostic technology for a range of clinical scenarios. When looking at limitations of CT, the image noise is dependent on the square root of the radiation dose and 3/2 power of the element used (pixel size). However, the axial resolution is highly dependent on object contrast, asymmetrical properties of the object, noise in the projected sample, and the reconstruction algorithm. In addition, several other factors will influence the spatial resolution in X-ray CT. These include the X-ray spot size, detector aperture, sample spacing, form of the convoluting filter, size of the reconstructed pixel, and contrast of the object. High resolution is required (on the micron scale) for most of the applications we are looking at. The definition of a high contrast object normally involves contrast above 10% from that of the background. With high contrast, the spatial resolution is generally between 0.5 and 1.0 mm. Low contrast is 0.5 to 3.0%, which results in spatial resolution between 2 and 4 mm. Clearly, for the applications we are pursuing, these resolutions are insufficient, with the potential application of screening for lung cancer. Of note, high contrast resolution is limited by the modulation transfer function of the system while low contrast sensitivity is limited by the radiation dose. An additional and probably more important source of error in CT images is independent of the system and is body or organ motion. Also, dense objects, such as prosthetics and bone, remain the source of significant artifacts.

There are multiple designs for CT scanners. An example is the fourth dimensional scanner that uses a circumferential collection of detector arrays with a moving X-ray source. This is a fan beam with a stationary circular detector. It will be important in two subtypes of CT that are discussed next.

Spiral or helical CT is now a common variant of CT. With spiral CT, the couch moves continuously at a constant speed while the tube and detector make a number of revolutions around the patient. The detectors are not directly wired but send out their signals through radio frequencies. The size of the slice depends on the speed at which the table is moved and the rate at which the source spirals. These data are reconstructed as a series of sliced images. The advantages of spiral CT are that it is faster and uses less contrast medium. The problem of slice misalignment is also overcome and it results in substantial cost reduction for a range of applications. The disadvantage is that the noise may be greater because a lower mA is used. There is also no cooling period between slices so that the heat loading is high within the tube. Also, a large amount of data is accumulated so computer power becomes critical. It is possible to generate images more frequently at 5 mm, but this results in heavy demands on tube heating and computing power. Some of these issues have been dealt with through multi-slice spiral CT.

A related technique is electron beam computer tomography (EBCT). With EBCT, a beam of electrons is generated from a stationary source swept at high speed across a tungsten arc of 210 degrees around the patient. The electron beam is generated from a gun that is steered through electromagnetic deflection to sweep around the anode. Therefore, the electron beam is moved and not the source. When the electron beam interacts with the tungsten ring, it results in an admission of a fan of X-rays. The circular sweep of the CT is similar to that produced with CT systems using moving parts. Images are obtained in 50 to 100 msec with a 3-mm slice thickness. Heating problems are reduced due to an integrated water system. The use of ionizing radiation represents a disadvantage over other CT techniques. With EBCT, the field is not symmetrical, so radiation exposure is greater in the supine rather than anterior positions. In addition, the design of EBCT is significantly more complex than conventional CT, which is among the reasons why so few devices are in use.

13.2.5 Light-Scattering Spectroscopy

Light scattering spectroscopy is similar to the technology known as elastic scattering.[11] It is a relatively low "tech" modality which could be advantageous, but has yet to be proven as useful as a diagnostic modality at high resolution. Light-scattering spectroscopy is based on the scattering spectrum of single scattered light. Scattering is due to the interactions with microstructures, generally nuclei and mitochondria. Since malignant cells exhibit a variety of morphologic abnormalities including nuclear enlargement, nuclear crowding, and hyperchromasia, the theory is that it will be able to distinguish malignant from normal regions. Unfortunately, to date, light-scattering spectroscopy remains largely untested *in vivo*. Due to this limitation, the author has elected not to discuss the technology in further detail.

13.3 SPECTROSCOPIC TECHNIQUES

Spectroscopic techniques are based upon alterations of the wavelengths of the emitted light from tissue rather than directional changes in the incident light directed toward tissue. While there is some clinical overlap between spectroscopic and structural imaging techniques, spectroscopic techniques assess the biochemical properties of the tissue while structural techniques assess the physical orientation of microstructure within tissue.

13.3.1   Fluorescence

Fluorescence can be divided into point measurements and imaging. The initial discussion will focus on point measurements, then we will move on to the most recent techniques of multiwavelength fluorescence imaging.

Fluorescence has become a very popular technique for studying cellular processes.[14,15] In large part, this is due to the wide range of high quality fluorescence probes and labeling protocols. However, its broad based clinical application to *in vivo* human imaging has been limited.

Fluorescence is an area of investigation that has been present for decades. It is the absorption of photons followed by emission of a photon at a longer wavelength. The theory behind fluorescence can best be illustrated in the Jablonski diagram shown in Figure 13.2. In this diagram, S_0 is the resting state while S_1 is the first excited state. These energy levels are different electronic states that are dictated by the rules of quantum mechanics. It should be understood that, for a given atom or molecule, the electrons (and rotational states) are only allowed to have discrete values. The electrons change in response to various stimuli that excite them to a raised energy state. Why these specific energy levels are allowed, yet others are not, can be explained through basic quantum mechanics, but is well beyond the scope of this work. Reference texts on the topic are listed for those interested.[3] In the Jablonski diagram, the S_1 and S_0 states are also subdivided into sublevels that, in this case, represent different vibrational levels of the molecule at the given electron excitation state. There is also a T_1 level, which corresponds to an electron transition to an energy level with a different electron spin state from S_0 and S_1, which are in the same spin state. Once again, the reason for the different energy levels of the electron spin transition fall under the rules of quantum mechanics, playing an important role in how absorbed energy is dissipated in cells.

When light normally interacts with a molecule, it is scattered on the order of 10^{-15} seconds, which for practical purposes can be considered essentially instantaneous. This corresponds to the normal scattering process of light (as outlined in Chapter 3). However, the light can undergo transient absorption, partially losing its energy into the sample. The light is emitted at different wavelengths. This is the process of fluorescence.

To begin with a simple example of the process of fluorescence, a quantum of light ($h\nu$) is absorbed. This results in the transition of the molecule from the S_0 state to the S_1 state. Since the S_1 state has various energy states that are closely spaced, the molecule will undergo internal conversion, where the electron drops to the lowest S_1 state, its lowest excited vibrational state. This event occurs on a picosecond interval. Therefore, since the electron is in the lowest S_1 state, all energy will be released from this lowest S_1 state.

There are three pathways by which the electron drops out of the S_1 state. These are emission (fluorescence), nonradiative relaxation, and transition to the T_1 state (triplet state).

With fluorescence emission, a photon is released but, since internal conversion has led to a loss of energy due to drop to the lowest S_1 level, the emitted energy is

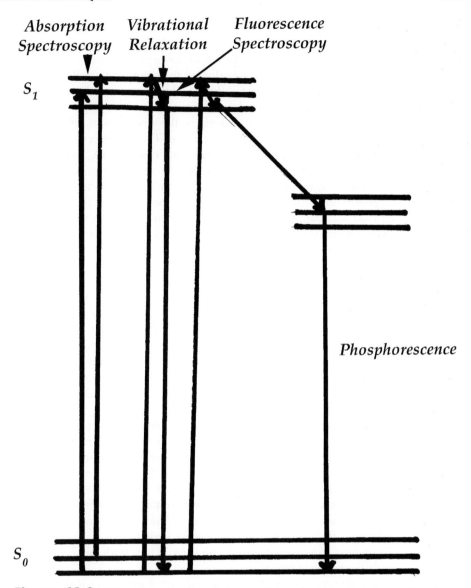

Figure 13.2 Schematic of different types of spectroscopic techniques. The lines demonstrate energy levels for electrons or bonds. The arrows represent transitions between different states. The different possesses are discussed in the text.

at a lower energy level than the incident light and therefore a longer wavelength. Fluorescence occurs when a photon is absorbed, then emission takes place on the order of 10^{-8} to 10^{-10} seconds. Among the properties of fluorescence that are of diagnostic value include the Stokes shift, shape of the emission spectrum, fluorescence lifetime, and to some degree the nonradiative losses.

As stated, the emitted photon will have a lower energy than the incident photon. This reduction in energy and frequency is known as the Stokes shift. When separated from the incident light and other fluorophores (fluorescence-producing molecules), it gives information about the evaluated fluorophore as well as the local environment. Environmental factors that influence fluorescence include ion concentration, pH, and the presence of molecular interactions.

The emission spectrum, which is the distribution of emission as a function of wavelength, also has diagnostic value. The fluorescence emission does not occur at a single wavelength but at the transition from S_1 to S_0, which occurs to the various ground state vibrational levels of S_0. As seen in Figure 13.2, once again the S_0 contains various sublevels due to different vibrational and rotational energy levels.

Besides the Stokes shift and emission spectrum, the fluorescence lifetime can also be used for diagnostic information. The average time for the molecule to go back to the ground state is known as the fluorescence lifetime. The fluorescence lifetime (τ) is given by the formula:

$$\tau = 1/(\alpha + \kappa) \tag{13.4}$$

Here α is the rate of transition from S_1 to S_0 and κ is the transition back to S_0 through nonfluorescence mechanisms. The fluorescence emission profile has a shape that is exponential in nature. While individual molecules have their own lifetimes, τ is also dependent on the local environment, which alters the lifetime. The fluorescence lifetime has several advantages. First, it can help distinguish molecules with similar Stokes shifts but different lifetimes. Second, the lifetime measurements are a property of the molecule, so they are predominately independent of probe concentration relative to fluorescence intensity.

Fluorescence emission, as stated, is not the only mechanism by which the molecule can drop to the S_0 state. This can occur through either nonradiative transitions or by crossover to T_1 states. Examples of nonradiative transitions include resonance energy transfer and collision quenching. Resonance energy transfer occurs when the molecule in the excited state transfers its energy to another nearby molecule, before emission can occur, through electromagnetic interactions. Collision quenching occurs when the excited molecule collides with another molecule capable of removing the absorbed energy before emission can occur. Molecular oxygen is a common example of a quenching molecule. Both of these mechanisms compete with fluorescence emission.

Another mechanism that competes with fluorescence is the crossover to the triplet state (T_1) rather than directly to S_0. A single electron has one of two spin states and T_1 represents a different spin state from S_0 and S_1, which poses a significant energy barrier for transition to S_0. While in the T_1 state, which is at a lower energy level than S_1, the molecule can release energy through emission (known as phosphorescence) or through the nonradiative mechanisms previously described. In general, phosphorescence occurs at a much slower rate and intensity than even fluorescence.

Another important mechanism for decreased fluorescence emission is photobleaching. When a fluorescent molecule has undergone repeated excitation, the molecule becomes irreversibly damaged and can no longer emit fluorescent light. This phenomenon is referred to as photobleaching and is highly dependent on the intensity of the light exposure, the probe type, and the environment. Photobleaching can have a detrimental effect on interpreting fluorescence data.

Fluorescence can be endogenous or can be from autofluorescence.[16] Autofluorescence occurs when compounds intrinsic to the tissue demonstrate fluorescence. Examples of molecules that exhibit autofluorescence include the pyridines/flavins (e.g., NAD, FAD, etc.), aromatic amino acids (e.g., tryptophan, tyrosine, and phenylalanine), structural proteins (such as collagen and elastin), and eosinophilic granules. Examples of exogenous fluorophores include 5-aminolevulinic acid (ALA), bis-carboxyethyl carboxyfluorescein, and porphyrins. Disadvantages to the presence of exogenous fluorophores are the side effect of photosensitivity with stimulation that can last for weeks, requiring avoidance of sunlight.

In addition to point measuring fluorescence techniques, which have been discussed to this point, there are also fluorescence-based imaging techniques. Point measurement techniques acquire data at a single point and may be impractical for use in screening large organs. To overcome this issue the technique of ratio fluorescence imaging (RFI) is attempted.[17] RFI consists of simultaneously measuring fluorescence in the green and red regions of the wavelength spectrum. The ratio between red and green is used to distinguish normal tissue from premalignant or malignant tissue. These data are present in pseudocolor and presented in real time. The normal mucosa when stimulated appears as green while dysplastic cells and carcinoma appear predominately red. One original commercial system, known as LIFE (light-induced fluorescence endoscopy), was developed by Xillix Technology (BC, Canada) to evaluate the airways for autofluorescent variables.[18]

While fluorescence can give information about tissue characteristics, the technology has been around for a considerable period of time and still has not found clear clinical indications. One reason for this is that large-scale clinical trials generally show unacceptably high sensitivities or specificities, depending on where the baseline is set. There are likely a variety of reasons for these results. One may be due to the fact that the biochemical baseline of tissue varies significantly from individual to individual. This includes not only differences in fluorophores, but also differences in local environments, nonradiative relaxation, and photobleaching. In addition, tissue architecture, absorption (particularly hemoglobin), scattering, the metabolic state of the tissue, and the biochemical environment can lead to serious variations in measurements. This all suggests that fluorescence as a diagnostic technique can be limited by variation from patient to patient, narrowing its usefulness as a diagnostic tool. Further research is needed to understand the mechanism behind the sensitivity or specificity issues.

13.3.2 Two-Photon Laser Scanning Microscopy

Two-photon laser scanning microscopy (TPLSM) is a technique that should be discussed because of its future potential and because it is a competitive technology to confocal microscopy.[19,20] With confocal microscopy, the incident light typically uses a wavelength at or near the UV region, which may be damaging to tissue. In addition, there is a relative lack of appropriate lens for use in the UV region. With two-photon fluorescence, the fluorophore absorbs two lower energy photons simultaneously, which typically but not always have the same wavelength. The two photons combine to excite the molecule to the S_1 level. The photons generally are in the red or near infrared regions, which have lower energy and are therefore safer for the tissue.

High laser intensities are required to induce this two-photon absorption and saturation and photobleaching will occur if a continuous wave source is used. Instead, short pulse lasers, on the order of femtoseconds, are used which have high peak powers but the same average power. The laser is therefore focused to a very small focal point (the diffraction limit) so that the laser intensity drops off rapidly both in the lateral and axial directions. Excitation volumes can be in the range of 0.1 μm^3. In general, imaging produces depths on the order of 300 to 600 μm.

TPLSM requires no pinhole like confocal microscopy. Furthermore, the galvano-meter mirror in a confocal microscope leads to some loss of returning light (half), but TPLSM does not contain this descanning mechanism, improving fluorescence collection. Ignoring tissue properties, the sensitivity to emitted photons with TPLSM is therefore dependent on the numeric aperture, throughput of the microscope, and the efficiency of the detectors.

The major disadvantage of TPLSM is the cost and size of the femtosecond lasers. Additionally, there is currently incomplete data on two-photon fluorophores and due to this, clinical usefulness has not yet been evaluated.

13.3.3   Near Infrared Absorption Spectroscopy

Absorption spectroscopy, which has been studied for well over half a century, is based on the fact that molecules absorb energy at specific wavelengths. In theory, the backreflection profile over a range of wavelengths will incorporate the absorption profile (lose backreflected light) of the tissue, and where specific wavelengths have been absorbed there will be no signal. In the near-infrared region, where most diagnostic absorption imaging is currently performed, absorption is occurring predominately by molecular bond transitions of vibrational and rotational modes.[1] For the purposes of illustration, atoms in a molecule can be viewed as connected to each through a spring. When electromagnetic radiation (EM) is applied to the bond, in this case infrared radiation, it leads to oscillation of the bond. This oscillation leads to a change in the displacement of the maximal and minimal distance between the two atoms in the bond or "spring." Usually, the vibrating bond acts like an oscillating dipole and reradiates the infrared light at the same wavelength that it was stimulated, but in different directions. This is the basis of scattering. However, these springs or bonds have certain specific frequencies, known as resonance frequencies, under which the EM photons are absorbed rather than reradiated. Each of these are specific frequencies to the type of bond and/or molecule. Once absorbed, de-excitation occurs when energy is lost through dissipation into the molecule (heat) or by emission of a less energetic photon. One of the biggest problems associated with absorption spectroscopy is that water makes up the largest constituent of tissue and its absorption spectrum overshadows that of virtually all other molecules. In addition, bond energies vary with environmental conditions. Some recent techniques using algorithms and statistical analysis have shown some promise for assessing tissue characteristics.[21] However, the acquisition rates remain relatively slow and the technology remains largely untested on a large scale.

13.3.4 Raman Scattering

Although the Raman effect involves a change in the wavelength of light, it is an instantaneous process and actually a form of scattering rather than traditional absorption. Therefore, it can occur at a wide range of wavelengths. It is based on changes induced in the rotational and vibrational states of the molecular bonds induced by infrared light. Most scattering between light and molecules is elastic; the energy of the photon striking the molecule is the same as the emitted light. About one in every million collisions is inelastic and involves a quantitative exchange of energy between the scattered and incident photon. With the Raman effect, monochromatic light is scattered by a molecule. A frequency shift then occurs above and below the incident light in a small fraction of the light scattered, which is known as the anti-Stokes and Stokes shift.[22] This shift is independent of the frequency of the incident light but the intensity varies with the fourth power of the frequency of the incident radiation. The Raman effect occurs when a beam of intense radiation passes through a sample that contains a molecule that can undergo a change in molecular polarizability as it vibrates. Raman is somewhat distinct from infrared absorption since changes in the polarizability are of more importance than that of the dipole moment. Polarizability is distinct from the classic dipole radiation. The electron cloud around the molecule elongates and contracts under the electromagnetic radiation in a manner distinct from the resting state or normal modes.

Symmetrical molecules have greater Raman effects than asymmetrical molecules, an effect which is opposite from traditional absorption and second harmonic generation (discussed in Chapter 9). The intensity of fluorescence produced may therefore be orders of magnitude higher than the Raman effect, completely obscuring the Raman spectrum. This is why Raman spectroscopy is performed in the near infrared. This frequency has been chosen to lie below most electron transitions and above the fundamental vibrational frequencies. The biggest disadvantage of Raman spectroscopy, in addition to its low sensitivity, is that the number of high-energy photons required may result in tissue damage and concurrently reduced penetration. It should also be noted that a relatively long time is required to obtain data, making its clinical viability questionable. It takes roughly 5 seconds to take a single Raman spectrum measurement with reasonably low signal to noise ratio. Therefore, Raman spectroscopy currently is an experimental technique which has yet to demonstrate clinical utility.

REFERENCES

1. Hecht, E. (1998). *Optics*. Third edition. Addison-Wesley, Reading, MA.
2. Lemasters, J. J., Quan, T., Trollinger D. et al. (2001). *Methods in Cellular Imaging*, A. Periasamy (ed.). Oxford Press, Oxford, UK, pp. 66–88.
3. Inoue, S. (1995). Foundations of confocal scanning imaging in light microscopy. *Handbook of Biological Confocal Microscopy*. Second edition, J. Pawley (ed.). Plenum Press, New York, pp. 1–17.
4. Goldie, R. G., Rigby, P. J., Pudney, C. J., and Cody, S. H. (1997). *Pulmon. Pharmacol. Ther.* 10, 175–188.
5. Jester, J., Petroll, P. M., Lemp, M. A., and Cavanagh, H. D. (1991). In vivo real-time confocal imaging. *J. Electron Microsc. Tech.* 18, 50–60.
6. Tearney, G. J., Webb, R. H., and Bouma, B. E. (1998). Spectrally encoded confocal microscopy. *Opt. Lett.* 23, 1152–1154.
7. Christensen, P. A. (1988). *Ultrasonic Bioinstrumentation*. First edition. John Wiley & Sons, New York.
8. Benkeser, P. J., Churchwell, A. L., Lee, C., and Aboucinaser, D. M. (1993). Resolution limitations in intravascular ultrasound imaging. *J. Am. Soc. Echocardiogr.* 6, 158–165.
9. van de Hulst, H. C. (1984). *Scattering of Light By Small Particles*. Dover, New York.
10. Patwari, P., Weissman, N. J., Boppart, S. A., Jesser, C. A., Stamper, D. L., Fujimoto, J. G., and Brezinski, M. E. (2000). Assessment of coronary plaque with optical coherence tomography and high frequency ultrasound. *Am. J. Cardiol.* 85, 641–644.
11. Mourant, J. R., Fiselier, T., Boyer, J. et al. (1997). Predictions and measurements of scattering and absorption over broad wavelength ranges in tissue phantoms. *Appl. Opt.* 36, 949–957.
12. Moore, J. and Zouridalis, G. (2004). *Biomedical Technology and Devices Handbook*. CRC Press, Boca Raton, FL, pp. 16-1 to 16-21.
13. Farr, R. F. and Allisy-Roberts, P. J. (1997). *Physics for Medical Imaging*. W. B. Saunders Company Ltd, London.
14. Richards-Kortum, R. and Sevick-Muraca, E. (1996). Quantitative optical spectroscopy for tissue diagnosis. *Annu. Rev. Phys. Chem.* 47, 555–606.
15. Berland, K. (2001). *Methods in Cellular Imaging*, A. Periasamy (ed.). Oxford University Press, Oxford, UK, pp. 5–20.
16. Harper, I. S. (2001). *Methods in Cellular Imaging*, A. Periasamy (ed.). Oxford University Press, Oxford, UK, pp. 20–40.
17. Andersson-Engels, S., Klintberg, C., Svanberg, K., and Svanberg, S. (1997). In vivo fluorescence imaging for tissue diagnostics. *Phys. Med. Biol.* 42, 815–824.
18. Lam, S., MacAuley, C., and Palcic, B. (1993). Detection and localization of early lung cancer by imaging techniques. *Chest*, 103, 12S–14S.
19. Denk, W., Strickler, J. H., and Webb, W. W. (1990). Two photon laser scanning fluorescence microscopy. *Science* 248, 73–75.
20. Manni, J. (1996). Two-photon excitation expands the capabilities of laser scanning microscopy. *Biophototon. Interact.* 44–52.
21. Moreno, P. R., Lodder, R. A., Purushothaman, K. R. et al. (2002). Detection of lipid pool, thin fibrous cap, and inflammatory cells in human aortic atherosclerotic plaques by near infrared spectroscopy. *Circulation* 105, 923–927.
22. Brame, E.G. and Grasselli, J. (1977) *Infrared and Raman Spectroscopy*. Marcel Dekker, New York, pp. 2–44.

14 INTRODUCTION TO CLINICAL RESEARCH DESIGN AND ANALYSIS

Kathy Jenkins, MD, PhD

Cardiologist, Director of Programs for Safety and Quality
Chairperson, Scientific Review Committee, Children's Hospital of Boston
Associate Professor, Harvard Medical School

Kimberlee Gauvreau, ScD

Biostatistician
Assistant Professor, Harvard School of Public Health

Chapter Contents

The goal of clinical research is to create information that increases our understanding of how to provide the best care to patients. The key to achieving this goal is in learning how to make *appropriate inference* — drawing appropriate conclusions — about the answer to a particular research question using the data collected from a clinical study. Usually the question we would like to answer is a general one. We begin by thinking about the research question in broad terms, then focus it to be more specific, develop a study design to address the question, and finally implement the study design as best as possible. The ability of an investigator to implement a study as designed allows us to *infer* "truth in the study." The adequacy of the design to address the broader question allows us to *infer* "truth in the universe."[1] Simply stated, the findings of the study are the findings. It is the broader interpretation of these findings that is dependent upon the study design and its implementation. The same process applies when we evaluate the literature or interpret data that have already been collected. We can only make appropriate inferences about whether a study has answered a particular question, or whether the results of the study can be generalized to non-studied populations, after examination and evaluation of the implementation and design.

As an example, suppose we would like to know: What factors predispose patients to postoperative arrhythmias? We must first narrow the universe — what patients, what operations, what arrhythmias, when? What factors predispose children to early postoperative tachyarrhythmias requiring treatment after Fontan surgery? is still a "truth in the universe" question. Theoretically, this question applies to every patient undergoing a Fontan procedure. What were the clinical risk factors for early postoperative tachyarrhythmias requiring treatment among patients undergoing Fontan procedures operated at Children's Hospital, Boston in 2003? is a question that can be studied more completely. It should be possible to determine an answer to this question, provided a reasonable design can be implemented. The study may also provide information about other Fontan patients, at other institutions and in different years, depending on its design. However, if even a well-designed study cannot be implemented suitably — if, for example, tachyarrhythmias are difficult to identify accurately using retrospective data — very little inference is likely to be possible. We will be unable to answer the question appropriately for patients undergoing the Fontan procedure at Children's Hospital in 2003, much less anywhere else. This chapter examines designs of clinical studies, sources of potential errors, and methods of analyzing the results. While some sections describe trials using pharmaceuticals and surgical procedures, conclusions can be extrapolated to imaging techniques.

14.1 ELEMENTS OF STUDY DESIGN

14.1.1 Choosing a Primary Research Question

The process of formalizing a research question is an important part of beginning an investigation. Although studies can often answer multiple questions at once, it is best to designate one question to be the main focus, and therefore the most important. This question is usually referred to as the *primary research question*. As we move through the study design and implementation, it is critical to evaluate the effect of any decisions made on our ability to answer this question. Ultimately, the success of the study will be determined based on an investigator's ability to make an inference about the primary question the study was designed to answer. Therefore, few if any compromises should be made that might make it hard to draw this inference. Similarly, the overall study design, including the sample size, should be chosen to ensure that the primary research question can be answered.

Finding questions that are interesting and important comes from a good knowledge of the current state of the art of one's field. However, if one is a novice, it is usually best to explore possible study questions with more experienced clinicians or researchers. It is also critical to review the literature carefully, to be aware of what studies have been done previously or are currently in progress. Postings by funding organizations, such as the National Institutes of Health (NIH), can be another useful source of information. A good working knowledge of basic principles of study design and analysis can be used to determine whether a question is feasible. We must be able to complete a study well enough to make an appropriate inference, using the patients and resources available, in a reasonable time frame. Otherwise the study is simply not worth doing.

14.1.2 Descriptive versus Analytical Studies

Although much literature has been generated from descriptive studies performed at single institutions, hypothesis-driven studies with a predetermined analytical plan and sample size will generally, by design, give much clearer answers to questions about best practice with the smallest possible sample size. The actual mechanism by which a study will be performed (e.g., a chart review of cases over a 10-year period) may be very similar whether a study is done to "describe outcomes for single ventricle patients at our institution between 1990 and 1999" or to "determine predictors of poor outcome after Fontan surgery performed between 1990 and 1999." More specific information about how to change practice is likely to be possible from the second study than from the first. Patient characteristics, such as age at Fontan or prior bidirectional Glenn operation, will be simple descriptive information in the first design, but will serve as potential predictors of poor outcome in the second. For example, study investigators may hypothesize that younger age at surgery is a risk factor for failing Fontan physiology, death, or transplant during follow-up.

An important aspect to hypothesis-driven analytical designs is limiting the study to patients who will specifically provide information regarding the research question. For example, patients who never underwent Fontan surgery might be included in a study describing outcomes of patients with single ventricle, but will be excluded from a study evaluating whether young age at Fontan is a predictor of poor outcome. The overall descriptive results presented in the first design may provide a confusing picture about outcome differences among patients operated on at different ages.

14.1.3   Threats to Validity

Our ability to infer an answer to a primary research question depends on the validity of the findings in the study. Different types of errors made during the design or implementation of a study can limit its validity, and therefore the inference that can be made about the research question. Prior to beginning an investigation, it is helpful to understand common types of threats to validity that lead to errors, so that we can attempt to avoid them. Different threats limit the validity of studies in different ways. Studies can be protected from mistakes in inference within the study design, the implementation, and in some cases, the analysis phase.

14.1.4   Random Variability

The random variability inherent in data collection and measurement can lead to mistakes in inference. Variation can arise from limitations in precision associated with making a measurement, or from natural variation in the clinical parameter itself, such as blood pressure. However, variation is also introduced when we attempt to make an inference about an entire population based on the data collected from a sample of individuals drawn from that population. This type of variation would occur even if we could achieve perfect measurement of an unchanging parameter. In general, protection against random error or sampling variability is achieved by increasing our sample size, thus increasing the precision of population-based estimates, and also by using analytic techniques that allow quantification of the extent to which random error may be responsible for the results obtained. The science of *biostatistics*, based primarily on determining the likelihood that certain data would have been obtained under specified theoretical conditions, provides numerous tools to help protect studies from mistakes in inference due to random variation in clinical data. Biostatistical techniques can quantify the degree to which random error may explain the findings observed in a study, and are almost always used as the basis for an inference from clinical research. It is important to remember that it is only possible to quantify the degree to which random error might explain the findings in a study if we use statistical tests that are appropriate for the type and amount of data available. If we use a test inappropriately, we will get an incorrect answer about the likelihood that a particular result could have been achieved by chance variability alone. It is impossible to make an "appropriate" inference from an "incorrect" answer.

14.1.5 Bias

Bias can be introduced at almost any point in the study process, and occurs whenever the details of the design or implementation make it more likely that the study will yield one particular conclusion rather than another. Bias is a general term; any aspect of a study that makes it more likely that the study will obtain one answer rather than another should be considered a form of bias. However, a number of common types have been identified. In general, most forms of bias cannot be corrected using analytical techniques.

Selection bias — When designing a study, we usually develop entry and exclusion criteria to define who is eligible. The entire population of patients meeting these criteria is the theoretical target population, and an investigator hopes to obtain an answer that pertains to all eligible subjects. *Selection bias* can be introduced if some subjects from the target population are more likely to be included in the study than others. Although a well-implemented study should still provide an answer to the research question for the patients that were actually included, this answer may not be true for the entire target population. For example, a study examining the prevalence of a cardiac diagnosis after a syncopal episode may provide an overestimation of this proportion if only those individuals referred to a cardiologist are studied. Also, a study may reach a correct conclusion about the effect of a drug in patients who comply with treatment, but not be generalizable to the population of all diseased patients. *Ascertainment bias* is a form of selection bias that occurs when a study has information about only a subgroup of the target population, so that certain subjects are more likely to be included in the study overall or in the group of patients with analyzable data. It may be nearly impossible to know anything about a population that is truly inaccessible. As an example, it may be very difficult or impossible to make an inference about groups with "silent" disease, such as, without undergoing a large-scale screening program in asymptomatic patients.

Bias in treatment allocation — Treatment allocation bias occurs if patients in different treatment groups are not comparable. For example, if investigators tend to assign sicker patients to one particular treatment rather than another a study may under- or overestimate the effect of that treatment. Sicker patients may have more potential benefit from any effective treatment, or be more likely to do poorly because of their initial worse status. Thus, errors in both directions are possible from biased treatment allocation.

Bias in outcome assessment — This occurs if outcomes are not assigned equivalently in different treatment groups. Investigators may assign outcomes following their prior assumptions, based on what he or she wishes or expects. Qualitative assignments are particularly vulnerable to this type of bias.

Misclassification bias — This can occur when there are errors in assigning patients to treatment or outcome categories. Subjects are not "classified" properly within the study. Random misclassification biases a study toward the hypothesis of no treatment effect and thus limits the power of the study. More patients will be necessary to show a true effect.

14.1.6 Confounding

Confounding can also threaten the validity of a study, because the results of the study might be misinterpreted due to the presence of the confounders. *Confounders* are factors that are associated with both the predictor of interest and with the study outcome. The effect of the confounder on the outcome is wrongly attributed to the predictor, and false conclusions are drawn.

A simple example would be a study that concludes that there is no difference in weight among 16 year olds in India and those in the United States based on measurements from 100 children, 50 from each country. A true difference in weight may be hidden if there are more males in the group of Indians than in the group from the United States. Within this study, gender is associated with both the predictor (being Indian) and with the outcome (weight). Confounding can be avoided within the study design by knowing that the groups are truly comparable. For example, a study restricted to males would not have reached the false conclusion described above. Confounding can also be dealt with by using analytic techniques; stratifying by gender would have avoided the problem. One of the most important roles for multivariable models is adjustment for multiple confounders. Of course, we must identify and collect data about the confounders in order to adjust for them later. Unmeasured or unmeasurable confounders can be especially problematic. Figure 14.1 shows an example of confounding in a study examining the effect of pulmonary vein size on outcome among infants with totally anomalous pulmonary venous connection.[2]

14.1.7 Effect Modification

Effect modification is a special type of confounding that can also lead to false conclusions, and occurs when a predictor has a different "effect" on the outcome in different subgroups of the study population. If the effects are in opposite directions, they can cancel each other out and lead to the false conclusion that there is no apparent effect of treatment on outcome. More commonly, the magnitude of the effect can be wrongly inferred. The only way to avoid errors from effect modification

A study was designed to determine whether patients with totally anomalous pulmonary venous connection (TAPVC) and small pulmonary veins died more frequently than those with larger veins. The study sample included patients with both isolated TAPVC and complex heterotaxy. The primary finding was that small vein size at diagnosis (defined as the sum of all pulmonary vein diameters measured by echocardiography, normalized to body surface area) was strongly associated with death.

Relationship of small veins to outcome:

	Small	Large
Died	15	3
Alive	2	15

Mortality rate 88% small versus 17% large, p < 0.001

However, heterotaxy syndrome was associated with both mortality (the outcome) and small vein size (the predictor).

Relationship of heterotaxy to outcome:

	Heterotaxy	Non-Heterotaxy
Died	13	5
Alive	3	14

Mortality rate 81% heterotaxy versus 26% non-heterotaxy, p = 0.001

Relationship of heterotaxy to small veins:

	Small	Large
Heterotaxy	11	5
Non-Heterotaxy	6	13

Rate of small veins 65% heterotaxy versus 28% non-heterotaxy, p = 0.02

In order to determine whether the original finding was merely due to patients with heterotaxy syndrome both having small veins and dying more frequently, i.e., to determine whether heterotaxy syndrome was a confounder in the relationship between vein size and mortality, a stratified analysis was performed. The relationship between small vein size and mortality was examined separately for patients with and without heterotaxy syndrome.

Relationship for patients with heterotaxy syndrome:

	Small	Large
Died	10	3
Alive	1	2

Mortality rate 91% versus 60%, p = 0.14

Relationship for patients without heterotaxy syndrome:

	Small	Large
Died	5	0
Alive	1	13

Mortality rate 83% versus 0%, p < 0.001

Thus, small vein size was significantly associated with mortality for patients without heterotaxy syndrome, and there was a trend towards significance in patients with heterotaxy syndrome. The original finding of small vein size at diagnosis could not be completely explained by small veins among heterotaxy patients.

Figure 14.1 Examples of confunding data.

is to identify it. If effect modification is present, analytic methods that do not account for it will lead to incorrect conclusions.

14.1.8 Evaluating Threats to Validity

Understanding and identifying threats to the validity of a study is clearly very important. It is not enough, however, to simply notice that a certain threat is present. To make an appropriate inference from data, we must also consider the magnitude and direction of the threat in relation to the study hypothesis or conclusion. For example, missing mortality outcome data from 10 out of 100 subjects would not be a major flaw in a study that concludes that mortality is very high, since determining that any of the missing subjects had died would only enhance the study conclusion. However, the same missing data would be a major threat to validity if the conclusion is that there was no mortality within the study group. The results of a study are not invalidated just because a flaw is present, however. As a rule, never criticize a study without carrying your concern about bias, confounding, or other problems to the next level of analysis. That is to say, never criticize a study without considering the magnitude and direction of the error in relation to the hypothesis. Only then can you determine whether the error requires that an investigator modify his or her conclusions.

14.2 CHOICE OF STUDY DESIGN

14.2.1 Randomized, Blinded, Controlled Clinical Trials

Randomized, blinded, controlled clinical trials are frequently regarded as the "gold standard" study design to definitively answer many research questions. This is because a well-implemented randomized, blinded, controlled trial automatically eliminates many common threats to validity. In fact, the only major threat to validity that remains in such a trial is making a false inference due to chance or random variation, and this threat can be eliminated by making certain that the study has an adequate sample size, and therefore adequate statistical power. Simply conducting a randomized trial does not, however, ensure that the study is of high caliber. In fact, since randomized trials are often burdensome and expensive, it is especially important that such studies address important research questions.

The individual "parts" of the randomized, blinded, controlled design protect the study in different ways. It is important to understand the protection that is provided by each individual component.

14.2.1.1 Protection by randomization

The primary protection provided by randomization is protection from confounding. By design, all variables that might potentially confound interpretation of the relationship between the predictor of interest and the primary outcome should be equally distributed between the study groups. The resultant study groups should then differ *only* with respect to the intervention. Of course, this does not always happen perfectly. With a simple randomized design, it is *still* possible for residual confounding to be present due to chance associations within the data, or because the sample size is small. In studies where such chance associations present a substantial threat to the validity of the study, including those where some clinical factors are very strong, independent predictors of the outcome, additional protection may be achieved through stratified randomized designs. Such designs ensure that key variables are equally distributed between the study groups. Randomization of treatment also prevents bias in treatment allocation.

14.2.1.2 Protection by blinding

The key protection provided by blinding is protection from bias. Blinding investigators and patients to treatment assignment during the study minimizes the bias that can be introduced if subjects are treated differently based on knowledge of treatment assignment. For example, an investigator might be more likely to maximize concomitant medications if he knows that a patient is randomized to placebo. This type of bias is due to differences in "co-interventions" between the study groups. Blinding is also critical during outcome assignment to avoid investigator bias. The more subjective (rather than objective) the outcome measure, the more important it is that it be assessed in a blinded fashion. For example, blinding is not very important in a trial examining mortality as an outcome, but it is critical in evaluating an improvement in school performance. Blinding of treatment status during outcome assignment can often be accomplished even in cases where blinding of treatment is not possible. For example, an independent assessment of left ventricular function by echocardiography can be made without knowing whether a patient recovering from myocarditis received gamma globulin or placebo. Some studies cannot be blinded. Such studies are particularly susceptible to threats to validity due to bias.

The terminology regarding blinding can often be confusing. "Double-blinding" can refer to blinding of both group assignment and assessment of outcome, or to blinding of both subjects and investigators to treatment assignment or outcome assessment. It is important to understand how the term is used in any particular context. In general, more blinding offers better protection.

14.2.1.3 Protection through use of an appropriate control group

Using an appropriate control or comparison group does not directly provide protection to the validity of a study, but it does greatly facilitate interpretation of the findings after a study is complete. For this reason, comparison groups must be chosen carefully, based on the particular research question. For studies assessing treatments, comparisons to placebo are commonly regarded as the gold standard. Placebos are used rather than "no treatment" to facilitate blinding and to minimize bias on the part of study subjects, particularly for subjective outcomes. Comparisons to standard treatment are also common, provided that a standard treatment exists.

Ethical or practical considerations sometimes exclude certain choices for comparison. This occurs even in cases where the answer to the research question is not known. Patients or physicians may be unwilling to enroll in trials when they have strong preferences about treatment alternatives or when comparisons will be made to treatments that are generally regarded to be ineffective.

14.2.1.4 Analytical issues

If a randomized, blinded, controlled study has been well designed and implemented, all possible measured and unmeasured confounding factors should be equally distributed between the groups, and no complex analytic techniques are necessary. For a dichotomous outcome variable, the final results from a lengthy trial can be summarized in a simple 2×2 table; for continuous outcomes, the results are summarized as a comparison of means between the groups. Despite this, baseline characteristics are often recorded at the beginning of a study, prior to randomization. This can sometimes be wrongly interpreted to indicate that complex analyses will be essential to make a proper inference. Although measurement of these variables is not required, it does allow post hoc documentation that the only difference between groups was, in fact, the intervention. It also allows statistical adjustment for any differences that remain despite the randomization (known as residual confounding).

Measurement of prerandomization variables also allows subgroup analyses. *Between group sub-analyses*, such as a comparison of placebo versus study drug in males only, are protected by the randomized design, but have less statistical power than the primary analysis. The sample size is smaller for such comparisons. *Within group sub-analyses*, such as a comparison of the effect of treatment on outcome in males versus females, are not protected from confounding by the randomized design.

When performing a randomized, blinded, controlled trial, the only type of analysis that is protected from threats to validity by the randomized design is an *intention to treat analysis*. This means that patients must be analyzed according to their initial treatment assignment, regardless of whether or not the treatment was actually administered. Other types of analyses can be performed, such as analyses in only those patients who actually received treatment, but these analyses are subject to the same threats to validity already described. They are not protected by the randomized design, and should not be reported as such. One example of a sentinel randomized, blinded, controlled clinical trial in pediatric cardiology is "The treatment of Kawasaki syndrome with intravenous gamma globulin" by Newburger et al.[3]

14.3   OTHER STUDY DESIGNS

Although randomized, blinded, controlled clinical trials offer many protections, some research questions are very difficult to study in this manner. Examples include studies where little is known about the question of interest, studies of rare outcomes, studies of conditions with long latency periods, and studies with ethical or practical issues that preclude randomization. These types of research questions can often be approached more efficiently using other designs. Although these studies are not inherently protected from threats to validity in the way that randomized, controlled, blinded clinical trials are, with appropriate attention, a valid inference can usually be made.

14.3.1 Longitudinal Studies

Longitudinal designs, or before/after studies, are a particularly efficient type of "experiment" involving a single group of subjects. Key variables are measured before and after an intervention occurs. Analyses typically involve paired data techniques, which have enhanced power, thus limiting sample sizes. Each subject serves as his or her own control, minimizing confounding. In addition to the usual threats to validity, this design is susceptible to several particular forms of bias. *Time-related confounding* occurs when a third factor that has an effect on the outcome varies throughout the time period of the study. Another threat can be due to carryover effects, where the effects of a treatment last beyond the treatment period. To protect against these problems, many longitudinal studies are designed as *crossover trials*, often with randomization to a first treatment followed by a "crossover" to a second treatment, with a washout period in between. An example of an important longitudinal study in pediatric cardiology is "Impact of inspired gas mixtures on preoperative infants with hypoplastic left heart syndrome during controlled ventilation" by Tabbutt et al.[4]

14.3.2 Cohort Studies

Cohort studies are a classic epidemiological design in which a group of individuals is followed over time to determine the incidence of a particular disease or condition. Unlike randomized trials, cohort studies can be used to address several research questions simultaneously. Simple descriptive studies are a straightforward type of cohort design, but more often studies are performed to determine the association between various predictor variables and an outcome. If the predictors are measured before the outcome occurs, the study is *prospective*. If the predictors are measured after the outcome, it is *retrospective*. Some studies include both prospective and retrospective features. For both types of cohort studies, the influence of various predictors on the outcome is inferred by an analysis comparing outcomes in subgroups of the cohort with different levels of the predictor.

Prospective cohort studies are frequently helpful in inferring causal explanations, since predictor attributes were clearly present prior to the development of the outcome. If such studies are well implemented, they can be as definitive as randomized, controlled trials in evaluating certain research questions. Confounding is almost always present, but can be dealt with using appropriate analytic techniques. The major problem with prospective cohort designs is that they are cumbersome and expensive; lengthy trials are also subject to the influence of multiple factors that are not stable over time, similar to longitudinal designs. Retrospective designs are frequently helpful in studying diseases with very long latency periods to avoid the necessity of following a cohort for an extended period. However, an inference from such trials can be limited by the quality of the predictor data that are available at the time the study is performed. One example of a prospective cohort study in the field of pediatric cardiology is "Pulmonary stenosis, aortic stenosis, ventricular septal defect: Clinical course and indirect assessment — Report from the Joint Study on the Natural History of Congenital Heart Defects" by Nadas et al.,[5] while an example of a retrospective cohort study is "The modified Fontan operation: An analysis of risk factors for early postoperative death or takedown in 702 consecutive patients from one institution" by Knott-Craig et al.[6]

14.3.3 Cross-Sectional Studies

In *cross-sectional studies*, data about predictor variables and outcomes for individual study subjects are collected at the same point in time. Since this type of study does not involve any follow-up, it can be more efficient than a cohort study for many types of questions. Cross-sectional studies are not nearly as valuable as prospective studies for inferring causality, however, since it will frequently be difficult to know for sure whether a certain factor is the cause or the result of another factor even if a study demonstrates a clear association between the two. Cross-sectional studies are useful for measuring the prevalence, but not incidence, of factors of interest, since there is no attempt to follow patients over time. They are also subject to *prevalence/incidence bias*, where the influence of a factor on disease duration leads to mistakes in inference about disease occurrence. An example of a cross-sectional study is "Fontan operation in five hundred consecutive patients: Factors influencing early and late outcome" by Gentles et al.[7]

14.3.4 Case-Control Studies

One of the most efficient designs for the study of rare problems is a *case-control study*. This type of study must be both well designed and well implemented, because it is subject to many sources of bias. Cases are chosen, frequently as a convenience sample, to include individuals who have the outcome of interest. Controls are selected as a sample of subjects who do not have the outcome. The investigator then compares the numbers of subjects with the predictor of interest for cases and controls. Thus, case-control studies are almost always retrospective, and the measurement of predictor variables can be subject to *recall bias*, where cases with a particular disease or condition recall exposures more readily than healthy controls.

Selection of controls is one of the most challenging aspects to the case-control design. Ideally, we would choose a control group that is representative of the target population (particularly in terms of the presence of the predictor of interest), at risk for the disease, and accessible to study investigators. Frequently, more than one control group is used to "compensate" for potential sources of bias in any one group. In addition, special analytic techniques must be used to make an appropriate inference. The results from a case-control design can generally be summarized in a 2×2 table, but cannot be used to estimate either incidence or prevalence of disease because the number of individuals in the study was chosen by the investigator, and therefore does not reflect the population at risk for disease. Matching is sometimes used to increase the similarities between cases and controls, particularly for key factors, leading to an increase in statistical power. However, matching must be used cautiously, as we cannot correct for "matching" in the analytic phase of the study. Serious mistakes can occur if we match on factors that are in the causal pathway between the predictor of interest and the outcome; this severely limits the statistical power to answer the study question. Also, matched designs require special analytic techniques. An example of a case-control study in pediatric cardiology is "Perspectives in Pediatric Cardiology, The Baltimore-Washington Infant Study" by Ferencz et al.[8]

14.3.5 Nested Case-Control Studies

A special type of design that eliminates problems associated with retrospective data collection and recall bias is a *nested case-control design*, where a case-control study on a limited number of subjects is conducted within the structure of a prospective cohort study. Here, the predictors are obtained prospectively, since all study subjects are part of the initial cohort. All cases of a rare disease are identified after the study is complete, and are compared to a random sample of controls selected from the remainder of the cohort without the outcome. An example of a nested case-control study is "Sudden death in patients after transcatheter device implantation for congenital heart disease" by Perry et al.[9]

14.3.6 Secondary Data

Using secondary data means analyzing data that were collected for another purpose to answer a different research question. Since primary data collection can be both burdensome and expensive, this can be a very efficient technique for reducing the costs of conducting a study, provided that an investigator is sufficiently cautious about limitations in the data. For example, it can be difficult to know for certain to what extent the data are accurate. Also, variables are frequently not included in the way an investigator might wish them to be. We are forced to make trade-offs and to use "proxies" in many cases. Despite these problems, valid conclusions can often be drawn from secondary data as long as we proceed with caution. An example of a study using secondary data is "In-hospital mortality for surgical repair of congenital heart defects: preliminary observations of variation by hospital caseload" by Jenkins et al.[10]

14.3.7 Diagnostic Test Evaluation

When studying the performance of a diagnostic test, as is the case with OCT, *sensitivity* and *specificity* are measured within a particular clinical context.

To accomplish this, it is necessary to precisely define whether patients do or do not have the disease in question; this requires the existence of a gold standard for the diagnosis. In cases where no gold standard exists, we must create a reasonable definition for the presence of disease. The study design is generally an "experiment," where both the test under investigation and the gold standard are universally applied to a cohort of patients so the performance characteristics of the test can be evaluated. Studies are usually designed to measure sensitivity and specificity with an acceptable level of precision. Since sensitivity is estimated only from diseased patients and specificity only from non-diseased patients, the study must include sufficient numbers of patients of each type.

The classification of patients as having a positive or negative test result is especially important. It is best to decide in advance the criteria that will be used to determine a positive test, and to design and implement a study where the assessment of the test result and disease status are completely independent and preferably blinded. It is especially important to determine how "borderline" test results will be handled. If sensitivity and specificity are estimated based only on tests that are easy to classify, these values will tend to be overestimated, since the cases eliminated are "on the margin" of the discrimination ability of the test. Studies of diagnostic test performance generally report *positive* and *negative predictive values* in addition to sensitivity and specificity. These values are directly influenced by the prevalence of disease in the studied population. Investigators sometimes report predictive values under varying conditions of disease prevalence to better understand the outcomes that would be likely to occur if the test was applied to a different population. An example of the evaluation of a diagnostic test is "Heart size on chest x-ray as a predictor of cardiac enlargement by echocardiography in children" by Satou et al.[11]

14.4   ELEMENTS OF DATA ANALYSIS

14.4.1   Types of Numerical Data

Every research study yields a set of data; the size of the data set can range from a few measurements to many thousands of observations. Before we can begin to interpret these measurements, however, we must first determine what kind of data we have. The simplest type of data is *nominal data*, in which the values fall into unordered categories or classes. Examples include gender, cardiac diagnosis, and postoperative survival status. Nominal data that take on only two distinct values (e.g., survival status alive or dead) are said to be *dichotomous* or *binary*. The categories of a nominal variable are often represented by numbers (males might be assigned the value 1 and females the value 2, for instance), but these numbers are merely labels. Both the order and the magnitude of the numbers are unimportant; we could just as easily let 1 represent females and 2 males. Numerical values are used mainly for the sake of convenience, making it easier for computers to perform complex analyses of the data.

When there is a natural order among the categories, the observations are referred to as *ordinal data*. One example is New York Heart Association (NYHA) classification, where a larger number represents poorer health status. While the order of the numbers used to represent the categories is important, we are still not concerned with their magnitudes. Furthermore, the difference between NYHA class I and class II is not necessarily the same as the difference between NYHA class III and class IV, even though both pairs of values are one unit apart. Together, nominal and ordinal measurements are called *categorical data*.

With *discrete data*, both order and magnitude are important. Numbers are actual measurable quantities restricted to taking on certain specified values, usually integers or counts. Examples include the number of beds available in a cardiac ICU, and the number of times a patient has been admitted to the hospital.

Continuous data are not restricted to specified values; fractional values are possible. Examples include age, weight, systolic blood pressure, and length of time a patient is on mechanical ventilation following cardiac surgery. With continuous data, the difference between any two values can be arbitrarily small; the accuracy of the measuring instrument is the only limiting factor.

Measurements that can vary from one subject to another such as cardiac diagnosis, NYHA classification, and age are called *variables* or *random variables*. At times we might require a lesser degree of detail than that afforded by a continuous variable; in this case, continuous measurements can be transformed into categorical ones. Instead of recording birth weight in grams, for example, we might categorize birth weight as <2500 grams versus ≥ 2500 grams. While the analysis is simplified, information is also lost. In general, the degree of precision required for a given variable depends on the research question.

14.4.2 Summarizing and Describing Data

Once we have determined the types of numerical data we have collected for our study, our next goal is to organize and summarize the measurements so we can identify the general features and trends. *Descriptive statistics* are techniques that are used for this purpose. Different descriptive techniques are more appropriate for summarizing different types of numerical data.

Tables or *frequency distributions* enable us to see how a set of measurements are distributed and are often used for nominal and ordinal data. To construct a table for discrete or continuous data, the observations must be divided into distinct, non-overlapping intervals if any useful information is to be obtained.

Graphs are pictorial representations of a set of values that quickly convey the general patterns and trends of the data. *Bar charts* can be used to display frequency distributions for categorical data, while *histograms* serve the same purpose for discrete and continuous data.[12] *Box plots* display only certain summaries of discrete or continuous measurements, including the 25th, 50th, and 75th percentiles and the minimum and maximum values.[12] The relationship between two continuous measurements can be depicted using a *scatter plot*.

Numerical summary measures are single numbers that quantify the important characteristics of an entire set of values, such as the center of a set of measurements, or the amount of variability among the different values. For example, the mean, median, and mode can all be used to describe a "typical" value, or the point about which the observations tend to cluster. The *mean* is the simple arithmetic average of the measurements; we sum up the values and divide by the number of observations. The *median* is the 50th percentile of the data, or the value such that 50% of the measurements lie below this value and 50% above. The *mode* is the observation that occurs most frequently. The best measure of the center of a data set depends on both the type of numerical data and the way in which the values are distributed. The mean can be used for discrete and continuous measurements. The median is used for discrete and continuous observations as well, but can also be used for ordinal data. For continuous measurements, the mean is most appropriate when the distribution of values is fairly symmetric; if the distribution is skewed, the median is probably a better choice since it is less sensitive to extreme data points. The mode is especially useful for nominal and ordinal data, where it identifies the category that occurs most frequently. Variability among values can be quantified using the range, the interquartile range, or the standard deviation. The *range* is the difference between the minimum and maximum values, and therefore emphasizes extreme values; the *interquartile range* is the difference between the 25th and 75th percentiles of the observations and is not sensitive to outliers. Either the range or the interquartile range is usually reported with the median. The *standard deviation* quantifies the amount of variability around the mean,[12] and is therefore reported with the mean. For categorical data, a table is probably more effective than any numerical summary measure.

14.4.3   Variability in Data

If the only goal of a research study is to summarize and describe a group of measurements, as might be the case for a purely descriptive study, we can limit the analysis to the descriptive statistics outlined above. However, we are usually interested in going beyond a simple summary and investigating how the information contained in a sample of data can be used to infer the characteristics of the population from which it was drawn. The process of drawing conclusions about an entire population based on the information in a sample is known as *statistical inference.*

Any group of observations randomly sampled from a population displays some amount of variability among the individual values; there is inherent biological variability among different subjects, and there can be variation due to the measurement process as well. Before we can make an inference, however, it is important to understand the ways in which the data values can vary across subjects. This can help us to decide which statistical technique to use to make an inference in a given situation.

Probability distributions are rules that describe the behavior of a variable such as systolic blood pressure or gender. For a nominal, ordinal, or discrete variable, we can write down each possible category or class for the data along with the probability that the category occurs. For a continuous variable, a probability distribution specifies probabilities associated with ranges of values. The *probability* of an outcome is its relative frequency of occurrence in a large number of trials repeated over and over again; it can be thought of as the proportion of times that a particular outcome occurs in the long run. If an outcome is sure to occur, it has probability 1; if it cannot occur, it has probability 0. An estimate of a probability may be determined empirically using a finite amount of data, or can be based on a theoretical model. Several theoretical probability distributions have been studied extensively and are very important in clinical research. For example, the *binomial distribution* is useful for describing the proportion of "successful" outcomes arising from dichotomous data, while the *normal distribution* is used to analyze the behavior of means of continuous variables.[12]

14.4.4 **Making an Inference from Data**

Methods of statistical inference fall into two general categories: estimation and hypothesis testing. With *estimation*, our goal is to describe or estimate some characteristic of a population of values — such as the mean white blood cell count among children with acute Kawasaki syndrome, or the proportion of these children who develop coronary artery abnormalities — using the information contained in a sample of data drawn from that population. Two methods of estimation are commonly used. *Point estimation* uses the sample data to calculate a single number to estimate the population parameter of interest; a point estimate is our "best guess" for the value of that parameter. For instance, for the dichotomous variable any coronary artery abnormality we might use the proportion in a sample of children with acute Kawasaki syndrome seen at a particular institution in the past five years to estimate the proportion for the entire population of all children with this disease from which the sample was drawn. The problem is that two different samples drawn from the same population, at different times, or from different hospitals, will have a different mix of patients and therefore different sample proportions; there is uncertainty involved in the estimation process. A point estimate does not provide any information about the inherent variability of the estimator; we do not know how close the sample proportion is to the true population proportion in any given situation. Consequently, a second method of estimation known as interval estimation is often preferred. *Interval estimation* provides a range of reasonable values that are intended to contain the parameter of interest with a certain degree of confidence. The range of values is called a *confidence interval*. A 95% confidence interval is constructed in such a way that if 100 random samples were selected from a population and used to construct 100 different confidence intervals, approximately 95 of the intervals would contain the true population parameter and 5 would not.

With *hypothesis testing*, we begin by claiming that the population parameter of interest is equal to some postulated value. For instance, we might claim that the mean IQ score for children born with hypoplastic left heart syndrome and who have undergone the Fontan procedure is equal to 100, the mean for the general population. This statement about the value of the parameter is called the *null hypothesis*. The *alternative hypothesis* is a second statement that contradicts the null hypothesis. Together, the null and alternative hypotheses cover all possible values of the population parameter; consequently, one of the two statements must be true. After formulating the hypotheses, we next draw a random sample from the population of interest. If we are conducting a test on a population mean, the mean of the sample is calculated. The observed sample mean is then compared to the value postulated in the null hypothesis to determine whether the difference between the sample mean and the postulated mean is too large to be attributed to chance or sampling variability alone. If there is evidence that the sample could not have come from a population with the postulated mean, we reject the null hypothesis. This occurs when the *p value* of the test (defined as the probability of obtaining a value of the sample mean as far from the postulated mean or even further away given that the null hypothesis is true) is sufficiently small, usually less than 0.05. In this case these data are not compatible with the null hypothesis; they are more supportive of the alternative. Such a test result is said to be *statistically significant*. Note that statistical significance does not imply clinical or scientific significance; the test result could actually have little practical consequence. By using this procedure, we would incorrectly reject a true null hypothesis 5 times out of 100. This probability is called the *significance level* of the hypothesis test. If there is not sufficient evidence to doubt the validity of the null hypothesis, we fail to reject it.

14.4.5   Common Statistical Tests for Comparing Two Groups

Rather than compare the mean of a single population to some postulated value, we often wish to compare the means of two different populations, neither of which is known. Most often we want to know whether the two means are equal to each other; the null hypothesis is that the means are in fact equal, and the alternative hypothesis is that they are not. For instance, we might wish to know whether mean white blood cell count is the same for children with acute Kawasaki syndrome who are treated with intravenous gamma globulin plus aspirin and those who are treated with aspirin alone.[3] We perform a test by selecting a random sample of observations from each population, and use the data in the two samples to determine which of the hypotheses is more likely to be true. If the two population means are equal, we would expect the sample means to be fairly close to each other. We therefore try to determine whether the observed difference in sample means is too large to be attributed to chance alone. When answering this question, the form of the test used depends on the type of data collected — measurements can come from either paired or independent samples.

The defining characteristic of paired data is that for each observation in the first group, there is a corresponding observation in the second group. For instance, right atrial pressure might be measured both before and after surgical intervention. If the mean pressure is identical at the two time points, we would expect the differences in values (measurement at time 2 minus measurement at time 1) to be centered around 0. If the mean difference is far from 0, this suggests that the population means are not in fact the same. The null hypothesis is evaluated using the *paired t test*.[12] Paired designs can be very powerful as they inherently control for extraneous sources of variation.

If the two underlying populations of interest contain unique study subjects and are independent of each other, the *two-sample t test* is used. There are two different forms of the two-sample *t* test; in the first, the standard deviations of the two populations are assumed to be equal to each other, and in the second they are not. Both tests evaluate the null hypothesis that the two population means are equal.

To use either the paired or the two-sample *t* test, the populations from which the samples of data are drawn must be approximately normally distributed. This means that a histogram of the data should display a symmetric, bell-shaped curve. If the populations are skewed, it is more appropriate to compare medians than means. In this case, nonparametric methods are used. *Nonparametric methods* make fewer assumptions about the distributions of the populations studied than the more traditional *t* tests. The general procedure is the same, however. We begin by making some claim about the underlying population in the form of a null hypothesis, then collect a random sample of data from the population and compare what is observed to what would have been expected if the null hypothesis were true, and finally either reject or do not reject the null hypothesis. If data are from paired samples, we use the *Wilcoxon signed-rank test*. If the samples are independent, the *Wilcoxon rank sum test* is used. Nonparametric methods may be applied to ordinal data as well as discrete and continuous data. If the assumptions underlying the *t* test are met, meaning that the underlying population of measurements is normally distributed, then the nonparametric tests are almost as powerful as traditional tests (they would require roughly the same sample size to reject a false null hypothesis); if these data are not normally distributed, the nonparametric tests are more powerful.

For nominal data, the number or proportion of times that a particular outcome occurs is the parameter of interest. Proportions from two independent samples can be compared using the *chi-square test* if the sample sizes are large and *Fisher's exact test* if they are not.[12] The two methods will give very similar *p* values when the sample size is large. When samples are paired, proportions are compared using *McNemar's test*.[12]

If the outcome variable of interest is the time between two events, such as the time from Fontan surgery to failure, defined as death or takedown of the procedure, distributions of time to failure can be compared for two independent samples using the *log-rank test*.[12]

14.4.6 Power and Sample Size

Before beginning any research study, we should have some idea about the number of subjects that will be needed to conduct our primary test of hypothesis.

It is important to note that with any hypothesis test, there are two kinds of errors that can be made. A *type I* error is made if we reject the null hypothesis when it is true. A *type II* error occurs if we fail to reject the null hypothesis when it is false and the alternative hypothesis is true. The *power* of a test is the probability of rejecting the null hypothesis when it is false; in other words, it is the probability of avoiding a type II error. The power may also be thought of as the likelihood that a particular study will detect a deviation from the null hypothesis given that one exists. It is extremely important that a study be designed to have adequate power; it is not enough to know that there is a small probability of rejecting the null hypothesis when it is true, there should also be a high probability of rejecting the null hypothesis when it is false. In most practical applications, a power less than 80% is considered insufficient. The most common way to increase the power of a test is to increase the sample size. In the early stages of planning a research study, investigators should reverse the calculations and determine the sample size that will be necessary to provide a specified power. An example of a sample size calculation is shown in Figure 14.2.

Research question: Does one-year change in full scale IQ differ for patients undergoing surgical closure of an atrial septal defect versus those undergoing device closure?

Outcome: The continuous variable change in full scale IQ, defined as IQ measured one year after defect closure minus IQ prior to procedure

Assumptions:

- Null hypothesis: mean change in IQ for patients in the surgical treatment group equals mean change in IQ for patients in the device group ($\mu_1 = \mu_2$, or $\mu_1 - \mu_2 = 0$)

- Statistical test used: two-sample t test for comparison of means

- Significance level of test: $\alpha = 0.05$

- Power of test: $1 - \beta = 0.80$

- Difference in mean change in IQ to be detected, if null hypothesis is not true: $|\mu_1 - \mu_2| = 5$

- Standard deviation of change in IQ in each treatment group: $\sigma = 6.7$

Calculation:

$$
\begin{aligned}
n &= \frac{(2\sigma^2)(z_{1-\alpha/2} + z_{1-\beta})^2}{(\mu_1 - \mu_2)^2} \\
&= \frac{(2)(6.7^2)(1.96 + 0.84)^2}{5^2} \\
&= 28.2
\end{aligned}
$$

Conclusion: 29 patients would be required in each of the two treatment groups, for a total of 58 patients

Figure 14.2   Sample size calculation for comparison of two means.

14.4.7 Evaluating Relationships among Continuous Measurements

Many clinical studies that measure a large number of continuous variables do not have a clearly defined research question. Instead, they measure the correlation between each pair of variables. *Correlation analysis* can be used to measure the strength of the association between two continuous measures. The goal is simply to establish whether or not a relationship exists, and to give some information about the strength of that relationship. The *Pearson correlation coefficient* is used if both variables involved are approximately normally distributed. If either variable has a skewed distribution or if there are outlying values, the *Spearman rank correlation* can be used instead.[12]

While a correlation coefficient might be an appropriate choice for some descriptive or exploratory studies, in many cases more powerful statistical methods are available. We often wish to go beyond merely determining whether a relationship exists. Like correlation analysis, *simple linear regression* can be used to explore the nature of the relationship between two continuous variables.[12] The main difference is that one variable is designated the outcome and the other the predictor. Correlation does not make this distinction; the two variables are treated symmetrically. Linear regression analysis makes it easier to explore transformations of one or both of the variables, including the use of cutpoints for the predictor variable. It allows us to draw conclusions about the relationship between the outcome and the predictor after controlling or adjusting for the effects of potential confounding variables. A regression model also enables us to predict the mean value of the outcome for a given value of the predictor.

14.4.8 Multivariate Models

In general, multivariate analysis is used to quantify the relationship between an outcome and two or more predictor variables. If the outcome is continuous, *linear* or *nonlinear regression* is used.[12] If it is dichotomous, *logistic regression* is appropriate.[12] *Proportional hazards models* can be used if the outcome is the time from some initial event until a subsequent failure event. Some of the predictor variables of interest might have a real effect on the outcome; others might have no influence at all. Multivariate modeling techniques help us to decide which of the variables are important in explaining the outcome, and to determine the nature and strength of these relationships.

REFERENCES

1. Hulley, S. B., Cummings, S. R., Browner, W. S. et al. (2001). *Designing Clinical Research*, 2nd ed. Lippincott Williams & Wilkins, Philadelphia, PA.
2. Jenkins, K. J., Sanders, S. P., Orav, E. J. et al. (1993). Individual pulmonary vein size and survival in infants with totally anomalous pulmonary venous connection. *J. Am. Coll. Cardiol.* 22, 201–206.

3. Newburger, J. W., Takahashi, M., Burns, J. C. et al. (1986). The treatment of Kawasaki syndrome with intravenous gamma globulin. *N. Engl. J. Med.* 315, 341–347.

4. Tabbutt, S., Ramamoorthy, C., Montenegro, L. et al. (2001). Impact of inspired gas mixtures on preoperative infants with hypoplastic left heart syndrome during controlled ventilation. *Circulation* 104(suppl. I), I-159–I-164.

5. Nadas, A. S., Ellison, R. C., and Weidman, W. H. (1977). Pulmonary stenosis, aortic stenosis, ventricular septal defect: Clinical course and indirect assessment — Report from the Joint Study on the Natural History of Congenital Heart Defects. *Am. Heart Assoc. Monogr.* No. 53.

6. Knott-Craig, C. J., Danielson, G. K., Schaff, H. V. et al. (1995). The modified Fontan operation: an analysis of risk factors for early postoperative death or takedown in 702 consecutive patients from one institution. *J. Thorac. Cardiovasc. Surg.* 109, 1237–1243.

7. Gentles, T. L., Mayer, Jr. J. E., Gauvreau, K. et al. (1997). Fontan operation in five hundred consecutive patients: Factors influencing early and late outcome. *J. Thorac. Cardiovasc. Surg.* 114(3), 376–391.

8. Ferencz, C., Rubin, J. D., Loffredo, C. A., and Magee, C. A. Eds. (1993). *Perspectives in Pediatric Cardiology, Volume 4*, Epidemiology of Congenital Heart Disease, The Baltimore-Washington Infant Study 1981–1989. Futura Publishing Company, Inc., Mount Kisco, NY.

9. Perry, Y. Y., Triedman, J. K., Gauvreau, K. et al. (2000). Sudden death in patients after transcatheter device implantation for congenital heart disease. *Am. J. Cardiol.* 85, 992–995.

10. Jenkins, K. J., Newberger, J. W., Lock, J. E. et al. (1995). In-hospital mortality for surgical repair of congenital heart defects: Preliminary observations of variation by hospital caseload. *Pediatrics* 95(3), 323–330.

11. Satou, G. M., Lacro, R. V., Chung, T. et al. (2001). Heart size on chest x-ray as a predictor of cardiac enlargement by echocardiography in children. *Pediatr. Cardiol.* 22, 218–222.

12. Pagano, M. and Gauvreau, K. *Principles of Biostatistics*, 2nd ed. Duxbury Brooks/Cole, Pacific Grove, CA.

APPENDIX 14-1

FDA Approval[1]

The Medical Device Amendments to the Federal Food, Drug, and Cosmetic Act (FD&C) allowed medical devices to be regulated by the Food and Drug Administration (FDA) in 1976. This authority was expanded with the Safe Medical Device Act of 1990 (SMDA). SMDA requirements are extensive but some examples include the manufacturers and distributors to report malfunction, serious injuries/illnesses and deaths. In addition, methods for tracking must be in place, postmarket surveillance, and FDA recall authority. What constitutes a medical device is discussed elsewhere.[2,3] Anyone who engages in the manufacture, preparation, propagation, compounding, assembly, or processing of a device for human use that meets the definition of a medical device is subject to regulations enforced by the FDA. The Class of the product, I, II or III, governs the level of regulation. Class I devices are subject to the least amount of controls which include manufacturing site registration, device listing, premarket notification, and Good Manufacturing Practices (GMP). Activities requiring registration include repackaging, relabeling, distributing of imported or domestic devices, and specifications development. "GMP regulation covers the methods, facilities, and controls used in preproduction design validation, manufacturing, packaging, storing, and installing medical devices. The GMP regulation identifies the essential elements required of the quality assurance program". The GMP designates two types of devices, critical and noncritical. Class II devices have additional or Special Controls such as labeling and mandatory performance standards. Class III devices are those of most interest with OCT. These devices can not be marketed unless they have one of two approvals, premarket approval (PMA) or 510(k).

The 510(k) or premarket notification basically states that a device is substantially equivalent to a predicate device that is already legally marketed. A 510(k) is also required when a product is modified or changed in a significant way, including its use, which may affect safety or efficacy. The 510(k) applications are generally significantly quicker than the PMA. For PMA, the rules are quite complex, but these are generally devices judged to be significantly dissimilar to all other devices on the market

Under the FD&C Act, manufacturers may receive an exemption to allow the manufacturing of devices intended solely for investigational use on humans. This involves the gathering of safety and effectiveness data for medical devices under the Investigational Device Exemption (IDE). How the IDE is handled is determined by the presence or absence of significant risk. Consult the FDA for more precise definitions but a significant risk device in general is one that:

1. Is implanted and represents a serious potential risk to the well being of the patient.
2. Is designed to support or sustain life and represents a serious potential risk to the well being of the patient.

3. "Is for use of substantial importance in diagnosing, curing, mitigating, or treating disease and represents a serious potential risk to the well being of the patient."
4. Otherwise presents a potential for serious risk to the health, safety, or welfare of the patient

If significant risk is determined, the applicant must submit specific information to the FDA, including that supervision will be performed by the institutional review board (IRB). For nonsignificant risk, IRB approval is required by not necessarily submission to the FDA.

REFERENCES

1. Added by Dr. Brezinski
   For more detailed information, the following WEB pages may be of use:
2. FDA: www.FDA.gov/
3. Devices: www.FDA.GOV/CDRH/DEVADVICE
4. Good Manufacturing Practice: www.fda.gov/cdrh/dsma/cgmphome.html

15 OCT IN CARDIOVASCULAR MEDICINE

Chapter Contents

15.1   GENERAL

As with subsequent chapters, clinical data on the application of OCT to the diagnosis of *in vivo* human pathology are still at an early stage and pioneering work in this area is ongoing. Therefore, the majority of this chapter will focus on the current state of the entire field as a reference for those interested in working in this area rather than just a review exclusively on what has currently been done with OCT.

Cardiovascular disease represents the leading cause of death worldwide and is among the most important applications of OCT. This chapter will deal with the application of OCT to vascular disease, particularly acute coronary syndromes, strokes, atrial fibrillation, and pulmonary hypertension. This chapter will also cover the clinical problems in detail, even when minimal OCT data are available, to provide an in-depth background for those interested in exploring these areas. In particular, the disease process, treatments, competing technologies, and socio-economic factors will be addressed.

15.2   ACUTE CORONARY SYNDROMES

15.2.1   General

Acute coronary syndromes (ACS) include myocardial infarctions (MI) and closely related unstable angina. Among the applications of OCT, the most important is the diagnosis and treatment of ACS and its precursor, coronary artery disease or coronary atherosclerosis (CAD). An MI or "heart attack" can be most easily described as death of heart tissue due to near or complete lack of blood flow to a given region of the heart. In general, myocardial (heart) cells can survive approximately 15 minutes with no blood flow before cells begin to die. Vessel occlusion is typically caused by the presence of atherosclerosis in the coronary arteries, the abnormal accumulation of materials in the heart vessel including lipid, fibrous tissue, calcium, blood, and/or inflammatory cells. How this atherosclerosis or plaque results in vessel occlusion will be discussed below. The dysfunctional myocardial tissue produced by an MI predisposes the patient to poor pumping function (heart failure) or potentially lethal rhythm disturbances, leading to its high mortality. With unstable angina, the vessel is dynamic with clots oscillating in size and at times sufficiently large to cause chest pain. The clot formation mechanism is essentially the same as that for MI and without treatment, the condition has a high likelihood of progressing to MI. Stable angina is generally exertional chest pain that does not worsen significantly over time. As will be noted later, most minimally invasive and invasive therapies are directed at stable angina, primarily because current diagnostics are limited in their ability to predict lesions which progress to ACS.

15.2.2 Epidemiology

In the United States, the prevalence of CAD is approximately 13,000,000 with MI at 7,000,000 in 2002.[1] Deaths from MI (not including heart failure, strokes, or arrhythmias) were approximately 600,000 in the same year, while all cancers together totaled 560,000, accidents 106,000, Alzheimer's 59,000, and HIV 14,000. When considering all cardiovascular diseases (particularly stroke, atrial fibrillation, and heart failure discussed in later sections), mortality is greater than the next five leading causes combined (Figure 15.1). Put another way, more lives will be saved by decreasing the MI mortality by 10% than preventing all breast (48,580 deaths in

Leading Causes of Death for All Males and Females
United States: 2002

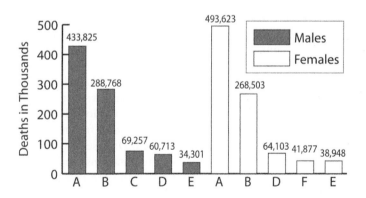

A Total CVD
B Cancer
C Accidents
D Chronic Lower Respiratory Diseases
E Diabetes Mellitus
F Alzheimer's Disease

Figure 15.1 Leading causes of death for all males and females (2002). Courtesy of American Heart Association (www.americanheart.org).

Deaths From Diseases of the Heart
United States: 1900–2002

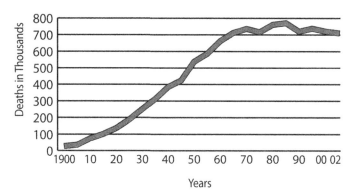

Figure 15.2   Deaths from heart diseases in the United States. (1900-2002). Courtesy of American Heart Association (www.american heart.org).

2004) or colorectal (56,730 deaths in 2004) cancer deaths. The probability at birth of eventually dying of cardiovascular disease is 47% compared with 22% for cancer.

Approximately one-third of all heart attacks present as sudden death. Of these, 50% of men and 64% of women who died suddenly of coronary heart disease (CHD) had no previous symptoms (most die outside the hospital). Within 6 years after a recognized MI, 7% of the patients will experience sudden death, and 22% of men and 46% of women develop disabling heart failure.[2] In 2005, the estimated direct and indirect cost of CHD was $142 billion.[1,3,4]

Cardiovascular disease has been the leading cause of death in the United States since 1918 (Figure 15.2). Contrary to popular belief, it is not only the leading cause of death in men, but also in women, and is on the rise in the latter (Figures 15.1 and 15.3). However, women tend to develop the disease at a slightly older age, approximately 8 years later than men, and frequently present with less predictable symptoms compared to those of men. Disease in women is more difficult to manage because of these often atypically presentations, and because it is more challenging to treat both because of the later onset and the fact that vessels tend to be smaller on average.

Cardiovascular disease is a world wide problem. Heart disease and stroke is expected to be the leading cause of death worldwide by 2020. The age-adjusted death rates vary worldwide with the highest rates in Eastern Europe (Russia 400 per 100,000 population) compared with Japan at 60 per 100,000.[5] According to the World Health Organization (WHO), cardiovascular disease represents one-third of all deaths globally. Furthermore, according to the WHO, in both developed and developing countries, 40 to 75% of all heart attack victims die before reaching the hospital. Cardiovascular disease is now more prevalent in India and China than

**Prevalence of Cardiovascular Diseases in Americans
Age 20 and Older by Age and Sex**
NHANES: 1999–2002

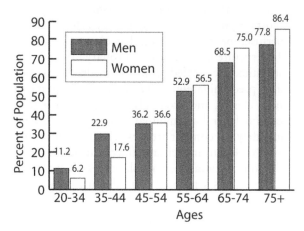

Figure 15.3 Cardiovascular disease by age in the United States.
Courtesy of American Heart Association (www.americanheart.org).

in all economically developed countries in the world combined. Cardiovascular disease is the leading cause of death in Europe, with over 4 million deaths per year (British Heart Foundation, European Cardiovascular Disease Statistics, 2000 edition). It is also on the rise in South America and the Middle East.

The major risk factors are increased blood pressure, nicotine abuse, diabetes, lipid levels, obesity, excessive alcohol consumption, a lack of physical activity, low daily fruit and vegetable consumption, family history, and psychosocial index.[6] However, a significant portion of heart attacks are in patients with none of these risk factors. Other potential risk factors include genetic abnormalities, high C-reactive protein levels, increased homocysteine levels, high fibrinogen levels, and the presence of certain infectious agents (e.g., Chlamydia pneumonia, *Helicobacter pylori*, and cytomegalovirus).[7]

15.2.3 Coronary Circulation

The circulation of the heart is through three major coronary arteries. These are the left anterior descending (LAD), circumflex, and right coronary artery (RCA). The LAD and circumflex come off a common but important conduit artery, the left main. Significant obstructions of the left main jeopardize such a large region of myocardium that it is extremely rare that it is treated through an intravascular approach. The left main and RCA come directly off the aorta and their orifices are just above the aortic valve. The LAD perfuses the major portion of the left ventricle while the RCA provides blood to the right ventricle and typically part of the posterior left. The circumflex perfuses the remainder of the left ventricle. Atherosclerotic disease in different arteries or different segments of the same artery can have different prognostic significance. The coronary arteries are unusual for arteries because they are perfused primarily during diastole and at a relatively lower pressure.

The structure of the normal coronary artery is shown in Figure 15.4. It consists of a thin intima, separated from the media by an internal elastic lamina. The media consists predominately of layers of smooth muscles that are intermingled with collagen, elastin, and proteoglycans. In elastic arteries such as the aorta and carotid, the central layer has minimal smooth muscle and high concentrations of elastin because they are conduit arteries. The outer adventitia of the coronary artery consists of loose collagen and elastin fibers, in addition to adipose tissue and fibroblasts. In addition, it contains small blood vessels (vasa vasorum), lymphatics, and nerves. The adventitia is separated from the media by the external elastic lamina.

15.2.4   Classification of Plaque

Atherosclerotic plaques have different compositions, with some more significantly predisposed to rupture than others. The major plaques are often classified using the American Heart Association Classification System, but a further subclassification will be discussed which focuses on vulnerable plaque.[8]

Type I through III lesions are very similar to each other and generally are not predisposed to rupture and the production of heart attacks. Type I lesions contain microscopic collections of lipid within their intima. These lesions are more frequent in infants and children. The lipid is primarily in the form of lipid-laden macrophages. The lipid tends to accumulate in regions of eccentric intimal thickening. It is clear that, while MI occurs at later ages, the process of atherosclerosis begins in the teens or earlier. A type II lesion is more organized in structure. It is predominately layers of stratified foam cells. Although most of the lipid is present in macrophages, it can also be found in smooth muscle cells and the extracellular space. A type III lesion is often referred to as a preatheroma. The lesions are characterized by microscopically significant collections of lipid and particles between increased layers of smooth muscle cells, typically with eccentric intimal hyperplasia (increased width of the intima).

Type IV lesions are known as atheromas and consist of a dense concentration of lipid in a *well-defined* region or lipid core. The lipid core may not result in the vessel impinging on the lumen since the vessel can remodel or push outward rather than inward. Therefore, lumen size can appear normal on an angiogram but the plaque size may be significantly large. The intima that lies between the lipid core and the lumen contains macrophages, collagen, and smooth muscle cells of varying concentrations potentially with interspersed lipid. The periphery of these lesions are particularly prone to rupture.

Type V lesions are classified as type Va (fibroatheroma), type Vb, and type Vc. Type V lesions are characterized by the presence of more fibrotic tissue than type IV.

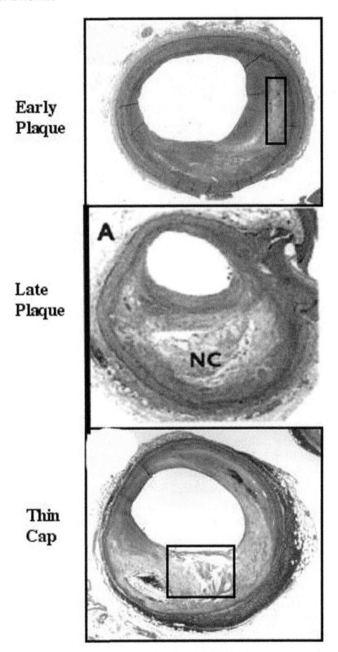

Figure 15.4   'Vulnerable Plaque' classified by in terms of thin capped fibroatheroma (TFCA). Squares demarcate diseased area and NC designated necrotic core. Courtesy Kolodgie, F.D., et al. (2003). *N.E.J.M.* 349, 2316-2325. [25] (modified)

In type Va, there is still a lipid core and the plaque now has developed one or more fibrous layer(s). In type Vb the lipid core now contains calcification. In type Vc, minimal lipid is present, including the lack of a defined lipid core. Type IV and V lesions are associated with the most significant acute coronary events such as MI, for reasons described below. When the surface of type IV and V lesions develop surface disruption, hematoma, and/or thrombus formation, they become type VI lesions. Again, the process of atherosclerosis plaque development begins relatively early in life. In one study 2% of 15 to 19 year olds and 20% of 30- to 34-year-old men had grade IV or V lesions.[9] For women of the same age group, the prevalence was 0% and 8%, respectively, consistent with the delayed development in women described above.

The American Heart Association classification of plaque uses broad criteria and has recently been further subclassified by Dr. Virmani and colleagues[10] based on the likelihood of rupture. The classifications specifically focus on plaques which lead to ACS, which will be the classification used for the remainder of this text. These are pathologic intimal thickening, early necrotic core, late necrotic core, and the thin cap fibroatheroma (TCFA), as shown in Figure 15.4. TCFAs are the most likely to progress to MI and these plaques are characterized by a thin fibrous cap (< 65 μm thickness), are infiltrated by macrophages and lymphocytes, have new microvessel growth, and have few supportive smooth muscle fibers and collagen. The cap usually overlays a hemorrhagic, necrotic core with substantial penetration of angiogenic blood vessels into the intima from the adventitia.

15.2.5   Unstable Plaque Histopathology

There are three major mechanisms that appear to lead to the vast majority of MIs. These are plaque rupture, plaque erosion, and near occlusive/calcified plaque.[11,12] Up until the 1980s, it was assumed that most heart attacks resulted from large plaque that, prior to the event, nearly occluded the vessel. It is now known that these near-occlusive plaques produce only a small portion of MIs. The most common microstructure for plaques resulting in MI are lipid-filled plaques (TFCA) that do not substantially impinge on the vessel lumen prior to rupture. These plaques contain a large lipid core. When rupture occurs blood is exposed to the vessel lumen and a clot forms. If the clot is sufficiently large, it reduces or eliminates blood flow to regions of the heart under the distribution of that vessel.

If the break in the cap or core is small, or local antithrombogenic factors are strong, the plaque may only marginally expand in size and therefore may not be symptomatic.[13] The plaque then becomes organized and endothelialized. This is the mechanism by which atherosclerosis is believed to progress. If the plaque rupture is large, then red cells, fibrin, and platelets fill the plaque core and lumen. Total occlusion is ultimately a race between the ability of thrombogenic factors to form the clot/expanding plaque against the antithrombogenic factors which break down the clot. The constituents may only partially occlude the lumen and become endothelialized, which results in an atherosclerotic plaque that, if hemodynamically compromising, leads to subsequent stable angina.

TCFAs are difficult to detect by conventional radiological techniques and, even when they are seen, their infrastructure, which determines their degree of instability, cannot be analyzed.[14,15] Furthermore, the large, firm near-occlusive plaques that are seen on angiography and produce the symptoms of stable angina serve predominately as markers (patients with large plaques are also likely to have plaque ruptures), rather than initiators of future acute syndromes. Even when these large plaques evolve to total occlusion, myocardial damage may be minimal due to the presence of collateral circulation that has developed.[16]

While these ruptured plaques are the most prominent mechanism for causing MIs, plaque erosion is another source. Plaque erosion occurs where the vessel has endothelial denudation and surface erosion, particularly on plaques that protrude into the lumen. Superficial thrombosis results as blood is exposed to tissue factor and collagen on the exposed surface. These are large scale regions generally associated with inflammation, including an accumulation of lipid-filled macrophages and T lymphocytes. Plaque rupture has been shown to be present in a ratio of 3:1 over surface erosion overall.[11] However, one study suggests plaque erosion tends to occur at a higher frequency in younger patients, particularly women, and is associated with higher incidence of sudden death.[12]

Thrombosis at the site of a hemodynamically significant lesion (large plaque), as stated above, was believed in earlier decades to be the primary mechanism for MI. However, it is now known that while this event does represent a small but significant portion of MIs, particularly highly calcified regions, plaque rupture and erosion remain the dominant events.

Remodeling does not directly lead to MI, but is a phenomenon indirectly related which is central to the development of atherosclerosis.[17] Positive remodeling is when a plaque grows outward as it develops rather than into the lumen of a vessel. It has been recognized that some plaques show no changes in luminal diameter initially as the plaque grows because the vessel is remodeling outward. Positive remodeling can be defined as a compensatory increase in vessel size in response to plaque burden. It is believed that this response is limited and eventually the plaque will encroach on the lumen of the blood vessel. Negative remodeling has also been described.[18] Paradoxical wall shrinkage has been described during the formation of atherosclerosis that cannot be attributed to plaque burden alone. There is some evidence that vessels that undergo positive remodeling are far more likely to be associated with vulnerable plaque.[19,20] This may explain why vulnerable plaques are more frequently seen in regions with only minor angiographic stenosis. It has been suggested that an increased wall stress can lead to vessel enlargement and an increased susceptibility to vessel constriction, increased shear stress, or myocardial contraction.

15.2.6 Triggers of Plaque Rupture

Identifying these high-risk plaques (prior to rupture) to improve patient risk stratification is particularly important in view of the high percentage of occlusions resulting in sudden death (approximately one-third).[1] However, to date no modality, including postmortem histopathology, has provided the level of resolution necessary to diagnose high-risk plaques prior to rupture.

The reason is that most ACS result from TCFA, but most TCFA do not lead to ACS. Therefore, since even pathology has not established which features indicate a TCFA will lead to ACS, then improved patient risk stratification techniques are needed to subclassify TCFA.[21–24] In one study, in up to 8% of patients dying of *non-cardiovascular* causes, ruptured plaques were noted in the coronary arteries.[21] In another intravascular ultrasound (IVUS) study, plaque rupture of target lesions occurred in 80 aqcute myocardial infarction (AMI) patients (66%) and in 31 stable angia pectoris (SAP) patients (27%) ($P < 0.001$).[22–23] Non-infarct-related or non-target artery plaque ruptures occurred in 21 AMI patients (17%) and 6 SAP patients (5%) ($P = 0.008$). Multiple plaque ruptures were observed in 24 AMI patients (20%) and 7 SAP patients (31%) ($P = 0.004$). As we will see, while OCT can identify plaques with intimal caps less than 70 μm (i.e., TFCA) and small lipid collections, which represents a substantial advance over other imaging modalities, a need exists for OCT advances to further risk-stratify these plaques beyond just the identification of TCFA. Therefore, when OCT imaging of plaque is discussed, identifying potential risk factors within TFCA will be a focus. It should also be pointed out that the vulnerable plaque referred to in the literature does not represent a specific known plaque histopathology but a clinical scenario. This is why the term vulnerable patient is frequently used, which emphasizes that most information is obtained post rupture.

The presumed general determinates of a plaque's vulnerability to rupture are the size/composition of the atheromatous hemorrhagic core (when the core is more than 40% of the plaque volume), the thickness of the fibrous cap covering the core, the collagen content/type of the cap, angiogenesis into the intima, and degree of inflammation within the cap.

The Virmani group has defined plaque vulnerability based on the actual cap thickness of fixed autopsy specimens. They found that in rupture, mean cap thickness was 23 ± 19 μm; 95% of the caps measured less than 64 μm within a limit of only two standard deviations.[26] Computer modeling, when the cap thickness is less than 100 μm, has also demonstrated that since the lipid core behaves as a mechanically soft medium, circumferential stress across the plaque becomes uneven during systole.[27] The stress on the soft core is redistributed onto the periphery of the cap, which is where thinning occurs. Since the current clinical technology with the highest resolution is high frequency ultrasound (40 MHz), which corresponds to an axial resolution of 80–90 μm or more, this suggests that these lesions are below the detection limit of current clinical imaging technologies. OCT has a greater than 20 μm resolution, which gives it the ability to measure thin cap thickness.

In addition to the thickness of the cap, the concentration of collagen type I in the cap determines its structural stability. Collagen depletion occurs in TFCA due to loss of smooth muscle, which synthesizes collagen, and the presence of protease from inflammatory cells. This renders the plaque wall mechanically weak. The ability of PS-OCT to assess collagen concentrations in caps could be a power tool in risk-stratifying TFCA that will be discussed later.[28,29]

Inflammatory cells, particularly macrophages, play a critical role in the development and progression of atherosclerosis.[30,31] These cells release metalloproteases, as well as lesser factors. This appears to be a central mechanism in the breakdown of the collagen matrix. The release of thrombogenic factors from local inflammation increases the likelihood of clot formation. While inflammatory cells play a pivotal role in the development of TCFA, their contribution to ACS is less clear. In patients with ACS, inflammation appears to be extensive throughout the arterial circulation, including both culprit and non-culprit regions.[32] This concept of ACS as a systemic inflammatory state is exemplified by elevations of markers such as the WBC counts or C-reactive protein. Yet, because in 50% of patients with MI the precipitating factors are acute (such as exercise and stress), it is difficult to envision inflammatory cells as a rupture trigger (although critical to progression of TFCA).[33] Studies demonstrating high concentrations of macrophages in ruptured TCFA have conducted these measurements only after ACS develops. It is, therefore, unclear whether increased concentrations of macrophages contribute to plaque rupture or migrate into the region (and the whole vascular system) subsequent to the rupture. If OCT is able to identify macrophage concentrations in plaques prior to rupture, it may be helpful in determining the precise role macrophages play in ACS.

In a large postmortem study, Geiringer, using 300 aorta and 100 coronary arteries, established that the normal human intima is devoid of blood vessels, which are limited to the adventitia.[34] Angiogenesis does not develop into the intima until the artery exceeds a critical thickness. Immunohistochemical studies of coronary atherectomy specimens, along with autopsy data, have provided evidence of an association between the presence of microvessels in the coronary atherosclerotic lesions and the development of unstable angina and MI.[23,35] Intraplaque rupture is presumed to occur from discontinuous endothelium of the microvessels without supporting smooth muscle cells. These new blood vessels are inherently weak. It is, therefore, postulated that they result in intraplaque hemorrhage, a sudden increase in plaque volume, and the development of plaque instability. Besides leading to plaque instability through rupture, some have suggested that red blood cell membranes may be a major (if not the largest) source of TFCA cholesterol.[25,36,37] Future studies should measure intimal angiogenesis with Doppler OCT or hemoglobin with absorption spectroscopy.

15.2.7 Interventions

Most invasive and minimally invasive therapies are directed at treating stable angina as treatments directed at preventing ACS are limited to some coronary artery bypass grafting (CABG) procedures and pharmaceuticals at this time. The major approaches to the treatment of CAD include CABG, percutaneous catheter-based coronary interventions, brachytherapy, and pharmaceuticals. It is important to discuss these therapies since OCT imaging could affect their application, cost, or effectiveness. It should be noted that for the last ten years, dramatic changes have occurred with these therapies and the field still remains very active. By the time this text is published, substantial changes in the approach to the patient with CAD may exist. Pharmaceutical therapy of ACS is an extensive topic on its own and will not be discussed here in detail, but it can be found in many texts and reviews on the topic.

A CABG is an invasive procedure that typically involves intubation of the patient, exposure of the heart by opening of the chest, perfusing the body with blood pumped by a bypass machine, and transposing/suturing vessels from parts of the body across coronary artery obstructions to improve flow. In addition to its invasiveness, CABG remains a very costly procedure. In the United States in 2003, an estimated 515,000 patients underwent CABG at an estimated average cost of $60,853 per procedure.[38] An imaging technology, such as OCT, that could potentially reduce the number of CABGs, could have a significant impact on healthcare costs.

There are a variety of catheter-based interventions that can be applied to coronary atherosclerosis. The three most common techniques are percutaneous transluminal coronary angioplasty (PTCA), coronary atherectomy, and stenting. There were approximately 1.5 million coronary angiograms performed in the United States in 2002, each at a cost of $18,000.[38] Approximately 660,000 coronary interventions were performed in the United States in the same year at a cost of $29,000 per procedure (www.hcup.ahrq.gov). Typically, more than one catheter (which are costly) is required during each procedure, such as an angioplasty balloon and stent catheter. A high resolution imaging technology, in either a catheter or guidewire design, could not only improve the efficacy of the procedure but also reduce cost by decreasing the number of devices required. Therefore, the development of an OCT guidewire could be a substantial advance in the field.

Lone PTCA is a procedure that has recently fallen out of favor, but its use may increase with the appropriate intravascular guidance. PTCA was pioneered in the 1960s and 1970s through the efforts of investigators including Gruentizg, Dotter, and Judkins.[39,40] It has been suggested that there are at least five possible mechanisms that cause this relatively simple procedure to demonstrate some efficacy. The PTCA catheter contains a balloon near its tip that, when inflated, works through either plaque compression, plaque fracture, medial dissection, plaque stretching, and/or stretching of the plaque free region.

With PTCA, the patient receives antiplatelet therapy. Typically the femoral region is anesthetized and a guide catheter is introduced up the aorta into the coronaries via a guidewire under fluoroscopic guidance. The patient is anti-coagulated and the PTCA catheter is introduced into the coronary artery via the guidewire through the guide catheter. This guidewire would already have been advanced across the lesion of interest. The balloon is inflated, typically at 4 to 8 atmospheres, under fluoroscopic guidance. The balloon inflation time varies from 30 seconds to 3 minutes, although in some instances, this time may be longer. Unless further inflations are needed, the balloon is removed and the patient is monitored for complications.

While PTCA is relatively inexpensive, several factors have led to dramatic reductions in its use. The two main factors are abrupt vessel closure and restenosis. With abrupt closure, sudden occlusion of the vessel occurs at or near the site of the intervention. It likely results from the extensive damage that occurs during the angioplasty such as fissuring into the media. The incidence is reported to be between 4.2 and 8.3%.[41] When this occurs outside the setting of the catheterization laboratory, an MI may result and emergent catheterization or CABG ensue.

Restenosis, a complication of PTCA, is the gradual occlusion of the vessel over a period of 6–12 months.[42] The patients usually present with progressive stable angina and only rarely do they present with MI. Restenosis rates are 30–60% and, as with other interventions, the most important predictor of restenosis is final luminal diameter. In addition, important predictors of success include the extent of plaque burden, ostial lesions, and diabetes. There are several likely etiologies for the existence of restenosis. Part of the loss of luminal dimension represents elastic recoil, which occurs very quickly after the procedure. Another source of late loss is intimal hyperplasia, which will be discussed in greater detail in the section about stents. Finally, some vessel spasm may exist. These complications have largely led to the replacement of PTCA by stenting.

Atherectomy is the catheter-based, physical removal of plaque from the site that could also potentially be guided by intracoronary imaging. Coronary atherectomy can be divided into two main classes of devices: directional coronary atherectomy (DCA) and rotational atherectomy (rotoblade). The classic DCA catheter contains a metal cylinder at its tip and an open window. Within the cylinder is a cup-shaped blade. On the side of the catheter opposite the window there is a balloon. To use the catheter, it is introduced into the coronary artery in a manner similar to the PTCA catheter. The window of the catheter is placed over the plaque of interest under fluoroscopic guidance. The balloon is then inflated to press the window on the plaque. The blade within the catheter is then pushed forward, cutting through the plaque and pushing the tissue forward into the hollow nosecone-collecting chamber of the catheter. After multiple cuts, sometimes at higher balloon pressures, the catheter is removed.

DCA probably works through two mechanisms. The first is the physical removal of tissue. The second, which may be the major source of improvement, is due to dilation by the balloon. However, DCA is plagued with similar restenosis and acute occlusion rates as PTCA, so its use has been significantly reduced, but there are clear clinical indications where DCA could benefit from intracoronary guidance.

With rotational atherectomy, the tip of the catheter now contains a head whose surface is coated with 10- to 40-μm diamond chips. The head rapidly rotates at approximately 200,000 rpm, which grinds the surface of the plaque. The rotoblade produces particles approximately 10 μm in diameter that embolize downstream and, at times, may lead to complications. The major advantage of rotational atherectomy is that it selectively removes calcified sections without removing normal tissue. However, while it does have a limited role, the relatively high restenosis rate and high incidence of complication, possibly by embolization, make it suboptimal for widespread use. The mechanisms of restenosis are similar to PTCA. However, since plaque burden is a predictor of restenosis, aggressive plaque debridement under intracoronary guidance prior to stent placement can result in a reduction in restenosis rates.

The use of conventional intracoronary stents has increased dramatically over the last 8–10 years, and it can safely be stated that it is the intracorornary procedure of choice to treat most symptomatic atherosclerotic coronary lesions.[43,44] These stents are metal scaffoldings that are expanded in the artery, which prevent the artery from recoiling. There are a variety of stent designs, but they are basically introduced the same way. The stents are placed on a balloon catheter and when in the appropriate site, the balloon is expanded to deploy the stent. The stent prevents elastic recoil via its mechanical support, but its initial restenosis rates were not dramatically better than angioplasty.

Because elastic recoil is essentially eliminated with most stents for mechanical reasons, intimal hyperplasia remains of concern with conventional stents. It has been well established that the amount of intimal hyperplasia is independent of vessel size, but the degree of hemodynamic compromise does depend on vessel size.[45–47] The same amount of intimal hyperplasia in a small vessel will result in more symptoms post procedure. Therefore, strategies that increase luminal diameter will, on average, reduce total vessel stenosis. The mechanism by which intimal hyperplasia occurs has been described as an inflammatory process resulting in prolonged cellular proliferation, monocytic invasions, and, ultimately, smooth muscle migration and collagen deposition. The process appears more severe in areas where the stent struts are imbedded in the lipid core and are in contact with damaged media compared with areas adjacent to fibrous plaque.

The use of high compression balloons has clearly reduced the restenosis rate by expanding vessels to larger diameters.[48] In one study, the cross-sectional diameter for the low (6 bars) and high pressure (up to 20 μm) were 2.55 ± 0.41 mm and 3.14 ± 0.37 mm, respectively. High frequency ultrasound has also been shown to reduce the restenosis rate as angiography is not as accurate in determining stent apposition to the wall.[49] High frequency ultrasound guidance has also been shown to reduce the need for aggressive anticoagulation in some patient populations.[50] Since plaque burden is linked to restenosis rate, applying an atherectomy catheter prior to stent placement could reduce the restenosis rate. The incidence of acute occlusion, now that anticoagulation schemes have been better defined, is lower than other interventional procedures. Even fissures noted at the edge of the stent have not been shown to increase the likelihood of a significant coronary event and, with the exception of brachytherapy (discussed below), are often repaired by 6 months.[51]

Pharmacologic agents, particularly antithrombotic agents, have had little impact in preventing the development of neointimal hyperplasia. In addition, their side effect profile includes thrombocytopenia, bleeding, neutropenia, dizziness, tremor, seizure, nausea, diarrhea, and nephrotoxicity. Therefore, ideally the drug delivery at the site of the lesion would lower the systemic drug distribution and have a better side effect profile.

Recently, drug eluting stents have been introduced to reduce the restenosis rates and lower the side effect profile.[52] These drug-coated stents generally consist of three components, the drug, the polymer coating, and the stent. The drugs often used are anticancer agents, immunosuppressive drugs, or organ anti-rejection agents.

The goal is to prevent smooth muscle cell proliferation, inhibit their migration, suppress inflammation/platelet activation, or promote wound healing. There are several drugs that have received considerable attention. Sirolimus has been used in the Ravel and Sirius trials.[53–55] Sirolimus (rapamycin, Rapamune) is an antimicrobial and an immunosuppressant agent (preventing inflammation). It causes the cell cycle to arrest at G_1/S transition. Early clinical studies have largely dealt with "ideal" arteries. For example, in the Ravel study, patients who demonstrated evidence of thrombi had ejection fractions <30%, and had ostial lesions, or LAD lesions not protected by bypass were studied. In this study, a 26% restenosis rate was seen in the control arm but no restenosis was seen in the drug-eluting stent group.

Paclitaxel is a drug used in the treatment of breast cancer. It inhibits smooth muscle cell proliferation. It works by acting on the microtubules of smooth muscle cells to prevent their migration. The early clinical trials with paclitaxel-eluting stents have included the TAXUS, ASPECT, ELUTES, and DELIVERY I studies.[56–58] Again, in the TAXUS study, at 12 months, major cardiac events were 3% in the treated group versus 10% in the control group. The ASPECT study showed an increase in subacute thrombosis. The DELIVER I study failed to show a significant reduction in target lesion revascularization. The long-term effects of the compounds used with drug-eluting stents is still under investigation. The first study examining real-world complex lesions reported a stent restenosis rate of 15%.[59]

A polymer coating is typically required because most drugs do not adhere to the stents. Concerns exist regarding increasing the inflammatory reaction and/or enhanced neointimal proliferation, both for biodegradable or non-biodegradable stents. The polymer can also be damaged by calcifications or overlapping stents. More promising coating agents are under investigation.

With regard to the stent itself, issues of concern include insufficient coverage, unequal drug delivery due to stent placement, and whether overlapping stents result in local toxicity. In addition, late stent malapposition (not abutting the wall) is an additional concern and is likely due to the failure of normal healing in the presence of traditional stents. The implications of these issues are the subject of ongoing research. Additionally, the importance of intravascular imaging in assuring approximation, stent position, and identification of sites for stent placement also remains a source of future study.

Brachytherapy has recently lost some popularity, but it is discussed here because OCT may be of some value in its successful application.[60] As stated, restenosis is a significant problem for catheter-based interventional procedures and is distinct from the process of atherosclerosis. Smooth muscle cells play an important role in this process. It is suggested that the source of the constituent material which leads to restenosis is proliferating smooth muscle cells, probably from the adventitia, that migrate into the neointima as described above. If these cells can be prevented from reproducing, from radiation-induced chromosomal damage with subsequent apoptosis, then in theory restenosis would be reduced. This is the theory behind the use of radiation to reduce restenosis or brachytherapy. For highly differentiated cells that do not divide, a very high dose of radiation is required to kill the cells (approximately 100 Gy).[61] However, for dividing cells, significantly lower doses can be used. Below, brachytherapy will be discussed as well as how OCT can be applied to the therapy.

Two types of radiation have been used to treat restenosis: beta (β) and gamma (γ) radiation. Gamma radiation is a high energy form of electromagnetic radiation. Beta radiation represents high energy electrons. When these high energy electrons are slowed down, particularly when colliding with nuclei, X-rays are emitted. Beta radiation operates over short distances while gamma radiation operates over much larger distances. The implications of this are discussed next.

In one study looking at stent implantation in 55 patients, the late luminal loss was significantly lower in the iridium-192 group (gamma radiation producer) than in the placebo group (0.38 ± 1.06 mm vs. 1.03 ± 0.97 mm).[57] Angiographically determined restenosis was found in 17% of the treated group and 54% of the placebo group. They reduced clinical events, specifically the frequency of revascularization of the target lesion, by 73%. In another smaller study of 23 patients undergoing PTCA using ^{90}Sr/Y (beta emitter), the restenosis rate at 6 months follow-up was 15%, significantly less than that for typical PTCA procedures.[62] In a larger more recent study of 130 consecutive patients undergoing stenting for in-stent restenosis, major cardiac events were reduced in the treated group, 29.9% versus 67.7%.[63] The incidence of angiographic restenosis was also reduced in these patients, 19% versus 58%.

There are several potentially serious complications to brachytherapy that may affect its application including the potential for aneurysms, high late occlusion rates, accelerated atherosclerosis (candy wrappers), and the potential for oncogenesis (cancer generation).[64] Because the radiation is changing the tissue composition of the artery and healing is impaired, there is the potential for aneurysm. In one study, 7 out of 8 patients treated with 25 Gy developed aneurysm-like features and one artery developed an aneurysm.[65] However, there is some question because some of the patients had received higher doses on the order of 50 Gy.[66] Although this has not been seen in other studies, it looms as a potential long-term complication. In a small number of patients treated with brachytherapy small fractures/fissures within the artery fail to heal.[67] There is little evidence this represents a serious risk, but one study does provide a concern. In this study, brachytherapy was associated with a 6.6% sudden occlusion rate 2 to 15 months after either angioplasty or stent placement, which could be related to the presence of fissures.[68] This is much higher than current rates with stenting and PTCA that are < 1.5%. If sudden occlusion rates are truly this high, then brachytherapy's relative value may be called into question. It has been shown that, in patients receiving radiation for other reasons, high dose radiation exposure has been shown to accelerate atherosclerosis. However, the relatively low doses of radiation used for brachytherapy are not likely to result in a risk higher than the underlying disease. The end of stent or candy-wrapper intimal hyperplasia seen may result from low dose radiation inducing proliferation at the periphery of the stent. Finally, how beta radiation effects sensitive tissue in the vicinity of the coronary artery such as sternal bone marrow and lymph nodes also needs to be evaluated.

If brachytherapy is to be utilized, several questions need to be addressed to allow widespread use of this technique. These include which emitter should be used, how to center the catheter, what is the optimal dose of radiation, and how to control exposure to catheterization laboratory personnel. Currently, whether or not beta or gamma emitters are better suited has not been established and neither are ideal emitters. Beta emitters operate over a very short distance; centering is therefore very critical. If the device is not centered, the artery will not have a relatively homogeneous exposure. Furthermore, if vessel size is large, sufficient penetration may not be achieved. Gamma emitters operate over a much larger distance so that centering is not as critical (energy fall off is slower). However, this means radiation penetrates deeper into other tissue, exposing that tissue, and more precautions need to be taken in the catheterization lab.

As stated, centering is important within the artery. Centering is particularly a problem in the right coronary that has rapid bends, making catheters ride along the wall rather than the center. Balloons on the catheter can be used in an attempt to center the device, but in general, the vessels are eccentric, making it much harder to implement. It is possible that intravascular image guidance, possibly with an OCT guidewire, is a potential solution to the problem.

The radiation-dosing regime has also not been standardized, with current doses ranging from 8 to 25 Gy.[64] If the dose is too low, it has been shown that it can actually increase proliferation. If the dose is to high, complications like aneurysms may become a problem.

With respect to exposure of hospital personnel, the use of radiation increases exposure above standard angiography and is worse with gamma emitters, which have deeper penetration. It has been estimated that the cardiologist will receive a total of four times the dose over a standard balloon angioplasty procedure.

OCT imaging, particularly delivered in the form of a guidewire, could again help eliminate some of the problems associated with brachytherapy. The real-time imaging of the technology could be used to both center the catheter, to assess plaque morphology, and evaluate stent placement. Still, the question of the ultimate utility of brachytherapy remains an issue.

Laser atherectomy has recently fallen out of favor. For those interested, references are provided.[69,70]

15.2.8 Imaging Modalities

Imaging modalities currently used to assess the coronary arteries include angiography, MRI, IVUS, and CT. Nuclear techniques such as SPECT and PET are used for assessing coronary flow, but would have limited utility for assessing vulnerable plaque or the local area after coronary intervention.[71]

Coronary angiography, which was introduced in 1959, is the fluoroscopic (X-ray) imaging of coronary arteries in the presence of IV contrast.[72–74] The contrast is generally injected into the coronary artery via catheters placed in either the left main or RCA (or arterial or venous bypass grafts). Angiography has been the gold standard for assessing the coronary arteries for decades and continues to be the most widely used imaging modality. It allows the identification of relatively large plaque and is useful for guiding current coronary interventions. However, it is a technology that lacks the resolution ($>500 \mu$m) necessary to identify the small plaques (vulnerable or unstable plaques). Recently, quantitative techniques have been developed, but performance remains less than that of technologies such as IVUS.

A detailed discussion of the properties and theory of MRI are discussed in Chapter 13. Magnetic resonance imaging (MRI) has recently been examined as a tool for assessing the coronary arteries.[75–77] However, it is limited by a lack of resolution, organ movement, and acquisition rate. Most MRI techniques have a resolution worse than 500μm. From discussions above, this makes it unlikely as a technique for assessing coronary vulnerable plaque although it has demonstrated potential for assessing the degree of hemodynamically significant CAD (similar to angiography). There are some systems used in small animals that achieve resolutions below 20μm, but they can only image small volumes and require high magnetic strength, which makes them impractical for coronary imaging.

MRI of the coronary arteries is particularly difficult due to the motion of the heart both from its rhythmic motion and lung displacement. In addition, the phrase "MRI" is used to describe a wide range of techniques, which include varying magnetic strengths and pulse sequences. Many of these techniques are not available to most institutions or cannot be used simultaneously.

To some degree motion can be controlled with EKG gating, diaphragm gating, and breath holding. The time window from the EKG is generally obtained in mid to late diastole, which lasts approximately 100 to 150 msec. Breath holding is usually possible for most patients for 20 seconds so that 20 heart beats can be obtained at a heart rate of 60 beats per minute. However, a significant portion of this patient population may not be able to hold their breath this long and have high heart rates and/or rapid respiratory rates. It should also be noted that fat suppression could be required since fat, which is in close approximation to the vessel, has a very strong signal making imaging of the coronaries difficult.

MRI techniques tried have included spin-echo (SE), gradient-echo, and spectroscopic imaging. With SE imaging, the pulse sequence is set up in such a way so that when the final pulse is performed, the blood is moved out of the field. Therefore, blood and the vessel appear black. With this technique 256 lines and 256 heart beats take approximately 3 minutes, allowing only proximal coronary arteries and bypass grafts to be evaluated.[78,79] Therefore, it currently cannot be useful for the general identification of atherosclerosis (replacing angiography) and is even less effective for the identification of high risk plaques.

Two-dimensional gradient-echo has also been used to perform imaging of the coronary artery, where blood appears bright.[80] An image of 128 lines can usually be obtained in one breath hold. Blood appears bright in these images. Turbulent blood flow will be less bright or even black in the image. This is one of the most important ways that distinct stenosis can be detected. However, inconsistent breath holding, which is the ultimate limit in the resolution, can induce misregistration artifacts that can appear as a false positive for occlusion or stenosis.

Three-dimensional, gradient-echo imaging can also be used with respiratory gating.[81,82] An MR navigator is used to monitor the diaphragm. This technique has relatively high resolution on the order of 1–2 mm. However, acquisition rates on the order of 30–50 minutes are required.

A newer technique that may be useful for imaging the coronary artery is echo-planar imaging.[83–85] With this technique, ultra-fast gradient switches are used to give multiple readouts of the echo during a single RF pulse. The speed is substantially greater than previously described techniques and a three-dimensional volume of the coronary tree can be obtained within one breath hold. However, image quality remains an issue. The resolution is ultimately limited by the signal to noise ratio and by breath holding.

Another technique is three-dimensional GE imaging within a breath hold.[86,87] High power gradients allow more images to be acquired at approximately 100 ms. This technique has the advantage; the course of coronary arteries is complex, but the volume can be angulated in any desired direction. With an approach called volume coronary angiography using targeted scans (VCATS), different scans are used for different sections. By segmenting, the problem of inconsistent breath holding in other methods is eliminated. The resolution in all directions is 1–2 mm. In one study, the identification of proximal coronary stenosis was in the range of 90%. The breath holding time ultimately limits the resolution. However, when looking for a technology with a resolution less than 50 μm, this resolution appears unfeasible for assessing high risk plaque.

Due to the described limitations, particularly resolution and cost, it is unlikely that MRI will be useful for the identification of unstable plaque in the near future. Ultimately, however, it may replace angiography under certain circumstances for screening for CAD. To do this, several limitations must be overcome. First, the sensitivity and specificity must increase to image not only the proximal arteries, but also the more distal vessels. Second, the cost must be reduced which may currently be prohibitive. One study concluded that society be willing to pay $50,000 per QLAY gained which would make costs currently ineffective.[88] Third, metal components such as metallic hemostat clips, sternal clips, sternal wires, and graft markers produce artifacts.

High frequency ultrasound has several potential roles as a method of intravascular imaging.[89] It has been shown to be superior to quantitative coronary angiography with respect to the identification of plaque and stent deployment. Most studies in the recent past have been performed with 30-MHz ultrasound catheters, which have an axial resolution of approximately 110 μm. The catheter is produced in two forms, a rotating transducer or a phase array device. The rotating transducer uses a speedometer cable to rotate a transducer at the tip of the catheter. The phase array does not rotate but consists of a series of small ultrasound transducers around the periphery of the catheter. Each has its own advantages and disadvantages. Specifically, rotational catheters are generally smaller, but array devices do not require speedometer cables and therefore nonlinear rotation.

Recently, a 40-MHz catheter has been introduced with an axial resolution of 80–90 μm. The diameter of the catheter is generally in the range of 1 mm in diameter. One of the applications where IVUS has shown usefulness is in guiding stent deployment. As stated, the Cruise trial has shown that IVUS has benefits in guiding stent deployment above the use of high pressure balloons. The final vessel diameter by IVUS was 7.78 ± 1.71 mm versus 7.06 ± 2.13 mm.[90] More important, the incidence of follow-up intervention in the IVUS group was reduced by 44%. Other studies have confirmed these results. Because of its low resolution, IVUS may not reach widespread use, but IVUS as an experimental tool has been a powerful asset in understanding the mechanisms of CAD and interventions on the disease. This has included the identification of the process of remodeling. Although remodeling does not identify vulnerable plaque, a recent series of studies have correlated positive remodeling with vulnerable plaque.[19,91] In one study, positive remodeling was found in 51.8% of patients with unstable angina but only 19.6% of patients with stable angina. Negative remodeling was found in 56.5% in stable angina but only 31.8% of those with unstable angina.

IVUS has several significant limitations for unstable plaque. First, the catheters are relatively large and likely will not be produced in the form of a guidewire in the near term. Second, and probably most important, IVUS is of low resolution, with a current axial resolution of 80–90 μm. This is likely why it has been unable to consistently identify vulnerable plaque.

Although the resolution of IVUS is limited, ultrasound elastography is currently under investigation to improve its diagnostic potential.[92,93] Ultrasound elastography looks at the elastic properties of tissue and is based on the modulation of ultrasound speckle or the RF modulation signal. At known changes in pressure, speckle modulation varies with the tensile properties of the tissue. It may be that someday, this modality will give insight into plaque vulnerability by assessing structural integrity. However, to date, it has largely gone unproven and its advantage over OCT elastography, which has a higher resolution, needs to be tested. IVUS and OCT elastography are discussed in Chapter 10.

Both electron beam (EBCT) and multi-slice (MSCT) computer tomography have been proposed as a method of identifying coronary disease.[94] EBCT can identify and quantify calcium in the coronaries. Generally, the calcium confirms the presence of atherosclerosis while its absence makes CAD unlikely. However, it has not proven useful for identifying unstable plaque. Furthermore, a majority of the American College of Cardiology/American Heart Association Consensus on the topic stated:

> The majority of the members of the Writing Group would not recommend EBCT for diagnosis obstructive CAD because of its low specificity (high percentage of false positive results), which can result in additional expense and unnecessary testing to rule out a diagnosis of CAD. The 1999 ACC/AHA Coronary Angiography Guideline Committee reached similar conclusions.[95]

To date EBCT has not established a clear role in CAD management.

MSCT has a limited ability to characterize plaque in the proximal coronary arteries, subclassifying them into calcific, fibrotic, and lipid-laden.[96] However, due to its limited resolution, slow acquisition rate, and inability to be easily combined with interventions it likely will not play a role in risk-stratifying TFCA in the near future, although it may have a role in global patient risk stratification.

There are a variety of new experimental procedures under investigation for assessing plaque including thermography, spectroscopy, and diffuse reflection. As stated, the intima of plaques that undergo rupture and possibly those that undergo erosion are characterized by inflammation. Since inflammation is associated with an increase in temperature, some have speculated that regional increases in temperature can be used to determine sites of coronary instability.[97–99] In one study of carotid artery plaques in patients undergoing carotid endarterectomy, investigators used a thermistor gauge needle-tip to measure temperatures. In this study, most regions had temperature differences that varied by 0.2 to 0.3°C but vulnerable plaque regions had substantially warmer temperatures. These regions correlated with increased inflammatory cell content ($p = 0.0001$). This technique has the potential for identifying inflammatory regions; however, it has yet to be applied *in vivo*. It will be of interest to see how the practicalities of *in vivo* assessments, such as variations in blood flow and the presence of IV contrast (which may change the temperature), influence the temperature measurement in the coronaries. In addition, the fact is that inflammation during ACS is a diffuse process, as described, and therefore may not be useful for distinguishing stable from unstable plaques.

A variety of spectroscopic techniques have been applied to the problem of diagnosing high risk plaques, but are still in experimental stages. These include Raman, near-infrared diffuse reflection spectroscopy, and IR spectroscopy. Raman spectroscopy, which utilizes a phenomenon described in the imaging section, has been applied *in vitro*. In these studies, Raman in the near-infrared was used to quantify cholesterol, cholesterol esters, triglycerides, phospholipids, and calcium salts in artery specimens.[100] The technology has not been used *in vivo* and has significant potential limitations: It requires direct contact with the surface of the artery, the source powers used are often well above tissue safety standards, and acquisition rates tend to be unacceptably long (10–100 seconds for one data point). Therefore, evidence that this technology will be practical for routine *in vivo* use in the near future is currently weak.

Recently, a technology known as near-infrared diffuse reflection spectroscopy has been applied to the analysis of *in vitro* plaques.[101] The technology analyzes gross infrared backreflection patterns from tissue followed by extensive post processing. Using a FT-spectrophotometer, a good correlation between cholesterol content and measurements (0.926 correlation coefficient) was demonstrated. The technology has several significant limitations: The measurement time is currently in the range of 30 seconds per scan, contact with the vessel is required, and it is unclear whether the standardization procedures used *in vitro* could be implemented across the general population *in vivo*.

In the early 1990s a group used near-infrared imaging that utilized a fiber-optic laser and a parallel vector supercomputer.[102] The authors successfully used this system *in vitro* to determine the various cholesterol constituents, including oxidized LDL. Unfortunately, little work has subsequently been performed on this approach. In addition, it is unclear from the group's manuscript what the source power and data acquisition rates were, in addition to whether direct contact is required.

15.2.8.1 Animal models

As markers distinguishing those TFCA that progress to ACS from those that don't, future animal studies are likely warranted to study this. In addition, animal models remain important in assessing the efficacy of pharmaceuticals. Currently, there is no accepted gold standard animal model for atherosclerotic plaque formation and rupture.[103–105] The rodent and rabbits are the two most important models used now. Some of the rodent models include the apoE −/− mice, the LDLR −/− mice, and the genetic hypersensitive-hyperlipidemic rats.[106,107] While catheter-based imaging is impractical, future OCT work may examine atherosclerosis in these models with a hand-held probe or OCT microscope. Rabbits, on the other hand, possess blood vessels with diameters that exceed 1 mm. The "classic" atherosclerotic rabbit model involves placing the animal on a high cholesterol diet, most commonly with concominant balloon angioplasty of a native vessel. A recently published study by investigators on this proposal has demonstrated an association of intraplaque hemorrhage and the size of the necrotic core with plaque instability in humans.[25] This same study proposed a rabbit model in which complex plaques were formed in rabbits fed a high cholesterol diet, whose iliac arteries were denuded and then injected with autologous RBC into the established plaque. Plaques injected with RBCs demonstrated an increase in macrophage infiltration (with iron deposits present), higher lipid content (including cholesterol crystals), and substantial reductions in smooth muscles and collagen matrix as compared with plaques that were not injected with RBCs. These plaques resemble the TCFAs in humans but do not progress to rupture. We are examining the use of vasoactive factors as well as pro-angiogenic agents to induce rupture.[108,109] Recently, another group has introduced a rabbit model where rupture was obtained taking advantage of p53 activation.[213]

15.2.9 OCT for Coronary Imaging

15.2.9.1 Historical development

In a discussion with the author in the early 1990s, Dr. Arthur Weyman, the Director of Echocardiography at Massachusetts General Hospital (MGH), stated that the problem with assessing vulnerable plaque was the limited resolution. At that time, the high resolution of optical technologies (or more precisely, those based on electromagnetic radiation) seemed to have the greatest potential for overcoming this problem. Most noteworthy was the modality referred to as OCT, pioneered by James Fujimoto Ph.D., Eric Swanson M.S., and David Huang from Massachusetts Institute of Technology (MIT), along with others in the field of low coherence interferometry. The technology had been applied to imaging cracks in fiber optics and the transparent tissue of the eye with unprecedented resolution, but had not been successfully applied to imaging nontransparent tissue.

With OCT, several significant limitations existed in adapting the technology to intravascular imaging. First, the penetration depth of OCT imaging at this time was 500 μm in nontransparent tissue, while a few millimeters was needed. Second, the technology was too slow for intravascular imaging, taking approximately 40 seconds to generate an image. Third, imaging had not been performed through either a catheter or endoscope. Fourth, registration of images with nontransparent tissue pathology, which is needed to verify the effectiveness of the technology, would be difficult at these resolutions. Finally, the system electronics would require significant modification. However, in spite of these limitations, the technology was ultimately adapted for intravascular imaging.

15.2.9.2 Early OCT work

In a collaboration of our group at MGH and the Fujimoto laboratory at MIT, pilot work was begun. Initial results in studies begun in 1994 were associated with only limited success, but ultimately, in a 1996 article in *Circulation* (and one shortly afterward), we were able to report significant positive results.[110,111] Several clinically relevant observations were noted. First, OCT images demonstrated a high correlation with histopathology. Some earlier data demonstrating structural imaging have been included in the next few paragraphs to orient the reader to the underlying basis of the proposal. In these studies, OCT images accurately displayed atherosclerotic plaque morphology when compared to the corresponding histopathology. In Figure 15.5, an OCT image of a heavily calcified aortic atherosclerotic plaque is seen with the corresponding histopathology.[110] A thin layer of intima of less than 50 μm in diameter, which is identified by the arrow, is overlying the plaque. In Figure 15.6, the top image shows an atherosclerotic plaque with a thin intimal cap, again less than 50 μm in diameter, seen (arrow) over a collection of lipid, a TCFA. In the bottom image, a fissure is seen at the intimal-medial border. Second, measurements made at both 830 and 1300 nm showed the superior penetration of the latter, due to the substantially reduced scattering and reduced absorption relative to the visible light region and water absorption peak. Therefore, mean wavelengths around 1300 nm resulted in maximal penetration. In Figure 15.7, imaging is performed of a human epiglottis at 830 and 1300 nm. The cartilage deep within the epiglottis is seen at 1300 nm but not at 830 nm. Finally, imaging was demonstrated through calcified plaque and of arterial fissures.

After these initial observations, the project took two distinct directions, comparison with high frequency IVUS and the advancement of the technology for *in vivo* intravascular imaging.[112–114] Initial comparisons of OCT with IVUS, done in collaboration with Neil Weissman, M.D., of Washington Hospital, validated the superior resolution compared to IVUS. In Figure 15.8, the OCT image of an *in vitro* coronary artery is on the left, the IVUS image is on the right.[114] The intimal hyperplasia (thickening of the inner layer) is clearly seen with OCT but not with IVUS. Quantitative measurements from OCT were also superior to IVUS. The axial resolution of both OCT and IVUS (30 MHz) was measured directly from the point spread function (PSF). The axial resolution of OCT was 16 ± 1 μm

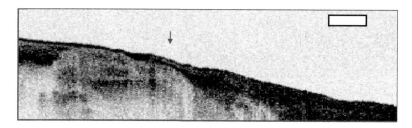

Reflectance

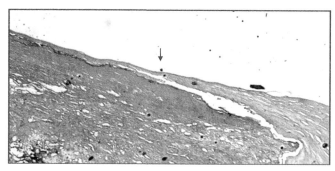

Figure 15.5 Thin intimal cap. The top is an OCT image of a heavily calcified plaque with minimal plaque while the bottom is the corresponding histopathology. The plaque is on the left and the relatively normal tissue on the right. The arrow denotes a thin intimal cap less than 50 μm in diameter at some points. Bar represents 500 μm. Image courtesy Brezinski, M.E., et. al. (1996). *Circulation.* 93(6), 1206–1213.

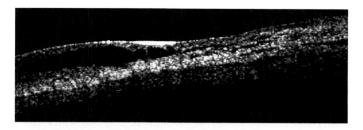

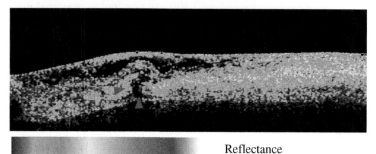

Reflectance

Figure 15.6 Unstable plaque. The top image is an unstable plaque with the black area on the left being lipid and the yellow arrow is the thin intimal cap. The bottom image is fissuring below the plaque surface shown by the arrows. Image courtesy Brezinski, M. E., et. al. (1996). *Circulation.* 93(6), 1206–1213.

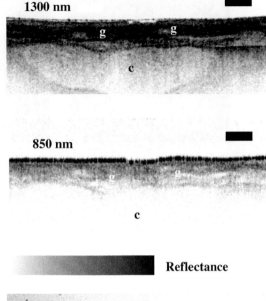

Figure 15.7 Wavelength dependency of imaging. The top image is of a human epiglottis at 1300 nm, the middle at 850 nm (used for imaging transparent tissue), and the bottom the corresponding histology. It can be seen that the cartilage (c) and glands (g) are far better defined at 1300 nm than at 850 nm due to an optical window (relatively low scattering and absorption). Image courtesy Brezinski, M. E., et. al. (1996). *Circulation*. 93(6), 1206–1213.

compared with 110 ± 7 μm for IVUS.[112] Similarly, when examining near-occlusive coronary plaques, as shown in Figure 15.9, both the ability of OCT to image through the width of the artery and the superior resolution compared with IVUS (40 MHz) are seen.[113] The OCT images are the three images on the top of the figure.

Significant technologic advances at this time, required for *in vivo* vascular imaging, included the development of an OCT catheter, a high speed data acquisition system to get imaging near video rate, noise reduction in the system electronics, and synchronization of the catheter and system components.[114,115]

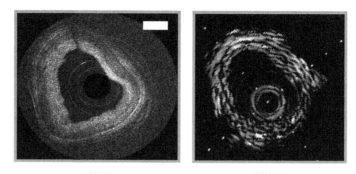

Figure 15.8   OCT and IVUS images of human coronary arteries. On the left is the OCT imag; on the right is the IVUS images. The three layered structure of this artery with intimal hyperplasia are clearly delineated with OCT but not by IVUS. Courtesy Tearney G. J., et al. (1996). *Circulation.* 94(11), 3013. [114]

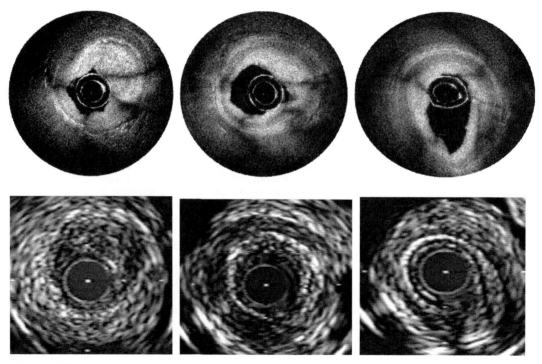

Figure 15.9   OCT and IVUS images of atherosclerotic human coronary arteries. The top three images are the OCT images, the bottom are IVUS images at 40 MHz. It can be seen that OCT sharply defines the plaque which is not seen in well in the IVUS images. Courtesy Patwari P, et al. (2000). *American Journal of Cardiology.* 85, 641–644.

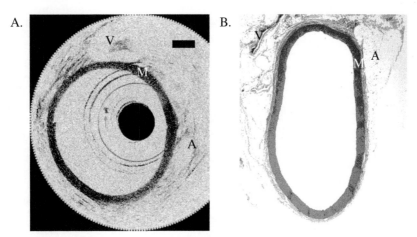

Figure 15.10 In vivo catheter based imaging of the rabbit aorta. The left is the OCT image; the right is the histology. In the OCT image, the media (M), inferior vena cava (V) and adventitia (A) are clearly defined. Imaging was performed in the presence of a saline flush. Courtesy Fujimoto JG, et al. (1999). *Heart.* 82, 128–133.

15.2.9.3 *In vivo* imaging

The next major milestone for OCT was to move to *in vivo* imaging. This came through three directions. In the first, imaging was performed *in vivo* in the normal rabbit aorta and structural detail was delineated at unprecedented resolution.[116] This is demonstrated in Figure 15.10. In Figure 15.11, a stent is imaged in a rabbit aorta and compared with 40-MHz ultrasound.[117] The second important advance toward *in vivo* intravascular human imaging was the commercialization of OCT through LightLab Imaging (formerly Coherent Diagnostic Technologies LLC) of Westford, MA.[118] LightLab Imaging was sold to Goodman Co. Ltd. in 2002. Figure 15.12 is an example of imaging in both the longitudinal (axial) and cross-sectional directions of *in vitro* coronary plaque during automated pullback. It can be seen in the axial image that multiple plaques are noted (arrows). In addition, *in vivo* human studies have been performed by several groups (particularly the one at MGH), whose work has spanned from assessing intracoronary stent position to examining plaque morphology.[119,120] To date, studies have been predominately observational and clear markers of those plaques that lead to ACS have yet to be established, as will be discussed below. Figures 15.13 and 15.14 are images generated *in vivo* in humans using a LightLab Imaging engine, courtesy of Professor Ishikawa (Kinki University Japan), and Joseph M. Schmitt Ph.D. (chief scientific officer, LightLab Imaging).[118] The OCT ImageWire (0.4 mm cross-sectional diameter) was delivered via an over-the-wire soft occlusion balloon catheter (Goodman Advantec, 4 F Proximal, 2.9 F at the distal tip). In Figure 15.13, this image of a human coronary artery shows a stent in a vessel with thick neointimal growth. Both the cross-sectional image (A) and axial image (B) are shown. The white line demonstrates where each image correlates. In Figure 15.14, a plaque is seen

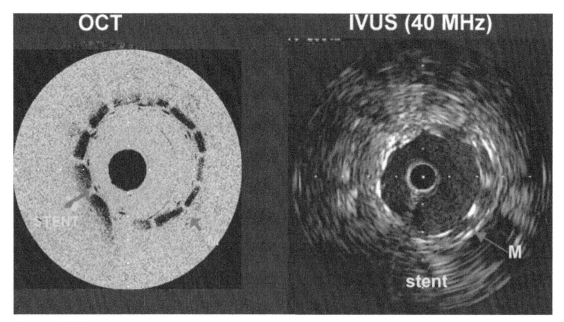

Figure 15.11   Comparison of stent imaging with OCT and IVUS. The left is an OCT image of a stent placed in a rabbit aorta imaged in the presence of saline flush. The right is IVUS imaging of the same stent again with a saline flush. Unpublished data courtesy of Li, X., J. G. Fujimoto, and M. E. Brezinski. [117]

which covers almost two quadrants in cross section (arrow in A). In the axial image (B), the arrows demonstrate plaques viewed along the catheter length. In Figure 15.15, also courtesy of LightLab Imaging, a relatively normal coronary artery is shown along with its corresponding angiogram. In Figure 15.16, contributed by LightLab Imaging (Professor Eberhard Grub, and Joseph M. Schmitt PhD), an intimal tear is noted *in vivo* in humans using a 0.017" imaging guidewire in the presence of a saline flush. In Figure 15.17, provided by the same group, neointimal hyperplasia is noted in the OCT image on the right, but is poorly defined in the IVUS image on the left.

15.2.10   Limitations

A major limitation of intravascular OCT imaging is that imaging is attenuated by blood. Approaches to overcome this limitation are saline flushes (possibly in association with increased acquisition rates), balloon occlusion, and index matching. Saline flushes have been used successfully both in animals and in patients *in vivo*.[118,119] However, concerns include fluid overloading the patient, which leads to pulmonary edema, and whether the field can be completely cleared in all clinical scenarios. Saline flushes may be reduced with higher data acquisition rates, possibly used with the newer Fourier-based approaches. Occlusion balloons have

Rapid Acquisition L-mode

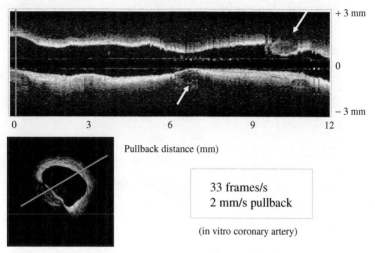

Figure 15.12 Axial imaging of an in vitro coronary artery. The images were generated with a Lightlab Imaging system. The unit allows imaging both along the plane of the catheter (upper) as well as in cross section (lower). Multiple TCFA are noted along the length of the catheters. Image courtesy of Lightlab, J. Schmitt PhD.

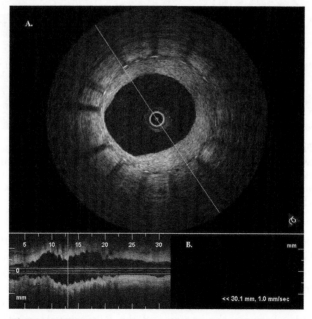

Figure 15.13 In vivo imaging of in stent restenosis in human coronary artery. Image courtesy of Professor Ishikawa (Kinki University Japan) and Joseph Schmitt PhD, Lightlab Imaging. Published in Brezinski, M. E. (2006). *Intl. J. of Cardiol.* 107, 154–165.

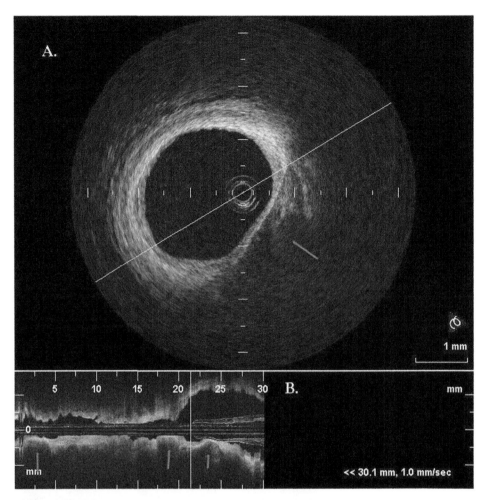

Figure 15.14   In vivo human imaging. In this image, a plaque is seen which covers almost two quadrants in cross section (line in A). In the axial image (b), the lines demonstrate plaques viewed along the catheter length. Image courtesy of Professor Ishikawa (Kinki University Japan) and Joseph Schmitt PhD, Lightlab Imaging. Published in Brezinski, M.E. (2006). *Intl. J. of Cardiol.* 107, 154–165.

been used *in vivo* by LightLab Imaging through the efforts of Professor Eberhard Grube and Professor Ishikawa. Future studies are required to assess the ultimate viability of this approach. An alternative technique to clearing the field is index matching.[121]

The source of the signal attenuation caused by blood is the difference in refractive index between the serum and the RBC cytoplasm. This is seen in Figure 15.18 from the study where saline, blood, and lysed blood are flowing over a reflector. Almost all the signal has returned from the lysed blood so that the reduced penetration in blood is not due to absorption or the cell membranes. By increasing

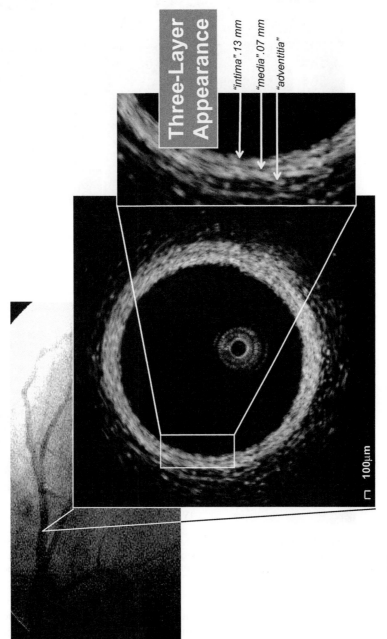

Figure 15.15   In vivo human coronary artery and comparison with angiography. Courtesy Dr. Grube, Siegburg (Heart Center, Siegburg, Germany) and J. Schmitt PhD, Lightlab Imaging.

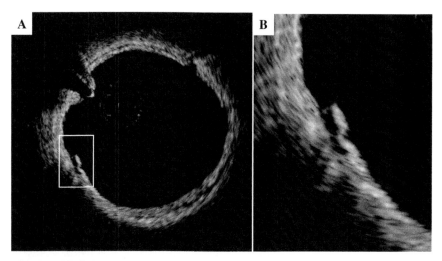

Figure 15.16   Presence of an in vivo intimal flap. In this patient, an intimal flap is identified with the OCT imaging guidewire. Courtesy Dr. Grube, Siegburg (Heart Center, Siegburg, Germany) and J. Schmitt PhD, Lightlab Imaging. Courtesy Stamper, D., et al. (2006). *J. Am. Coll. Cardiol.* 47(8), C69–79.

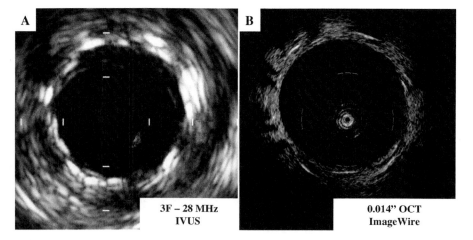

Figure 15.17   Less severe in-stent restenosis. In this patient, in-stent restenosis is noted as imaged with the 0.014 inch imaging guidewire. Courtesy Dr. Grube, Siegburg (Heart Center, Siegburg, Germany) and J. Schmitt PhD, Lightlab Imaging. Courtesy Stamper, D., et al. (2006). *J. Am. Coll. Cardiol.* 47(8), C69–79.

the serum refractive index to equal that of the cytoplasm, blood would be almost completely transparent. In Figure 15.19A, the effects of increasing the serum refractive index with dextran and IV contrast are noted. Dextran resulted in a $69 \pm 12\%$ increase in image intensity relative to the saline control while IV contrast resulted in a $45 \pm 4\%$ increase. Both effects were highly significant ($p < 0.005$). This

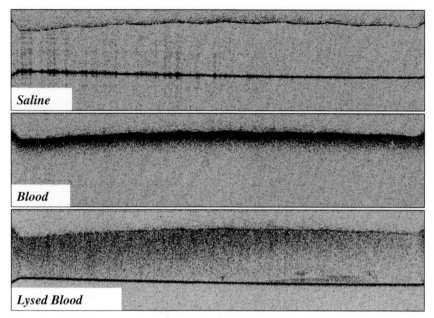

Figure 15.18 Qualitative proof of mismatch as the source of scattering. In the top image, saline is being circulated over a reflector. The top line is the inner layer of the tubing. The second line is the reflector. Saline is between them. In the middle image, blood is being circulated rather than saline through the system. The reflector is no longer seen. In the bottom image, blood has been removed, lyzed, and reintroduced into the system. The signal off the reflector returned to values not significantly different from control subjects. Reprinted, with permission, from Brezinski, et al. (2001). *Circulation.* 103, 2000–2004.

was not due to changes in hematocrit or RBC concentration as shown in Figure 15.19B. While dextran and IV contrast are not viable for *in vivo* use, current research is focusing on synthesizing more potent, clinically safe compounds for index matching.

A second potential limitation of OCT is that its penetration through the arterial wall is in the range of 2–3 mm. This is generally sufficient for the imaging of most arteries, but with some large necrotic cores, the entire length of the core cannot be imaged. Since the decay is exponential, increasing power (at constant SNR and dynamic range) on the sample within the range that does not induce tissue damage will likely not have a large effect on penetration. Approaches that may lead to improved penetration include increasing the dynamic range, use of a parallel ultrasound beam, and image processing techniques. A variety of approaches are being applied to potentially increase dynamic range including alternatives to time domain OCT (discussed in this chapter), such as spectral radar and source swept OCT, which are discussed in Chapter 5. Work on these new techniques is preliminary and future efforts are necessary for comparisons with traditional OCT for assessing arterial pathology. In addition, as most research groups are utilizing sources with median wavelengths above 1300 nm, some of the reduced penetration

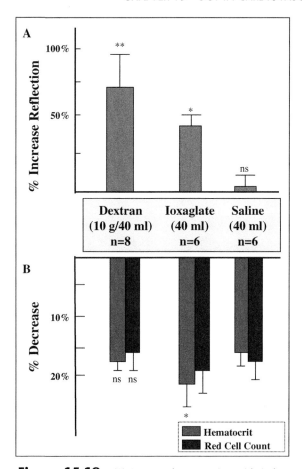

Figure 15.19 (A) Increased penetration with index matching. In this figure, the difference in penetration after addition of compounds or controls is seen. The saline control had no significant effect (7 ± 3%). Dextran results in a 69% ± 12% increase in penetration. Intravaenous (IV) contrast resulted in a 45 ± 4% increase. Both were significantly different than the control. (B) Influence of compounds on hemocrit and red cell count. Therefore, the improved penetration was not due to either a change in cell size or number. There was a slight but significant difference in red cell size after IV contrast compared with saline control. This might have contributed slightly to the effect. Reprinted, with permission, from Brezinski, et al. (2001). *Circulation.* 103, 2000–2004.

may be due to water absorption and a movement toward sources slightly below 1300 nm (1250–1280 nm) would likely be beneficial. Also, median wavelengths around 1700 nm deserve further investigation due to their relatively low scattering and absorption.[122]

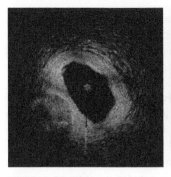 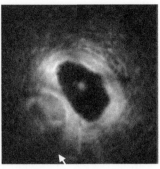 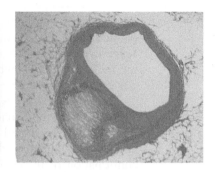

Figure 15.20 Image processing to improve assessment of image structural detail. The OCT image on the left is unprocessed. The image in the middle has been processed using the STICKS technique (as discussed in chapter 11) which allows improved definition of the plaque. The right is the corresponding histopathology. Rogowska J, and Brezinski, M. E. (2000). *IEEE Trans Med Imaging.* 19(12), 1261–1266.

Multiple scattering reduces resolution and therefore image penetration through a reduction in contrast. By the use of a parallel ultrasound beam, it has been shown that multiple scattering can be substantially reduced and ultimately may allow deeper imaging in tissue.[123] This approach is still under investigation.

Image processing may improve the ability to identify deeper structures within tissue. One example that has been used by our group is the use of the STICKS technique, which has been shown to enhance imaging in coronary arteries.[124] An example series of images of processed data is shown in Figure 15.20, where the plaque in the center image is more clearly defined.

15.2.11 Current State of Intravascular OCT Imaging

OCT has been developed for high resolution intravascular imaging and the identification of TFCA. Current clinical efforts focus on the identification of OCT markers of plaque vulnerability to improve patient risk stratification, in addition to technology development. Work focusing on overcoming the attenuation from blood and increasing penetration depth has been addressed in the previous section.

15.2.12 Current OCT Markers for *In Vivo* Studies

As OCT is now being introduced for *in vivo* human imaging at a resolution higher than any current imaging technology, allowing the identification of TCFA, studies are underway to further define OCT plaque markers which aide in patient risk stratification. This is because most ACS results from TCFA, but most TCFA do not lead to ACS, resulting in a need to further risk-stratify TFCA, as discussed above.

The presumed general determinates of a plaque's vulnerability to rupture are the size/composition of the atheromatous hemorrhagic core (when the core is more than 40% of the plaque volume), the thickness of the fibrous cap covering the core, the collagen content/type of the cap, angiogenesis into the intima, and the degree of inflammation within the cap. Currently, three OCT markers are used to risk-stratify plaques *in vivo* in human studies: cap thickness, cap-intimal borders, and granularity suggestive of macrophage/foam cell content. The first is an accurate measurement of intimal cap thickness, which goes largely undisputed and has been demonstrated by our group and others.

The second marker is the ability of OCT to identify plaque macrophages.[125,126] This is an intriguing concept put forward by a group at MGH and is an important marker to be evaluated in future studies. However, there is only one study that attempts to quantify this and because of several concerns discussed below, confirmation of these results seems appropriate. First, the authors neither used a "training set" nor predetermined criteria for what constituted a positive or negative predictive value in the OCT image, but determined the criteria after examining the measured data set.[125] They established the value for raw OCT data to be between 6.15 and 6.35%, giving 100% sensitivity and specificity. It may be useful to examine a different statistical approach. It is of note that they did not appear to use these values for interpreting their subsequent *in vivo* study, which did not have histological validation.[126] Second, they performed a median filter over a 3×3 square kernel that corresponds to an effective axial resolution for macrophages worse than 30 μm. Therefore, the macrophages would either need to be densely packed or greater than 30 μm in diameter. Third, they used the mean background noise for their calculation. The problem with this is best illustrated with a simulation, where we take the case of two different background variations. In the first case, the background has low pixel variations, between 1 and 3. In the second case, the pixels vary from 1 to 9. If we make the concentration of macrophages on both images the same and macrophages have a constant value of 50, the normalized standard deviation (NSD) on the first image will be 50.31%, and on the second 46.77%, meaning that the concentration of macrophages on image 1 is higher than on image 2 and this is not true. Finally, when the same group applied the approach *in vivo*, there was minimal difference between the culprit plaque and remote regions (5.54 ± 1.48 vs. 5.38 ± 1.56) and between culprit lesions in unstable versus stable plaque (5.91 ± 2.06 vs. 4.21 ± 11.74).[126] Particularly when no training set has been established with histologic correlations and the concern over variation due to background noise, it is unclear how these small variations will be diagnostic when spread over a large population. Furthermore, as discussed above, it is unclear whether macrophage concentrations increase in the high risk plaque before or after rupture. While the work on OCT assessments of macrophages remains intriguing, future work in this area is needed to confirm the results and determine the ultimate clinical value.

The third proposed OCT marker of current *in vivo* studies is the ability of OCT to distinguish fibrous, fibrocalcific, and lipid-laden plaque by the backreflected signal, the latter as plaques of interest.[119] In a recent study, the three plaques were, respectively, differentiated as homogeneous, signal-rich region; well-delineated, signal-poor region with sharp borders; and signal-poor region with diffuse borders. Therefore, fibrocalcific and lipid-laden plaques are distinguished by the sharp versus diffuse borders. However, this study was associated with a high false positive rate

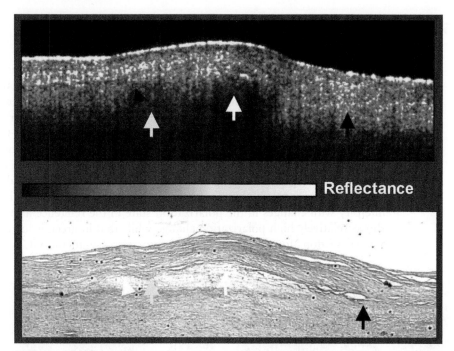

Figure 15.21 Examining intima boundary sharpness as a marker for lipid. In this figure, the yellow arrows demonstrate the cap-lipid interface that is diffuse and is covered by a highly scattering cap (yellow reflections). The white arrow, which is also over lipid, identifies cap with lower scattering and the cap-lipid interface is sharply defined. Finally, the black arrow shows the intimal-elastic layer interface (no lipid present) that is diffuse with an intima that is highly scattering. The reason for the diffuse nature of the intimal-elastin border is likely that multiple scattering is obscured due to high scattering within the intima. It is unclear, therefore, whether the lipid interface is diffuse because of the core comp[osition, as previously suggested, or multiple scattering from the cap. *Courtesy of Circulation* (1996). 93, 1206–1221 (modified).

with fibrous caps diagnosed as lipid-laden plaques (sensitivity 71 to 79%). The difficulty in identifying plaque by the diffuse nature of the plaque border is exemplified by Figure 15.21 from our initial paper on plaque rupture.[110] In this figure, the yellow arrows demonstrate the cap/lipid interface that is diffuse and is covered by a highly scattering cap (yellow reflections). The white arrow, which is also over lipid, identifies a cap with lower scattering and the cap-lipid interface is sharply defined. Finally, the black arrow shows the intimal-elastic layer interface (no lipid present) that is diffuse with an intima that is highly scattering. The reason for the diffuse nature of the intimal-elastin border appears to be that multiple scattering by the intima/cap that is obscuring the border (which would also explain a more rapid local exponential decay) due to high scattering within the intima/cap. It is unclear, therefore, whether the lipid interface is diffuse in some OCT images because of the core composition, as previously suggested, or because of multiple scattering from the cap, the latter of which could reduce its predictive power. Future studies are required to establish the mechanism that makes some plaques have diffuse cap/core interfaces.

15.2.13   Potential Additional OCT Markers

While the above markers are currently being examined as part of *in vivo* studies, additional markers are demonstrating diagnostic potential. These include polarization sensitive OCT (PS-OCT) for assessing collagen, OCT elastography, dispersion measurements to assess lipids, and OCT Doppler to assess angiogenesis.

Collagen concentration in the cap is one of the most critical factors in determining plaque stability.[127–129] We have recently utilized single detector PS-OCT to identify organized collagen in arteries as well as in other tissue.[130,131] In Figure 15.22, TCFA is examined by trichrome blue (D and F), picrosirius (C), and OCT at two orthogonal polarization states (A and B). The OCT images show variable polarization sensitivity, which means that different portions of the image change in backreflection intensity with changes in the polarization state in the reference arm. The red arrow shows relatively high polarization change while that in green is low. The picrosirius stained section (for organized collagen) correlates with the OCT image, showing

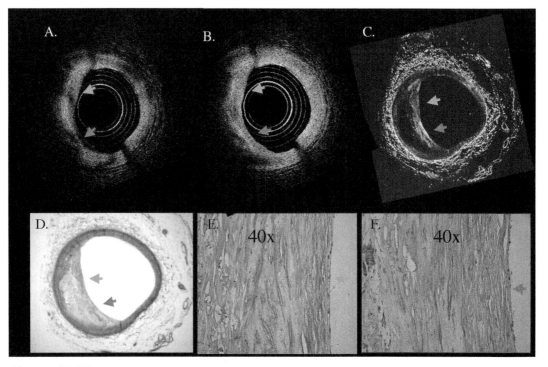

Figure 15.22   Assessing plaque collagen with PS-OCT. Single Detector PS-OCT changes the polarization state in the reference arm and allows collagen assessment. A and B demonstrate OCT images taken at two different polarization states. The image changes more in some regions than in others due to the higher concentration of collagen in the former. In C, the picrosirius section shows reduced collagen with the upper arrow but increased collagen at the lower arrow. In D, E, and F, trichrome stained sections are shown which is notable for a lack of inflammatory cells, the significance of which is discussed in the text. Unpublished data based on Giattina, S. D., et al. (2006). *Int. J. Cardiol.* 107, 400–409.

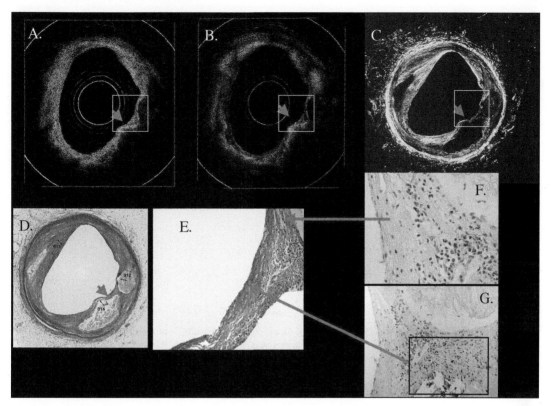

Figure 15.23 Thin walled plaque with minimal organized collagen. In the figure, a plaque with a wall thickness at points less than 60 çm in diameter is imaged. In parts A and B, the white box demonstrates an area of thin intima with almost no organized collagen. This is confirmed in the picrosirus stained section. Some inflammatory cells are noted in deeper sections, but have minimal presence near the surface. Courtesy Giattina, S.D., et al. (2006). *Int. J. Cardiol.* 107, 400–409.

areas of high and low birefringence. In the trichrome section, the images are relatively homogeneous, but the lack of cellularity is relatively striking in spite of collagen loss. This adds to the concern that measuring macrophage activity may not correlate with plaque vulnerability. This is seen again in Figure 15.23 where the OCT image of another small plaque shows different changes in backreflection intensity with polarization changes, with the arrow showing minimal change. The picrosirius section (C) confirms this region lacks organized collagen. The trichromes are notable for a lack of CD68 staining (macrophages), except on G (square). Strong correlation was noted between collagen concentrations in the cap measured by PS-OCT and collagen as assessed by picrosirius staining. The largest OCT changes had a positive predictive power for collagen of 0.889 while the smallest change was 0.929 for a negative predictive power. Furthermore, the study was performed with only a partially polarized source. Superior results are anticipated when the source is completely polarized. This effect was not dependent on whether the collagen was

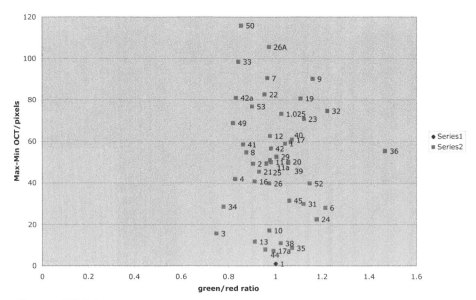

Figure 15.24 Green/red ratio. Since color by picrosirus correlates with fiber size/type, its relationship to OCT polarization changes was examined. With this technique, fiber types were not distinguished which is the source of future work. Giattina, S.D., et al. (2006). *Int. J. Cardiol.* 107, 400–409.

type I or type III, as in Figure 15.24. With single detector PS-OCT, banding in the image or changes in each A-scan of the image are measured with changes in reference arm polarization as opposed to dual detector approaches that measure the absolute value of the backreflected light polarization state. The key is that relative changes are measured and not absolute values. The advantages of the single detector approach (discussed in Chapter 8) include that it is not altered by catheter bending or by catheter compression. Catheter bending or compression essentially occurs with all interventional procedures. With the dual detector approach, as the catheter rotates or is pulled back, the polarization state changes making uniform tissue appear heterogeneous by OCT imaging.

Ultrasound elastography has been applied to arteries for over a decade to assess their tensile strength. However, due to its relatively low resolution, ultrasound elastography is not likely to be practical for the identification of high risk plaque. Recently, OCT elastography has been introduced for evaluating the elastic properties of tissue.[132] Due to its high resolution, OCT elastography has the potential of quantifying the cap mechanical strength as discussed in chapter 9.[133,134] An example elastography image of human aorta is shown in Figure 15.25, where the direction of the arrows represents strain direction while the length represents degree of displacement. However, many technique obstacles need to be overcome before elastography can be applied effectively *in vivo*.

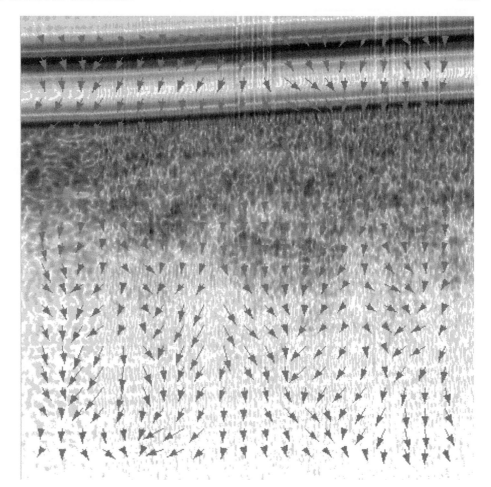

Figure 15.25 OCT elastography of human atherosclerotic arteries. This is discussed in Chapter 9 in detail. The arrows indicate displacement direction and intensity, which various dependent on the regional mechanical properties of this coronary artery. Courtesy Rogowska, J. et al. (2006). *Br. J. Rad.* (in press).

As discussed above, angiogenesis into plaque, with the subsequent rupture of its thin capillaries, may be an important mechanism of plaque instability. OCT Doppler has been demonstrated to identify blood flow in capillaries within tissue. OCT Doppler has the potential to identify angiogenesis in plaque and, while no current publications have been produced in this area, it is highly likely to be the subject of future investigations. Measurement of plaque hemoglobin may also be of benefit for similar reasons.

Although most absorption techniques are difficult to implement with OCT *in vivo*, as outlined in Chapter 9, dispersion analysis for lipid may be possible to implement. This may potentially be achieved through monitoring changes in the PSF, chirping, or photon correlations.[135–137]

15.2.14 Technological Investigations

Although a wide range of technological advances are being pursued by multiple research groups, two areas of study are of particular importance for intravascular imaging: microelectrical mechanical devices or MEMs-based catheters,[138,139] and the development of an OCT guidewire. MEMs technology would allow the elimination of the speedometer cable, preventing the associated nonlinear rotation. The development of an OCT guidewire would have a huge impact on the field, allowing for a single imaging device to be kept in place while different interventional devices are passed over it. Among other things, this would allow a reduced number of exchanges and procedure costs.

15.2.15 Other Coronary Areas

There are a whole variety of areas where OCT could be potentially applied in the coronary including monitoring Kawasaki disease, to opening chronic total occlusions.[140–142,210] However, more work is needed to evaluate the efficacy in these areas.

15.3 STROKE

15.3.1 General

OCT has the potential to both identify plaques that lead to stroke and guide to interventional procedures in the cerebral circulation. Stroke (cerebral infarct) is the loss of blood flow to a region of the brain resulting in death to that brain tissue. It can result from a range of processes including blood vessel disease (primarily atherosclerosis), nonatherogenic embolism, decreased blood flow to the brain (e.g., cardiac arrest), or vessel rupture. The first three are generally referred to as ischemic strokes while the final is known as a hemorrhage stroke. Brain cells are probably the cells of the body most vulnerable to loss of blood flow. In animal studies, total occlusion of the distal blood vessels results in irreversible neuronal death in 3 minutes. The brain only repairs itself by forming a fibrogliotic scar in damaged sites, since neurons do not multiply in adulthood. Because different sections of the brain perform different functions, the symptoms of a stroke will be different depending on the artery occluded. In general, the term transient ischemic attack (TIA) refers to the transient loss of blood flow to the brain where neurologic symptoms resolve within 24 hours. TIAs have a high incidence of progressing to ischemic stroke if left untreated. Amaurosis fugax, the transient painless loss of vision in one eye, is basically a type of TIA. The blindness lasts for a few seconds or minutes, and then recovers. Again, this is a serious condition and immediate medical care should be sought out to prevent permanent blindness.

15.3.2 Epidemiology

In the United States, the prevalence of stroke is 5,400,000.[1] Each year 700,000 people experience new or recurrent strokes, with 500,000 the first attack and 200,000 a recurrent attack. It is the third leading cause of death after MI and cancer at approximately 275,000. Of all strokes, 88% are ischemic, 9% are intracerebral hemorrhage, and 3% are subarachnoid hemorrhage. This accounts for 1 in 15 deaths. About 8–12% of ischemic strokes and 37–38% of hemorrhagic strokes result in death within 30 days.[143,144] After their initial stroke, 22% of men and 25% of women will die within 1 year. Of those having a stroke before the age of 65, 51% of men and 53% of women will die within 8 years.[145] Each year about 40,000 more women then men have a stroke. The prevalence of silent cerebral infarction between the ages of 55 and 64 is approximately 11% while 43% above the age of 85.[146–147] The incidence among African Americans is twice that of whites. The incidence is also higher among Hispanic Americans.

It is estimated that in 2005, direct and indirect costs associated with stroke will exceed 56 billion dollars in the United States.[4] In terms of disability, 50–70% of stroke survivors remain functional while 20% are institutionalized and 15–30% permanently disabled.[4] Stroke is an international problem. The WHO estimates that 15 million people each year suffer strokes and 5 million are permanently disabled. In 2002, it was estimated that 5.5 million deaths occurred from stroke worldwide.

15.3.3 Risk Factors

In 2000, 70% of respondents correctly named at least one established stroke warning sign and 72% correctly named at least one established risk suggesting that prevention is possible in a significant number of patients.[148] Major risk factors for stroke include age, race, hypertension, smoking, carotid artery stenosis, diabetes, hyperlipidemia, TIA, and cardiovascular abnormalities, such as atrial fibrillation and valvular heart disease.[149] Other potential risk factors are migraines, oral contraceptive use, alcohol abuse, hypercoaguable states, elevated homocysteine levels, and illegal drug use (such as cocaine).

Four risk factors deserve further attention because they generally should be easily controlled. Blood pressure is a powerful determinate of stroke, with patients having a blood pressure less than 120/80, the risk over a lifetime compared with hypertension patients is dramatically reduced. Therefore, blood pressure assessment should be routine at clinic visits and hypertension treated when identified. Atrial fibrillation is another stroke-independent risk factor with a 2- to 10-fold increase (1–10 strokes per year) depending on risk factors, which are discussed in a later section.[1,150] Smoking increases the risk of stroke approximately 1.5 times with it occurring in a dose-dependent fashion. After 3 to 5 years of cessation of smoking, the risk begins to decline. TIAs need to be addressed aggressively as they are associated with a high risk of progressing to stroke in the first 6 months, but the exact risk depends on the source of TIAs. Controlling hyperlipidemia and diabetes is also important for preventing strokes.

15.3.4 Cerebral Circulation

The anterior and middle of the brain receives its blood supply from the two internal carotid arteries (that are primarily extracranial), which divide into the anterior and middle cerebral arteries, in addition to the ophthalmic artery. The posterior sections of the brain are supplied by the two vertebral arteries. These major arteries are elastic consisting of a thin intima, large tunica media consisting mostly of elastic tissue, and adventitia. The four major arteries (carotids and vertebrals) are connected via the Circle of Willis, which allows some redundancy in blood supply. This is important because in many instances the redundancy allows one carotid to be occluded temporally when necessary without risk of stroke.

15.3.5 Carotid Plaque

Likely the most important application of OCT in preventing stroke is in the assessment and treatment of plaque in the carotid arteries and the prevention of embolization to distal sites. Carotid disease is thought to make up 20–30% of all strokes.[149] It is typically treated with either carotid endarterectomy, intraarterial catheter-based interventions, and medical therapy. With carotid endarterectomy, blood flow to the arterial section is halted and plaque is scraped off the surface of the artery.[151]

A stroke risk of 26% at 24 months exists in patients with ischemic neurologic symptoms and internal carotid artery stenosis greater than 70%.[152] Therefore, in patients with high grade stenosis, significant symptoms, and who are good surgical candidates, the decision to proceed with surgical treatment is typically straightforward. However, since not all patients with stenotic lesions develop stroke, improved techniques for patient risk stratification may reduce the number of patients needing surgery.

This is particularly true with asymptomatic stenotic lesions, and likely requires improved patient risk stratification. The Asymptomatic Carotid Atherosclerosis Study (ACAS) showed that for patients with ultrasound detected stenosis of at least 60%, carotid endarterectomy had a statistically significant advantage over medical management in preventing ipsilateral hemispheric stroke.[153] The 5-year stroke rate was 11% in medically treated patients compared with 5.1% for those with carotid endarterectomy. However, several concerns exist about surgery in asymptomatic patients who may be aided by improved plaque characterization. First, the 53% risk reduction was associated with only a 5.9% decrease in absolute risk reduction. Controversy exists then in both the cost-effectiveness and surgical risks associated with 50–80 surgeries preventing one stroke per year.[154] Second, it is unclear that the results can be generalized to the whole population. The surgical risk in this study was lower than previously published as surgeons appeared to be more skilled then average. As pointed out in one review, if the surgical risk was increased by only 2% (more consistent with the community), the positive results of the study would disappear.[155] In addition, the patient population may be of lower risk then the general population of interest. Third, medical therapy used at the time was aspirin and newer approaches (such as aspirin and clopidogrel) may be more effective, changing the interpretation of the data. Finally, the outcome of disabling stroke and stroke death was not statistically different between the two groups (same). Therefore, improved plaque characterization and potentially different treatment approaches would be beneficial with respect to these parameters.

Characteristics of plaque instability are under investigation for reasons analogous to the coronary plaque characterization described above. Pathologic investigations have not conclusively identified those microstructural features that result in plaque-producing embolism, although plaque rupture appears the most likely and which plaques ultimately rupture. In one study, plaque rupture was seen in 74% of symptomatic patients and only 32% of asymptomatic patients. Fibrous cap thinning was seen in 95% of symptomatic patients while 48% in asymptomatic.[156] Plaque stability likely is a balance between infrastructure destruction and the ability of vascular smooth muscle cells to proliferate and lay collagen. This lack of understanding is limited by the fact that histopathologic assessments are only obtained at one time point (either at endarterectomy or autopsy), limiting the predictive power relative to techniques that follow plaques as a function of time. Certain features that may play a role in plaque rupture include intraplaque hemorrhage, cholesterol content, increased calcium, intraplaque fibrin, inflammation, decreased collagen, and ulceration. Hemorrhage was the first characteristic difference noted between symptomatic and asymptomatic plaques, although this opinion is not universally accepted.[157] This is consistent with the increase in angiogenesis noted in carotid plaque, similar to the coronary.[158,159] High lipid content also may be an important microstructural feature. Plaques removed from symptomatic patients have more extracted lipid and cholesterol than asymptomatic patients. Usually the most stenotic portion contains more cholesterol, more calcium, and less collagen. In addition, the lipid plaques with low levels of collagen are associated with plaque ulceration, subintimal hemorrhage, and ischemic neurologic symptoms.[160] However, other investigators have failed to confirm differences between symptomatic and asymptomatic plaques.[161,162] Therefore, a focus of current research uses noninvasive or minimally invasive techniques to follow plaques as a function of time to improve patient risk stratification.

Of note is that at least one distinction may exist between plaque ruptures in the coronary and carotid circulations. In one study, ruptures were found in the midportion of the carotid cap rather than the shoulder.[163] Some difference between the two should not be surprising for several reasons. First, the carotids are perfused at high pressure during systole, while the coronary is perfused primarily at low pressure during diastole. In the coronary, then, rupture may occur during systole by pressure at the top of the plaque from myocardial contraction rather then high velocity blood flow. In the carotid, the plaque is more likely to rupture from shear stress during systole. Second, the coronaries are muscular arteries while the carotids are elastic arteries, which may affect how angiogenesis and vessel breakdown occurs. Therefore, it cannot be assumed without further proof that coronary and carotid disease progress in an identical manner.

15.3.6 Imaging Approaches

Among the most important techniques used for assessing carotid plaques include transcutaneous ultrasound, MRI, CT, angiography, radionuclear imaging, and IVUS. Digital subtraction angiography (DSA) has been generally considered the gold standard for assessing the carotids in the past. However, it has an approximately 1% risk of permanent neurologic deficits with minor asymptomatic infarcts also occurring at high frequency.[164,165] In addition, the vessel is imaged in silhouette rather than cross section, and information of plaque microstructure is not available. Furthermore, angiography does a poor job in assessing ulceration as determined by the North American Symptomatic Carotid Endarterectomy Trial, with a sensitivity and specificity for ulcerative plaques that was 45.9% and 74.1%, respectively. Therefore, other technologies have now moved to the forefront.

External duplex ultrasound is currently the primary screening tool for carotid disease. It is noninvasive, relatively inexpensive, has respectable accuracy for detecting significantly stenotic lesions, and can be performed in a relatively short period of time. However, it is limited by resolution, amount of tissue that needs to be penetrated, and angle with respect to the lumen (which affects assessments of luminal diameter). There are some studies that suggest an ability to at least partially characterize plaque, with one study of heterogeneous plaque by ultrasound having a greater risk for developing cerebral symptoms and infarction than luminal narrowing.[166] However, effective plaque characterization will almost undoubtedly require a technique with superior resolution.

Because of the limitations of ultrasound and angiography, magnetic resonance angiogram (MRA) and contrast enhanced MRA are increasingly used. In a recent meta-analysis of 63 publications, MRI demonstrated 95% sensitivity and 90% specificity for distinguishing lesions above and below 70% cross-sectional diameter.[167] In contrast, the sensitivity and specificity of ultrasound was 86 and 87%, respectively. It has also demonstrated some ability to characterize plaque.[168-170] In addition, the use of intravenous ultrasmall particles of iron oxide (USPIOs) may be used to identify plaque macrophages.[171] However, its ability to define plaque microstructure is currently limited by resolution, it is not performed in real time, and it is relatively expensive. Nevertheless, because of limited motion compared to coronary imaging and the potential to use higher strength magnets (3 Tesla or greater) over this shorter distance, future studies are needed to determine the value of MRI-based techniques for risk-stratifying carotid plaque.

Some authors have suggested that both single slice spiral CT and MSCT can distinguish lipid core and fibrous cap.[172,173] However, this needs to be confirmed, particularly in view of the limited resolution of CT. In addition, MRI has also been shown superior to CT in assessing plaque, particularly when calcium levels are high.[174]

Intravascular ultrasound and nuclear imaging have had only limited application to the carotids to date.[175-177] IVUS has the same limitations as in the coronary, in addition to stroke risk, but potentially has some role in guiding placement. While nuclear techniques can allow assessments of features such as oxidized LDL, inflammation, or thrombus formation, it does so at a resolution insufficient for characterizing plaque.

15.3.7 Treatment

There are several approaches to treating diseases of the carotid, which include carotid endarterectomy, intravascular stenting, and medical therapy.

15.3.7.1 Carotid endarterectomy

Carotid endarterectomy (CE) is the most common method for treating significant stenosis. With CE, the artery is occluded and the brain is perfused through the opposite carotid. Alternatively, the blood may be shunted around the region occluded during the procedure. The artery is then opened and the inner portion of the vessel is "peeled" away, often in one piece. If necessary, a venous graft can be placed into the carotid so it can be widened or repaired. The artery is then closed and blood allowed to flow through it. Missed plaque may be dislodged and result in an embolism. It is estimated that over 134,000 procedures were performed in the United States in 2002, and likely near 155,000 in 2005.[178] In general, the procedure is associated with a low perioperative stroke risk of approximately 2–4% in otherwise younger patients, but on the order of 19% for patients over 80.[152] In addition, certain patients are considered as having unexpectedly high risk. These include those with severe CAD, recent MI, contralateral carotid disease, the need for major organ transplant, significant heart failure, dialysis-dependent renal failure, uncontrolled diabetes, previous radical neck surgery, previous neck irradiation, spinal immobility, plaque burden not completely respectable (e.g., at ostium of the aorta or high bifurcation lesions), contralateral laryngeal nerve paralysis, or severe restenosis after prior endarterectomy.

15.3.7.2 Stent and angioplasty

Since CE is a heavily studied procedure with well-documented benefits, any procedure being evaluated to replace it will receive considerable scrutiny, particularly intravascular-based techniques. Therefore, intracarotid stenting has been extensively evaluated for approximately a decade,[179,180] but is currently limited to patients at high risk for developing CE complications. However, for the reasons discussed above, intravascular guidance for better plaque characterization and stent placement may increase its utilization. Current approaches typically utilize devices to prevent embolization by distal occlusion during the procedure, which includes complaint silicone balloons.[181] The occlusion reduces distal embolization of dislodged plaque fragments. Major studies examining the efficacy of stenting include: The Stenting and Angioplasty with Protection in Patients at High Risk for Endarterectomy (SAPPHIRE), ACCULINK to Revasculation of Carotids in High Risk Patients Registry (ARCHeR), and Registry Study to evaluate the Neuroshield Bare Wire Cerebral Protection System and X-Act Stent in patients at high risk for Carotid Endarterectomy (SECuRITY).[182] These studies suggest that stenting has a lower complication risk than surgery in the high risk group of patients. In a recent survey, the technical success of this procedure was greater than 95%, the complication rate was approximately 5%, and the restenosis rate was 2.4% at 3 years.[183] However, poor stent approximation has been noted and intravascular guidance could be beneficial in improving success rates, as well as better risk stratification of patients to perform stent placement.[184] Of note, if the procedure is to be performed by a non-cardiologist, the doctor should be familiar with managing carotid body induced asystole and hypotension or post-procedure hypertension to avoid cerebral hemorrhage.

15.3.7.3 Medical treatment

Medical therapy for carotid disease, like that for coronary disease, will not be discussed here in detail. First, aggressive treatment of risk factors such as hypertension, diabetes, and hypercholesterolemia should be performed when possible. Second, various antiplatelet or antithrombotic therapies are being examined which include aspirin, clopidogrel, ticlopidine, dipyridamole, and coumadin.[185–187] Their application and in which patient population they are used can be found in current medical texts and reviews.

15.3.7.4 OCT and carotid artery disease

Carotid artery disease was discussed in detail because it is an area in which OCT could represent a powerful diagnostic modality. However, minimal OCT research has been performed in this area to date and it represents an important future application of the technology. An OCT image of a human intracerebral artery is shown in Figure 15.26. There has been significant imaging of the aorta, which is an example of another elastic artery like the corotid that develops atherosclerosis under similar conditions to those of the carotid. An OCT image of the aorta is shown in Figure 15.27.

15.3.7.5 Other potential cerebrovascular applications

There are other disorders in the cerebral circulation that may be amenable to OCT guidance. These include vertebral artery stenosis, AV malformations, unruptured aneurysms (cerebral or subarachnoid), and intracranial artery stenosis.[188] Treatment of vertebral artery stenosis is particularly attractive because it is difficult to approach surgically and stenting is currently associated with a high restenosis rate.[189] These areas will likely be the source of future important OCT research.

15.4 PULMONARY HYPERTENSION

Pulmonary hypertension (PTN) is another potential area where OCT can have a clinical impact. PTN is generally defined as an increase in pulmonary systolic pressure above 30 mmHg or an increase in the mean pressure above 20 mmHg. It is associated with a high mortality rate due to right ventricle failure, arrhythmias,

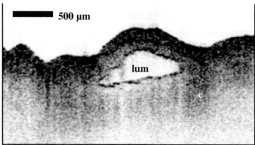

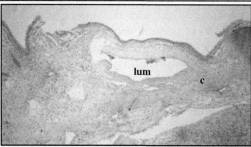

Figure 15.26 Imaging cerebral artery. This figure demonstrates an OCT imaging of a human cerebral artery. The top is the OCT image, the bottom is the histology. The lumen (lum) is clearly delineated in addition to the layers of the arteries.

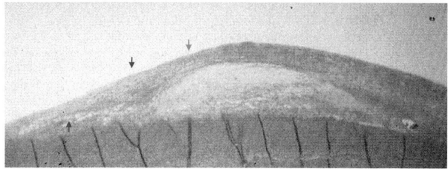

Figure 15.27 OCT imaging of an atherosclerotic elastic artery. In this image, multiple plaques are noted within the artery. The arrows identify lipid collections in addition to the plaque identified by OCT. Courtesy Brezinski, M. E., et al. (1997). *Heart. 77*, 397–404.

pneumonia, or thromboemboli. Currently, the etiology and progression of PTN is diagnosed and followed with right-sided pressures (intracardiac catheterization), angiograms, and lung biopsies, which are discussed below. However, these procedures have limitations that may be overcome with high resolution intravascular imaging.

The pulmonary circulation operates under much lower pressures than the systemic circulation. The normal peak systolic pulmonary pressure of the pulmonary circulation is 22 mmHg with a mean of 14 mmHg. The normal pulmonary circulation has a relatively low resistance and can sustain relatively large variations in cardiac output (volume). The latter is achieved through recruitment, where vessels in the cranial portion of the lung expand as the output into the lung increases. However, while the right side of the heart can withstand large variations in volume, it cannot withstand high pressures. A relatively normal right ventricle begins failing at around 50 mmHg.

PTN is a leading cause of mortality world wide. It can be primary or secondary. Among the common causes of secondary pulmonary hypertension are left-sided heart failure, valvular heart disease (e.g., mitral stenosis or regurgitation), intracardiac shunts (e.g., ventricular septal defect), pulmonary embolism, connective tissue diseases, drug-related disorders, and emphysema. The disease typically presents as exertional shortness of breath and fatigue. The causes of death associated with pulmonary hypertension again are emboli, arrhythmia, right ventricular failure, and pneumonia.[190] The histologic changes associated with pulmonary hypertension, which result in vessel narrowing and reduced compliance, were described by Heath and Edwards. They tend to be the same independent of the secondary cause, and are defined by a classification system which bears their name[191]:

Grade I: Hypertrophy of media in arteries and arterioles

Grade II: Intimal hypertrophy and cellular proliferation

Grade III: Intimal fibrous, diffuse obliteration of arterioles, small muscular pulmonary arteries

Grade IV: Plexiform lesions of small arteries and arterioles that are recannulized thrombi

Grade V: Development of vein like branches of hypertrophied muscular arteries that have a thin wall and have a hyalinization of the intima

Grade VI: Development of necrotizing arteritis

Grade III likely represents the beginning of irreversible changes, the importance of which is discussed below.[192,193] It has also been suggested that medial thickening may be useful in determining irreversibility, thereby predicting poor surgical outcome.[192–194] The disease most likely progresses because of chronic vasoconstriction resulting from hypoxia, acidosis, increased pulmonary blood flow, increased shear stress, or dysfunction of the vascular endothelium.

Primary pulmonary hypertension is a devastating disease where there is no apparent secondary cause. Median age at diagnosis is in the mid-20s and the onset of symptoms to diagnosis is 2 years.[192] Once the diagnosis is made, survival is between 12 and 48 months.[195] Except for the childhood variant, the disorder occurs predominately in females with a ratio of 5:1.[196] The disorder is currently believed to result from endothelial cell dysfunction. A variant has now also been identified in association with HIV, where pulmonary hypertension is 25 times higher than the general population.[194]

There are multiple reasons for evaluating the vascular morphology in patients with pulmonary hypertension, the most important are probably assessing irreversibility and guiding therapy. Irreversibility is the point past which the course of the disease (vascular disease) is no longer reversible. It is a particularly important issue in patients undergoing surgical repair. If patients with underlying irreversibility of pulmonary vascular changes undergo heart transplant, for example, mortality is very high (greater than 50%). In this case, heart transplant is contraindicated for two reasons: since heart donors are scarce, a more appropriate recipient would be indicated and, if surgical treatment is deemed necessary, heart-lung transplant would be the procedure of choice. Currently, the vascular bed is evaluated by hemodynamics, angiography, and open lung biopsy. Hemodynamic measurements are useful for making the diagnosis of pulmonary hypertension, but there is no good correlation between pressures and irreversibility.[195] Angiography is useful for the identification of large emboli and regional disease, but does not establish irreversibility and can induce heart failure in patients with elevated pulmonary artery pressures. Open lung biopsy remains the gold standard, but the high mortality and susceptibility to sampling error from regional variations raises the need to search for alternative diagnostic techniques.[197] Recently, high frequency ultrasound has been introduced for the evaluation of the pulmonary microstructure. Unfortunately, the limited resolution at 30 MHz does not allow consistent identification of pathology.[198] The three-layered appearance is often only definable in large vessels.[199] Furthermore, it does not allow examination of the periphery of the lung or vessels below 1 mm in diameter. Better catheter design is needed, such as a guide balloon, since current catheters poorly maneuver through the circulation.[200]

In addition to identifying irreversibility, high resolution imaging could aid in monitoring the therapeutic response, and possibly differentiating primary from secondary disease. Many disorders that lead to PTN, such as emphysema and connective tissue diseases, require systemic therapy with compounds such as steroids. The ability to monitor the effects of therapy on the progression of vascular disease would be powerful in managing these patients.

To date, OCT imaging of the pulmonary artery has been minimal. However, this represents a particularly attractive area for its application.

15.5 ATRIAL FIBRILLATION

Pulmonary vein ablation is a therapy that has shown great promise for the treatment of atrial fibrillation, of which OCT guidance may play a role. Atrial fibrillation is the most common sustained arrhythmia.[201] Its incidence increases with age and is found in about 5% of the population over the age of 65.[202] It is the change in the normal rhythmic beating of the atria to chaotic electrical excitation. Although it may be picked up by chance on a routine visit in elderly patients, younger patients typically present with palpitation, dyspnea, fatigue, lightheadedness, or embolism (including stroke). It is associated with a high incident of stroke (untreated, approximately 1–10% per year depending on risk factors) and difficulty in controlling heart rate. The most feared complication of atrial fibrillation is stroke since the atria are no longer contracting and therefore are fertile for thrombus formation. This means that lifetime anticoagulation, usually coumadin, and frequent blood tests monitoring the level of anticoagulation are necessary. Therefore, in addition to its high morbidity and mortality, and in view of the incidence of the disease, this represents a significant burden on the healthcare system.

Therefore, other approaches have been pursued including antiarrhythmics, surgery, and catheter-based procedures. Antiarrhythmic drugs may reduce the time to first recurrence of atrial fibrillation, but are of limited long-term efficacy. Current drugs do not typically bring about a cure of the disease process and patients who respond may require suppressive antiarrhythmic drugs for many decades. Antiarrhythmic drugs also carry a risk of ventricular proarrhythmia (life-threatening arrhythmias). The surgical maze procedure involves the creation of surgical incisions and cryolesions in the right and left atria after placing the patient on cardiopulmonary bypass.[203] Advantages of this procedure include a high cure rate and removal of the left atrial appendage (the site of most thrombus formation). This procedure is best suited to patients who are undergoing cardiopulmonary bypass for concomitant mitral valve surgery or other indications. It is more difficult to justify in patients with lone atrial fibrillation due to the risks and costs associated with the procedure.

Catheter ablation of atrial fibrillation involves two approaches — the catheter maze procedure and more recently, focal catheter ablation within the pulmonary veins. The catheter maze procedure is aimed at replicating the surgical maze procedure by the creation of linear atrial lesions with RF ablation catheters.[203] The advantage of the catheter procedures include the avoidance of a thoracotomy and cardiopulmonary bypass. However, with the conventional ablation used with the maze, the creation of these long lesions in the left atrium has been associated with a risk of thromboembolic stroke and pulmonary vein stenosis making it relatively unattractive.

Pulmonary vein ablation has recently been developed for the treatment of both new onset and paroxysmal atrial fibrillation, attacking its trigger.[204–206] It has been found that the pulmonary veins may be a source of rapid discharging foci. Catheter ablation of such foci may result in a freedom from atrial fibrillation recurrence in a large percent of patients with atrial fibrillation. In one *New England Journal of Medicine* study, 94% of patients had triggers in the pulmonary veins.[207]

The risk associated with pulmonary vein ablation includes the development of pulmonary vein stenosis.[208] If RF ablation is too extensive, stenosis will result. If too little, atrial fibrillation is not prevented. The procedure is currently monitored via surface temperature, which has been shown to be unreliable.[209] The finding of

| **Histology** | **OCT** |

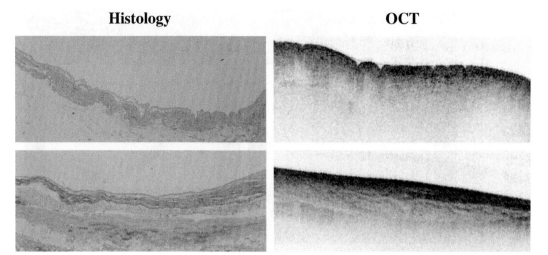

Figure 15.28   OCT imaging of normal goat pulmonary vein. Shown here are pulmonary vein histology on the left and OCT images on right. These examples are taken from the junction of the vein (at left) with the left atrium (at right). The cell surface interfaces and subendocardial myocyte layer are noted. The muscular layer is thicker to the right, and thins out to the left of the image, as the vein leaves the atrium. Unpublished data from D. Keane and M. Brezinski.

a focal source of atrial fibrillation within the pulmonary veins at first may be surprising, however, *in utero* the pulmonary veins receive an investiture of atrial myocardium that is capable of firing rapidly and triggering paroxysms of atrial fibrillation in later life.[209] It seems likely that the optimal time to ablate such focal sources may be early in the clinical course of paroxysmal or new onset atrial fibrillation, before the patient develops the atrial substrate for chronic atrial fibrillation (through the process of electrical and structural remodeling). OCT has demonstrated some potential for monitoring pulmonary vein ablation. *In vitro* studies in our lab have shown that ablation can be monitored in real time. An *in vitro* image of a pulmonary vein and the corresponding histology is shown in Figure 15.28. OCT sharply differentiates the media from the adventitia. In Figure 15.29, *in vitro* cryoablation is performed of *in vitro* pulmonary vein at 40 seconds. Structural changes are obvious in the image. Future *in vivo* work is needed to establish its utility for guiding ablation.

15.6   CONCLUSION

Cardiovascular applications of OCT remain among the most important for the technology. Potential cardiovascular applications are not limited to those discussed above, but these remain among the most promising applications in the near future.

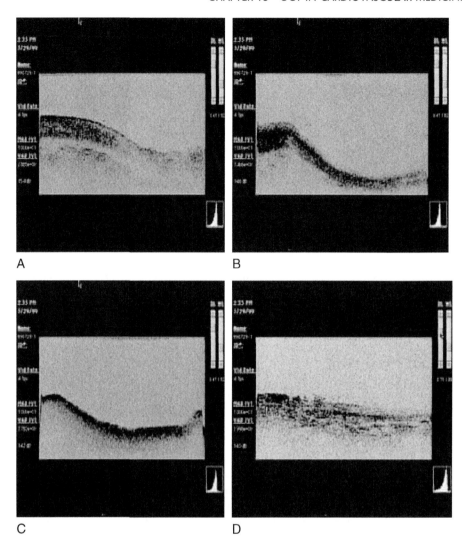

Figure 15.29 Cryoablation of pulmonary vein. A goat pulmonary vein is cryoablated from 0 to 40 second. It can be seen that changes in the vessel are noted through the ablation process. Unpublished data from D. Keane and M. Brezinski.

REFERENCES

1. Heart Disease and Stroke Statistics, www.americanheart.org
2. Hurst, W. (2002). *The Heart, Arteries and Veins*, 10th ed. McGraw-Hill, New York.
3. National Health Expenditures Amounts, and Average Annual Percent Change, by Type of Expenditure: Selected Calendar Years 1990–2013.
4. www.cms.hhs.gov.

5. Yusuf, S., Reddy, S., Ounpuu, S., and Anand, S. (2001). Global burden of cardiovascular disease. *Circulation* 104, 2746–2753.

6. Yusuf, S., Hawken, S., Ounpuu, S. et al. (2004). Effect of potentially moditiable risk factors associated with myocardial infarction in 52 countries. *Lancet*, 364, 937–952.

7. O'Conner, S., Taylor, C., Campbell, L. A. et al. (2001). Potential infectious etiologies of atherosclerosis: A multifactorial perspective. *Emerg. Infect. Diseases* 7, 780–788.

8. Stary, H. C., Chandler, A. B., Dinsmore, R. E. et al. (1995). A definition of advanced types of atherosclerotic lesions and histological classification of atherosclerosis. A report from the Committee on Vascular Lesions of the Council on Arteriosclerosis, American Heart Association. *Circulation* 92, 1355–1374.

9. McGill, H. C., McMahan, A., Zieske, A. W. et al. (2000). Association of coronary heart disease risk factors with microscopic qualities of coronary atherosclerosis in youth. *Circulation* 102, 374–379.

10. Kolodgie, F. D., Burke, A. P., Farb, A. et al. (2001). The thin cap fibroatheroma: A type of vulnerable plaque. *Curr. Opin. Cardiol.* 16, 285–292.

11. Farb, A., Burke, A. P., Tang, A. I. et al. (1996). Coronary plaque erosion without rupture into a lipid core: a frequent cause of coronary thrombosis in sudden death. *Circulation* 93, 1354–1363.

12. Virmani, R., Burke, A. P., Farb, A., and Kolodgie, F. D. (2002). Pathology of unstable plaque. *Prog. Cardiovasc. Dis.* 44, 349–356.

13. Burke, A. P., Kolodgie, F. D., Farb, A. et al. (2001). Healed plaque ruptures and sudden death. Evidence that subclinical rupture has a role in plaque progression. *Circulation* 103, 934–940.

14. Ambrose, J. A., Tannenbaum, M. A., Alexopoulos, D. et al. (1988). Angiographic progression of coronary artery disease and the development of myocardial infarction. *Am. Coll. Cardiol.* 12(1), 56–62.

15. Little, W. C., Constantinescu, M., Applegate, R. J. et al. (1988). Can coronary angiography predict the site of a subsequent myocardial infarction in patients with mild to moderate coronary artery disease? *Circulation* 78(5), 1157–1166.

16. Habib, G. B., Heibig, J., and Forman, S. A. (1991). Influence of coronary collateral vessels on myocardial infarct size in humans. *Circulation* 83(4), 739–746.

17. Glagor, S., Weisenberg, E., Zaris, C. K. et al. (1987). Compensatory enlargement of human atherosclerotic arteries. *N. Engl. J. Med.* 316, 1371–1375.

18. Pasterkamp, G., Wensing, P., Post, M. et al. (1995) Paradoxical arterial wall shrinkage may contribute to luminal narrowing of human atherosclerotic femoral arteries. *Circulation* 91, 1444–1449.

19. Schoenhagen, P., Ziada, K. M., Kapadia, S. R. et al. (2000). Extent and direction of arterial remodeling in stable versus unstable coronary syndromes. *Circulation* 101, 598–603.

20. Pasterkamp, G., Schoneveld, A. H., Haudenschild, C. C. et al. (1998). Relation of arterial geometry to luminal narrowing markers for plaque vulnerability: The remodeling paradox. *J. Am. Coll. Cardiol.* 32, 655–662.

21. Falk, E. (1983). Plaque rupture with severe pre-existing stenosis precipitating coronary thrombosis. *Br. Heart J.* 50, 127–131.

22. Goldstein, J. A., Demetriou, D., Grines, C. L. et al. (2000). Multiple complex coronary plaques in patients with acute myocardial infarction. *N. Engl. J. Med.* 343, 915–922.

23. Rioufol, G., Finet, G., Ginon, I. et al. (2002). Multiple atherosclerotic plaque rupture in acute coronary syndrome: three-vessel intravascular ultrasound study. *Circulation* 106, 804–808.

24. Asakura, M., Udea, Y., Yamaguchi, O. et al. (2001). Extensive development of vulnerable plaques as a pan-coronary process in patients with myocardial infarction. *J. Am. Coll. Cardiol.* 37, 1284–1288.

25. Kolodgie, F. D., Gold, H. K., Burke, A. P. et al. (2003). Intraplaque hemorrhage and progression of coronary atheroma. *N. Engl. J. Med.* 349, 2316–2325.

26. Virmani, R., Kolodgie, F. D., Burke, A. P. et al. (2000). Lessons from sudden coronary death. *Arterioscler. Thromb. Vasc. Biol.* 20, 1262–1275.

27. Loree, H. M., Kamm, R. D., Stringfellow, R. G., and Lee, R. T. (1992). Effects of fibrous cap thickness on peak circumferential stress in model atherosclerotic vessels. *Circ. Res.* 71, 850–858.

28. Herrmann, J. M., Pitris, C., Bouma, B. E, Boppart, S. A., Jesser, C. A., Stamper, D. L., Fujimoto, J. G., and Brezinski, M. E. (1999). High resolution imaging of normal and osteoarthritic cartilage with optical coherence tomography. *J. Rheumatol.* 26, 627–635.

29. Giattina, S., Courtney, B., Herz, P., Harman, M., Shortkroff, S., Stamper, D. L., Liu, B., Fujimoto, J. G., and Brezinski, M. E. (2006). Measurement of coronary plaque collagen with polarization sensitive optical coherence tomography (PS-OCT). *Int. J. Cardiol.*, 107: 400–409.

30. Shak, P. K., Falk, E., Badimon, J. J. et al. (1995). Human monocyte derived macrophage collagen breakdown in fibrous caps of atherosclerotic plaques, potential role of matrix degrading metalloproteinases and implications for plaque rupture. *Circulation* 92, 1565–1569.

31. Libby, P. (2001). Current concepts in the pathogenesis of the acute coronary syndromes. *Circulation* 104, 365–372.

32. Buffo, A., Biasucci, L. M., and Liuzzo, G. (2002). Widespread coronary inflammation in unstable angina. *N. Engl. J. Med.* 347, 5–12.

33. Tofler, G. H., Stone, P. H., Maclure, M. et al. (1990). Analysis of possible triggers of acute myocardial infarction. *Am. J. Cardiol.* 66, 22–27.

34. Geiringer, E. (1951). Intimal vascularization and atherosclerosis. *J. Pathol. Bacteriol.* 63, 201–211.

35. Tenaglia, A. N., Peters, K. G., Sketch, M. H. et al. (1998) Neovascularization in atherectomy specimens from patients with instable angina. *Am. Heart J.* 135, 10–14.

36. Pasterkamp, G. and Virmani, R. (2002). The erythrocyte: a new player in atheromatous core formation. *Heart* 88, 115–116.

37. Arbustini, E., Morbini, P., D'Armini, A. M. et al. (2002). Plaque composition in plexogenic and thromboembolic pulmonary hypertension: the critical role of thrombotic material in pultaceous core formation. *Heart* 88, 177–182.

38. www.hcup.ahrq.gov.

39. Dotter, C. T. and Judkins, M. P. (1964). Transluminal treatment of arteriosclerotic obstruction: Description of a new technique and a preliminary report of its application. *Circulation* 30, 654–661.

40. Gruentzig, A. R., Senning, A., and Seigenthaler, W. E. (1979). Non-operative dilatation of coronary artery stenosis-percutaneous transluminal coronary angioplasty. *N. Engl. J. Med.* 301, 61–68.

41. Roubin, G. S., Cannon, A. D., Agrawal, S. K. et al. (1992). Multicenter investigation of coronary stenting to treat acute or threatened closure after percutaneous transluminal coronary angioplasty: clinical and angiographic outcomes. *J. Am. Coll. Cardiol.* 22, 135–143.

42. Popma, J. J., Califf, R. M., and Topol, E. J. (1991). Clinical trials of restenosis after coronary angioplasty. *Circulation* 84, 1426–1436.

43. Dotter, C. T. (1969). Transluminally placed coil spring endarterial tube grafts. Long-term patency in canine popliteal artery. *Invest. Radiol.* 4, 329–334.

44. Goy, J., Sigwart, U., Vogt, P. et al. (1991). Long term follow-up of the first 56 patients treated with intracoronary self-expanding stents. *Am. J. Cardiol.* 67, 569–575.

45. Pearce, B. J. and McKinsey, J. F. (2003). Current status of intravascular stents as delivery devices to prevent restenosis. *Vasc. Endovasc. Surg.* 37, 231–237.

46. Farb, A., Sangiorgi, G., Carter, A. J. et al. (1999). Pathology of acute and chronic coronary stenting in humans. *Circulation* 99, 44–52.

47. Barth, K. H., Virmani, R., Froelich, J. et al. (1996). Paired comparison of vascular wall reaction to palmaz stents. *Circulation* 93, 2161–2169.

48. George, G., Haude, M., and Ge, J. (1995). Intravascular ultrasound after low and high inflation pressure coronary artery stent implantation. *J. Am. Coll. Cardiol.* 26, 725–730.

49. de Jaegere, P., Mudra, H., Figulla, H. et al. (1998). Intravascular ultrasound guided optimized stent deployment. *Eur. Heart J.* 19, 1214–1223.

50. Colombo, A., Hall, P., Nakamura, S. et al. (1995). Intracoronary stenting without anticoagulation accomplished with intravascular ultrasound guidance. *Circulation* 91, 1676–1688.

51. Sheris, S. J., Canos, M. R., and Weissman, N. J. (2000). Natural history of intravascular ultrasound detected edge dissections from coronary stent deployment. *Am. Heart J.* 139, 59–63.

52. van der Hoeven, B. L., Pires, N. M., Warda, H. M. et al. (2005). Drug-eluting stents: Results, promises, and problems. *Int. J. Cardiol.* 99, 9–17.

53. Marx, S. O. and Marks, A. (2001). Bench to bedside: the development of rapamycin and its application to stent restenosis. *Circulation* 104, 852–855.

54. Morice, M. C., Serruys, P. W., Sousa, J. E. et al. (2002). A randomized comparison of sirolimus-eluting stent with a standard stent in patients with stenosis in native coronary artery. *N. Engl. J. Med.* 346, 1773–1780.

55. Schofer, J., Schluter, M., Gershlick, A. H. et al. (2003). Sirolimus eluting stents in patients with long atherosclerotic lesions in small coronary arteries. *Lancet* 362, 1093–1099.

56. Gershlick, A.H. (2002). European evaluation of paclitaxel eluting stent, ELUTES. *Circulation* 105, E39.

57. Park, S. J., Shim, W. H., Ho, D. S. et al. (2003). A paclitaxel eluting stent for the prevention of coronary restenosis. *N. Engl. J. Med.* 348, 1537–1545.

57a. Grude, E., Silber, S., Hauptmann, K. E. et al. (2003). TAXUS I. *Circulation* 107, 38–42.

58. Colombo, A., Drzewiecki, J., Banning, A. et al. (2003). Randomized study to assess the effectiveness of slow and moderate release polymer based paclitaxel eluting stent. *Circulation* 108, 788–794.

59. Lemos, P., Saia, F., Lightart, J. M. et. al. (2003). Coronary re-stenosis after sirolimus-eluting stent implantation. *Circulation* 108, 257–260.

60. Iteirstein, P. S., Massullo, V., Jani, S. et al. (1997). Catheter based radiotherapy to inhibit restenosis after coronary stenting. *N. Engl. J. Med.* 336, 1697–1703.

61. Bacq, Z. and Alexander, P. *Fundamentals of Radiobiology, 2ⁿᵈ Edition, Pure and Applied Biology: Modern Trends in Physiological Sciences.* Butterworths, London.

62. King, S. B., Williams, D. O., Chougule, P. et al. (1998). Endovascular beta radiation to reduce restenosis after coronary balloon angioplasty. *Circulation* 97, 2025-2030.

63. Waksman, R., White, L., Chan, R. C. et al. (2000). Intracoronary gamma radiation therapy after angioplasty inhibits recurrence in patients with in-stent restenosis. *Circulation* 101, 2165–2171.

64. Bernard, O. F., Mongrain, R., Lehnert, S. et al. (1997). Intravascular radiation therapy in atherosclerotic disease: promises and premises. *Eur. Heart J.* 18, 1385–1395.

65. Condado, J. A., Gurdiel, O., Espinoza, R. et al. (1996). Late follow-up after percutaneous transluminal coronary angioplasty and intracoronary radiation therapy (IRCT). *Discoveries in Radiation for Restenosis,* S. B. King, R. Waksman, and I. R. Crocker (eds.). Emory School of Medicine Press, Atlanta, GA, p. 105.

66. Waksman, R. (1996). Local catheter-based intracoronary radiation therapy for restenosis. *Am. J. Cardiol.* 78 (Suppl. 3A), 23–28.

67. Meerkin, D., Tardiff, J. C., Bertrand, O. F. et al. (2000). The effects of intracoronary brachytherapy on the natural history of postangioplasty dissections. *J. Am. Coll. Cardiol.* 36, 59–64.

68. Costa, M.A., Sabate, M., van der Giessn, J. et.al. (1999). Late coronary occlusion after intracoronary brachytherapy. *Circulation* 100, 789–792.

69. Giri, S., Ito, S., Lansky, A. et al. (2001). Clinical and angiographic outcome in the laser angioplasty for restenotic stents (LARS) multicultural registry. *Cathet. Cardiovasc. Intervent.* 52, 24–34.

70. Mehran, R., Dangas, G., and Mintz, G. (2000) Treatment of in-stent restenosis with excimer laser coronary angioplasty versus rotational atherectomy. *Circulation* 101, 2484–2489.

71. Davies, J. R., Rudd, J. F., Fryer, T. D., and Weissberg, P. L. (2005). Targeting the vulnerable plaque: The evolving role of nuclear imaging. *J. Nucl. Cardiol.* 12, 234–246.

72. Ambrose, J. A., Tannenbaum, M. A., Alexopoulos, D. et al. (1988). Angiographic progression of coronary artery disease and the development of myocardial infarction. *Am. Coll. Cardiol.* 12(1), 56–62.

73. Little, W. C., Constantinescu, M., Applegate, R. J. et al. (1988). Can coronary angiography predict the site of a subsequent myocardial infarction in patients with mild to moderate coronary artery disease? *Circulation* 78(5), 1157–1166.

74. Jost, S., Deckers, J. W., Nikutta, P. et al. (1994). Evolution of coronary stenosis is related to baseline severity — a prospective quantitative angiographic analysis in patients with moderate coronary stenosis. INTACT Investigators. *Eur. Heart J.* 15(5), 648–653.

75. van Geuns, R., Wielopolski, P. A., de Bruin, H. G. et al. (1999). Magnetic resonance imaging of the coronary arteries: Techniques and results. *Prog. Cardiovasc. Dis.* 42, 157–166.

76. Toussaint, J., Lamuraglia, G. M., Southern, J. F. et al. (1996). Magnetic resonance images lipid, fibrous, calcified and thrombotic components of human atherosclerotic in vivo. *Circulation* 94, 932–938.

77. Worthley, S., Helft, G., Fuster, V. et al. (2000). Serial in vivo MRI documents arterial remodeling in experimental atherosclerosis. *Circulation* 101, 586–589.

78. Paulin, S., von Schulthess, G. K., Fossel, E. et al. (1987). MR imaging of the aortic root and proximal coronary arteries. *Am. J. Roentgenol.* 148, 665–670.

79. Alfidi, R. J., Masaryk, T. J., Haacke, E. M. et al. (1987). MR angiography of the peripheral, carotid, and coronary arteries. *Am. J. Roentgenol.* 149, 1097–1109.

80. Edelman, R. R., Manning, W. J., Burstein, D. et al. (1991). Breath-hold MR angiography. *Radiology.* 181, 641–643.

81. Li, D., Kaushikkar, S., Haacke, E. M. et al. (1996). Coronary arteries: Three-dimensional MR imaging with retrospective respiratory gating. *Radiology* 201, 857–863.

82. Botnar, R. M., Stuber, M., Danias, P. G. et al. (1999). Improved coronary artery definition with T2-weighted, free-breathing, three-dimensional coronary MRA. *Circulation* 99, 3139–3148.

83. van Geuns, R. J. M., Wielopolski, P. A., Rensing, B. et al. (1998). Clinical evaluation of breath-hold MR coronary angiography using targeted volumes. *Proc. Int. Soc. Magn. Reson. Med.* 1, 34, (abstract).

84. Wielopolski, P. A., Manning, W. J., Edelman, R. R. et al. (1995). Single breath-hold volumetric imaging of the heart using magnetization-prepared 3-dimensional segmented echo planar imaging. *J. Magn. Reson. Imaging 5*, 403–409.

85. Wielopolski, P. A., Oudkerk, M., de Bruin, H. et al. (1996). Breath-hold 3D MR coronary angiography. *Radiology* 210(P), 273 (abstract).

86. Wielopolski, P. A., van Geuns, R. J. M., deFeyter, P. et al. (1998). Breath-hold coronary MR angiography with volume targeted imaging. *Radiology* 209, 209–219.

87. van Geuns, R. J. M., Wielopolski, P. A., Rensing, B. et al. (1998). Clinical evaluation of breath-hold MR coronary angiography using targeted volumes. *Proc. Int. Soc. Magn. Reson. Med.* 1:34 (abstract).

88. Maria, G. M,, Hunink, M. D., Kuntz, K. M. et al. (1999). Noninvasive imaging for the diagnosis of coronary artery disease: focusing the development of new diagnostic technology. *Ann. Intern. Med.* 131, 673–680.

89. Hibberd, M. G., Vuille, C., and Weyman, A. E. (1992). Intravascular ultrasound: basic principles and role in assessing arterial morphology and function. *Am. J. Card. Imaging* 6(4), 308–324.

90. Fitzgerald, P., Oshima, A., Hayese, M. et al. (2000). Final results of the Can Routine Ultrasound Influence Stent Expansion study (CRUISE). *Circulation* 102, 523–530.

91. Nissen, S. E. (2002). Applications of intravascular ultrasound to characterize coronary artery disease and assess the progression of atherosclerosis. *Am. J. Cardiol.* 89, 24B–31B.

92. de Korte, C. L. and van der Steen, A. F. (2002). Intravascular ultrasound elastography: An overview. *Ultrasonics* 40(1–8), 859–865.

93. van der Steen, A. F. W., de Korte, C. L., and Cespedes, E. I. (1998). Intravascular elastography. *Ultraschall. Med.* 19, 196–201.

94. Schoepf, U. J., Becker, C. R., Ohnesorge, B. M., and Kent Yucel, E. (2004). CT of coronary artery disease. *Radiology* 232, 18–37.

95. O'Rouke, R. A., Brundage, B. H., Froelicher, V. F. et al. (2000). American College of Cardiology/American Heart Association expert consensus document on EBCT for the diagnosis and prognosis of coronary artery disease. *J. Am. Coll. Cardiol.* 36, 326–340.

96. Schroeder, S., Kopp, A. F., Baumbach, A. et al. (2001). Noninvasive detection and evaluation of atherosclerotic coronary plaques with multislice computed tomography. *J. Am. Coll. Cardiol.* 37, 1430–1435.

97. Casscells, W., Hathorn, B., David, M. et al. (1996). Thermal detection of cellular infiltrates in living atherosclerotic plaques. *Lancet,* 347, 1447–1449.

98. Stefanadis, C., Toutouzas, K., Tsiamis, E. et.al. (2001). Increased local temperature in human coronary atherosclerotic plaques: an independent predictor of clinical outcome in patients undergoing a percutaneous coronary intervention. *J. Am. Coll. Cardiol.* 37, 1277–1283.

99. Courtney, B.K., Nakamura, M., Tsugita, R. et.al. (2004). Validation of a thermographic guidewire for endoluminal mapping of atherosclerotic disease: An in vitro study. *Cathet. Cardiovasc. Intervent.* 62, 221–229.

100. Weinmann, P., Jouan, M., Dao, N. et al. (1999). Quantitative analysis of cholesterol and cholesteryl esters in human atherosclerotic plaques using near infrared Raman spectroscopy. *Atherosclerosis* 140, 81–88.

101. Jaross, W., Neumeister, V., Lattke, P. et al. (1999). Determination of cholesterol in atherosclerotic plaques using near infrared diffuse reflection spectroscopy. *Atherosclerosis* 147, 327–337.

102. Cassis, L. and Lodder, R. A. (1993). Near-IR imaging of atheromas in living arterial tissue. *Anal. Chem.* 65, 1247–1258.

103. Bennett, M. R. (2002). Breaking the plaque: Evidence for plaque rupture in animal models of atherosclerosis. *Arterioscler. Thromb. Vasc. Biol.* 22, 713–714.

104. Rekhter, M. D. (2002). How to evaluate plaque vulnerability in animal models of atherosclerosis? *Cardiovasc. Res.* 54, 36–41.

105. Nakashima, Y., Plump, A. S., Raines, E. W. et al. (1994). ApoE-deficient mice develop lesions of all phases of atherosclerosis throughout the arterial tree. *Arterioscler. Thromb.* 14, 133–140.

106. Ishibashi, S., Goldstein, J. L., Brown, M. S. et al. (1994). Massive xanthomatosis and atherosclerosis in cholesterol-fed low density lipoprotein receptor-negative mice. *J. Clin. Invest.* 93, 1885–1893.

107. Kumai, T., Oonuma, S., Kitaoka, Y. et al. (2003). Biochemical and morphological characterization of spontaneously hypertensive hyperlipidaemic rats. *Clin. Exp. Pharmacol. Physiol.* 30, 537–544.

108. Nakamura, M., Abe, S., and Kinukawa, N. (1997). Aortic medial necrosis with or without thrombosis in rabbits treated with Russell's viper venom and angiotensin II. *Atherosclerosis* 128, 149–156.

109. Celletti, F. L., Waugh, J. M., Amabile, P. G. et al. (2001). Vascular endothelial growth factor enhances atherosclerotic plaque progression. *Nature Med.* 7, 425–429.

110. Brezinski, M. E., Tearney, G. J., Bouma, B. E., Izatt, J. A., Hee, M. R., Swanson, E. A., Southern, J. F., and Fujimoto, J. G. (1996). Optical coherence tomography for optical biopsy: Properties and demonstration of vascular pathology. *Circulation* 93, 1206–1213.

111. Brezinski, M. E., Tearney, G. J., Bouma, B. E., Boppart, S. A., Hee, M. R., Swanson, E. A., Southern, J. F., and Fujimoto, J. G. Imaging of coronary artery microstructure with optical coherence tomography. *Am. J. Cardiol.* 77(1), 92–93.

112. Brezinski, M. E., Tearney, G. J., Weissman, N. J., Boppart, S. A., Bouma, B. E., Hee, M. R., Weyman, A. E., Swanson, E. A., Southern, J. F., and Fujimoto, J. G. (1997). Assessing atherosclerotic plaque morphology: comparison of optical coherence tomography and high frequency intravascular ultrasound. *Heart* 77, 397–404.

113. Patwari, P., Weissman, N. J., Boppart, S. A., Jesser, C. A., Stamper, D. L., Fujimoto, J. G., and Brezinski, M. E. (2000). Assessment of coronary plaque with optical coherence tomography and high frequency ultrasound. *Am. J. Cardiol.* 85, 641–644.

114. Tearney, G. J., Brezinski, M. E., Boppart, S. A., Bouma, B. E., Weissman, N. J., Southern, J. F., Swanson, E. A., and Fujimoto, J. G. (1996). Catheter based optical imaging of a human coronary artery. *Circulation* 94(11), 3013.

115. Tearney, G. J., Brezinski, M. E., Bouma, B. E., Boppart, S. A., Pitris, C., Southern, J. F., and Fujimoto, J. G. (1997). In vivo endoscopic optical biopsy with optical coherence tomography *Science* 276, 2037–2039.

116. Fujimoto, J. G., Boppart, S. A., Tearney, G. J., Bouma, B. E., Pitris, C., and Brezinski, M. E. (1999). High resolution in vivo intra-arterial imaging with optical coherence tomography. *Heart* 82, 128–133.

117. Li, X., Gold, H., Weissman, N. J., Saunders, K., Pitris., C., Fujimoto, J. G., and Brezinski, M. E. Assessing stent approximation with oct and comparison with IVUS in an in vivo rabbit model (unpublished).

118. Lightlab Imaging Inc. (Westford, MA), www.lightlabimaging.com.

119. Yabushita, H., Bouma, B. E., Houser, S. L. et al. (2002). Characterization of human atherosclerosis by OCT. *Circulation* 106, 1640–1645.

120. Jang, I. K., Tearney, J. G., and Bouma, B. E. (2001). Visualization of tissue prolapse between coronary stent struts by optical coherence tomography. *Circulation* 104, 2754–2759.

121. Brezinski, M., Saunders, K., Jesser, C., Li, X., and Fujimoto, J. (2001). Index matching to improve OCT imaging through blood. *Circulation* 103, 1999–2003.

122. Bouma, B. E., Nelson, L. E., Tearney, J. G., Jones, D. J., Brezinski, M. E., and Fujimoto, J. G. (1998). Optical coherence tomographic imaging of human tissue at 1.55 mm and 1.8 mm using Er- and Tm-doped fiber sources. *J. Biomed. Opt.* 3, 76–79.

123. Schenck, J. and Brezinski, M. (2002). Ultrasound induced improvement in optical coherence tomography. *Proc. Natl. Acad. Sci.* 99, 9761–9764.

124. Rogowska, J. and Brezinski, M. E. (2001). Evaluation of the adaptive speckle suppression filter for coronary optical coherence tomography imaging. *IEEE Trans. Med. Imaging* 19, 1261–1266.

125. Tearney, G. J., Yabushita, H., Houser, S. L. et al. (2003). Quantification of macrophage content in atherosclerotic plaques by optical coherence tomography. *Circulation* 107, 113–119.

126. Macneil, B. D., Jang, I. K., Bouma, B. E. et al. (2004). Focal and multi-focal plaque macrophage distributions in patients with acute and stable presentations of coronary artery disease. *J. Am. Coll. Cardiol.* 44, 972–979.

127. Bartos, F. and Ledvina, M. (1979). Collagen, elastin, and desmosines in three layers of bovine aortas of different ages. *Exp. Gerontol.* 14, 21–26.

128. Katsuda, S., Okada, Y., Minamoto, T. et al. (1992). Collagens in human atherosclerosis. *Arterioscler. Thromb.* 12, 494–502.

129. Rekher, M. D., Hicks, G. W., Brammer, D. W. et.al. (2000). Hypercholesterolemia causes mechanical weakening of rabbit atheroma. *Circ. Res.* 86, 101–108.

130. Giattina, S. D., Courtney, B. K., Herz, P. R. et al. (2005). Measurement of coronary plaque collagen with polarization sensitive optical coherence tomography (PS-OCT). Cardiovascular Revascularization Therapies Mtg., Washington, D. C., p. 609.

131. Giattina, S., Courtney, B., Herz, P., Harman, M., Shortkroff, S., Stamper, D. L., Liu, B., Fujimoto, J. G., and Brezinski, M. E. (2006). Measurement of coronary plaque collagen with polarization sensitive optical coherence tomography (PS-OCT). *Int. J. Cardiol.*, 107, 400–40.

132. Schmitt, J. M. (1998). OCT elastography: imaging microscopic deformation and strain in tissue. *Opt. Express* 3(6), 199–211.

133. Rogowska, J., Patel, N., Fujimoto, J., and Brezinski, M. E. (2004). Quantitative OCT elastography technique for measuring deformation and strain of arterial tissue. *Heart* 90(5), 556–562.

134. Rogowska, I., Patel, N., and Brezinski, M. E. Quantitative optical coherence tomography elastography, method for assessing arterial mechanical properties. *B. J. Radiology* (in press).

135. Liu, B., Macdonald, E. A., Stamper, D. L., and Brezinski, M. E. (2004). Group velocity dispersion effects with water and lipid in 1.3 μm OCT system. *Phys. Med. Biol.* 49, 1–8.

136. Nasr, M. B., Saleh, B. E., Sergienko, A. V., and Teich, M. C. et. al. (2003). Demonstration of dispersion cancellation quantum optical tomography. *Phys. Rev. Lett.* 91, 083601-1 to 083601-4.

137. Brezinski, M. E. (2004). Optical Coherence Tomography NIH Optical Imaging Workshop. National Institutes of Health. Bethesda, MD., Sept. 20, 2004.

138. Tran, P. H., Mukai, D. S., Brenner, M., and Chen, Z. In vivo endoscopic optical coherence tomography by use of a rotational microelectromechanical system probe. *Opt. Lett.* 29, 1236–1238.

139. Zara, J. M., Izatt, J. A., Rao, K., Tazdanfar, S., and Smith, S. W. (2002). Scanning mirror for optical coherence tomography using an electrostatic MEMs actuator. IEEE Int. Symp. Biomed. Imaging, 297–300.

140. Cordero, H., Warburton, K., Underwood, L. et al. (2001). Initial experience and safety in the treatment of chronic total occlusions with fiber optic guidance technology: optical coherent reflectometry. *Cathet. Cardiovasc. Intervent.* 54, 180–187.

141. Suero, J., Marso, S., Jones, P. et al. (2001). Procedural outcomes and long-term survival among patients undergoing percutaneous coronary intervention of a chronic total occlusion in native coronary arteries: A 20-year experience. *J. Am. Coll. Cardiol.* 38, 409–414.

142. Puma, J., Sketch, M., Tchieng, J. et al. (1995). Percutaneous revascularization of chronic coronary occlusions: An overview. *J. Am. Coll. Cardiol.* 26, 1–11.

143. Rosamond, W. D., Folsom, A. R., Chambless, L. E., Wang, C., McGovern, P. G., Howard, G., Copper, L. S., and Shahar, E. (1999). Stroke incidence and survival among middle-aged adults: 9-year follow-up of the atherosclerosis risk in communities (ARIC) cohort. *Stroke* 30, 736–743.

144. Woo, D., Gebel, F., Miller, R., Kothari, R., Brott, T., Khoury, J., Salisbury, S., Shukla, R., Pancioli, A., Jauch, E., and Broderick, J. (1999). Incidence rates of first-ever ischemic stroke subtypes among blacks: A population-based study. *Stroke* 30, 2517–2522.

145. Hurst, W. (2002). *The Heart Arteries and Veins.* 10th edition. McGraw-Hill, New York.

146. Howard, G., Wagenknecht, L. E., Cai, J., Cooper, L., Kraut, M. A., and Toole, J. F. (1998). Cigarette smoking and other risk factors for silent cerebral infarction in the general population. *Stroke* 29, 913–917.

147. Bryan, R. N., Wells, S. W., Miller, T. J., Elster, A. D., Jungreis, C. A., Poirier, V. C., Lind, B. K., and Manolio, T. A. (1997). Infarct like lesions in the brain: prevalence and anatomic characteristics at MR imaging of the elderly — data from the Cardiovascular Health Study. *Radiology* 202, 47–54.

148. Schneider, A. T., Pancioli, A. M., Khoury, J. C., Rademacher, E., Tuchfarber, A., Miller, R., Woo, D., Kissela, B., and Broderick, J. P. (2003). Trends in community knowledge of the warning signs and risk factors for stroke. *JAMA* 289, 343–346.

149. Ingall, T. J. (2000). Preventing ischemic stroke. *Postgrad Med* 107, 34–50.

150. Wolf, P. A., Abbott, R. D., and Kannel, W. B. (1991). Atrial fibrillation as an independent risk factor for stroke: the Framingham Study. *Stroke* 22, 983.

151. Debakey, M. E. (1975). Successful carotid endarterectomy for cerebrovascular insufficiency. *JAMA* 233, 1083–1085.

152. Hallett, J.W., Pietropaoli, J. A., Ilstrup, D. M. et al. (1998). Comparison of North American Symptomatic Carotid Endarterectomy Trial and population-based outcomes for carotid endarterectomy. *J. Vasc. Surg.* 27, 845–851.

153. Executive Committee for the Asymptomatic Carotid Atherosclerosis. Endarterectomy for asymptomatic carotid artery disease. *JAMA* 273, 1421–1428.

154. Warlow, C. (1995). Endarterectomy for asymptomatic carotid stenosis? *Lancet* 345, 1254–1255.

155. Dodick, D. W., Meissner, I., Meyer, F. B., and Cloft, H. J. (2004). Evaluation and management of asymptomatic carotid artery stenosis. *Mayo Clin. Proc.* 79, 937–944.

156. Carr, S., Farb, A. Pearce, W. H. et al. (1996). Atherosclerotic plaque rupture in symptomatic carotid artery stenosis. *J. Vasc. Surg.* 23, 755–766.

157. Imparato, A. M., Riles, T. S., Mintzer, R., and Baumann, F. G. (1988). The importance of hemorrhage in the relationship between gross morphologic characteristics and cerebral symptom in 376 carotid artery plaques. *Ann. Surg.* 197, 195–203.

158. McCarthy, M. J., Loftus, I. M., Thompson, M. M. et al. (1999). Angiogenesis and atherosclerotic carotid plaque: An association between symptomatology and plaque morphology. *J. Vasc. Surg.* 30, 261–268.

159. Jezlorska, M. and Wooley, D. E. (1999). Local neovascularization and cellular composition within vulnerable regions of atherosclerotic plaques of human carotid arteries. *J. Pathol.* 188, 189–196.

160. Seeger, J. M., Barratt, E., Lawson, G. A. et al. (1995). The relationship between carotid plaque composition, plaque morphology, and neurologic symptoms. *J. Surg. Res.* 58, 330–336.

161. Hatsukami, T. S., Ferguson, M. S., Beach, K. W. et al. (1997). Carotid plaque morphology and clinical events. *Stroke* 28, 95–100.

162. Bassiouny, H. S., Davis, H., Massawa, N. et al. (1989). Critical carotid stenosis: Morphologic and chemical similarity between symptomatic and asymptomatic plaques. *J. Vasc. Surg.* 9, 202–212.

163. Carr, S., Farb, A., Pearce, W. H. et al. (1996). Atherosclerotic plaque rupture in symptomatic carotid artery stenosis. *J. Vasc. Surg.* 23, 755–766.

164. Cloft, H. J., Joseph, G. J., and Dion, J. E. (1999). Risk of cerebral angiography: A meta-analysis. *Stroke* 30, 317–320.

165. Bendszus, M., Koltzenburg, M., Burger, R. et al. (1999). Silent embolism in a diagnostic cerebral angiography and neurointerventional procedures: A prospective study. *Lancet* 354, 1594–1597.

166. Leahy, A. L., McCollum, P. T., Feeley, T. M. et al. (1988). Duplex ultrasonography and selection of patients for carotid endarterectomy: Plaque morphology or luminal narrowing? *J. Vasc. Surg.* 8, 558–562.

167. Nederkoorn, P. J., van der Graff, Y., and Hunink, M. G. (2003). Duplex ultrasound and magnetic resonance angiography compared with digital subtraction angiography in carotid stenosis. *Stroke* 34, 1324–1332.

168. Hatsukamai, T. S., Ross, R., Polissar, N. L., and Yuan, C. (2000). Visualization of fibrous cap thickness and rupture in human atherosclerotic carotid plaque in vivo with high resolution magnetic resonance imaging. *Circulation* 102, 959–964.

169. Trivedi, R. A., Graves, M. J., Horsley, J. et al. (2004). MRI derived measurements of fibrous cap and lipid core thickness: the potential for identifying vulnerable carotid plaques in vivo. *Neuroradiology* 46, 738–743.

170. Yaun, C., Kerwin, W. S., Ferguson, M. S. et al. (2002). Contrast-enhanced high resolution MRI for atherosclerotic carotid artery tissue characterization *J. Magn. Reson. Imaging* 15, 62–67.

171. Kooi, M. E., Cappendijk, V. C., Cleutjens, K. B. et al. (2003). Accumulation of ultrasmall superparamagnetic particles of iron oxide in human atherosclerotic plaques can be detected by in vivo magnetic resonance imaging. *Circulation* 107, 2453–2458.

172. Estes, J. M., Quist, W. C., Costello, P. et al. (1998). Noninvasive characterization of plaque morphology using helical computed tomography. *J. Cardiovasc. Surg.* 39, 527–534.

173. Oliver, T. B., Lammie, G. A., Wright, A. R. et al. (1999). Atherosclerotic plaque at carotid bifurcation: CT angiographic appearance with histologic correlation. *Am. J. Neuroradiol.* 20, 897–901.

174. Randoux, B., Marro, B., Koskas, F. et al. (2001). Carotid artery stenosis: Prospective comparison of CT, three dimensional gadolinium enhanced MR, and conventional angiography. *Radiology* 220, 179–185.

175. Weissman, N. J., Mintz, G. S., Laird, J. R. et al. (2000). Carotid artery intravascular ultrasound: Safety and morphologic observations during carotid stenting. *J. Am. Coll. Cardiol.* 35 (Suppl.), 10A.

176. Clark, D. J., Lessio, S., O'Donoghue, M., Schainfeld, R., and Rosenfield, K. (2004). Safety and utility of intravascular ultrasound guided carotid artery stenting. *Cathet. Cardiovasc. Intervent.* 63, 355–362.

177. Davies, J. R., Rudd, J. F., Fryer, T. D., and Weissberg, P. L. (2005). Targeting the vulnerable plaque: The evolving role of nuclear imaging. *J. Nucl. Cardiol.* 12, 234–246.

178. Centers for Disease Control, www.cdc.gov.

179. Dierthrich, E. B., Ndiaye, M., Reid, D. B. et al. (1996). Stenting in the carotid arteries, initial experience in 110 patents. *J. Endovasc. Surg.* 3, 42–62.

180. Yaday, J. S., Roubin, G. S., Iyer, S. et al. (1997). Elective stenting of the extracranial carotid arteries. *Circulation* 95, 376–381.

181. Albuquerque, F. C., Teitelbaum, G. P., Lavine, S. D. et al. (2000). Balloon protected carotid angiography. *Neurosurgery* 46, 918–923.

182. Gray, W. A. (2004). A cardiologist in the carotids *J. Am. Coll. Cardiol.* 43, 1602–1605.

183. Wholey, M. H. and Al-Mubarek, N. (2003). Updated review of the global carotid artery stent registry. *Cathet. Cardiovasc. Intervent.* 60, 259–266.

184. Clark, D. J., Kessio, S., O'Donoghue, M. et al. (2004). Safety and utility of intravascular ultrasound guided carotid artery stenting. *Cathet. Cardiovasc. Intervent.* 63, 355–362.

185. Schellinger, P. D. and Hacke, W. (2005). Stroke: Advances in therapy. *Lancet Neurol.* 4, 2.

186. Ling, G. S. and Ling, S. M. (2005). Preventing stroke in the older adult. *Cleve. Clin. J. Med.* 72 (Suppl. 3), S14–S25.

187. Hankey, G. J. (2005). Preventable stroke and stroke prevention. *J. Thromb. Haemost.* 3, 1638–1645.

188. Doerfler, A., Becker, W., Wanke, I. et al. (2004). Endovascular treatment of cerebrovascular disease. *Curr. Opin. Neurol.* 17, 481–487.

189. Wehman, J. C., Hanel, R. A., Guidot, C. A. et al. (2004). Atherosclerotic occlusive extracranial vertebral artery disease. *J. Interven. Cardiol.* 17, 219–232.

190. Rich, S., Dantzker, D., Ayres, S., Bergofsky, E. et al. (1987). Primary pulmonary hypertension. A national prospective study. *Ann. Intern. Med.* 107, 216–223.

191. Heath, D. and Edwards, J. E. (1958). The pathology of hypertensive pulmonary disease. *Circulation* 18, 533–547.

192. Bush, A., Haworth, S. G., Hislop, A. A., Knight, W. B., Corrin, B., and Shinebourne, E. A. (1988). Correlations of lung morphology, pulmonary vascular resistance, and outcome in children with congenital heart disease. *Br. Heart J.* 59, 480–485.

193. Friedli, B., Kidd, B. S. L., Mustard, W. T., and Keith, J. D. (1974). Ventricular septal defect with increased pulmonary vascular resistance: Late results of surgical closure. *Am. J. Cardiol.* 33, 403–409.

194. Petitpretz, P., Brenot, F., Azarian, R., Parent, F., Rain, B., Herve, P., and Simmoneau, G. (1994). Pulmonary hypertension in patients with HIV. *Circulation* 89, 2722–2727.

195. Wagenvoort, C. A. (1975). Pathology of congestive pulmonary hypertension. *Prog. Resp. Res.* 9, 195–202.

196. Moraes, D. and Loscalzo, J. (1997). Pulmonary hypertension: Newer concepts in diagnosis and management. *Clin. Cardiol.* 20, 676–682.

197. Davies, L., Dolgin, S., and Kattan, M. (1997). Morbidity and morality of open lung biopsy in children. *Pediatrics* 99, 660–664.

198. Day, R. W. and Tani, L. Y. (1997). Pulmonary intravascular ultrasound in infants and children with congenital heart disease. *Cathet. Cardiovasc. Diagn.* 41, 395–398.

199. Ishii, M., Kato, H., Kawano, T., Akagi, T., Maeno, Y., Sugimura, T., Hashino, K., and Takagishi, T. (1995). Evaluation of pulmonary artery histopathologic findings in congenital heart disease: An in vitro study using intravascular ultrasound. *JACC* 26, 272–276.

200. Day, R. W. and Tani, L. Y. (1997). Pulmonary intravascular ultrasound in infants and children with congenital heart disease. *Cathet. Cardiovasc. Diagn.* 41, 395–398.

201. Braunwald, E. (2001). *Heart Disease: A Textbook of Cardiovascular Medicine*. W. B. Saunders Company, Philadelphia, PA.

202. Kopecky, S. L., Gersh, B. J., McGoon, M. D. et al. (1987). The natural history of lone atrial fibrillation: A population based study over three decades. *N. Engl. J. Med.* 317, 669–674.

203. Cox, J. L. and Ad, N. (2000). New surgical and catheter-based modifications of the Maze procedure. *Sem. Thorac. Cardiovasc. Surg.* 12, 68–73.

204. Jais, P., Haissaguerre, M., Shah, D. et al. (1997). A focal source of atrial fibrillation treated by discrete radiofrequency ablation. *Circulation* 95, 572–576.

205. Haissaguerre, M., Jais, P., Shah, D. et al. (1998). Spontaneous initiation of atrial fibrillation by ectopic beats originating in the pulmonary veins. *N. Engl. J. Med.* 339, 659–666.

206. Hsieh, M., Chuen-Wang, C., Wen, Z. et al. (1999). Alterations of heart rate variability after radiofrequency catheter ablation of focal atrial fibrillations originating from pulmonary veins. *Circulation* 100, 2237–2243.

207. Haissaguerre, M., Jais, P., Shah, D. et al. (1998). Spontaneous initiation of atrial fibrillation by ectopic beats originating in the pulmonary veins. *N. Engl. J. Med.* 339, 659–666.

208. Robbins, I. M., Colvin, E. V., Doyle, T. P. et al. (1998). Pulmonary vein stenosis after catheter ablation of atrial fibrillation. *Circulation* 98, 1769–1775.

209. Chauvin, M., Shah, D., Haissaguerre, M. et al. (2000). The anatomic basis of connections between the coronary sinus musculature and the left atrium in humans. *Circulation* 101, 647–654.

210. Patel, N. A., Li, X., Stamper, D. L., Fujimoto, J. G., and Brezinski, M. E. (2003). Guidance of aortic ablation using optical coherence tomography. *Int. J. Cardiovasc. Imaging* 19(2), 171–178.

211. Staper, D., Weismann, N. J. and Brezinski, M. E. (2006). Plague characterization with optical coherence tomography. *J. Am. Coll. Card.* 47(8), C69–79.

212. Brezinski, M. E. (2006). Optical coherence tomography for identifying unstable coronary plague. *Int. J. Card.* 107, 154–165.

213. Chen, W. Q., Zang, Y., Zhang, M. et al. (2004). Induction of atherosclerotic plaque instability in rabbits after transfection of human wild type p53 gene. *Chin. Med. J.* 117, 1293–1299.

16 OCT IN THE MUSCULOSKELETAL SYSTEM

Chapter Contents

There are a large number of potential applications for OCT in the musculoskeletal system, but likely the most important are in the early diagnosis and monitoring of osteoarthritis, monitoring of non-mechanical arthritic disorders, and the assessment of ligament and tendon disorders.

16.1   OSTEOARTHRITIS

16.1.1   General

Osteoarthritis or degenerative joint disease (DJD) is the most common joint disease, affecting an estimated 43 million Americans in 2002.[1] It is the leading cause of disability in the United States. Its cure or effective management could be one of the largest breakthroughs for the healthcare system, resulting in improved quality of life to an ever aging population. The disorder primarily affects the knee, hip, vertebral, temporomandibular, and distal interphalangeal (finger) joints. Although all tissue within the osteoarthritic joint exhibits morphologic changes, the hallmark of the disease is cartilage degeneration.

A need exists for an imaging technique to identify early disease or subtle changes in moderate disease so that the effectiveness of therapeutics can be monitored. This is not achievable with current diagnostic techniques. The need for a method to assess the course of therapeutics for osteoarthritis (OA), emphasizing its importance as a healthcare problem, has recently lead the NIH to introduce the OA initiative.[2]

Risk factors for OA include age, gender (women), weight/height ratio, muscle imbalance (such as weak hamstrings relative to the quadriceps), abnormal gait, weight, injury to a joint, and likely genetics.[1]

16.1.2   Cartilage Structure

Most articular cartilage consists of hyaline cartilage, a highly organized form of cartilage well designed for its function in the joint. Joint cartilage without evidence of OA consists of approximately 70–80% water.[3] Collagen makes up 50% of the dry weight of cartilage, while proteoglycans (PG) constitute an additional 10%.[4] Collagen is a fibrous protein that has a rod-like shape and is insoluble in water.[5] OCT assessment of collagen is among its greatest diagnostic attributes. Collagen's fundamental unit is tropocollagen, a molecule with three peptide chains, in a helix, of about 1000 residues each. Collagen molecules can associate and crosslink to form larger fibers. Articular cartilage is composed of primarily type II collagen fibers with types V, VI, IX, and XI present to a lesser degree. There are little to no type I fibers present, which is the predominate collagen in the body. While the exact arrangement varies from joint to joint and person to person, Benninghoff's model has generally become widely accepted. In this model organization of these fibers is arranged throughout the cartilage into zones. In knee cartilage in Figure 8.3, the most superficial zone, the lamina splendens, accounts for 5–10% of the total thickness of the cartilage.[6] Here the collagen fibrils are oriented parallel to the articular surface. The transitional zone, the largest percent width, contains tangential fibers that blend with radial fibers. The deep zone is tangential to the surface. As the fibers enter the deepest area, the calcified zone, it is believed that their orientation is rotated and intertwined.

The PG consist of glycosaminoglycans (GAGS) bound to protein. The most prevalent GAGS are chondroitin and keratan sulfate and small but significant quantities of hyaluronic acid.[7] The high water content is believed to provide the cartilage with its elastic resistance to compression while the collagen is critical in maintaining the structural integrity or form of the tissue.[8,9]

16.1.3   Osteoarthritis Cartilage

In joints eventually progressing to clinically evident OA, biochemical changes in the cartilage occur prior to macroscopic abnormalities such as fibrillation. These include an increase in water content, a loss in orientation of the surface collagen fibers, an increase in the proportion of chondroitin sulfate, and a relative reduction in keratan sulfate.[3,9,10] However, there is some evidence that microfractures in the subchondral bone are among the first changes. Because of the collagen loss with subsequent increase in water content, the cartilage frequently is thickened followed later by progressive thinning. The first macroscopic changes include erosion and flaking of the cartilaginous surface, replication of chondrocytes creating small clonal clusters, and depletion of PG.[3] With advance of the disease, clefts and fibrillations form, focal cystic areas (filled with fibrous tissue) appear within the subchondral bone, and blood vessels from subchondral bone penetrate into the articular cartilage. With further progression, the cartilaginous layer is more deeply eroded or entirely denuded in focal areas. This loss of cartilage accounts for the radiographic thinning of the joint space. At these later stages, osteophytes or bony outgrowths also develop at the margin of the articular surface and often into extra-articular structures.

16.1.4   Knee Anatomy

A significant percentage of the work performed will be on humans or animal knees, so an overview of the human knee anatomy is presented. The anatomy of the knee is shown in Figure 16.18. Knee menisci, the C-shaped fibrocartilage located between the femur and tibia cartilage, are important for load-bearing and shock-absorption in the joint and for keeping the joint stable. The anterior cruciate ligament (ACL) prevents the femur from sliding excessively forward. The medial collateral ligament (MCL) prevents the femur from moving medially across the joint.

16.1.5 Current Imaging Modalities

Recent reports in animals suggest that the progression of articular cartilage damage in OA may be modified. However, their efficacy has not been confirmed in humans because there is no method of effective evaluation. Symptoms such as pain correlate poorly with disease progression since cartilage lacks nerves and symptoms are occurring from supportive tissue. Similarly, current imaging techniques are insufficient for this purpose. Therefore, a true clinical need exists for better methods to image early changes in cartilage and monitor the progression of changes at all stages.[11–13] A variety of noninvasive and minimally invasive methods have been applied to the assessment of cartilage pathology including radiography, ultrasound, computed tomography (CT), magnetic resonance imaging (MRI), and arthroscopy, but they have not been effective for high resolution assessment of cartilage.[14]

It should be emphasized that for imaging techniques, registration of location from one examination to the next remains critical, as this is essential for monitoring the effectiveness of therapeutics. It becomes particularly difficult when the joint changes at different time points, such as due to swelling or alterations in the joint space. The National Institute of Arthritis and Musculoskeletal and Skin Disease (NIAMS) has recently reviewed registration as part of the Osteoarthritis Initiative.[15]

Conventional radiography is the most common imaging method for assessing the integrity of joints and plays a major role in the field. It is relatively inexpensive and is able to identify only gross changes in the joint space (severe disease). However, as cartilage is only a few millimeters thick, it is not a technique for assessing more subtle changes required to monitor therapeutics. The relatively low resolution and the inability to assess many areas of the cartilage surface, in addition to microstructural changes such as altered water content and collagen breakdown, prevent its general application to the assessment of early cartilage changes. Both digital and microfocal radiography have been used to improve the sensitivity with only limited success.[14] When conventional radiography is used quantitatively, standardized protocols (registration) are needed at each time point including the use of devices which maintain a constant joint position.[2]

Externally applied "low frequency" (<10 MHz) sonography has not played a large role in the evaluation of joint spaces due to its relatively low resolution (greater than 500 μm) and access to only limited areas of the joint.[16] Furthermore, measurements of joint space narrowing obtained from radiographs appear more reproducible and precise than estimates of cartilage thickness obtained from external sonography.[17] High frequency intra-articular sonographic imaging has the potential to substantially improve resolution. Current 25-MHz probes have a resolution in the range of 125 μm.[17] However, high frequency ultrasound requires the presence of a transducing medium, has a limited penetration, the transducers are relatively large, and imaging typically requires an operating room. Furthermore, even the potential application of a 40-MHz transducer will not yield resolutions greater than 80–90 μm and calcium attenuation becomes a concern.

Though CT is superior to conventional radiography in assessing joint changes such as soft tissue and osseous abnormalities, it also has a resolution greater than 500 μm, making it unfeasible for evaluation of early cartilage changes.[18] It is also expensive for routine screening and time-consuming to implement.

MRI represents a powerful tool for evaluating joint abnormalities, particularly when using coils, contrast, magnetic strength, and various sequences.[19,20] The many different protocols used for cartilage are reviewed elsewhere and will not be discussed here in detail. It is more sensitive than CT or radiography for assessing the extent of osteoarthritic changes such as subchondral sclerosis, meniscal degeneration, and osteophytosis. MRI has had some success in identifying cartilage softening, thinning, collagen breakdown, and swelling.[21–24] It is currently one of the most promising noninvasive techniques for evaluating cartilage. However, despite its advantages, MRI has not achieved widespread use for the detection of chondral abnormalities. Difficulties include its high cost, especially if serial screening (of multiple joints) is required, relatively low resolution, and the long time necessary for data acquisition. Furthermore, it has yet to be demonstrated that cartilage width could be measured with sufficient precision (and effectively registered) to monitor the effects of ongoing therapy. Finally, as pointed out in previous chapters, MRI involves a variety of magnetic strengths, as well as different protocols and contrast agents, a fact that should be kept in mind when comparing techniques in the literature. To include all these techniques, the time for acquisition and interpretation of data would be prohibitive. Furthermore, many of these techniques may not be available to the average hospital. Therefore, while MRI is an excellent technique for giving a whole picture of the joint, it probably will not be the technique of choice for routinely following therapy.

Direct visualization of intra-articular structures is achieved with the techniques of standard and needle arthroscopy.[25,26] The advantage of arthroscopy is that it provides a direct, magnified view of the articular surfaces of the joint as well as other supportive structures such as the menisci. Direct visualization is limited by an inability to image below the tissue surface, which prevents the precise determination of cartilage thickness. Furthermore, arthroscopy is currently an operating room technique that makes it impractical for routine screening. There is a less invasive needle arthroscope, which uses a small-diameter fiber scope ($\sim$2 mm), that could potentially be applied in an office setting.[29] However, it is relatively invasive and registration is difficult.

Micro-CT and diffraction enhanced X-ray are two techniques that recently have shown high resolution in small samples, but their cost, risk, and practicality need to be examined.[27]

16.1.6 Serum and Urine Markers

Serum and urine markers of OA have been examined over recent years, and these include type II collagen degradation, keratan sulfate epitopes, and cytokines. However, they likely serve as a marker of a systemic arthritis burden rather than as a method for managing problems associated with a single (few) joint(s). It is unclear how one or two joints can be managed by following the body's total arthritic activity. Hopefully, though, these markers will demonstrate a role in identifying patients in a higher risk group. This topic has been reviewed extensively at www. niams.nih.gov/ne/oi/oabiomarwhipap.htm, the NIH for the osteoarthritis initiative (NIAMS).

16.1.7 Chondroprotection and Cartilage Repair

Essentially all current clinically used therapeutic agents are directed at pain relief, not chondroprotection or cartilage repair. This is problematic as the ultimate goal is to protect or repair cartilage. Chondroprotection is a method that prevents cartilage from breaking down. There are also cartilage repair techniques that attempt to replace damaged cartilage.

16.1.7.1 Methods of chondroprotection

A variety of pharmaceuticals have been suggested as chondroprotective agents. A few will be mentioned below including tetracyclines, GAGS, polysulfuric acid, and pentosan polysulfate.[28]

In addition to their antimicrobial effect, it has been suggested that tetracyclines inhibit tissue metalloproteinases, which breakdown cartilage.[29,30] This may be due to their ability to chelate calcium and zinc ions. Doxycycline, a derivative of tetracycline, has been shown to inhibit collagenase and gelatinase activity in articular cartilage and also to reduce the severity of OA in animal models.

Glycosaminoglycan polysulfuric acid (GAGPS) influences collagen breakdown via its effect on collagenase.[31] It is a highly sulfated GAG. In animals, medial femoral condyle lesions have been shown to be reduced with GAGPS treatment. In a five-year trial in humans, improvement in multiple parameters, such as time to return to work, has been shown but again following symptoms is an unreliable method for assessing progression of disease. In addition, adverse reactions included allergic reactions and heparin-like effect.

Pentosan polysulfate is an extract of beech hemicellulose. It can inhibit the release and action of granulocyte elastase. Animal studies have demonstrated its ability to preserve cartilage PGs.[32-33]

Other potential chondroprotection agents include glycoaminoglycan-peptide complex, growth factors, hyaluronic acid, and cytokines.[34]

16.1.7.2 Methods of cartilage repair

There are a variety of methods that are currently being evaluated for repairing cartilage. These include changing the load on the joint, penetration of the sub-chondral plate (which will be referred to here as the microawl procedure), tissue transplantation, mesenchymal stem cell transplant, and chondrocyte transfer. Some of the techniques are more effective on small defects while others are useful for large lesions.

The theory behind changing the load is to mechanically realign the joint so that reaction force is directed away from diseased cartilage and toward the remaining healthy joint surface.[3] However, changing the load has fallen out of favor because, although some patients report dramatic improvement, the relief after these procedures is usually partial and temporary, with satisfactory results in a majority of patients lasting between 3 and 12 years.

The microfracture technique is probably the most common procedure used in an attempt to regenerate full thickness defects.[35,36] The microfracture technique is used for degenerative lesions <3 cm^2 in humans. Microfracture surgical technique is performed using a surgical awl (i.e., device to make small holes) to make multiple small holes (microfractures) through the subchondral bone plate of focal full thickness chondral defects. The holes are separated 1–2 mm apart. Meticulous care is given to ensure the most peripheral of the lesions is penetrated by the awl to aid the healing of reparative cartilage tissue to the surrounding articular surface. The microfractured area provides a rough surface for the adherence of blood clot from bleeding through the bone perforations, which contains undifferentiated mesenchymal cells from the subchondral bone and leads to cartilage formation. However, the criteria that lead to procedure success (e.g., presence of adjacent normal tissue, clot adhesion to cartilage, and complete perforation of the subchondral bone) and the long-term outcome of the procedure (whether hyaline cartilage ultimately develops) have not been established.

Cartilage transplant is another procedure currently used in an attempt to repair cartilage. The current practice is to transplant the cartilage with the underlying bone to capitalize on bone to bone healing, since the cartilaginous cap by itself will not bond to the recipient bone or cartilage (it is avascular). Various biological adhesives are under investigation to further improve bonding. The two types of tissue transplantation are autologous grafts (sample from patient) and allografts (sample from another person).[37–39]

Studies have shown that autologous grafts can last for more than a decade. However, because of the small number of possible donor sites from which osteochondral autologous grafts may be obtained, such as the patella, use of these grafts has been limited to selected localized regions of damaged articular cartilage. In addition, other concerns are donor site morbidity and limitations in treatment size. Allografts are attractive because they have greater availability (postmortem or post-resection) and because they can be prepared in any size. One study looking at 92 patients has found that 75% were viable at 5 years, 64% at 10 years, and 63% at 14 years. More research is required to further improve success rates, such as with potential pharmaceutical therapies.

The limited ability of other techniques has led investigators to seek methods of transplanting cells that can form cartilage into chondral and osteochondral defects.[40] Experimental work has shown that both chondrocytes and undifferentiated mesenchymal cells placed in articular cartilage defects survive and produce a new matrix. Recently, the use of chondrocytes and undifferentiated mesenchymal cells in gels on patients in small studies resulted in hyaline cartilage in 50–75% of patients. An attractive aspect of the procedure is that it can be used over very large defects. However, the inherent disadvantages of the technique are adhesions and an extremely prolonged recovery time.

16.1.8 *In Vitro* and *In Vivo* **Human Data**

The potential role of OCT in OA is threefold. First in assessing the extent of OA. Second, with chondroprotective agents the focus is on assessing their effectiveness in treating OA. Third, with techniques for cartilage repair, the focus is on assessing their efficacy. For example, with the microawl procedure, OCT imaging would have several important roles. This would include confirming that the edges of "normal" tissue assessed visually actually contain no significant OA, that micropenetrations are actually through the subchondral plate, and that the development of the new cartilage is monitored.

Data currently demonstrating the potential of OCT for article cartilage imaging are shown.[41] Figures 16.1 and 16.2 are OCT images of normal cartilage. Figure 16.1 is a human patella where the red arrows identify the bone cartilage interface. Cartilage width is defined at 10 μm resolution. In Figure 16.2, a series of images of interphalangeal joints are shown where cartilage (C) and bone (B) are seen. The supportive tissue(s) is also noted.

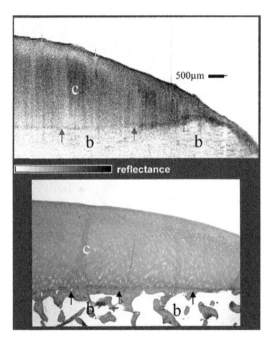

Figure 16.1 OCT imaging of normal human cartilage. The top is the OCT image, the bottom is the histopathology. Cartilage (c) and bone can be clearly identified in vitro along with the bone cartilage interface in this normal patella. Figure courtesy Herrmann, J. et al. (1999). *J. Rheumatolog.* 26(3), 627–635. [41]

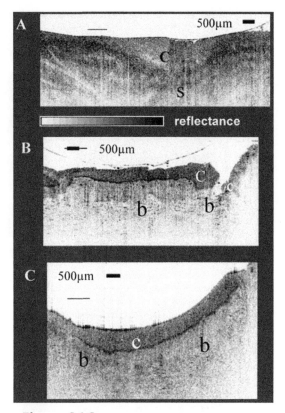

Figure 16.2 OCT imaging of normal metatarsal cartilage. The three images were obtained of the articular surfaces of the metatarsal joints, emphasizing the ability of OCT to image small joints such has the hands and feet. In the images, cartilage (c), bone (b), and ligaments (s) are noted. Image courtesy Herrmann, J. et al. (1999). *J. Rheumatolog.* 26(3), 627–635. [41]

Figure 16.3 demonstrates pathologic articular cartilage that has become fibrocartilage. The fibrous bands are clearly delineated (F). Figure 16.4 is an osteoarthritic femoral head where a fibrous band has developed on the surface (f), the cartilage is thinned to the left, and the bone cartilage interface is disrupted (nb).

An important research focus for assessing cartilage is the application of PS-OCT.[42] In Figure 16.5, the normal tissue is on the left while the osteoarthritic tissue is on the right. The tissue is imaged at two polarization states and the polarization sensitivity of imaging is lost in the diseased tissue. The polarization effect in the normal tissue is shown by a series of bands in the tissue that result from rotation of the polarization state due to tissue birefringence. The position of the bands changes with changing incident polarization.

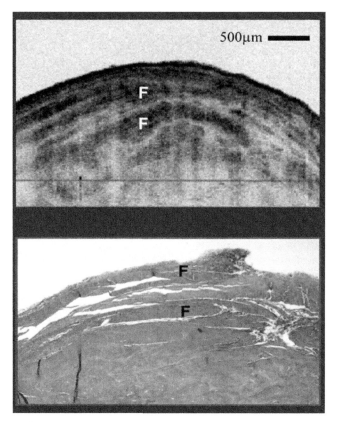

Figure 16.3 OCT imaging of fibrocartilage. Fibrocartilage is one form of joint degradation in osteoarthritis. In the top OCT image, fibrous bands are noted (f). Image courtesy Herrmann, J. et al. (1999). *J. Rheumatolog.* 26(3), 627–635. [41]

In Figure 16.6, the polarization effect has been removed in normal cartilage by treating the tissue with collagenase. The untreated tissue is on the left of the figure, the collagenase treated sample is on the right.

Studies were then extended to examine the phenomena with comparisons with picrosirius stained tissue. In Figure 16.7, images a, b, and c demonstrate the polarization-sensitive changes in this section of normal cartilage (each at different polarization state). In Figure 16.7d the cartilage is grossly normal, and in Figure 16.7e there is a section stained with picrosirius red (polarization sensitive) that demonstrates organized collagen. Positive birefringence is shown as a brightly stained area.

In Figure 16.8, images a, b, and c demonstrate polarization-sensitive changes that are more irregular compared to Figure 16.7. In the picrosirius stained section, the staining is also irregular, consistent with the OCT images.

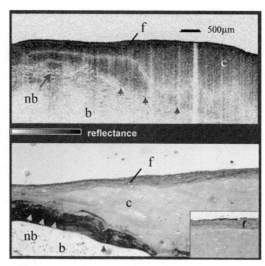

Figure 16.4 OCT imaging of severely diseased cartilage of the femoral head. In this image, severe factors are noted. First, the cartilage is thinned on the left. Second, on the left identified by nb is disruption of the bone-cartilage interface. Third, a fibrous band is noted at the surface of the cartilage. Image courtesy Herrmann, J. et al. (1999). *J. Rheumatolog.* 26(3), 627–635. [41]

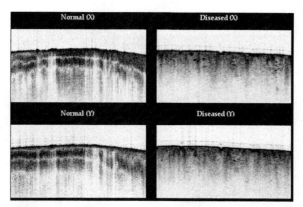

Figure 16.5 Polarization sensitive imaging of cartilage. This represents our first work suggesting that osteoarthritic cartilage losses its polarization sensitivity (which will be confirmed in the data to follow). On the left is normal cartilage with the characteristic banding pattern. On the right is osteoarthritic cartilage where the banding pattern has been lost. Image courtesy Herrmann, J. et al. (1999). *J. Rheumatolog.* 26(3), 627–635. [41]

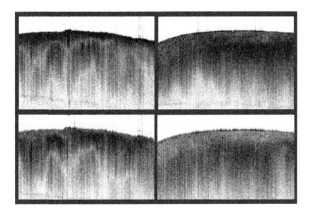

Figure 16.6 Use of collagenase to remove PS-OCT sensitive imaging. On the left, OCT imaging is performed of mildly diseased cartilage where this banding pattern is essentially intact. Mildly diseased cartilage need to be used since the collagenase enzyme was too large to penetrate normal cartilage significantly. After exposure to collagenase, the banding pattern is lost, suggesting it was secondary to organized collagen. Unpublished data from M. Brezinski and J.G. Fujimoto.

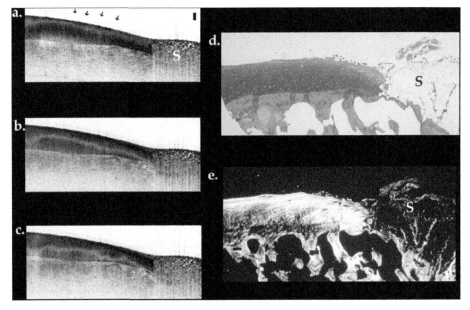

Figure 16.7 PS-OCT sensitivity of normal cartilage. On the left (a, b, and c), cartilage is imaged at three different polarization states which results in shifts in the well defined bands. In e, the picrosirius stained section shows homogeneously bright cartilage which suggests strongly organized collagen. Image courtesy of Drexler, W., et al. (2001). *J. Rheumatol.* 28, 1311–1318. [42]

Figure 16.8 PS-OCT imaging of mildly diseased cartilage. In the OCT images (a, b, and c), the banding pattern is present but not as organized as in figure 16.8. In e, organized collagen is present by picrosirius but areas of disorganization are also noted. Image courtesy Drexler, W., et al. (2001). *J. Rheumatol.* 28, 1311–1318. [42]

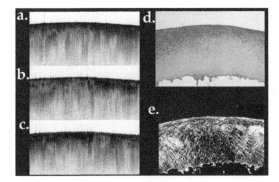

Figure 16.9 PS-OCT imaging of severely osteo-arthritic cartilage. In the OCT images, no banding pattern is noted. In the picrosirius stained section, there is no evidence of organized collagen. However, in figure d, the cartilage appears to be of normal thickness. Therefore, osteoarthritis is detected by OCT prior to cartilage thinning. Image courtesy Drexler, W., et al. (2001). *J. Rheumatol.* 28, 1311–1318. [42]

In Figure 16.9a, b, and c, polarization sensitivity is lost in the cartilage. While the cartilage is thick and shows no gross fibrillations or erosions, it has lost its birefringence as determined by picrosirius staining. This is consistent with the hypothesis that OCT could be used as an early indicator of OA by identifying collagen disorganization.

The correlation between OCT polarization sensitivity and birefringence by picrosirius staining was assessed quantitatively.[42] Twenty-four *in vitro* samples were examined. The OCT images and histology were graded with a three-point system by a blinded investigator. The mean score for OCT images was 0.5 +/− 0.2. The scores for picrosirius stained samples read by a blinded investigator were 0.5 +/− 0.2 mean. There was only one sample that differed in score between the OCT images and the blinded investigator, thus there was no significant difference between these groups.

In Figure 16.10, *in vivo* images of human osteoarthritic joints are shown using a hand-held imaging probe on the patient's knees just prior to joint resection. On the left of each image is relatively normal tissue while that on the right is significantly more diseased. The diseased sections are less dense and the birefringence is substantially weaker. An *in vivo* OCT image of knee cartilage with the corresponding histology is shown in Figure 16.11. The right side shows no cartilage while the cartilage on the left is heterogeneous due to disease. Clinical trials are underway to examine the ability of OCT to predict long-term outcomes after microawl procedures and meniscectomy.

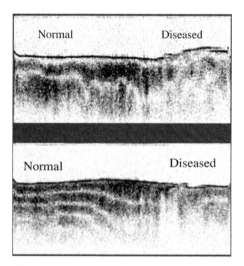

Figure 16.10   In vivo imaging of human knee cartilage. OCT imaging was performed of a human knee during open knee surgery. The left of the image is relatively normal cartilage, by the banding pattern, while the right is more significantly diseased and lost its PS-OCT sensitivity. This confirmed, among other points, that the banding pattern was not an in vitro phenomena due to, for instance, dehydration or postmortem distortion. Image courtesy Li, X., et al. (2005). *Arthritis Res. Ther.* 7, R318–323.

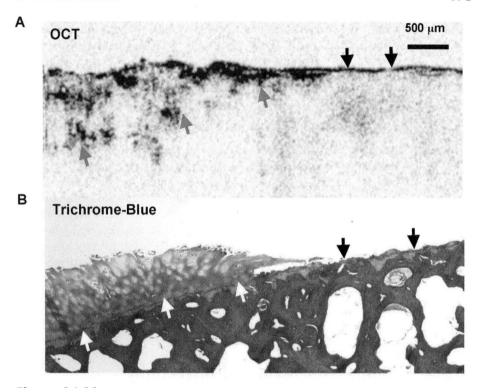

Figure 16.11 Severely osteoarthritic in vivo knee. In this section of tissue, cartilage remains on the left but is grossly inhomogeneous with poor resemblance to the normal cartilage seen above. On the right, no significant cartilage exists. Image courtesy Li, X., et al. (2005). *Arthritis Res. Ther. 7*, R318–323.

16.2 CURRENT OA ANIMAL MODELS

Animal models are important in developing therapeutic approaches for OA. A variety of modalities, ranging from cartilage transplant to pharmaceuticals, have demonstrated considerable promise as methods for altering the progression of OA, as previously discussed. Currently, animal models using rabbits, rats, guinea pigs, mice, and dogs involve sacrificing the animal at given time points. This has several drawbacks. First, rather than following the time course of the disease in a given animal, data for that animal are only obtained at a single time point, making it difficult to take into account heterogeneity between animals. Second, since data are only obtained at one time point, large numbers of animals are required for time course studies. Therefore, experiments tend to be expensive due to the costs of the large number of animals and the cost associated with tissue processing. Third, only limited quantities of the given therapeutic may be available. Finally, data analysis is limited to sites selected for histologic analysis. A technology capable of identifying changes in osteoarthritic joints in real time, near the resolution of histology, and without the need for animal sacrifice could be a powerful tool for assessing the efficacy of therapeutic agents. It will be demonstrated that OCT, combined with rat models of OA, can be used to assess potential therapeutics.

16.2.1   Canine Anterior Cruciate Ligament Transection Model

Anterior cruciate ligament transection (ACLT) in the dog is historically a popular model of OA and has been studied by many investigators.[44] It is attractive to some because the large animal size is well suited for surgical procedures. There are, however, conflicting studies regarding the progression of the disease and the extent of the degenerative changes. Some groups have found consistent fibrillation of the medial tibial plateau by 8 weeks, whereas others reported less severe changes even after 27 weeks.[45,46] Recent studies showing full-thickness loss of articular cartilage 54 months after ACLT, which included a process of hypertrophic repair, is persuasively similar to human OA.[47] However, weight, age, and breed of their animals as well as surgical technique, cage activity, and exercise in these studies are considerably different. In addition to this controversy, utilizing the dog limits large-scale studies because of the high cost of purchase and daily board (54 months in duration).

16.2.2   Rabbit Partial Medial Meniscectomy

Early rabbit models used partial medial meniscectomy to induce OA.[48] Histological and biochemical changes in this model are somewhat similar to human OA. This model might be useful, but the progression of the osteoarthritic lesion is relatively slow and potentially limited. Some groups have reported that these lesions never progressed to the stage of severe destructive OA.[49] The lesions also tended to be focal and were limited to the medial compartment. It was noted that regeneration of the meniscus could interfere with the progression of osteoarthritic change and add variability to the model.

16.2.3   Rabbit ACLT

The rabbit unilateral ACLT model has recently become popular.[50] In this model, no full thickness ulcerations were seen at 4 weeks but were seen at 12 weeks.[51] The percent change in cartilage area and thickness was decreased in almost all regions. The relatively short time to severe OA makes this model attractive for the study of chondroprotective agents. Furthermore, rabbit models have the additional advantage over the dog in that they are small and inexpensive, making them more attractive when large numbers of animals are required for studies of chondroprotective agents. A major limitation of the rabbit model is that it is still a relatively large animal model, requiring often a relatively large amount of experimental therapeutics and expense.

16.2.4 Guinea Pigs

With partial medial meniscectomy in guinea pigs, changes in the superficial and middle layers of the articular cartilage can be seen as soon as 24 hours post surgery. The progression of the disease is dependent upon the surgical procedure. Partial medial meniscectomy results in a more rapid onset and progression than transection of the lateral collateral ligament with the meniscus left intact. However, full thickness lesions often take more then 90 days to develop. This would be an attractive animal model to examine in combination with OCT, if not for the success in the rat.

16.2.5 Rats

OA has been induced in rats both via chemical and mechanical insult.[52,53] Injection of either mono-iodoacetate or polyethyleneimine results in a very rapid degradation of the articular cartilage, with changes seen as early as 3 days post-injection. In some cases, exposure of the subchondral bone occurred in as little as 15 days, but typically in the range of 3 to 4 weeks. Mechanical induction of OA has a slower time course and is more similar to human OA in terms of its distribution. With a medial meniscal tear and MCL transect, surface fibrillations can occur in a week. However, exposure of the subchondral bone takes up to 6 weeks. In studies where the ACL was transected, but the meniscus left intact, surface fibrillations did not occur until 3 weeks post surgery. Our current model is transection of the ACL and medial meniscus, which results in study completion in about 4 weeks. How OCT imaging is performed was changed as studies progressed. While OCT imaging through an arthroscope was initially considered in the rat, it was found that the most convenient method is to open the joint at one-week time periods and then image with a hand-held probe.

16.2.6 Mice

A number of studies have also indicated mice to be a possible model for OA. However, due to their small size it may be difficult to effectively utilize these animals for the assessment of early OA, particularly when examining cartilage repair.

16.2.7 Assessing Cartilage OA with OCT in the Rat

Histology remains the gold standard for assessing changes in cartilage, but sacrificing animals and performing standard biopsies is unattractive for the reasons discussed. Imaging technologies have been applied with only limited success. Although traditional methods of imaging cartilage such as conventional radiography, CT, and transcutaneous ultrasound have applications in humans, their usefulness in animal models are limited.

MRI using a superficial coil can be used in midsize animal models with approximately 100 μm resolution. However, MRI has not achieved widespread use in animals. Difficulties include insufficient resolution, high cost (especially if serial screening is required), the need for a high powered magnet, and the time necessary for data acquisition.[54] Furthermore, it has yet to be demonstrated that cartilage width could be measured with sufficient precision to monitor the effects of ongoing therapy even in humans.

We have successfully demonstrated the ability of OCT to monitor the induction and progression of OA in both rat and rabbit animal models.[55–57] Due to the many advantages of the rat, particularly size, we have focused our efforts on this model. In our initial studies, experimental OA was induced in male Wistar rats (180–200 g) by injecting sodium iodoacetate dissolved in 50 μl of sterile saline into the knee through the patellar ligament. OCT imaging was performed at various time intervals to assess the articular cartilage. Figure 16.12 depicts images obtained of the medial femoral condyle one week post-injection. It can be seen that polarization sensitivity has been lost in the cartilage exposed to sodium iodoacetate but not in the contralateral untreated knee. At 2 weeks (Figure 16.13) both polarization sensitivity and bone cartilage interface have been lost. By 4 weeks (Figure 16.14), all cartilage has been lost in the treated knee. At this late time interval it also appears

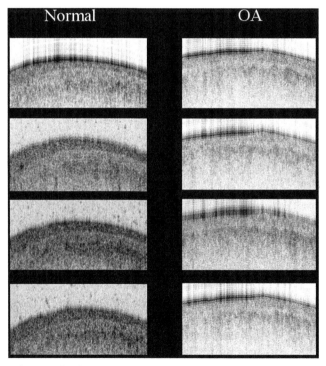

Figure 16.12 Images of the femoral condyle 1 week of OA was induced by Sodium Iodoacetate. Polarization sensitivity has been lost in the treated knee but not the control knee. Courtesy Patel, N.A., et al. (2005). *IEEE Trans. Med. Imaging.* 24(2), 155–159.

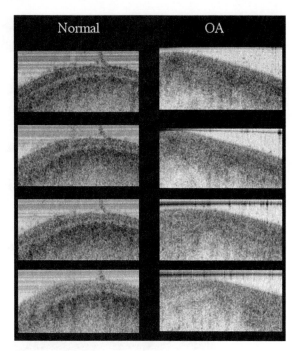

Figure 16.13 Images of the femoral condyle 2 week of OA was induced by Sodium Iodoacetate. The bone cartilage interface has now been lost in the treated knee but not the control knee. Courtesy Patel, N. A., et al. (2005). *IEEE Trans. Med. Imaging.* 24(2), 155–159.

that articular cartilage thinning is occurring in the untreated knee, which has been previously seen. This thinning may be due to either a lack of use or abnormal use of the contralateral leg in these anesthetized animals.

Additional data have been obtained in a mechanical model of OA, where an ACL was transected along with a medial meniscal tear. Rats were then sacrificed at various time points (1, 2, or 3 weeks) following treatment. Figure 16.15 indicates representative changes found at each time frame. The left images are the OCT images. Histological analysis is provided for structural comparison in the middle images (Masson's trichrome) and collagen organization in the right images (picrosirius red). Over time, there is a progressive thinning of the articular cartilage (green arrows) and attenuation of the bone-cartilage interface with a complete loss in some areas at 3 weeks (black arrows). Surface fibrillations (red arrows) can be detected as early as 1 week with significant erosion seen at 3 weeks. The changes seen in the OCT images correspond well with those seen in histological preparations. Furthermore, picrosirius red staining indicates a progressive loss in the highly organized pattern of collagen fibers (as noted by the darker areas in the histology) and a decrease in the diameter of the collagen fibers (an increase in green coloration likely representing type III collagen).

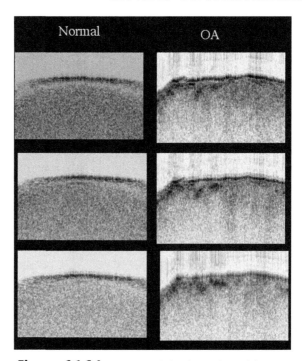

Figure 16.14 Images of the femoral condyle at 4 weeks posttreatment. While the treated knees have lost all detectable cartilage, the control knee now have developed cartilage thinning, likely due to mechanical misuse of the joint. OA was induced by sodium iodoacetate. This emphasizes that separate control animals are necessary and not just the use of the controlateral leg. Courtesy Patel, N. A., et al. (2005). *IEEE Trans. Med. Imaging.* 24(2), 155–159.

The results of this study indicate that OCT is able to detect progressive structural changes in the cartilage (thickness, surface fibrillations, etc.) as well as changes in the bone-cartilage interface. In addition, consistent with previous results, polarization sensitivity was lost when collagen disorganization occurred. An example image of disruption of the bone-cartilage interface (lower, Figure 16.16) with initial apparent increase in the thickness of the cartilage is seen.

We then initiated an *in vivo* study allowing us to sequentially monitor the changes in articular structures of individual animals over time. In this study, OA was chemically induced with sodium iodoacetate (as described above). An incision was made lateral to the patella, including penetrating the capsule, and the patella was dislocated to allow the femoral condyle to be imaged using a hand-held probe. After the imaging was completed, the incision site was sutured and the animals re-imaged at different time points. Each knee (both treated and control) was re-imaged at the following time points: 1, 2, 3, 4, and 8 weeks. Figure 16.17 demonstrates the

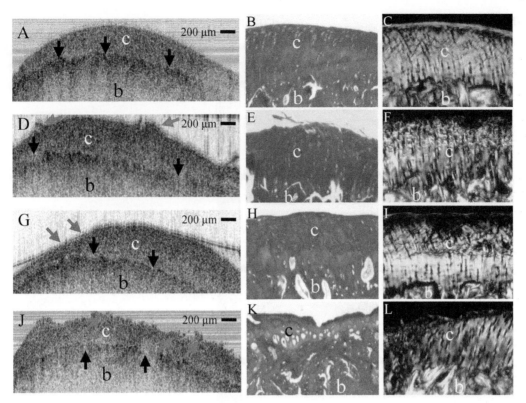

Figure 16.15 Mechanically induced OA. In this study, OA was produced by transection of the medial collateral ligament and full thickness artificial meniscal tear. The left are the OCT images at 0 (A), 1 week (D), 2 weeks (G), and 3 weeks (J) respectively. The second column is the trichome stained section while the third is the corresponding picrosirius. Progressive degeneration is noted in each. Courtesy Roberts, M.J. (2003). *Anal. Bioanal. Chem.* 377, 1003–1006.

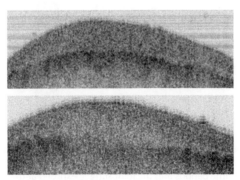

Figure 16.16 OCT image of the normal and OA induced knees at 3 weeks. Again OA was introduced mechanical as in figure 16.15. In b, the OA knee, the cartilage have acutely thickened (a sign of early OA) and the bone cartilage interface has become diffuse. Courtesy Roberts, M.J. (2003). *Anal. Bioanal. Chem.* 377, 1003–1006.

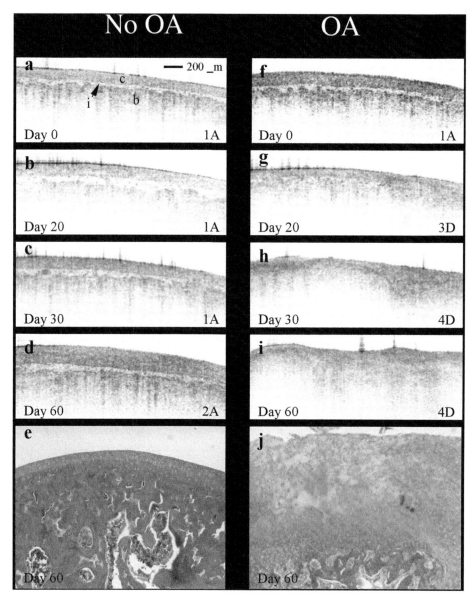

Figure 16.17 OCT images of the same locations on the control (a-e) and iodoacetate treated (f-j) lateral condyles of the same animal followed sequentially in time. The same rats were imaged at different time periods and the joint surface exposed at these times. The articular cartilage (ac), bone (b), and subchondral bone (sb) are identified. Data are compared with histology at the time of sacrifice (day 60). The surface is shown by the red arrow, the bone cartilage interface by the green arrows. Treated condyles demonstrated progressive loss of cartilage/bone interface, articular cartilage, and subcondral bone with time, while the controls retained the interface and a smooth cartilage surface. Histology at day 60 showed the degree of cartilage degradation in the treated joint (Masson's trichrome). Image courtesy Adams, S. B., et al. (2006). *J. Orth. Res.* 24, 1–8.

progressive changes seen in both the treated and control leg of a representative rat. In the treated leg, there is a progressive loss in the bone-cartilage interface as well as a thinning of the cartilage. However, the bone-cartilage interface remains well defined in the control leg. The thickness of the cartilage in the control leg appears to increase with time. This increase may be due again to an increase in water content and may represent early changes indicative of the onset of OA in the control leg. Another disadvantage of the chemical model, along with those discussed, includes the leakage of the chemical agent through the suture. This is one of the reasons to move to a mechanical model.

In all of our studies utilizing the rat model, a number of the control legs started to show some indication that they were developing OA after 3 weeks. This is in spite of the fact the animals are heavily anesthetized. As stated, future studies will require the incorporation of control animals in which there is no induction of OA in either leg.

16.3 TENDON/LIGAMENTS

Abnormalities of tendons and ligaments can lead to significant morbidity, including, but not limited to, pain, rupture, and tears.[58–63] The vulnerability of tendons and ligaments to pathology development has been attributed to the relative avascularity of the tissue, mechanical trauma, and subsequent collagen degradation. For example, Achilles and ACL rupture has been linked to underlying prior histopathologic abnormalities of collagen: Collagen fibers become nonparallel, decreased collagen, and bundle dimensions become irregular. In addition, the pathophysiology of many disorders remains unknown, such as the chronic pain in the patella tendon that often occurs following ACL replacement. Although biopsy is a viable diagnostic option, it may be hazardous, create additional pathologies, or be non-yielding if the correct location is not sampled. MRI, ultrasound, and arthroscopy all have major roles in assessing tendons and ligaments, but these technologies are not optimal at assessing pathologic changes at a histological level. The ability to analyze these structures on a micron scale could be a powerful tool both for experimental analysis to gain better insight into the pathophysiology of these disorders as well as a potential diagnostic modality.

16.3.1 General Ligaments and Tendons

Tendons are a form of connective tissue that consists primarily of dense, parallel collagenous bundles. The collagen is arranged in specific patterns to handle the high mechanical forces it experiences. The fibers are somewhat flexible but have great resistance to pulling forces. The primary cells are the fibroblasts, which are present in small concentrations. Muscle-tendon units connect to bone, which stabilizes, or moves under often considerable force. Ligaments are similar to tendons except

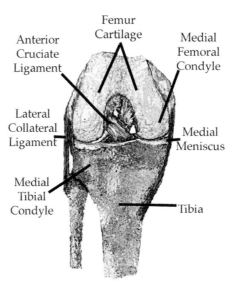

Figure 16.18 Major structures of the knee. The major structures of the knee are discussed as this is the most common joint examined in the text.

the bundles are not as regularly arranged. This likely results because tendons withstand high force in one direction, but ligaments protect the joint from a variety of directions. A ligament typically stretches across joints and helps to reduce excessive movement in the joint. Ligaments in general alone cannot sustain control of the joint and if muscle weakness or failure occurs, damage to the ligament or joint will occur.

16.3.2 OCT Imaging of Tendons and Ligaments

OCT images of a ligament and tendon are shown in Figure 16.19: the top is a ligament, the bottom a tendon.[64] Imaging of the biceps tendon, as well as its corresponding picrosirius stained sections, is shown in Figure 16.20 for illustrative purposes. In Figure 16.21, we see the unruptured section of the author's ACL, which demonstrates normal birefringence by both OCT and picrosirius staining. In Figure 16.22, the ruptured section is shown with loss of birefringence.

The most relevant tendons and ligaments to OCT imaging are the ACL, the Achilles tendon, and the ligaments/tendons making up the rotator cuff. The Achilles tendon and rotator cuff will be discussed below, while the application to the ACL is not straightforward and will be the subject of future work.

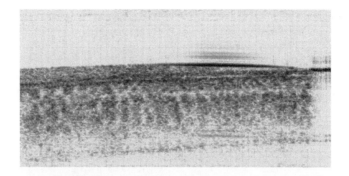

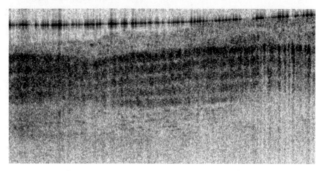

Figure 16.19 PS-OCT ligament and tendon. The top image is of ligament; the bottom is tendon. Distinct birefringence patterns are noted, which is not surprising in view of their different organizations of collagen. Unpublished data from M. Brezinski.

16.3.3 Achilles Tendinopathy

The Achilles tendon (tendo calcaneus) connects the posterior ankle muscles (gastrocnemius, plantaris, and soleus) to the posterior calcaneum (heel). As a reminder of the power of this complex, the average person can stand on their toes for an extended period of time, sustaining their entire body weight with the complex. Therefore, the normal function of the tendon is critical in body mobility, particularly since it needs to respond to relatively intense spontaneous pressure changes. In view of its critical role in lower extremity dynamics, injury or rupture of the tendon can lead to lifelong disability.

Complete or partial tear of the tendon is readily diagnosed by conventional techniques, but rupture prevention is the long-term goal.[65] While excessive muscle strength relative to tendon strength/length can lead to some ruptures, which is why those engaging in strenuous activity should prepare adequately, structural abnormalities in the tendon likely predispose to many ruptures. The Achilles tendon is often referred to as undergoing tendonitis but what typically occurs is the

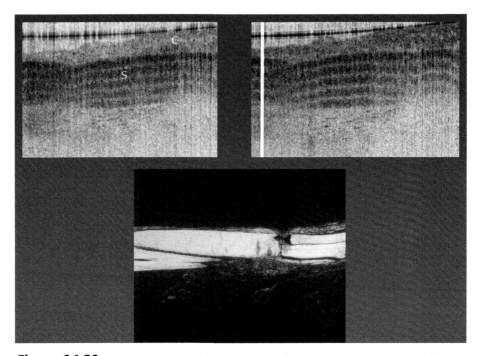

Figure 16.20 PS-OCT imaging of biceps tendon. The top two OCT images are of biceps tendons imaged at two different polarization states. OCT effectively demarcates the layers of the tendon. The banding pattern is from the polarization sensitivity of the organized collagen. The plane of polarization is rotating as the light is passing through the tissue. The pattern changes with change in the polarization state. The supportive tissue around the collagen shows minimal polarization sensitivity. In the third image of histology, the sample is stained with picrosirus and demonstrates that organized collagen is present in the birefringent region (bright region). In the figure S is supportive tissue while c is collagen. Martin, S.D., et al. (2003). *Int. Orthop.* 27, 184–189.

process of chronic tendinosis.[66] Tendonitis is the inflammation of a tendon whereas tendinosis is "death" of a region of the tendon. Diagnosing tendinosis in patients with ankle pain may identify those predisposed to rupture. While we have imaged normal Achilles tendon, as in Figure 16.23, no work has currently been done in patients with tendinosis at the time of this publication.

16.3.4 Shoulder Pain

One area which does deserve considerable attention is shoulder. The shoulder is the most mobile of all joints, having relatively minimal contact between the humerus and the scapula (and clavicle) through direct articulations. The joint's position and integrity are therefore achieved primarily via muscles (glenohumeral, thoraco-humeral, and biceps groups), ligaments, and tendons of the shoulder, neck, back,

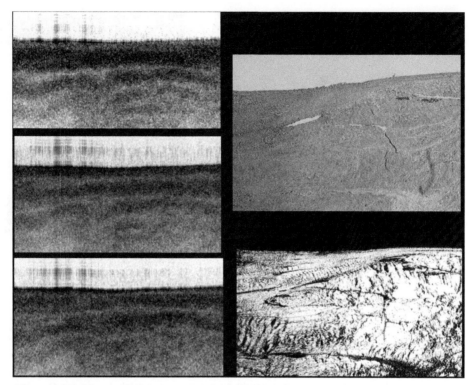

Figure 16.21 An unruptured section of a resected ACL is seen with histology and picrosirus staining. Again, a banding pattern is seen and highly organized collagen found in the pricrosirius stained section. Courtesy Martin, S. D., et al. (2003). *Int. Orthop. 27*, 184–189.

chest, and arms. Therefore, this minimal dependency on a large synovial joint makes the shoulder both extremely mobile and vulnerable to a wide range of injuries, often difficult to correctly define. The reader is encouraged to review the shoulder anatomy, which will not be addressed in detail here. Among the most important areas are the rotator cuff and the long head of the biceps tendon. The rotator cuff consists of the tendons of the subscapularis, supraspinatus, infraspinatus, and teres minor muscles, which are fused with the capsule of the shoulder joint. The rotator cuff plays an important role in stabilizing the shoulder joint. Subtle damages in one of these tendons can lead to substantially reduced shoulder function, yet be undetected by conventional imaging techniques.

The origin on the scapula and insertion in the radius leaves the long head of the biceps branchii muscle with interactions over both shoulder and elbow joint. An OCT image and the corresponding histology of a biceps tendon are shown in Figure 16.20. Biceps tendonitis can result from pathologies including impingement syndrome (regional swelling from overuse) or slipping of the tendon from the groove. Irrespective, management can be extremely difficult both for the professional and recreational athlete. Inability to control symptoms can lead to the end of an

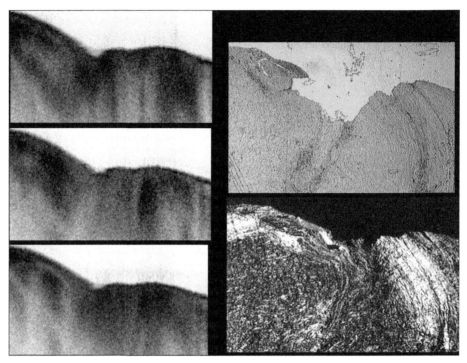

Figure 16.22   The figure demonstrates the ruptured end of the ACL from Figure 16.21. It can be seen that polarization sensitivity has been completely lost in the OCT images, and in the picrosirius stained region, it can be seen that collagen organization is almost completely lost. Courtesy Martin, S. D., et al. (2003). *Int. Orthop.* 27, 184–189.

athletic career. Techniques are needed to evaluate the biceps tendon, as well as other structures of the shoulder joint. To date, no imaging has been performed of pathologic bicep samples.

16.4   OTHER POTENTIAL APPLICATIONS

The potential number of OCT applications in musculoskeletal disease is enormous, some of which include guiding spinal surgery, assessing tumor margins, and guiding nerve repair. However, one disorder that is uncommon with devastating effects on both the patient and family is osteogenesis imperfecta (OI).[67] OI is a genetic disorder where bones become fragile and bone mass is low. It varies in severity from death in the perinatal period to very mild fracturing. In most cases, the disorder is caused by a mutation of one of two genes for the alpha chains of collagen type I. In general, medical therapy is unsuccessful. However, recently, bisphosphates, which are used to treat osteoporosis, have been applied to the problem. These are potent antiresorptive drugs that have shown some effectiveness in severe cases. However, concerns

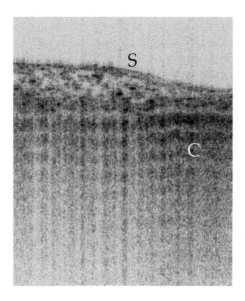

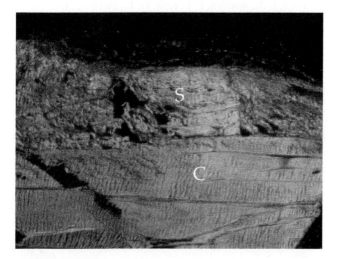

Figure 16.23 In this figure, an OCT of an Achilles tendon is seen along with the corresponding histology (picrosirius stain). The birefringence of the normal collagen region (C) is seen in addition to the loose supporting connective tissue (S). Courtesy Martin, S. D., et al. (2003). *Int. Orthop. 27, 184–189.*

over side effects are considerable, particularly the long-term effects of decreased bone growth and diminished bone remodeling. For patients with less severe disease, such as those having a few noncritical fractures per year, the decision to treat can be agonizing. The ability to quantify the state of collagen in these patients with OI could substantially improve the ability of physicians and families to make the decision to use these potentially hazardous medications. What is particularly interesting is the disease may possibly be followed through OCT imaging of either the teeth or gum line.

16.5   CONCLUSION

OCT represents an attractive new technology for musculoskeletal imaging. In addition to osteoarthritis, a wide range of potential applications exist.

REFERENCES

1. www.cdc.gov
2. www.niams.nih.gov/ne/oi/index.htm
3. Brandt, K. D. (1989). Pathogenesis of osteoarthritis. *Textbook of Rheumatology*. 3rd edition, W. N. Kelly, E. D. Harris, Jr., S. Ruddy, and C. B. Sledge (eds.). WB Saunders, Philadelphia, pp. 1480–1500.
4. Tyler, J. A., Watson, P. J., Koh, H. L., Herrod, N. J., Robson, M., and Hall, L. D. (1995). Detection and monitoring of progressive degeneration of osteoarthritic cartilage by MRI. *Acta. Orthop. Scand.* 66(Suppl. 266), 130–138.
5. Myers, S. L., Dines, K., Brandt, D. A., Brandt, K. D., and Albrecht, M. E. (1995). Experimental assessment by high frequency ultrasound of articular cartilage thickness and osteoarthritic changes. *J. Rheumatol.* 22, 109–116.
6. Jeffery A. K., Blunn, G. W., Archer, C. W. et al (1991). Three dimensional collagen architecture in bovine articular cartilage. *J. Bone Joint Surg.* 73B, 795–801.
7. Muir, H., Maroudas, A., and Wingham, J. (1969). The correlation of fixed negative charge with glycosaminoglycan content of human articular cartilage. *Biochim. Biophys. Acta* 177, 494–503.
8. Kempson, G. E., Muir, H., Pollard, C., and Tuke, M. (1973). Correlations between the compressive stiffness and chemical constituents of human articular cartilage. *Biochim. Biophys. Acta* 215, 70–78.
9. Venn, M. and Maroudas, A. (1977). Chemical composition and swelling of normal and osteoarthritic femoral head cartilage. *Ann. Rheum. Dis.* 36, 121–129.
10. Mankin, H. J. and Thrasher, A. Z. (1975). Water content and binding in normal and osteoarthritic human cartilage. *J. Bone and Joint Surg.* 57A, 76–80.
11. Lozada, C. J. and Altman, R. D. (1999). Chondroprotection in osteoarthritis. *Bull. Rheum. Dis.*, 46, 5–7.
12. Cole, A. D., Chubinskaya, S., Luchene, L. J. et al (1994). Doxycycline disrupts chondrocyte differentiation and inhibits cartilage matrix degradation. *Arthritis Rheum.* 32, 1727–1734.
13. Shimizu, C., Kubo, T., Hirasawa, Y. et al. (1998). Histomorphometric and biochemical effect of various hyaluronans on early osteoarthritis. *J. Rheumatol.* 25, 1813–1819.
14. Abadie, E., Ethgen, D., Avouac, B. et al. (2004). Recommendations for the use of new methods to assess the efficacy of disease modifying drugs in the treatment of OA. *Osteoarthritis Cartilage* 12, 263–268.
15. www.niams.nih.gov/ne/oi/imaging.htm

16. Aisen, A. M., McCune, W. J., and MacGuire, A. (1984). Sonographic evaluation of the cartilage of the knee. *Radiology* 153, 781–784.

17. Agemura, D. H., O'Brien, Jr., W. D., Olerud, J. E., Chun, L. E., and Eyre, D. E. (1990). Ultrasonic propagation properties of articular cartilage at 100 MHz. *J. Acoust. Soc. Am.* 87, 1786–1791.

18. Chan, W. P., Lang, P., Stevens, M. P., Sack, K., Majumdar, S., Stoller, D. W., Basch, C., and Genant, H. K. (1991). Osteoarthritis of the knee: Comparison of radiology, CT, and MR imaging to assess extent and severity. *AJR* 157, 799–806.

19. Eckstein, F. and Glaser, C. (2004). Measuring cartilage morphology with quantitative magnetic resonance imaging. *Sem. Musculoskeletal Radiol.* 8, 329–353.

20. Peterfy, C. G., Guermazi, A., Zaim, S. et al. (2004). Whole Organ Magnetic Resonance Imaging Score (WORMS) of the knee. *Osteoarthritis Cartilage* 12, 177–190.

21. Yulish, B. S., Montanez, J., Goodfellow, D. B., Bryan, P. J., Mulopulos, G. P., and Modic, M. T. (1997). Chondromalacia patellae: Assessment with MR imaging. *Radiology* 16, 763–766.

22. Konig, H., Aicher, K., Klose, U., and Saal, J. (1990). Quantitative evaluation of hyaline cartilage disorders using Flash sequence. Clinical applications. *Acta. Radiol.* 31, 377–381.

23. Hayes, C. W., Sawyer, R. W., and Conway, W. F. (1990). Patella cartilage lesions: In vitro detection and staging with MR imaging and pathologic correlation. *Radiology* 175, 479–483.

24. Hodler, J., Trudell, D., Parthria, M. N., and Resnick, D. (1992). Width of the articular cartilage of the hip: Quantitation by using fat-suppression spin-echo MR imaging in cadavers. *AJR* 159, 351–355.

25. Burks, R. T. (1990). Arthroscopy and degenerative arthritis of the knee: A review of the literature. *Arthroscopy* 6, 43–47.

26. Ike, R. W. and O'Rouke, K. S. (1993). Detection of intra-articular abnormalities in osteoarthritis of the knee. A pilot study comparing needle arthroscopy with standard arthroscopy. *Arthritis Rheum.* 36, 1353–1363.

27. Batiste, D. L., Kirkley, A., Laverty, S. et al. (2004). High resolution MRI and micro-CT in an ex vivo rabbit ACL. *Osteoarthritis Cartilage* 12, 614–626; Mollenhauer, J., Aurich, M. E., Zong, Z. et al. (2002). Diffraction enhanced X ray imaging of articular cartilage. *Osteoarthritis Cartilage* 10, 163–171.

28. Lozada, C. J. and Altman, R. D. (1999). Chondroprotection in osteoarthritis. *Bull. Rheum. Dis.* 46, 5–7.

29. Cole, A. D., Chubinskaya, S., Luchene, L. J. et al. (1994). Doxycycline disrupts chondrocyte differentiation and inhibits cartilage matrix degradation. *Arthritis Rheum* 32, 1727–1734.

30. Brandt, K. D. (1995). Modification by oral doxycycline administration of articular cartilage breakdown in osteoarthritis. *J. Rheumatol.* 22(Suppl. 43), 149–151.

31. Altman, R. D., Dean, D. D., Muniz, O. E. et al. (1989). Prophylactic treatment of canine osteoarthritis with glycosaminoglycan polysulfuric acid ester. *Arthritis Rheum.* 32, 759–766.

32. Golding, J. C., Ghosh, P. et al. (1983). Drugs for osteoarthrosis: I. The effects of pentosan polysulphate (SP54) on the degradation and loss of proteoglycans from articular cartilage. *Curr. Ther. Res.* 32, 173–184.

33. Smith, M. M., Gosh, P., Numata, Y. et al. (1994). The effects of orally administered calcium pentosan polysulfide on inflammation and cartilage degradation. *Arthritis Rheum.* 37, 125–136.

34. Rejholec, V. (1987). Long term studies of antiosteoarthritic drugs: An assessment. *Sem. Arthritis Rheum.* 17(Suppl 1), 35–53.

35. Steadman, J. R., Rodney, W. G., Singleton, S. B. et al. (1997). Microfracture technique for full thickness chondral defects. Technique and clinical results. *Op. Tech. Orthop.* 7, 200–205.

36. Rodrigo, J. and Steadman, J. R. (1994). Improvement of full-thickness chondral defect healing in the human knee after debridement and microfracture using continuous passive motion. *Am. J. Knee Surg.* 7, 109–116.

37. Outerbridge, H. K., Outerbridge, A. R., and Outerbridge, R. E. (1995). The use of a lateral patellar autologous graft for the repair of large osteochondral defect in the knee. *J. Bone Joint Surg.* 77A, 65–72.

38. Buckwalter, J. A. and Mankin, H. J. (1998). Articular cartilage: Degeneration and osteoarthritis repair, regeneration, and transplantation. *AAOS Instr. Course Lect.* 47, 487–504.

39. Beaver, R. J., Mahomed, M., Backstein, D. et al. (1992). Fresh osteochondral allografts for post-traumatic defects in the knee. *J. Bone Joint Surg.* 74B, 105–110.

40. Jackson, D. W. and Simon, T. M. (1996). Chondrocyte transplantation. *Arthroscopy* 12, 732–738.

41. Herrmann, J., Pitris, C., Bouma, B. E., Boppart, S. A., Fujimoto, J. G., and Brezinski, M. E. (1999). High resolution imaging of normal and osteoarthritic cartilage with optical coherence tomography. *J. Rheumatol.* 26(3), 627–635.

42. Drexler, W., Stamper, D., Jesser, C., Li, X. D., Pitris, C., Saunders, K., Martin, S., Fujimoto, J. G., and Brezinski, M. E. (2001). Correlation of collagen organization with polarization sensitive imaging in cartilage: Implications for osteoarthritis. *J. Rheumatol.* 28, 1311–1318.

43. Li, X., Martin, S. D., Pitris, C., Ghanta, R., Stamper, D. L., Harman, M., Fujimoto, J. G., and Brezinski, M. E. (2005). High-resolution optical coherence tomographic imaging of osteoarthritic cartilage during open knee surgery. *Arthritis Res. Ther.* 7, R318–R323.

44. Johnson, R. G. (1986). Transection of the canine anterior cruciate ligament: Concise review of experience with this model of degenerative joint disease. *Exp. Pathol.* 30, 209–213.

45. McDevitt, C. A. and Muir, H. (1976). Biochemical changes in the cartilage of the knee in experimental and natural osteoarthritis in the dog. *J. Bone. Joint Surg.* 58B, 94–101.

46. Johnson, R. G. (1986). Transection of the canine anterior cruciate ligament: concise review of experience with this model of degenerative joint disease. *Exp. Pathol.* 30, 209–213.

47. Brandt, K. D., Myers, S. L., Burr, D., and Albrecht, M. (1991). Osteoarthritic changes in canine articular cartilage, subchondral bone, and synovium fifty-four months after transection of the anterior cruciate ligament. *Arthritis Rheum.* 34, 1560–1570.

48. Moskowitz, R. W., Davis, W., Sammarco, J. et al. (1973). Experimentally induced degenerative joint lesions following partial meniscectomy in the rabbit. *Arthritis Rheum.* 16, 397–404.

49. Shapiro, F. and Glimcher, M. J. Induction of osteoarthritis in the rabbit knee joint: Histologic change following meniscectomy and meniscal lesions. *Clin. Orthop.* 147, 287–295.

50. Vignon, E., Bejui, J., Mathieu, P. et al. (1987). Histological cartilage changes in a rabbit model of osteoarthritis. *J. Rheumatol.* 14(Suppl 14), 104–106.

51. Yoshioka, M., Coutts, R. D., Amiel, D., and Hacker, S. A. (1996). Characterization of a model of osteoarthritis in the rabbit knee. *Osteoarthritis Cartilage* 4, 87–98.

52. Dunham, J., Hoedt-Schmidt, S., and Kalbhen, D. A. (1993). Prolonged effect of iodoacetate on articular cartilage and its modification by an anti-rheumatic drug. *Int. J. Exp. Pathol.* 74, 283–289.

53. Troop, R., Buma, P., Krann, P. R., Hollander, A. P., Billinghurst, R. C., Poole, A. R., and Berg, W. B. (2000). Differences in type II collagen degradation between peripheral and central cartilage of rat stifle joints after cranial cruciate ligament transection. *Arthritis Rheum.* 43, 2121–2131.

54. Batiste, D. L., Kirkley, A., Laverty, S. et al. (2004). High resolution MRI and micro-CT in an ex vivo rabbit ACL. *Osteoarthritis Cartilage* 12, 614–626.

55. Roberts, M. J., Adams, S. B. Jr., Patel, N. A. et al. (2003). A new approach for assessing early osteoarthritis in the rat. *Anal. Bioanal. Chem.* 377, 1003–1006.

56. Patel, N. A., Zoeller, J., Stamper, D. L., Fujimoto, J. G., and Brezinski, M. E. (2005). Monitoring osteoarthritis in the rat model using optical coherence tomography. *IEEE Trans. Med. Imaging* 24(2), 155–159.

57. Adams, S. B., Herz, P. R., Stamper, D. L. et al. (2006). High resolution imaging of progressive articular cartilage degeneration, *J. Ortho. Res.* 24, 1–8.

58. Movin, T., Gad, A. et al. (1997). Tendon pathology in long-standing achillodynia. *Acta Orthop. Scan.* 68, 1701–1775.

59. Astrom, M. and Rausing, A. (1995). Chronic Achilles tendinopathy. *Clin. Orthop. Related Res.* 316, 151–164.

60. Kartus, J., Movin, T. et al. (2000). A radiographic and histologic evaluation of the patella tendon after harvesting its central third. *Am. J. Sports Med.* 28, 218–226.

61. Kartus, J., Magnusson, L. et al. (1999). Complications following arthroscopic anterior cruciate ligament reconstruction. *Knee Surg. Sports Traumatol. Arthrosc.* 7, 2–8.

62. Azangwe, G., Mathias, K. J. et al. (2000). Macro and microscopic examination of the ruptured surfaces of the anterior cruciate ligaments of rabbits. *J. Bone Joint Surg.* 82B, 450–456.

63. Wojtys, E., Huston, L., Lindenfeld, T. et al. (1998). Association between the menstrual cycle and anterior cruciate ligament injuries in female athletes. *Am. J. Sports Med.* 26, 614–619.

64. Martin, S. D., Patel, N. A., Adams, S. B. et al. (2003). New technology for assessing microstructural components of tendons and ligaments. *Int. Orthop.* 27, 184–189.

65. Brandser, E. A., El-Khoury, G. Y., and Saltzman, C. L. (1995). Tendon injuries: Application of magnetic resonance imaging. *Can. Assoc. Radiol. J.* 46, 9–18.

66. Astrom, M. and Rausing, A. (1995). Chronic Achilles tendinopathy. *Clin. Orthop. Related Res.* 316, 151–164.

67. Rauch, F. F. and Glorieux, H. (2004). Osteogenesis imperfecta. *Lancet* 363, 1377–1385.

17   OCT IN ONCOLOGY

Chapter Contents

17.1   GENERAL

The clinical information in this chapter as with other clinical chapters is designed to aid in OCT development and is not intended to be recommendations on current patient care.

A neoplasm (new growth) may be benign or malignant (cancer). Neoplasms have the common feature that cell growth continues without regulation by the external environment. In other words, they continue to grow unabated. A cancer has the capacity to invade tissue and potentially metastasize (move) to distant sites through the blood or lymphatic system, forming new cancer sites. Cancers generally occur because of abnormal DNA either from replication errors or carcinogens, usually combined with a genetic predisposition. These errors result in either activation of proto-oncogenes or the inactivation of tumor suppressor genes. Cancers are less common in tissue that does not divide, such as cardiac myocytes, and more common in tissue with rapid turnover, such as the gastrointestinal tract. In this chapter disorders will not be grouped in terms of organ systems, but rather where significant OCT research has been performed.

17.1.1   Epidemiology of Cancer

Cancer as a whole is the second major cause of death worldwide.[1] According to the National Cancer Society, the 2004 estimates for the top 5 cancers prevalent in men in the United States are prostate (33%), lung (13%), colorectal (11%), bladder (6%), and melanoma (4%). For women, it is breast (32%), lung (12%), colorectal (11%), uterine (6%), and ovarian (4%). In terms of cancer deaths in men, the most common causes are lung (32%), prostate (10%), colorectal (10%), pancreatic (5%), and leukemia (5%). For women, it is lung (25%), breast (15%), colorectal (10%), ovarian (6%), and pancreatic (6%).

17.1.2 Need for a Diagnostic Imaging Technique

The identification of early malignant changes remains a central objective of clinical medicine since, once widely metastatic, most cancers become incurable. Excisional biopsy and cytology, with subsequent histologic processing, remains a cornerstone of early diagnostics. Both have had a substantial impact on the management of many neoplasms such as cervical and skin cancer. However, in many instances, diagnostics based on excisional biopsy or cytology are ineffective due, for instance, to sampling errors. A technology capable of performing optical biopsy (imaging at a resolution comparable to histopathology without the need for tissue removal) could substantially improve the ability of clinicians to identify malignancies at curable stages. Large areas could be screened with the imaging system at high resolution and relatively low cost. The clinician would no longer be relying on a "few static views" to make the diagnosis but could use a continuum of information, obtained at different orientations, in a manner analogous to 2D echocardiography or ultrasound. This will be the central approach to applying OCT for the early diagnosis of malignancy. In addition, OCT will play a role in assessing tumor margins during surgical resection, the efficacy of therapy (e.g., radiation) in promoting tumor death, and of blocks of tissue being assessed by pathologic processing (discussed at the end of the chapter).

17.1.3 Epithelium

In general, OCT imaging will likely demonstrate its greatest potential as a method for the early detection of epithelial cancers. It will be assessing somewhat subtle changes in the structure of the epithelium that are needed to identify early cancers. Therefore, a basic understanding of epithelium is indicated. Epithelia are a diverse group of tissues that predominately cover body surfaces, cavities, and tubes. Usually, they represent barriers between the internal body tissue and external environment. The basement membranes are important structures of the epithelium which separate it from the underlying supportive tissue. Penetration of the basement membrane by cancers is often when a tumor metastasizes. In addition, thickness of the epithelium, which may be useful in assessing early cancers, is often determined by the distance between the surface and basement membrane.

Epithelia are traditionally classified by four morphological parameters, the number of cell layers, the shape of the component cells, the presence of surface specialization, such as cilia or keratin, and the presence of glands (structures used to secrete fluids of various composition from the epithelium). Four common epithelia are shown in Appendix 17-1. To detect early malignancy, among the tissue properties which OCT could potentially measure include width of the epithelium, nuclear size, nuclear variation, concentration of mitotic figures, integrity of the basement membrane, nuclear to cytoplasmic ratio, organization of glands, and presence of abnormally superficial nucleated cells. *This is perhaps the greatest obstacle in developing OCT for early cancer detection.* Structural OCT currently lacks the resolution or contrast to identify many of these factors. Therefore, other adjuvant techniques need to be explored such as speckle analysis, polarization sensitivity, and elastography.

17.2   ESOPHAGEAL CANCER

17.2.1   General

The luminal gastrointestinal tract is divided into the mouth, esophagus, stomach, small intestine, large intestine, and colorectal region. This is excluding exocrine organs such as the pancreas and liver. The gastrointestinal tract layers consist of the mucosa, submucosa, muscularis propria, and adventitia. The mucosa is further subdivided into the epithelium, a supportive lamina propria, and a thin smooth muscle layer known as the muscularis mucosa, which produces local movement and folding. As shown in Figure 17.1, there are generally four types of mucosa in the GI tract.[2] The first is the protective layer of the mouth, anorectal region, and esophagus, which is generally stratified squamous epithelium and designed to prevent substances from entering the body (i.e., primarily serves as conduit). The esophagus normally has an alkaline environment. The stomach contains a large number of glands, but only a few substances are absorbed through it (e.g., alcohol), and is responsible for breaking food down into smaller components. The environment is acidic. The small intestine is designed for absorption. The mucosa consists of finger-like projections called villi which are used to maximize surface area. Secretions enter the small intestine from the biliary and pancreatic ductile system. Glands are also present for further digestion. The large intestine also contains villi but their primary purposes are water regulation and holding undigested food components.

Gastrointestinal malignancies are among the most important where OCT can potentially play a role. These include esophageal, gastric, biliary, and colorectal malignancies. Esophageal cancer will be discussed here and other GI applications will be discussed later in the chapter.

There are two major forms of esophageal cancer, squamous cell carcinoma and adenocarcinoma. Probably no other cancer has shown more variability with respect to region and race. The incidence of the adenocarcinoma form is high and rising in the United States but low in the Far East. The incidence of squamous cell carcinoma is high in China, Iran, and North France but comparably low in the United States.

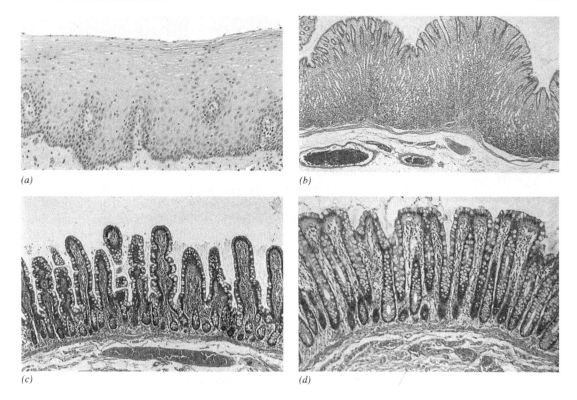

Figure 17.1 This figure demonstrates generally the four types of mucosa. The top image is a protective mucosa which is commonly found in the oral cavity, pharynx, esophagus, and anal cannal. It is stratified squamous (Appendix 17-1). The second type of mucosa is secretory. It commonly occurs in the stomach. It consists of long, closely placed glands. The third type of mucosa is absorptive. This type of mucosa is found throughout the small intestine. The mucosa contains finger like projections called villi. These villi contain short glands called crypts. The fourth type of intestinal mucosa is that of the large intestine. The surface is contains straight, packed tubule glands. These cells are responsible for water absorption and secreting goblet cells which lubricate the feces for movement through the large intestine. Courtesy Young, B., and Heath, J.W. (2000). *Functional Histology*, fourth edition, 251. [2]

17.2.2 Adenocarcinoma of the Esophagus

17.2.2.1 Epidemiology

Barrett's disease, sometimes referred to as columnar-lined esophagus, is the largest predisposing condition for adenocarcinoma of the esophagus and the most important screening population. Barrett's esophagus is the replacement of the normal squamous epithelium of the distal esophagus with columnar-like epithelium, which is described below.[3] The problem with this simple definition is that distinguishing the gastric-esophageal junction (gastric tissue vs. short segment Barrett's) may be difficult, particularly in cases of hiatal hernia, so that extension greater than 2–3 cm from the presumed GE junction is often needed to make the diagnosis confidently. Barrett's is common among patients with chronic gastroesophageal reflux or "heartburn." The intestinal like mucosa appears to be better suited to handle the acid refluxed from the stomach rather than the normal squamous epithelium of the esophagus. Approximately 80% of all patients with Barrett's disease have described symptoms of reflux. Barrett's carries a 30 to 40 times increased risk of developing adenocarcinoma.[3] Obesity, particularly in those of below average height, may be emerging as an additional risk factor with at least a threefold increase. In addition, some links have been seen with smoking and alcohol use.

The mean age for developing Barrett's is 40, but generally it is not diagnosed until after age 60. However, there are a significant number of cases in children, usually in patients with predisposition to reflux, such as neurogenic disease or atresia of the esophagus. The prevalence of Barrett's within the United States is between 0.3% and 2%.[4] Among patients with reflux, the prevalence is 8–20%.[5,6] The incidence of adenocarcinoma in Barrett's is approximately 7–9%.[7] Without early diagnosis, esophageal adenocarcinoma can be considered universally fatal.[8,9] It should be noted that some controversy exists over these numbers, which may be somewhat biased due to sampling issues.

Barrett's disease is generally a white male disease. The ratio of the disease in Caucasian to African American is 20:1. The prevalence is 2:1 in males versus females. Once considered a rare tumor of the esophagus, adenocarcinoma now exceeds squamous cell carcinoma as the most common malignancy of the esophagus. Similarly, Barrett's disease has increased dramatically from 1970 to 1990. However, some have suggested that at least part of this increased risk can be attributed to increasing numbers of gastroenterologists performing endoscopic procedures. As the number of gastroenterologists performing endoscopy has now leveled, it will be of interest to see how the rate will rise in the future.

Limited data exist on Barrett's disease and adenocarcinoma in other countries. As a generalized statement, this is a disease of Western society. However, in Japan, where both have been historically low, significant increases have recently been noted. In the UK, incident rates are between 9.6 to 18.3 per 100,000 and have been increasing at a rate similar to the United States.[10]

There are three distinct types of Barrett's esophagus that can occur separately or in combination: specialized columnar epithelium, fundus, and junctional (Figure 17.2).[3] They each have mononucleated cells that are present due to inflammation. Specialized columnar epithelium is the most common. It consists of a villiform surface with crypt-like intestinal mucosa. The crypts are glandular. The epithelial lining consists of columnar cells and Goblet cells. This is incomplete metaplasia since the cells are not functional. The fundus type resembles the gastric mucosa of the body and fundus. The surface is pitted with villi. The deep glands contain mucus. The junctional type resembles the gastric cardia. The surface is pitted with mucus-secreting cells.

17.2.2.2 Screening

The diagnosis of Barrett's is generally made endoscopically. It is usually apparent by visual inspection, except when short segment disease is present. However, in cases

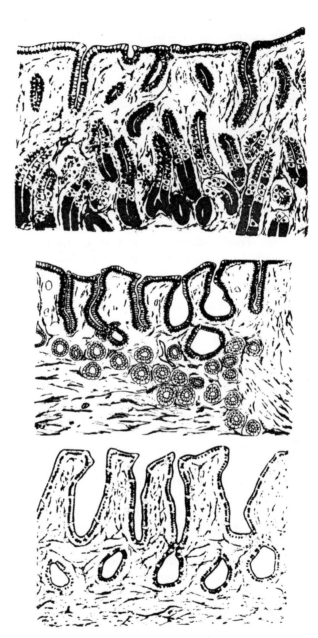

Figure 17.2 Types of Barrett's esophagus. Knowing what epithelium that needs to be imaged is important in developing technology for image. The three types of Barrett's esophagus are specialized columnar, fundus, and junctional epithelium. The different epithelium are described in the text. Courtesy Phillips, R. W., and Wong, R. K. (1991). *Gastroenterol. Clin. North Am. 20,* 791–815. [3]

where the diagnosis is in question, dyes can be used to aid assessment. Toluidine blue identifies metaplastic but not squamous mucosa, although it is weaker with dysplasia. Lugol's iodine will identify squamous epithelium but not columnar epithelium. These are discussed in more detail below.

The progression to adenocarcinoma is believed to go from columnar metaplasia, low grade dysplasia, high grade dysplasia, then adenocarcinoma. There is a lot of controversy surrounding how to screen and when and how to intervene. Low grade dysplasia is characterized by preserved architecture, enlarged hyperchromatic nuclei not exceeding half the cell's size, and rare mitosis. High grade dysplasia is characterized by distortion of gland architecture with branching and budding of crypts, villous appearance of mucosal surfaces, enlarged hyperchromatic nuclei exceeding 50% of the cell size, irregularly stratified nuclei, and frequent mitotic figures. It is generally defined as intraepithelial neoplasia that has not infiltrated the basement membrane. Frank adenocarcinoma is characterized by grossly abnormal cell morphology, large quantities of glands in close approximation, and necrosis. It is well recognized that significant interobserver variability exists in grading specimens needed for clinical decision making.[11] The risk of lymph node metastasis reaches 20–25% in the presence of submucosal infiltration.[12] Markers are actively studied to address this issue. Few animal models exist to improve our understanding of the pathophysiology.

Based on the size of the population at risk, significant controversy exists over whom to screen. The ideal screening procedure would be low cost, low risk to the patient, highly sensitive and specific, acceptable to the patient, inexpensive, and have the potential to be applied to large numbers of patients. There currently are no modalities meeting all these criteria, which is why OCT demonstrates such significant potential. As an example protocol, in a patient known to have Barrett's, endoscopic surveillance of the esophagus is performed every 12–18 months looking for high grade dysplasia or adenocarcinoma.[13,14] Endoscopic screening currently involves biopsies every 2 cm. However, screening has its problems.[15,16,24] One problem is that after 6 to 8 biopsies, bleeding can become heavy and the esophagus becomes difficult to visualize. Another problem is that the procedure is expensive. A recent study using a Markov simulation model found that the cost-effectiveness was extremely sensitive to the true prevalence of Barrett's in the population and the real probability that Barrett's will progress to adenocarcinoma.[17] Currently, there are conflicting data on both these numbers. As pointed out above, the large intraobserver variation in the interpretation of histology is an additional problem. Finally, sampling error remains a problem. The technique for the most part relies on blind biopsies. In a study of 30 patients, adenocarcinoma only represented 1.1 cm^2 of the esophagus and therefore can be easily missed.[18] Furthermore, foci of unsuspected adenocarcinoma were found in 40% of patients with high grade dysplasia after surgical resection.[19]

17.2.2.3 Treatment

Controversy exists for treating Barrett's disease and high grade dysplasia. Esophagectomy is still considered the standard of care by many for the treatment of high grade dysplasia. There are several reasons for this. First, studies have shown that a large number of patients with the preoperative diagnosis of high grade dysplasia already have invasive carcinoma when the esophagus is examined postoperatively. Second, many believe that most of the patients with high grade dysplasia will eventually develop adenocarcinoma. Third, reconstructive procedures using stomach or colon have an excellent functional recovery (morbidity about 30%) and a mortality of less than 3%.[20] However, there are counter arguments to not performing surgery for high grade dysplasia. First, it should be pointed out that the above argument assumes that the interpretation of high grade dysplasia by the pathologist is correct. Second, many others would argue that most high grade dysplasia will not progress to adenocarcinoma. Third, some feel that the adenocarcinoma is slowly progressive and that if closely followed, even if it develops, the patient can be caught at a curable stage.

Treatment with proton-pump inhibitors is associated with a substantial reduction in the onset of dysplasia in patients with Barrett's esophagus.[21] However, with respect to the treatment of Barrett's disease, although cases have been documented where acid suppression has resulted in regression of the disease, these cases are rare and it should not be considered curative. Three additional endoscopic approaches are photodynamic therapy (PDT), laser therapy, and argon-plasma coagulator to treat adenocarcinoma-severe dysplasia other than esophagectomy. PDT has been proposed, which again is light-induced activation of an administered photosensitizer, which results in local injury by the production of singlet oxygen. Although it has certain advantages, problems associated include persistent Barrett's esophagus in approximately 50% of patients after treatment, approximately 5% chance of subsquamous islands of Barrett's esophagus after treatment, and persistent molecular abnormalities.[22] Lasers are used primarily for palliation, but seem to be gradually being replaced by stents. For use with Barrett's esophagus, it may not provide homogeneous ablation of the mucosa and glands may persist. Perforation is also a concern. An argon-plasma coagulator is a non-contact electrocoagulation device that involves high frequency monopolar current conducted to the tissue by a flow of ionized argon gas.[23] The depth of injury can reach 6 mm. Concerns are similar to those of laser ablation.

17.2.2.4 Diagnostic modalities other than traditional endoscopy

High frequency ultrasound — It is superior to CT or MRI for detecting esophageal wall penetration and for detecting regional lymph node involvement. High frequency ultrasound can be linear or circular. However, it lacks the ability to identify dysplasia and early cancer. Probe frequencies vary from 7.5 to 20 MHz.[26–28] It can pick up the superficial mucosa, deep mucosa, submucosa, muscularis propria, and adventitia. It may be useful for surgical staging, which is controversial from a cost-effective basis based on relatively high false negatives, although it does not appear to have a role in identifying dysplasia or directing biopsies.[26–28] The cost-effectiveness is dealt with elsewhere.[28]

Chromoendoscopy — It employs chemical staining agents applied to the GI tract to identify specific mucosal subtypes or to highlight surface characteristics of the epithelium. The agents generally are not used alone for the identification of dysplasia but do serve certain adjuvant roles. In patients with Barrett's, the goal of chromo-endoscopy is to increase the detection of localized intestinal metaplasia or the esophagus/gastric border, not dysplasia or adenocarcinoma. Tissue stains are applied directly on the mucosal surface during endoscopy using a spray catheter and after stain application. Vital stains include Lugol's solution and methylene blue.[29,30] Lugol's solution stains normal esophageal mucosa and can aid in highlighting the squamocolumnar junction. Methylene blue is taken up by absorptive epithelium, primarily of the small bowel and colon. Its effectiveness with Barrett's is variable. In addition, there is a question of methylene blue resulting in DNA damage when used excessively.[31] These dyes may play their largest role in use with high magnification endoscopy.

High magnification endoscopy — The concept of using a combination of staining with either high resolution or magnification endoscopy to allow accurate identifica-tion of Barrett's disease as well as dysplasia is currently under examination. Metaplasia was detected in areas with a prominent tubular or villous pattern.[32,33] Although pilot studies appear promising, future work is needed.

Brush and balloon cytology — Cytology offers the promise of sampling a large area for dysplastic tissue. Endoscopic brushing is promising for Barrett's as a variety of biomarkers have been identified as precursors.[34] However, cytology interpretation of dysplasia is even more difficult than histologic interpretation. Balloon techniques are discussed below and have primarily been used in screening for squamous cell carcinoma.

Fluorescence, reflectance, and light scattering spectroscopy — These are relatively simple techniques that have been applied in an attempt to diagnose early neoplasia with only limited success. They generally are associated with inadequate sensitivity or specificity. The potential advantage of these techniques for diagnosing dysplasia is that they are not imaging techniques but look at subcellular processes, albeit complex in nature. Recently, the three techniques have been combined with some success, but significant future work is needed.[35]

Optical coherence tomography – We initially demonstrated OCT's capability of imaging normal GI tissue and tissue with Barrett's disease in the late 1990s.[36,37] This is seen in Figure 17.3 where normal esophagus, Barrett's disease, and adenocarci-noma are shown. In another study, distinct layers were identified as the muscularis propria in the presence of propylene glycol (Figure 17.4).[167] In contrast, the presence of gland- and crypt-like morphologies and the absence of a layered structure were observed in Barrett's.[38] Similar results were found by Jackle et al.[39] Layers from epithelium to the muscularis propria were identified in normal squamous epithelial. The most obvious sign of Barrett's was absence of the layered structure. Additional work has confirmed the ability to differentiate normal squamous from Barrett's.[40] OCT generally images five layers: the squamous epithelium, lamina propria, muscu-laris mucosa, submucosa, and muscularis propria. In the mid esophagus, layers imaged are the squamous epithelium, lamina propria, submucosa, and muscularis

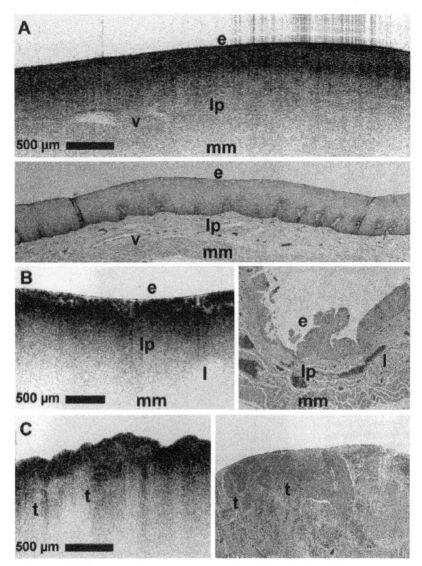

Figure 17.3 In vitro Barrett's and adenocarcinoma. An OCT image of normal esophageal mucosa and the corresponding histology are shown in A with the lamina propopria and muscularis mucosa is clearly defined. The middle image (B) is a section of Barrett's mucosa where glandular structure is noted and the layered appearance lost. The lower image (C) shows adenoocarcinoma where all structure is lost. Courtesy of Pitris, C., et al. (2000). *J. Gastroenterol.* 35, 87–92.

propria. There is poor imaging of the muscularis mucosa in the mid esophagus. As stated, OCT has shown a feasibility for imaging Barrett's, although dysplasia is much more difficult to image and will be discussed below. It can easily be envisioned that the small OCT catheter can be introduced through the nose or mouth in an

PROPYLENE GLYCOL AS A CONTRASTING AGENT

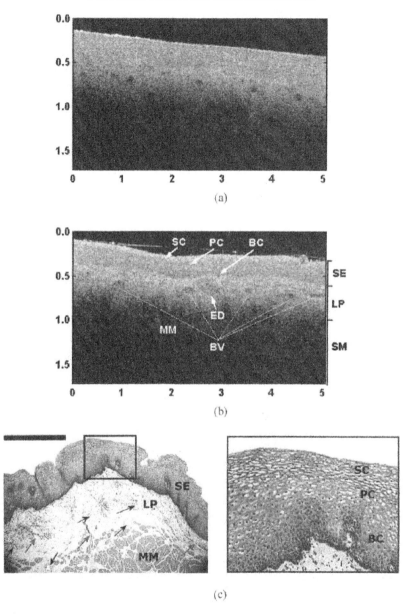

Figure 17.4   OCT images in the presence and absence of propylene glycol. The top is the OCT image of human esophagus in the absence of propylene glycol, a contrast agent. The second is in its presence. Improved identification stratified squamous epithelium (SE), lamina propria (LP), and basal cells (BC). Courtesy of Wang, R.K. and Elder, J.B. (2002). *Lasers in Surg. Med.* 30, 201–208.

office-like procedure to diagnose Barrett's. This would alleviate the need for an endoscope and general anesthesia, in addition to reducing cost and likely patient reluctance. OCT to date has not demonstrated an ability to consistently diagnose dysplasia. This is not surprising in view of how difficult the diagnosis is to make with histopathology. Nevertheless, other approaches are under examination including increased resolution, speckle analysis, and algorithms analyzing scattering. Imaging can be performed radially, along the length of the catheter (axial), or forward from the tip of the catheter. Balloons are sometimes used to expand the esophagus and can sometimes allow for spiral imaging up the esophagus, similar to spiral CT. The size of the esophagus makes it more difficult to image compared with the coronary system. The lumen size is much greater than mucosa, so that radial imaging requires, in general, the acquisition of a large amount of data. Forward and axial imaging only allows small areas to be sampled. An optimal catheter design is yet to be developed. In conclusion, OCT represents an attractive technology for the diagnosis of Barrett's esophagus. However, its role in the diagnosis of dysplasia has yet to be established.

17.2.3 Squamous Cell Carcinoma

17.2.3.1 Epidemiology

Another cancer of the esophagus, which is almost equal in incidence in the United States to adenocarcinoma, is squamous cell carcinoma.[41] In the United States, there is not a well-defined risk factor like Barrett's disease for adenocarcinoma. In regions outside the United States such as China, Iran, and Africa, the incidence is very high, warranting screening.[42] In China, for example, esophageal cancer represents 22% of all deaths.[43] The incidence of squamous cell carcinoma in Japan is much higher than that in most developed countries, but lower than that in China.[44] In high risk populations, esophagitis and atrophy can be seen as early as age 25, dysplasia at 35, and with carcinoma generally occurring over the age of 60.[45]

Squamous cell carcinoma is two to four times higher in males than in females, and is more common for blacks than whites.[46,47] The incidence ranges from less than 5 cases per 100,000 among whites in the United States to 100 per 100,000 in high risk areas of China and Iran.[47]

In the United States, the disease tends to be higher in the northeast and one of the strongest differences is between the high rate in urban areas compared to reduced rates in rural areas.[48] Among urban areas, rates in Washington D. C. among blacks are the highest.[49]

The prognosis of esophageal cancer is poor if not detected early, particularly if lymph node involvement exists. Between 1983 to 1990, the 5-year survival rate in the United States was 9.2%.[50] If the tumor involves either the lamina propria mucosae layer or the upper third of the submucosa, the chance of lymph node metastasis is 5–25%, so some suggest preoperative lymph node evaluation.[51–55] If the tumor has penetrated more than one-third into the submucosa, metastasis to lymph nodes is between 20 to 60%.[53–55]

17.2.3.2 Risk factor

In the United States, the greatest risk factors are alcohol and nicotine abuse.[56] The risk of esophageal cancer in large cohort studies is five times higher among cigarette smokers and tenfold among heavy cigarette smokers.[57] This is partly related to increased alcohol consumption among smokers.[58] However, even in patients who do not drink excessive alcohol, the risk still remains high. This increased risk extends to cigar smokers.[58]

In a study in France, the difference between high and low alcohol intake ranges from 20- to 50-fold.[59] Combined with drinking, smoking results in a 100-fold increase. The increase appears higher for hard liquor but does also occur with beer and wine. The risks for heavy consumers of alcohol are inversely proportional to the nutritional status.[60]

Probably the most important risk factor for squamous cell carcinoma world wide is poor nutrition. This is suspected to be the etiology in high risk groups in Iran and China.[61,62] In particular, diets low in fruit and vegetables have been associated with increased squamous cell carcinoma even in high risk regions of developing countries. Specific important nutrients include vitamin A, vitamin C, magnesium, iron, riboflavin, and zinc.[63] In a study in China, 30,000 villagers were given vitamin and mineral supplements.[61,62] After 5.25 years, cancer rates were lower by 13%. A benefit appeared among those who received beta-carotene, vitamin E, and selenium, but the effects were even greater for gastric cancer. Thermal injury can also predispose to esophageal cancer.[64,65] In individuals who routinely drink hot drinks, the risk of esophageal cancer is increased, particularly in Asia and South America.

Abrasions of the esophagus can also lead to squamous cell carcinoma. In regions where water is scarce and the diets are high in rough bulky food, the risk is increased. Chronic esophagitis is the most frequent lesion found in populations at risk for esophageal cancer and may be considered by some a precursor lesion.[66,67] In an autopsy study performed in Japan, 58% of patients with esophageal cancer had esophagitis.[68] In another series in northern Iran, 86% of the subjects had esophagitis.[69]

Atrophy is another risk factor for squamous cell carcinoma.[69] In northern Iran, 13% of men and 8% of women have thinning esophageal mucosa. Similar increases were seen in black populations in New Orleans.[66]

The World Health Organization (WHO) classified dysplasia in three distinct categories, mild, moderate, or severe dysplasia, but some variation is found in the literature on classification.[70] Atypical cells localized to the basal zone are indicative of mild dysplasia while immature cells in more than three quadrants of the epithelium represent severe dysplasia. Dysplastic tissue is generally thinned, which may be useful when evaluated by high resolution imaging. Since OCT can accurately measure the thickness of epithelium to micron scales, it is promising for the diagnosis of dysplasia. With carcinoma *in situ*, the entire mucosa is involved. In China, in high risk populations, the incidence may be as high as 28% while in the general population of northern Iran, the incidence may as high as 5%.[71] In autopsies studied in

Japan, 12% of patients had dysplasia while a similar group in the United States had a rate of 0.2%.[68,72]

Similar to the situation in Barrett's disease, the identification of dysplastic lesions is difficult endoscopically. They often cannot be detected and in other cases the lesions resemble esophagitis.[73]

In one study in China, of 105 patients with mild dysplasia, 15% progressed to severe dysplasia, 40% remained unchanged, and 45% regressed to normal.[74] Of the 79 patients with severe dysplasia, 27% progressed to cancer, 33% essentially stayed the same, and 40% regressed to mild dysplasia or normal.[74] In another study in the United States, severe dysplasia was not an indicator of outcome as assessed by balloon cytology.[75] The reason for the difference may be that, while the prevalence of dysplasia in the populations studied is similar, the incidence of squamous cell carcinoma is much higher in China than in the United States. However, similar to dysplasia in Barrett's disease, many investigators recommend intervention when dysplasia is detected.[73]

17.2.3.3 Detection of early cancer

Screening is clearly needed for high risk populations, particularly overseas. Some potential techniques include esophageal balloon-mesh cytology (EBC), ultrasound, endoscopy with Lugol's solution, and ultrasonic tactile sensor.

With EBC, a balloon is introduced through the mouth. Nylon mesh is placed around the balloon to serve as an abrasive surface to collect superficial cells. The balloon typically has two lumens, one for inflation and the other for aspiration. The 5-year survival for patients with cancer detected early by EBC in China is 90% compared with an overall survival rate in the United States of 5–7%.[75–79] However, in a study in the United States, the balloon was far less effective. First, inflammation at times was confused with dysplasia.[75] The authors ultimately concluded that while the technique was valuable in China, its usefulness in the United States was questionable. They felt, based on their experience, exophytic lesions or invasion were required for EBC to be successful. This was because of the low incidence of carcinoma; high incidence of esophagitis in the population; inability to distinguish inflammatory from dysplastic changes; and a less predictable dysplasia to cancer pattern in the United States.

Endoscopic ultrasound has been shown to be useful for staging tumors, but with the exception of an isolated abstract, studies have demonstrated its inability to identify dysplasia and early carcinoma reliability.[80]

Lugol's solution is very useful for the identification of some dysplastic lesions. This solution stains normal epithelium, which produces glycogen that interacts with the iodine. In one study, 23 out of 25 invasive carcinomas, 6 out of 7 carcinoma's *in situ*, and 5 out of 7 cases of severe dysplasia were unstained.[81] On the other hand, mild to moderate dysplasia was not well detected. In another study done in China, the sensitivity for identifying high grade dysplasia or cancer was 96% while specificity was 63%.[82] However, while Lugol's solution is useful for identifying dysplasia, it is not reliable for assessing margins accurately, an important consideration when performing endoscopic mucosal resection.

One technique that has been used in an attempt to distinguish tumors limited to the mucosa from deeper penetration is ultrasonic tactile sensor (UTS).[83] With UTS, stiffness differences between tumor states is measured. The ultrasound probe has a primary frequency. When the probe is in contact with tissue, frequency shifts occur that are dependent on the stiffness of the tissue. Statistical significant differences were noted between limited mucosal disease and submucosal disease. Current problems are sensitivity to temperature and pressure in addition to a limited ability to image the posterior wall of the esophagus. Furthermore, larger studies are needed to establish sensitivity and specificity.

Fluorescence has not been extensively examined as a method for assessing dysplasia. There is one paper examining 48 patients with malignancy, but dysplastic tissue was not looked for and it is unclear what percentage of patients were squamous cell and what percentage were adenocarcinoma.[84]

17.2.3.4 Treatment

Endoscopic mucosal resection (EMR) is a method for treating early esophageal malignancies.[85] A typical procedure begins by passing a loop snare through the lesion and coagulation applied to the snares. If the lesion is large, the process is repeated. Since the advent of EMR, preoperative evaluation of tumor invasive depth has become important. If the tumor has penetrated more than one-third into the submucosa, metastasis to lymph nodes is between 20 to 60%, which excludes patients from EMR.[85-87] A role for high resolution guidance is therefore envisioned.

If the tumor involves either the lamina propria mucosae layer or the upper one-third of the submucosa, the chance of lymph node metastasis is 5–25%, so some suggest preoperative lymph node evaluation.[85-87]

Esophagectomy remains the most common treatment for either type of esophageal cancer.[86-89] Typically, either intestine is used to replace the esophagus or for direct attachment with the stomach, although a gastrostomy tube may be used. In the United States, the mortality was 2.5% in hospitals with high surgical volume and 15% with low surgical volumes. Complications include pneumonia, cardiac arrhythmia, anastomotic leakage, and infection. In addition 60–80% of the patients have regurgitation, aspiration, and cough. Techniques which could reduce the need for esophagectomy would therefore be of considerable clinical value.

17.3   GASTRIC CANCER

17.3.1   Epidemiology

The stomach consists of different sections with distinct morphology and function. The section closest to the esophagus is the cardia and the next section is the fundus. The next and largest section is the body or corpus. The final section before the duodenum is the pylorus. Globally, the layers of the stomach, which differ in composition from section to section, are the mucosa, muscularis mucosa, and submucosa followed by three muscle layers. The inner bend is called the lesser curvature and the outer bend is the greater curvature.

Gastric adenocarcinoma is the most important tumor of the stomach, comprising 90 to 95% of all stomach cancers. It is the second most common cancer in the world.[90] Its incidence in Japan, Russian, and China is four- to sixfold more common than its incidence in the United States, UK, and Canada. In the world, the incidence is approximately 900,000 while deaths are in the range of 650,000. In the United States, there are approximately 24,000 cases of gastric cancer with 14,000 dying of the disease (www.cancer.org). In Japan, it accounts for 20 to 30% off all incident cancers. World wide, it exceeds lung cancer in incidence.[91] The male to female ratio is two to one. In 1940, it was the most common cause of cancer death, but its overall incidence has decreased dramatically.[92–94] Interestingly, while cancer of the distal stomach has decreased, like adenocarcinoma of the esophagus, adenocarcinoma of the proximal stomach is on the rise. This suggests that the mechanism behind carcinogenesis in the different sections of the stomach is not the same. The 5-year survival of gastric cancer remains less than 20% overall.

There are two general types of mucosal gastric cancer histologically.[94] One type has an intestinal morphology, resembling colon. It forms bulky tumors composed of large glandular structures. The well-polarized columnar cells often have brush borders. Intestinal type is regional and is likely due to the ingestion of exogenous toxins. The second type is the diffuse type, whose incidence is relatively constant world wide. Where the incidence of type I cancer is low, this tends to be the most common cancer. The diffuse type extends widely with no distinct margins and the glandular structures are rarely present. The cells tend to be present in small clusters or individual cells. Individual cells infiltrate and thicken the stomach wall without forming a discrete mass. These cancers tend to occur in the cardia, in younger patients, and have a worse prognosis. It is generally found in areas where the incidence of gastric cancer is high.

The distribution of cancer within the stomach is 30% in the pylorus, 30% in the body, 30% in the cardia and fundus, and the remaining 10% of cancers are diffuse.[95] Tumors are more common on the lesser curvature but when an ulceration is present on the greater curvature, it is more likely to be malignant.

17.3.2 Risk Factors

Nitrate metabolites, either produced from poorly digested food or abnormal metabolism in the stomach, remain the greatest risk factor.[96–98] Additional risk factors include *Helicobacter pylori* infection, chronic atrophic gastritis, intestinal metaplasia, pernicious anemia, high fat diets, low fruit/vegetable intake, smoking, genetics, and occupation. Occupations associated with increased gastric cancer include miners, painters, fishermen, agricultural workers, and textile workers.

17.3.3   Staging

When gastric cancer is superficial, it typically does not present with symptoms. The most common symptoms on presentation of more extensive cancer are weight loss and abdominal pain, followed by nausea and anorexia. Gastric cancer can be grossly staged as early and advanced gastric cancer. Advanced cancer is defined as spread of the cancer below the submucosa into the muscular layer. In Japan, there is a major screening program underway and 30% of all cancers are found at an early stage.[91] In the United States, where screening is not cost-effective, only 10 to 15% are found at an early stage.[95]

In the current staging system, stage 0 is limited to the mucosa, which has a 5-year 90% survival rate. Stage IA has negative nodes but invades into the lamina propria or submucosa, with a 5-year survival rate of approximately 60%. Type IB has negative nodes with invasion into the muscularis propria, which has a 5-year mortality of less than 50%.

The goal of current diagnostic techniques is to differentiate mucosal from submucosal disease, guide minimally invasive surgery, and identify tumor margins. The primary techniques for assessing the extent of gastric cancer are endoscopy and barium studies. They should be viewed as complementary and not competitive techniques. The improvement of double contrast barium techniques over the last two decades has improved the ability of radiologists to assess gastric cancer and ulcers. The double contrast technique allows more detailed assessment of the gastric mucosa and lack of distensibility from diffuse carcinoma. Advantages of barium are its low cost, low percentage of side effects, and patients tolerate it well.[97] It should be noted that less than 5% of gastric ulcers represent carcinoma. Therefore, some gastroenterologists recommend 6 weeks of acid suppression followed by repeat barium contrast. Overall though, its limitation is its resolution, which is generally not sufficient to diagnose disease localized to the mucosa.

17.3.4   Diagnosis

Endoscopy is considered the most specific and sensitive technique for identifying gastric cancer.[98] It allows direct visualization as well as the ability to perform biopsy or cytologic brushings. New scopes result in a more comfortable examination, as they are thinner, more maneuverable, and safer for the patient. However, endoscopy does not allow imaging below the surface to distinguish mucosal versus submucosal disease.

MRI, CT, ultrasound, and biomarkers (e.g., CEA, CA-19.9, CA-50) have their role in staging, but their utility in the early diagnosis of cancer is questionable.

OCT represents an attractive alternative for assessing early gastric adenocarcinoma.[99-101] The optimal goal is again threefold — identifying patients with disease limited to the mucosa, margins, and guiding endoscopic mucosal resection. Figure 17.5 demonstrates OCT images of the human GI tract with propylene glycol as a contrast agent. It can be seen that the lamina propria and muscularis mucosa are clearly defined by OCT. Additional OCT images of the stomach have been

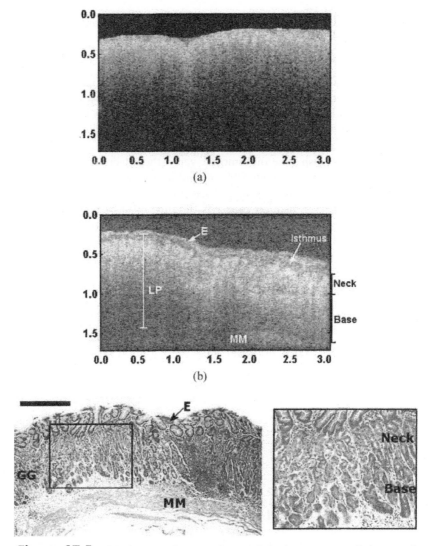

Figure 17.5 OCT images of a normal stomach in the presence and absence of propylene glycol. The top image is in the absence but the second is in the presence. The epithelium, neck, and base are more clearly defined. Courtesy of Wang, R. K. and Elder, J. B. (2002). *Lasers in Surg. Med.* 30, 201–208.

demonstrated by other groups. OCT has demonstrated considerable potential as a method to diagnose early gastric malignancies.

The treatment for gastric carcinoma is variable depending on the stage, and the focus here is mucosal resection in the appropriate screening population. Here lies OCT's greatest potential advantage. In addition, as with other organ systems, endoscopes can be made relatively small and inexpensive.

17.4 BLADDER CANCER

17.4.1 Epidemiology

The bladder holds urine prior to excretion. It is covered by transitional mucosae designed almost exclusively to allow stretch and prevent toxicity from urine. When the bladder is relaxed, it normally appears five to six cells thick, but in the stretched state it may appear only two or three cells thick. Below this, the wall consists of three loosely arranged layers of smooth muscle and elastic fibers, which are often difficult to distinguish. The outermost layer is the adventitia which contains arteries, veins, and lymphatics. About 50% of bladder cancers are believed to be related to cigarette smoking.[102] Other agents include aniline dyes and *Schistosoma* (parasite) in developing countries. Bladder cancer is the fifth most common cancer in the United States, representing 6% of all male cancers and 3% of all female cancers.[1] Since well-described risk factors/signs (e.g., hematuria) occur in 80% of all cases, in general it is a preventable disease. Bladder cancer is staged using either the American Joint Committee on Cancer (AJCC) or TNM staging systems.[103] In AJCC stage 0, noninvasive papillary carcinoma or carcinoma *in situ* is present. In stage I, the tumor has invaded the subepithelial connective tissue. In stage II, the tumor has invaded the muscle. In stage III, the tumor has entered the perivesical fat, prostate, uterus, and/or vagina. In stage IV, invasion of the pelvic or abdominal wall occurs or metastasis to lymph nodes or distal sites. The tumor type is transitional cell carcinoma approximately 93–94% of the time. Overall, 75% of tumors present as superficial lesions. Low grade transitional tumors that are papillary tend to be less likely to be invasive while carcinoma *in situ* is more of a precursor to muscular invasion. There is a major need for improved imaging and guided interventions due to the high rate of recurrence and the multifocal nature of the disease, which suggests a general molecular abnormality of the bladder.

17.4.2 Staging

Most cases present as type 0 and I (approximately 33 and 29%, respectively). Five-year survival rates for stages 0, I, II, III, and IV are 90.2, 86.6, 65.1, 47.9, and 23.2%, respectively. Therefore, it is critical to identify the disease before there is muscle invasion.[104]

The treatment strategy depends on the stage of the tumor. The percentage undergoing transurethral resection (TUR) for transitional cell carcinoma (TCC) are 94, 92.4, 63.8, 40.9, and 44.7% for stage 0, I, II, III, and IV, respectively. Patients not undergoing TUR generally undergo cystectomy. The patients are monitored as a 50% recurrence rate exists.

The ability to distinguish tumors limited to the mucosa from those penetrating the lamina propria (into the muscle) is critical, particularly in view of the high rate of recurrence and the development of disease in other regions. If a reliable method can be devised for distinguishing the two types, it can dramatically influence treatment. If it is confidently known that the tumor is limited to the mucosa, laser treatment or simple electrocautery can be performed. This becomes particularly important in view of the high incidence of impotence when more aggressive treatment is needed.

17.4.3 Diagnosis

Cystoscopy, the direct visualization of the bladder surface via optical fiber bundles, is the current gold standard for the diagnosis of TCC.[107] Under cystoscopic evaluation, those tumors raised above the bladder surface can be readily detected. Furthermore, cystoscopy can be used to play some role in the TUR of superficial tumors.[108] However, as alluded to above, some clinically relevant pathology, which appears benign or stage 0 by direct visualization, may go unobserved. This would include multiple site involvement at the time of local resection or the presence of recurrences during subsequent screening. In addition, cystoscopy does not allow subsurface microstructural information to be obtained, such as the extent of tumor invasion into the bladder wall. Blind biopsy or cytology may be used in conjunction with cystoscopy, but they are only of limited value.[109]

The limitations of cystoscopy have led investigators to examine other methods for interrogating the bladder wall. MRI, transabdominal ultrasound, and CT are powerful imaging technologies for a wide range of medical applications, including the assessment of distant metastasis.[108–110] Unfortunately, their relatively low resolution (>500 μm) prevents assessments of microstructural changes within the bladder wall, with the mucosa only a few 100 μm thick. High frequency endoscopic ultrasound (20 MHz) has been applied experimentally to the assessment of bladder carcinomas. However, its axial resolution of greater than 150 μm is also not sufficient for reliably identifying layers within the bladder wall, which has prevented its adoption into routine clinical use.[110] Both diffuse reflectance and fluorescence techniques have been explored, but high false positive rates have raised questions about their ultimate utility.[111] An imaging technology capable of assessing the bladder wall near the resolution of histopathology would be a powerful tool to overcome these limitations in the management of TCC.

Several pilot studies have also been performed *in vitro* and *in vivo* on the urinary tract. In one study, both *in vitro* imaging of human bladder pathology and *in vivo* imaging of normal rabbit bladder were performed.[112] This included *in vitro* studies on transitional cell tumors post-resection. Normal bladder is seen in Figure 17.6 while TCC is seen in Figures 17.7. *In vivo* imaging performed on a rabbit bladder has also been included to demonstrate limitations in current endoscope-based systems; specifically radially imaging catheters are not ideal due to the large size of the bladder. In the normal bladder, the mucosal-submucosal interface is identified by the arrows, in addition to superficial blood vessels. In Figure 17.8, structure is lost in transitional carcinoma *in vivo*.[113] Studies have been performed combining OCT with fluorescence.[114] However, the advantage of adding fluorescence is unclear, particularly in view of the added complication of the procedure.

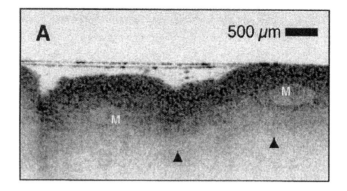

Figure 17.6   Normal bladder. The top is the OCT image of normal bladder while B is the corresponding histology. The upper arrows are the mucosal/submucosal interface and the bottom arrows are the submucosal/muscularis interface. Courtesy of Jesser, C. A., et al. (1999). *Br. J. Radiol.* 72, 1170–1176.

17.5   LUNG CANCER

17.5.1   Epidemiology

In the United States, lung cancer is the second most common cancer diagnosed in men and women, but the leading cause of cancer mortality. It represents 14% of all cancers diagnosed.[1] In Western Europe, lung cancers cause more deaths than colorectal, cervical, and breast cancer combined. In the middle of the 20th century this was predominately a disorder of the Western industrialized countries, but the disorder has now spread to other regions, such as Eastern and Southern Europe, paralleling increased nicotine consumption.

Survival rates from lung cancer are poor. Because of the high incidence and poor prognosis, in the United States one in three deaths from cancer in men are due to lung cancer and one in four in women. Current survival rates, which are between

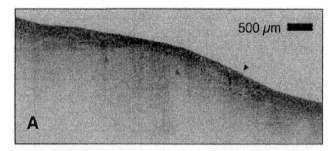

Figure 17.7 Transitional cell carcinoma. The top in the OCT image of invasive transitional cell carcinoma. Distinct layers, boundaries, and capillaries are no longer noted. Courtesy of Jesser, C. A., et al. (1999). *Br. J. Radiol. 72*, 1170–1176.

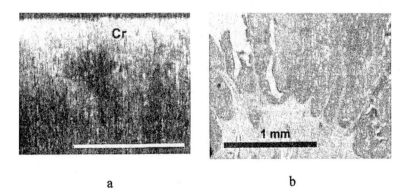

a   b

Figure 17.8 In vivo imaging of transitional cell carcinoma. Courtesy of Zagaynova, E. V., et al. (2002). *J. Urol. 167*, 1492–1496.

7 and 13%, have not changed significantly in 30 years. Due to the large reserve of the lung, symptoms do not occur until late and less than 20% are diagnosed at an early stage (stage I).

World wide, lung cancer is more common in men (37.5 new cases per 100,000) compared with women (10.8 per 100,000). However, in the United States a surge in women smokers has occurred over the last few decades. Between 1973 and 1996, age-adjusted incidence rates declined by 2.5% in men but increased by 123% in the women. Mortality rates increased only 8.9% among men, but 148.3% in women over the same period.

The ages with the greatest incidence are between 65 and 79, which account for more than half the annual incidence. In the United States, mortality rates for lung cancer demonstrate substantial regional variability. In the last few decades, for white men smoking rates are high in the southeast and south central regions, and relatively low in the northeast. Among white women, relatively high rates appear on the Atlantic and Pacific coasts. For both sexes, rates were low in the mountain and plains states. For blacks, rates are low in the south and high in the northeast.

17.5.2   Risk Factors

The most important risk factor for lung cancer appears to be nicotine abuse.[115] Some have stated that tobacco smoking accounts for between 75% and 90% of lung cancer risk. The association between lung cancer and smoking was raised as early as the end of World War I. After World War II, the number of reports linking the two increased dramatically. Finally, in 1964, the U. S. Surgeon General stated that "... cigarette smoking is causally related to lung cancer... (and) the magnitude of the effect far outweighs all other factors."

The lifetime risk of lung cancer among male smokers is 14.6 and 8.3% among female smokers.[116] However, the lag between starting cigarette smoking and the development of lung cancer is 30 or more years and a higher risk for women may be noted in subsequent decades. In a recent prospective study, women, in spite of smoking at a later age and smoking fewer and lighter cigarettes than men, appear to be more likely to develop lung cancer. Of note, since the lifetime risk among smokers is less than 20%, then other factors, possibly genetics, must be involved.

Cessation of smoking appears to reduce the risk, but does not eliminate it. This is particularly important since as many as 50% of all adults in the United States are former smokers. Current smokers have as much as a 20-fold greater risk than nonsmokers. The risk after 2–10 years of smoking cessation is eightfold. After 10 years, the risk is 2.2-fold. Half of the new cases of lung cancer occur in ex-smokers. Median age of onset for former smokers is slightly later than that for current smokers.

Of the remaining 10–15% of lung cancers not directly related to smoking, associations include radon, secondhand smoke, and occupational exposure. Radon is an inert gas formed from radium during the decay of uranium.[117] Most of the radon exposure is indoors and from soil. Radon may lead to lung cancer because it breaks down and releases alpha particles. Data from uranium miners have established a relationship between radon exposure and lung cancer. The U. S. EPA has established that approximately 15,000 annual cases of lung cancer are due to radon exposure. Ten thousand radon attributed deaths are estimated to occur among smokers and 5,000 deaths among nonsmokers.

Secondhand smoke appears to be an independent risk factor for lung cancer. It was noted that nonsmoking women whose husbands smoke carry an increased risk of lung cancer.[118,119] In 1986, the U. S. Surgeon General and National Research Council reported that secondhand smoke is a significant cause of human lung cancer. Although there is some disagreement as to how much it contributes to lung cancer incidence, it probably increases the risk by about 60%. This would represent about 2000–4000 cases in the United States.

17.5.3 Histopathology

Lung cancer does not refer to a single histologic malignancy.[120] Histologically, 20% of all lung cancers are small cell. The remaining 80% are classified as non-small cell lung cancer (NSCLC). All NSCLCs are classified together because they commonly coexist together, the histology can often be difficult to differentiate, and they have dissimilar treatment regimes and prognoses relative to small cell. Of the NSCLCs, 40% are adenocarcinomas, 30% are squamous cell carcinoma (also called epidermoid), and 10% are large cell undifferentiated.

Small cell carcinomas are the most aggressive of the lung tumors. They grow rapidly and metastasize early. This is why, unlike the NSCLCs, they are generally treated with chemotherapy rather than surgical resection. Even though they carry a poorer prognosis, they are more responsive to chemotherapy and tend to occur in the major bronchi.

Squamous cell was once the most common lung cancer. However, adenocarcinoma has now become more common. Squamous cell carcinoma typically arises from major bronchi. In its earliest form, carcinoma *in situ*, the normal delicate columnar epithelium is replaced with thick stratified squamous epithelium. The carcinomas may grow to a relatively large size before metastasizing.

Adenocarcinomas account for about 40–45% of all lung cancers. Their origin tends to be in the peripheral bronchi rather than the major bronchi, which makes it difficult to detect by bronchoscopy.

Bronchioloalveolar carcinoma grows along the alveoli like a scaffolding, with little involvement of pulmonary parenchyma. The radiographic appearance resembles that of pneumonia and, while metastasis to lymph nodes is common, death usually occurs via respiratory insufficiency.

17.5.4 Staging System

The staging systems here are generalities, and an oncology textbook should be consulted for more exact staging.[121] Non-small cell (NSC) and small cell cancers (SC) are essentially staged separately, as curative surgery with chemotherapy and radiation remain the focus for NSC and for SC life extension with chemotherapy and radiation are the focus. The goal is to find NSC at stages IA, IB, and IIA as these hold the best prognosis of 5-year survival. Even then, mortality is high with stage IB greater than 3 cm with no lymph node involvement with a 5-year prognosis of approximately 55%. It is unclear whether carcinoma *in situ* or dysplasia should be treated. The time scale is poorly understood and it is found in a high percentage of patients. It is necessary to know if preinvasive lesions advance and how their treatment affects mortality. A simple two-stage system is used with SC. With limited stage disease the tumor is confined to one hemisphere and possibly regional lymph nodes. Extensive disease, which is far more common, extends beyond these parameters. The median survival without treatment is approximately 10 weeks; with treatment it is 50–60 weeks.

17.5.5   Treatment

Treatments for lung cancer include surgery, chemotherapy, PDT, and radiation. For small cell carcinoma, as stated, the treatment of choice is generally chemotherapy. For those NSCLCs where distant metastasis has not occurred, the patient may be a candidate for surgery. In general, outcome is poor unless caught at an early stage.

If the cancer is detected early, significant survival rates have been obtained. Unfortunately, less than 20% are detected at stage I. The survival for stage I is 70% at 5 years and 60% at 10 years. The operative mortality is 4% for lobectomy and 1.6% for pneumonectomy.

Lung cancer prevention has also been attempted in chemoprevention trials, without significant success.[122] In an example large randomized trial in over 29,000 male smokers between the ages of 50 and 69 in Finland, the vitamins alpha-tocopherol and beta-carotene (ATBC) were given to patients. Patients receiving vitamins had a statistically significant increased risk of lung cancer (18%). In the CARET study, a randomized, double-blind, placebo-controlled trial, vitamin A and beta-carotene were given to 14,420 smokers and over 4,000 asbestos-exposed workers. A 28% increase was noted in lung cancer in the vitamin-treated group.

PDT has the potential to eliminate dysplasia or carcinoma *in situ*.[123] It is not likely to be an effective therapy for larger neoplasia because the maximal therapeutic depth is no more than 2 cm and is probably less than 1 cm. It was tested in one study where 38 patients underwent treatment with a hematoporphyrin derivative. A complete response occurred in 13 patients with 3 of these exhibiting recurrences. A complete response occurred in patients whose tumors were radiographically negative and appeared superficial by bronchoscopy. Because dysplasia and even carcinoma *in situ* appears common in smokers and the natural history is not well understood, it may ultimately be found that treatment is not warranted and if treatment is deemed needed, PDT may be of use. In one study, cancer developed in 42.9% of smokers with marked dysplasia and 10.6% in moderate dysplasia by 10 years. Therefore, a better understanding of the natural history of dysplasia is needed. The major complications of PDT are hemoptysis from large tumors and airway obstruction from edema. It should be noted that OCT may be found useful in guiding the application of PDT as well as detecting early disease.

17.5.6 Screening

Early detection should have at least six criteria: (1) there must be a phase where the disease is not yet terminal, (2) the technology should be capable of identifying this state, (3) the appropriate screening group should be identified, (4) an intervention must exist to treat the disorder at this state, (5) screening must be safe, and (6) there should be a cost-effectiveness to the procedure.[124]

Three major National Cancer Institute studies in the 1970s (from the Mayo clinic; Johns Hopkins Oncology Center; and Memorial Sloan Kettering, MSKCC cancer center in addition to a smaller Czechoslovakia study) failed to demonstrate a benefit for screening smokers with X-rays. In the MSKCC and Johns Hopkins study, the addition of cytology was of no benefit. Using mathematical modeling, one study felt that X-rays could detect stage I tumors in at most 16% of the patients. Many physicians still do not agree and continue to screen.

Possible appropriate screening groups will need to be evaluated. This includes pack-years, years after cessation of smoking, age, and the presence of chronic obstructive disease.

It is estimated that a tumor 1 cm in diameter, which is the smallest lesion that can be detected on a plain chest film, has undergone 30 volume doubling. Small cell, which is a rapidly growing carcinoma, may take 2.5 years to reach this size while non-small cell may take 7–13 years.

Conventional CT was examined for screening in the 1970s.[124,125] However, it was slow, used relatively high radiation, and was not cost-effective. In the 1990s spiral CT was introduced for screening with the advantage of high speed and being relatively inexpensive.[124] Recently, fast, low dose (50 mA), spiral CT which produces radiation 1/6 that of conventional CT and only 2–10x that of a conventional chest X-ray was introduced. It takes approximately 20 seconds, images can be obtained on a single breath hold, no contrast is used, and it costs only slightly more than a chest X-ray. To date though, no randomized clinical trials have been published and most are observational. Spiral CT has the advantage that it also looks at the silent areas. It also gives evaluation of lymph nodes, pleura, and vessels. Since the incidence of adenocarcinoma is increasing, which tends to be in the periphery, CT has this additional advantage.[124]

An early CT study was the early lung cancer action project (ELCAP) (1992). This study looked at 1000 symptom-free volunteers, aged 60 years or older, with at least 10 pack-years of cigarette smoking, no previous cancer, and medically fit to undergo thoracic surgery. Chest X-rays and low dose CT were used on these patients. Noncalcified nodules were found in 23% of patients by CT but only 7% by chest X-ray. Malignant disease was found in 2.7% by CT but only 0.7% by X-ray. Stage I disease was seen in 2.3% of patients by CT versus 0.4% by chest X-ray. Of the 27 patients demonstrated to have malignancy by CT, 26 were resectable. Of the 27 malignancies, 18 were adenocarcinomas and none were small cell, a concern described below. Technical advances in spiral (or helical) CT scan have an estimated potential of picking up stage I disease between 80 and 85% of the time.

Several studies have examined spiral CT for screening. In the Matsumoto study (1996), a mobile, low dose helical CT was used. The 3967 patients were smokers and nonsmokers aged 47–74 years. In this study, 1% had suspicious lesions, 2% had indeterminate lesions, and 2% had suspected cancers. Of the 19 cancers picked up (0.48%), 16 were stage I and three were stage IV. Peripheral adenocarcinomas were found in 12 of the 19 patients.

Several problems have been noted with spiral CT studies. In a Japanese study, lung cancer rates were 0.46% in nonsmokers compared with 0.52% in smokers.[125] Second, there was a predominance of adenocarcinoma, above historical value, and lack of small cell (expected incidence of 20%), raising the possibility that clinically irrelevant, slow-growing tumors are being picked up. Third, there is a high false positive rate for tumors as a whole in the order of 10 to 20%.[126] This can result in an increased number of unnecessary procedures, including thoracotomy as well as increased morbidity and excessive emotional stress on these patients. Many of these patients would likely die of other causes. Finally, spiral CT is less sensitive for detecting proximal tumors and 70% of all lung cancers are in the large proximal airways, most amenable to bronchoscopy. Ultimately, another screening technique such as OCT may be used or spiral CT may need to be combined with another modality. While some authors have suggested using PET in combination with spiral CT, this is not likely to be practical. Sputum- or serum-based techniques are being explored.

Fluorescence is a technique that has been used for early lung neoplasia, such as dysplasia or carcinoma *in situ*.[127] The principles behind fluorescence have been described in Chapter 13. The most commonly used system is the laser-induced fluorescence endoscope (AFB-LIFE or LIFE, Xillix, Richmond, BC, Canada). This system excites the tissue with an incident wavelength of 442 nm. The recovered light is filtered by two filters of different wavelength, one in the red region (630 nm and longer) and the other in the green (480 to 520 nm). The data are processed via a nonlinear function which combines the red and green data intensity values, creating a single number. Special imaging algorithms have been generated, based on data of *in vitro* and *in vivo* analysis and Monte Carlo simulation modeling. Other systems that have been used to a lesser degree are the Storz (Karl Storz, Tuttlingen, Germany) and SAFE (Pentex, Asahi, Optical, Tokyo, Japan).

Autofluorescence appears to be the approach of choice over exogenous probe addition. This technique uses differences in the autofluorescence characteristics of normal and neoplastic epithelia to localize lesions. In one trial, a 47% absolute increase in the sensitivity of detecting early neoplasm (dysplasia and carcinoma *in situ*) in high risk patients compared to white light bronchoscopy was noted.[127]

However, it has several problems. First, the bronchoscope can only get as far as the fourth generation bronchus, representing only a small portion of the tree. Second, the procedure is time-consuming. Third, the technique primarily identifies dysplasia and carcinoma *in situ*. However, the natural history of dysplasia, unlike stage I, is not well understood. It appears to be a far more common and diffuse process than cancer as one study showed 40% moderate dysplasia and 12% severe dysplasia in smokers. For ex-smokers, 25% had moderate dysplasia and 6% had severe dysplasia. Finally, it does not distinguish between dysplasia from carcinoma *in situ* well.

Sputum cytology has been examined as a method for identifying early malignancies. It can potentially be used either as a direct method for screening or in conjunction with techniques such as spiral CT (to reduce the false positive rate). Current tests are more sensitive to squamous cell carcinoma rather than adenocarcinoma, possibly because the latter is more peripheral. Among the factors examined are immunostaining with heterogeneous nuclear ribonuclear protein (hnRNP) A2/B1, quantitative image cytometry by looking at malignant-associated changes (MAC) using DNA texture analysis of the cell nuclei, and molecular biological markers.[128] Currently, not a single set of biomarkers has emerged that can be used for accurate prediction of the development of lung cancer in any particular individual. For example, although promising, one recent study in miners using hnRNP A2/B1 has shown 80% sensitivity in those with lung cancer primaries.[127] The low sensitivity in combination with the large screening population puts into question its ultimate utility.

Another potential method is the measurement of volatile organic compounds, such as alkanes and benzene derivatives, which are present in the breath of cancer patients. However, these techniques remain experimental and the source of future investigations.

We initially performed OCT imaging of the *in vitro* tracheobronchial system in 1998.[130] This work examined *in vitro* imaging of the human epiglottis, vocal cords, trachea, and bronchi. An image of the epiglottis was also seen in the original 1996 *Circulation* publication. Good correlation was seen with histology. In Figure 17.9, OCT images are seen of the bronchus and corresponding histology.[131] The mucosa (I) and cartilage (C) are seen as well as glands (g). These features are confirmed by the histology. The first *in vivo* studies were performed in the rabbit, where the trachea, main bronchi, and secondary bronchi were assessed.[132] Imaging was performed at 8 frames per second at 15-μm resolution. An example image is in Figure 17.10 of a rabbit trachea. Here cartilage (c) and muscle (m) are identified. In this study, the epithelium, glands, and their ducts were also identified. Recently, OCT imaging has been performed *in vivo* in humans.[133] A Pentex/LightLab system was used at 30 frames per second at 16-μm resolution. In Figure 17.11a, the bronchoscopic image of an early squamous cell carcinoma is seen where the tumor is designated by A. The mucosa is also well differentiated. In Figure 17.11b the corresponding histology is seen which correlates well. OCT therefore demonstrates considerable potential as a method for the diagnosis of early neoplasms. In addition, it could be

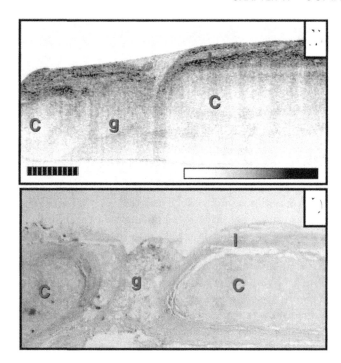

Figure 17.9   Human epiglottis. In vitro OCT image of human epiglottis (A) and associated histology (B). The epithelial (e) and lamina propria (l) are clearly identified as well as cartilaginous (c) and glandular (g) structures. Courtesy Pitris, C., et al. (1998). *Am. J. Respir. Crit. Care Med.* 157, 1640–1644.

a powerful tool for improving our understanding of the natural history of dysplasia and carcinoma *in situ*.

17.6   SCLEROSING CHOLANGITIS AND CHOLANGIOCARCINOMA

17.6.1   Epidemiology

The primary function of the biliary system is to deliver digestive enzymes to the small bowel. These enzymes are primarily from the liver, gallbladder, and pancreas (pancreatic duct). The diameter of the biliary tree is approximately similar to those of the coronary arteries and is challenging to reach through upper endoscopy, although performed. Sclerosis (sclerosing cholangitis) can occur which can be intrahepatic and/or extrahepatic. It is a disorder that occurs in over 70% of patients with inflammatory bowel disease.[134] Patients with sclerosing cholangitis present with jaundice, purities, right upper quadrant pain, and/or acute cholangitis.

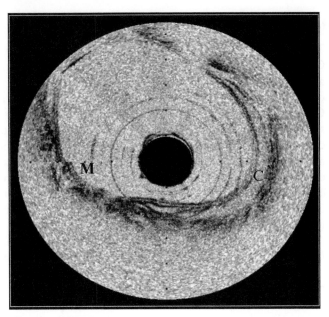

Figure 17.10 In vivo rabbit trachea. In vivo imaging was performed of a rabbit trachea through an OCT endoscope. The cartilage (c) and trachealis muscle (m) are noted. Tearney, G. J., et al. (1997). *Science.* 276, 2037–2039.

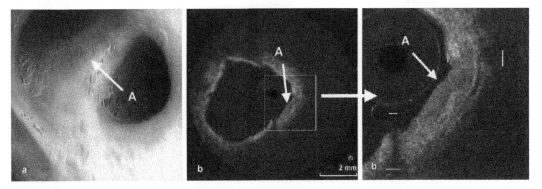

Figure 17.11 In vivo imaging of squamous cell carcinoma. In a, subtle changes in the bronchus are noted. In b (and its enlargement), the tumor in the mucosa and submucosa are seen, as well as loss of submucosal glands and structural definition. Courtesy of Tsuboi, M. et al. (2005). *Lung Cancer.* 49, 387–394.

As the disease generally is segmental, a "beaded" appearance in the biliary system is seen by radiographic techniques. The treatment of choice for severe sclerosing cholangitis is often stenting, surgery, or dilatation when localized obstruction exists.

17.6.2   Diagnosis

The techniques used to assess the biliary system include hepatobiliary ultrasound, CT, MR cholangiopancreatography, percutaneous transhepatic cholangiogram, and endoscopic retrograde cholangiopancreatogram, in addition to direct biopsy or brushings. All are limited in their ability to diagnose malignant transformation. Cholangiocarcinoma, an adenocarcinoma generally of the extrahepatic biliary system, is a complication of sclerosing cholangitis (approximately 8%). Brush cytology and biopsy (approximately 1 mm deep) can be highly variable in their accuracy. Furthermore, when the tumor is beneath desmoplasia it is even more difficult to diagnose. The prognosis is poor but early detection and liver transplant can significantly improve it. Prognosis is 1 to 2 years but if caught early and treated, can reach 30% at 5 years.[135–137] Therefore, a high resolution imaging would likely improve the prognosis in these patients.

Two studies have examined OCT in the biliary system. In the first we performed *in vitro* imaging.[138] In Figure 17.12 the mucosa, submucosa, muscularis, and adventitial layers are well visualized. In addition, glands (arrow) are noted in the biliary tree. This is confirmed in the corresponding histology (C). The second study was performed *in vivo* in humans.[139] The penetration with OCT is sufficient to visualize the adventitia in the extrahepatic duct and hepatic parenchyma in the intrahepatic duct. In Figure 17.13, an *in vivo* OCT image of cholangiocarcinoma is demonstrated with the loss of normal epithelial structure and presence of glandular

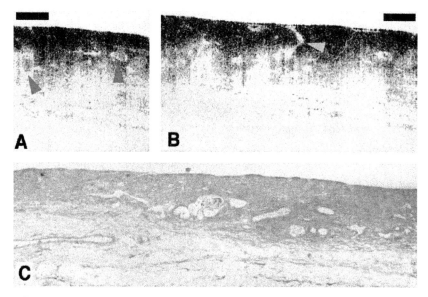

Figure 17.12   Imaging of normal biliary system. A and B are in vitro images of the common bile duct. The arrows in A are glands containing secretions while the one in B is a duct invaginating from the luminal surface. Tearney, G. J., et al. (1998). *Dig. Dis. Sci.* 43, 1193–1199.

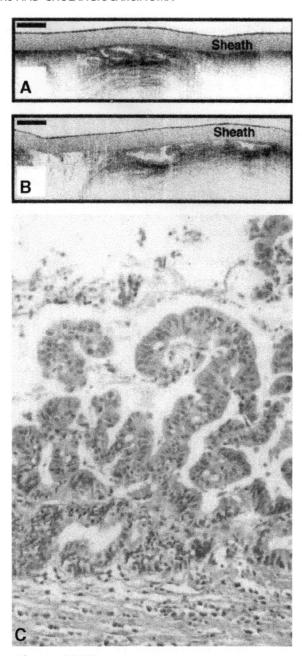

Figure 17.13 In vivo imaging of cholangiocarcinoma. A and B are OCT images of cholangiocarcinoma. The villiform papillary morphology is noted. Figure C is the histology, confirming the diagnosis. Poneros, J. M., et al. (2002). *Gastrointest. Endo. 55*, 84–89.

structure. Villiform papillary morphology is noted indicative of the cancer. The catheter sheath is also noted in the image. The ability to diagnose cholangiocarcinoma in the setting of sclerosing cholangitis could be a powerful tool in improving patient morbidity.

17.7   CERVICAL CANCER

OCT unlikely has a role in the general screening for cervical cancer in view of the effectiveness and cost considerations associated with the Pap smear. However, it may have a role in assessing patients with positive or inadequate Pap smears, focusing on intraepithelial neoplasia or microinvasion.

17.7.1   Epidemiology/Histology

The uterine cervix protrudes into the vagina (exocervix) and contains an endocervical canal that links the vagina with the uterus. The endocervical canal is lined by cells similar to those of the uterus — a single layer of tall, columnar, mucus-secreting cells. In areas where the cervix is exposed to the hostile environment of the vagina, the cervix is lined by thick stratified squamous epithelia. At the point where the endocervical canal opens, the junction between the two types of epithelium exists. The squamocolumnar junction is an area of rapid cell turnover and is typically the site of oncologic transformation.

While invasive cervical cancer is common world wide, it is not common in the United States. The incidence of cervical cancer is about 12,000 yearly with a death rate of approximately 5000. The relative low incidence is due to the effectiveness of Pap smears. Since the 1930s, rates of cervical cancer have decreased by 70%. The mortality rate reached a plateau into mid-1980s likely because of increased sexual activity at a younger age, increased prevalence of human papillomavirus (HPV), and reduced compliance to screening in high risk groups. Of women who are diagnosed with invasive cervical cancer, 65% have not had a Pap smear in the past five years. Risk factors for cervical cancer include smoking, sexually transmitted disease, low socioeconomic status, multiple lifetime partners, increased sexual activity, HPV infections, and immunosuppression (e.g., HIV). HPV is of particular interest because it is often measured in parallel with Pap smears to assess early cervical cancer. The high risk subtypes of HPVs include 16, 18, 31, 33, and 35. The HPV may integrate into the genome or may release a protein that binds and inactivates tumor suppressor genes p53 or Rb. The relative risk of malignant transformation of the cervix to high grade squamous intraepithelial lesion and invasive cancer has been reported as high as 296 for high risk patients.[140]

Squamous cell carcinoma represents over 80–90% of all cervical cancer and will be the focus of this work. The remaining carcinomas are adenocarcinoma, adenosquamous carcinoma, mixed, and metastatic, which all generally carry a worse prognosis.

17.7.2 Staging

The grading system for cervical cancer is stage 0, which is cervical intraepithelial neoplasia (CIN) I, II, or IIII (described below). Stage I is divided into stage IA with invasion limited to <5 mm into the stroma and <7 mm width and stage IB which is further subdivided into IB1 and IB2. With IB1 the lesion is less than 4 cm in size but greater than a IA and IB2 which is greater than 4 cm but still within the cervix. Type II extends beyond the cervix but not into the pelvic wall, and the carcinoma involves the vagina but not as far as the lower one-third. Stage III extends to the pelvic wall or lower third of the vagina, and stage IV is when the tumor invades the mucosa of the bladder or rectum or extends beyond the true pelvis. The 5-year survival of each stage of cervical cancer is stage IA 94%, IB 79%, stage II 39%, stage III 33%, and stage IV 7%.[102]

The focus with OCT is on CIN and microinvasion (stage IA) into the basement membrane. Premalignant lesions (CIN I, II, and III) may remain noninvasive for 20 years. CIN I is characterized as having nuclei atypia and perinuclear vacuolization, with one-third of the surface. CIN II is progression of the atypia to two-thirds of the epithelium. There is normal mitosis and differentiation above the basal layer. CIN III is carcinoma *in situ* and has diffuse loss of differentiation with marked cell/nuclear variation in size, hyperchromasia, and increased mitosis. The next step is invasive cancer, which penetrates the basement membrane.

The mean age at diagnosis of grade II and III CIN is decreasing because of increased sexual activity at younger ages. The age-specific incidence for CIN III now peaks between 25 and 29 years. The peak incidence of invasive cervical cancer is about 10–15 years later.

17.7.3 Diagnosis

The most common test to screen for cervical neoplasia is a Pap smear. The test swabs the cervix and cells removed are placed on a slide and assessed by microscopy. The mean sensitivity of Pap tests for detecting cervical abnormalities is 58% in screening populations and 66% in follow-up populations. The reason the Pap smear is effective in spite of these low sensitivities is because these lesions may remain noninvasive for up to 20 years.[141] Because cervical cancer is a slow growing tumor and low grade lesions often regress, serial testing with Pap smears is effective. In the United States, women are generally tested yearly. Approximately 50,000 positive or indeterminate Pap smears are found each year, which is the population of interest to apply OCT. It has been concluded that if people are screened more than every 3 years then this exceeds $50,000 per life year saved.[142,143] For this reason, in Europe, women are often screened as far apart as 3 years.

As stated, 65% of all women diagnosed with type I or greater have not had a Pap smear within 5 years. The remainder of the cases are false negatives. About two-thirds of all false negatives are due to sampling errors, the cells are not collected or not transferred to the slide. This may result from either poor technique in obtaining/processing specimens or physiologic changes (e.g., infection or neoplasia).

Older women are more likely to have acellular specimens while younger women are more likely to have smears obscured by cells. The most common sampling error though is lack of obtaining cells from the cervical transformation zone. The false negative rate is between 20 and 45%. There is also a wide range of inadequate smears from 2.9 to 20% between different laboratories. The number of inadequate smears is increasing nationally due to increased pressure to scrutinize smears. A current guideline is to recall patients following inadequate smears and immediate referral for colposcopy following three inadequate smears. The incidence of CIN in patients with inadequate smears after biopsy was only 9.7% in one study. Therefore, inadequate smears lead to anxiety and increased healthcare cost for a relatively low yield of CIN. Inadequate smears remain a major reason to develop techniques such as OCT for cervical use.

Intraepithelial lesions cannot be identified with the naked eye so colposcopy is used as a diagnostic technique. A colposcopy is the magnification of the view of the cervix in combination with the use of a light source. It is generally used in the setting of an abnormal or indeterminate Pap smear. It can be used to identify mucosal abnormalities characteristic of CIN or invasive cancer. Before imaging, a dilute acetic acid solution can be used which gives zones of high nucleic acid an opaque, white color. We will see that OCT has shown considerable potential for aiding colposcopy in identifying normal from abnormal tissue in the cervix.

The cost-effectiveness of HPV testing is of questionable value. As pointed out, most cervical cancers in the United States occur in patients who have not had a Pap smear in 5 years and not from false negatives. The cost-effectiveness of a yearly Pap smear is called into question so the addition of a second test to pick up less than 5000 cases in a screening population of greater than 40 million is questionable. HPV testing has a cost of $60 dollars per test, the Pap smear has a cost between $20 and $40 dollars per test. Even in high risk populations, greater than 70% of those with a positive HPV will also not have detectable disease. In addition, the presence of high risk HPV is equivalent to a precancerous lesion of the cervix does not appear to be true since high rates of transient infections have been observed in young women. However, HPV may someday be effective in identifying low risk groups.

Fluorescence has been used experimentally to assess cervical cancer and studies are currently ongoing.[144,145] Studies have been done both with autofluorescence and with added fluoroprobes. To date, there is not sufficient data to appreciate how effective this technology will be. Among the limitations of exogenously added fluoroprobes, direct application can result in inhomogeneous distributions while systemically added probes require careful control of dosage and are only useful at specific (and potentially variable) time periods. In addition, current systems cannot access the endocervix because of probe size and cervical mucus may hinder the correct diagnosis. For systems using autofluorescence, direct contact is required at each point, making their usage impractical. In addition, some studies failed to distinguish inflammation from CIN. Finally, specificities have been disappointing as in other organ systems.

17.7.3.1 OCT for assessing cervical malignancies

The cervix and cervical cancer was first examined by our group in 1999.[146] In this study, both cervical and uterine tissue were examined. Imaging was performed with a relatively slow OCT system *in vitro*. In Figure 17.14, normal *in vitro* human cervix was imaged with OCT. Figure 17.14A shows the ectocervix with the corresponding histology. The epithelium (e) and basement membrane (b) are clearly defined. Some groups have later asserted that the basement membrane can not

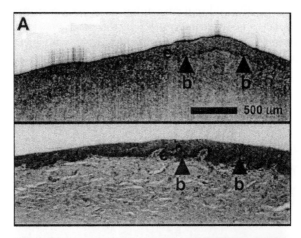

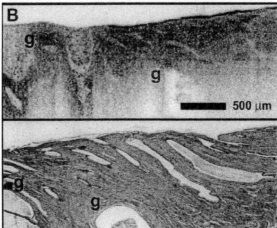

Figure 17.14 Human cervix imaged in vitro. A demonstrates ectocervix with the epithelial layer and b is the basal membrane. (B) The microstructure is OCT images of ectocervix. Here g represents deep endocervical glands, some of which fluid filled cysts. The epithelial layer has been partially denuded. Courtesy of Pitris, C., et. al. (1999). *Obstet. Gynecol.* 93, 135–139.

be detected with OCT. This may be due to either a lower resolution system used in the study or the fact that polarization sensitivity of the basement membrane was not taken into account. In Figure 17.14B, the endocervical canal is imaged (OCT image on the top and histology on the bottom). The deep endocervical glands are noted.

In Figure 17.15 from the same work, CIN and invasive cancer are examined. In Figure 17.15A, CIN is noted with thick irregular epithelial layer and thickening of the basement membrane. In Figure 17.15B, the tissue no longer has a layered appearance and is heterogeneous with no discernable basement membrane.

Three other subsequent studies have been performed.[147–149] The *in vivo* study is the one of greatest interest.[149] In this study, 50 patients (*in vivo*) were examined, 18 with CIN II or III, 14 with CIN I, 13 with metaplasia/inflammation, and two with invasive disease. The most striking finding was that normal epithelia appeared normal by OCT 100% of the time, suggesting a role in focusing biopsies.

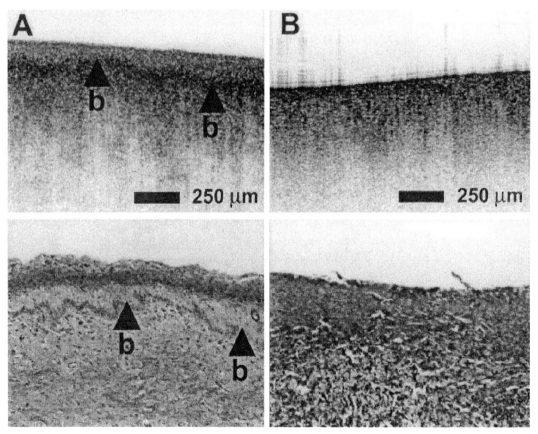

Figure 17.15   Cervical diseases imaged in vitro. (A) Carcinoma in situ with the basement membrane (b) intact. (B) Invasive carcinoma is shown with the corresponding loss of structure. Courtesy of Pitris, C., et. al. (1999). *Obstet. Gynecol.* 93, 135–139.

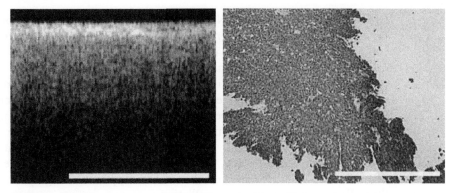

Figure 17.16 In vivo imaging of invasive carcinoma of the cervix. Courtesy of Escobar, P. F., et al. (2004). *Inter. J. Gynecol. Cancer.* 14, 470–474.

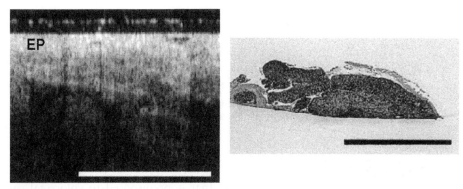

Figure 17.17 In vivo imaging of carcinoma in situ. Here the squamous epithelium (EP) is noted. Courtesy of Escobar, P. F., et al. (2004). *Inter. J. Gynecol. Cancer.* 14, 470–474.

Invasive disease was characterized by loss of structure, consistent with the original *in vitro* study. Although absolute criteria were not found for CIN, which awaits future work, a trend toward irregularity and thickening of the epithelium was noted. In Figure 17.16, invasive carcinoma is seen with the loss of layered structure. In Figure 17.17, CIN is noted where the epithelium is irregular in structure. Again, OCT demonstrates its greatest potential in diagnosing early CIN and the etiology of cellular atypia.

17.7.4 Treatment

The need for early, accurate diagnosis is evident both based on survival data and the treatments. In general, patients with CIN I require no further treatment other than follow-up since the lesions tend to resolve. There are currently two approaches to patients with CIN II and CIN III. The first approach comprises conservative out-patient treatment with cryotherapy, laser vaporization, and the electric loop excision procedure. The second approach is called cervical conization or cone biopsy that involves resection of an area including part of the endocervical canal. The morbidity associated with cone biopsy is primary hemorrhage, discomfort, readmission for secondary hemorrhage, and cervical stenosis. Conservative treatment is considered when the entire lesion can be visualized by colposcopy, there is no evidence of invasive disease, Pap smear and biopsy closely correlate, and there is no evidence of endocervical involvement. If microinvasion has occurred (stage I), treatment is either simple hysterectomy, radiation, or careful observation after cone biopsy. Stage II and higher are managed with radiation, possible surgery, or chemotherapy. Stage IV treatment is generally palliative.

17.8   OTHER POTENTIAL APPLICATIONS

17.8.1   General

There are other potential applications that exist for OCT in the field of oncology, but they will only be mentioned here briefly as currently they show less potential than those discussed. These include screening for colon cancer in patients with ulcerative colitis, uterine cancer screening, guiding laparoscopy, diagnosing early breast cancer, diagnosing skin cancer, and assessing the oral mucosa.

17.8.2   Colon Cancer

Ulcerative colitis is a condition that is predisposed to colorectal cancer and these patients require aggressive screening, often of mucosa that appears abnormal at baseline. Patients with pancolitis have a risk of colon cancer as high as 42% at 24 years.[150,151] A technology capable of identifying areas of early neoplasm of this relatively slow growing tumor could substantially impact on mortality. OCT has demonstrated feasibility for imaging the colon and inflammatory bowel disease.[36] However, implementation is challenging due both to the large amount of tissue to be imaged and the amount of blood in the field.

17.8.3   Postmenopausal Bleeding and Endometrial Carcinoma

OCT has the potential for the early diagnosis of early endometrial cancer. Five percent of all gynecological visits are for postmenopausal bleeding (PMB), and this number is expected to increase with current trends of hormonal replacement and prophylaxis to prevent breast cancer.[152] A feared cause of PMB is endometrial carcinoma, the most common gynecological malignancy. An estimated 40,320 cases of cancer of the uterine lining occurred in 1994, with 7,000 deaths (www.cancer.org). High exposure to estrogen increases the risk, which includes estrogen replacement therapy, tamoxifen for breast cancer prophylaxis, early menarche, late menopause, and never having children.

Although historically dilatation and curettage (D&C) has been used to evaluate PMB, its high cost and requirement of general anesthesia make it unattractive for screening this large population.[155] Other approaches under investigation include blind endometrial biopsy, transvaginal ultrasound, and transvaginal hysterosonography, which are intended as office procedures.[154–158] However, these techniques have their limitations. We have previously demonstrated a potential for OCT to assess the uterine mucosa and identify uterine cancer,[146] but there are several important limitations to its application. First, the uterine mucosa can be as thick as 5 mm. Second, the amount of deaths compared to the number of cases is relatively low because neoplasms present early due to PMB. Third, PMB is associated with a relatively large amount of bleeding, reducing the effectiveness of OCT imaging.

17.8.4 Breast Cancer

Theoretically, OCT may play a role in the assessment of early breast cancer, although limited data are currently available to support it. Approximately 216,000 new cases of invasive breast cancer occurred in 1994.[1] More relevant is that approximately 60,000 new cases of ductal carcinoma *in situ* (DCIS) are diagnosed each year. This number has increased in recent years likely due to more aggressive screening with mammography. When noninvasive, DCIS has a 5-year survival in excess of 97%. This is opposed to stage II, greater than 2 cm in diameter but less than 5 cm, which has a prognosis between 65 and 82%. Determining the presence of neoplasia, as well as the extent of DCIS and the potential of invasion, need to be determined to establish treatment. While some groups have advocated OCT guided biopsy, the most intriguing approach is passing an OCT imaging wire into the nipple and through the ductal system to allow large areas of DCIS to be examined for invasion.[159] This diagnostic wire approach awaits future assessment.

17.8.5 Skin Cancer

Skin cancer is the most common cancer, with more than 1 million cases per year. Other than malignant melanoma, the cancers are highly curable. Of the 10,000 deaths per year from skin cancer, approximately 8,000 will be due to malignant melanoma. Although OCT has been applied to the assessment of skin cancer, it does not follow the general principles outlined for OCT application in earlier chapters (i.e., biopsy can be easily performed).[160–163] The possible exception is the Mohs microsurgical procedure, used to treat aggressive forms of basal cell carcinoma greater than 2.0 cm in diameter.[163] With this procedure, large surgical margins are needed so there is a balance between assuring clean margins and disfigurement from excessive tissue removal. Therefore, the surgeon gradually removes tissue in small increments and checks frozen sections at each interval to define clean margins. This is both time-consuming and expensive. Therefore, OCT guidance of resection to clean margins can be an important application of the technology.

17.8.6   Oral and Pharyngeal Cancers

An estimated 28,000 cases of oral and pharyngeal cancers occur each year, with an estimated 7,000 deaths in 2004. These disorders are typically squamous cell carcinoma and are usually associated with nicotine or alcohol abuse. If not detected early, these neoplasms have a high mortality (43% at 5 years). Imaging with OCT has been performed of both the oral mucosa and larynx with OCT, including the epiglottis seen in the original 1996 *Circulation* publication.[130,131,132,164,166] Future work is needed to ultimately define its role in these areas.

17.8.7   Assessing Pathological Samples

Very often, as in the case of tissue removed after biopsy, only small numbers of histologic slides are generated relative to the amount of tissue available. This raises the possibility of providing inadequate staging as lesions of interest are not identified. An example is samples of DCIN of the breast where invasion is missed. OCT offers a method whereby large amounts of tissue can be examined at a histologic level and relatively low expense. To date, this an area of OCT research which has received only minimal attention.

17.9   CONCLUSION

OCT represents a promising technology for a wide range of early malignancies. However, the fact that early neoplasias have changes on a very small scale may make realization of this potential challenging.

REFERENCES

1. www.cancer.org
2. Young, B. and Heath, J. W. (2000). *Functional Histology, A Text and Color Atlas*, fourth edition. Churchill Livingstone, Edinburgh, Scotland, p. 251.
3. Phillips, R. W. and Wong, R. K. H. (1991). Barrett's esophagus: Natural history, incidence, etiology, and complications. *Gastroenterol. Clin. North Am.* 20, 791–815.
4. Taylor, A. L. (1927). The epithelial heterotopias of the alimentary tract. *J. Pathol. Bacteriol.* 30, 415–449.
5. Sarr, M. J., Hamilton, S. R., and Marrone, G. C. (1985). Barrett's esophagus: Its prevalence and association with adenocarcinoma in patients with symptoms of gastroesophageal reflux. *Am. J. Surg.* 149, 187–194.
6. Naef, A. P., Savary, M., and Ozzello, L. (1975). Columnar-lined lower esophagus: An acquired lesion with malignant predisposition. Report on 140 cases of Barrett's esophagus with 12 adenocarcinomas. *J. Thorac. Cardiovasc. Surg.* 70, 826–834.

7. Cameron, A. J., Ott, B. J., and Payne, W. S. (1985). The incidence of adenocarcinoma in columnar lined (Barrett's) esophagus. *N. Engl. J. Med.* 313, 857–859.

8. Rozen, P., Baratz, M., Fefer, F., and Gilat, T. (1995). Low incidence of significant dysplasia in a successful endoscopic surveillance program of patients with ulcerative colitis. *Gastroenterology* 108, 1361–1370.

9. Axon, A. T. (1994). Cancer surveillance in ulcerative colitis — A time for reappraisal. *Gut* 35, 587–589.

10. Froelicher, P. and Miller, G. (1986). The European experience with esophageal cancer limited to the mucosa and submucosa. *Gastrointest. Endosc.* 32, 88–90.

11. Melville, D. M., Jass, J. R., Morson, B. C. et al. (1996). Observer study of the grading of dysplasia in ulcerative colitis: Comparison with clinical outcome. *Hum. Pathol.* 20, 1008–1014.

12. Nigro, J. J., Hagen, J. A., DeMeester, T. R. et al. (1999). Prevalence and location of nodal metastasis in distal esophageal adenocarcinoma confined to the wall: Implications for therapy. *Cardiovasc. Surg.* 117, 16–25.

13. Sampliner, R. E. (1998). Practice guidelines on the diagnosis, surveillance, and therapy of Barrett's esophagus. The Practice parameter committee of the American College of Gastroenterology. 93, 1028–1032.

14. Falk, G. W., Rice, T. W., Goldblum, J. R. et al. (1999). Jumbo biopsy forceps protocol still misses unsuspected cancer in Barrett's esophagus with high grade dysplasia. *Gastrointest. Endosc.* 49, 170–176.

15. Spechler, S. J. (2002). Barrett's esophagus. *N. Engl. J. Med.* 346, 836–842.

16. Falk, G. W., Catalano, M. F., Sivak, M. V. et al. (1994). Endosonography in the evaluation of patients with Barrett's esophagus and high grade dysplasia. *Gastrointest. Endosc.* 40, 207–212.

17. Beck, J. R. and Pauker, S. G. (1999). The Markov process in medical prognosis. *Med. Dec. Making* 3, 419–458.

18. Caneron, A. and Carpenter, H. (1997). Barrett's esophagus, high grade dysplasia and early adenocarcinoma: A pathological study. *Am. J. Gastroenterol.* 92, 586–591.

19. Falk, G., Rice, T., Goldblum, J. et al. (1999). Jumbo biopsy protocol still misses unsuspected cancer in Barrett's esophagus with high grade dysplasia. *Gastrointest. Endosc.* 49, 170–176.

20. Daly, J. M., Frey, W. A., Little, A. G. et al. (2000). Esophageal cancer: Results of an American College of Surgeons Patient Care Evaluation Study. *J. Am. Coll. Surg.* 190, 562–573.

21. El-Serag, H. B., Aquirre, T. V., Davis, S. (2004). Proton pump inhibitors are associated with reduced incidence of dysplasia in Barrett's esophagus. *Am J. Gastroenterol.* 99, 1877–1883.

22. Siersema, P. D. (2005). Photodynamic therapy for Barrett's esophagus: Not yet ready for the premier league of esophagus interventions. *Gastointest. Endosc.* 62m, 503–507.

23. Conio, M. A., Cameron, A., Blanchi, S., and Filberti, R. (2005). Endoscopic treatment of high grade dysplasia and early cancer in Barrett's oesophgus. *Lancet Oncol.* 6, 311–321.

24. Farugi, S. A., Arantes, V., and Bhutani, S. et al. (2004). Barrett's esophagus: Current and future role of endoscopy and optical coherence tomography. *Dis. Esophagus* 17, 118–123.

25. Spechler, S. J. (1997). Esophageal columnar metaplasia. *Gastrointest. Endosc. Clin. North Am.* 26(7), 1–18.

26. Srivastava, A. K., Vanagunas, A., Kamel, P. et al. (1994). Endoscopic ultrasound in the evaluation of Barrett's esophagus: A preliminary report. *Am. J. Gastroenterol.* 89, 2192–2195.

27. Adrian, A., Cassidy, M. J., Schiano, T. D. et al. (1997). High resolution endoluminal sonography is a sensitive modality for the identification of Barrett's metaplasia. *Gastrointest. Endoscopy.* 46, 147–151.

28. Kinjo, M., Maringhini, A., Wang, K. K. et al. (1994). Is endoscopic ultrasound cost effective to screen? (Abstract). *Gastointest. Endosc.* 51, 467–474.

29. Gangarosa, L., Halter, S., and Mertz, H. et al. (2000). Methylene blue staining and ultrasound evaluation of Barrett's esophagus with low grade dysplasia. *Dig. Dis. Sci.* 45, 225–229.

30. Connor, M. and Sharma, P. (2003). Chromoendoscopy and magnification in Barrett's esophagus. *Gastrointest. Endosc. Clin. North Am.* 13, 269–277.

31. Oliver, J. R., Wild, C. P., Sahay, P. et al. (2003). Chromoendoscopy with methylene blue and associated DNA damage in Barrett's esophagus. *Lancet* 362, 373–374.

32. Stevens, P. D., Lightdale, C. J., Green, P. H. et al. (1994). Combined magnification endoscopy with chromoendoscopy for the evaluation of Barrett's esophagus. *Gastrointest. Endosc.* 40, 747–749.

33. Sharma, P., Weston, A. P., Topalvoski, M. et al. (2003). Magnification chromoendoscopy for the detection of intestinal metaplasia and dysplasia in Barrett's esophagus. *Gut* 52, 24–27.

34. Sharma, P. (2004). *Aliment. Pharmacol. Ther.* 20 (suppl. 5), 63–70.

35. Georgakoudi, I., Jacobson, B. C., Van Dam, J. et al. (2001). Fluorescence, reflectance, and light scattering for dysplasia in patients with Barrett's esophagus. *Gastroenterology* 120, 1620–1629.

36. Tearney, G. J., Brezinski, M. E., Southern, J. F., Bouma, B. E., Boppart, S. A., and Fujimoto, J. G. (1997). Optical biopsy in human gastrointestinal tissue using optical coherence tomography. *Am. J. Gastroenterol.* 92, 1800–1804.

37. Pitris, C., Jesser, C. A., Boppart, S. A., Stamper, D. L., Brezinski, M. E., and Fujimoto, J. G. (2000). Feasibility of optical coherence tomography for high resolution imaging of human gastrointestinal tract malignancies. *J. Gastroenterol.* 35, 87–92.

38. Li, X. D., Boppart, S. A., Van Dam, J. et al. (2000). Optical coherence tomography: Advanced technology for the endoscopic imaging of Barrett's esophagus. *Endoscopy* 32, 921–930.

39. Jackle, S., Gladkova, N., Feldchtiech, F. et al. (2000). In vivo endoscopic optical coherence tomography of esophagitis, Barrett's esophagus, and adenocarcinoma of the esophagus. *Endoscopy* 32, 750–755.

40. Zuccaro, G., Gladkova, N., Vargo, J. et al. (2001). Optical coherence tomography of the esophagus and proximal stomach in health and disease. *Am. J. Gastroenetrol.* 96, 2633–2639.

41. Ries, L. A. G., Miller, B. A., Hankey, B. F. et. al. (1994). SEER cancer statistics review, 1973–1991: Tables and graphs. National Cancer Institute, Bethesda, MD. NIH Pub. No. 94-2789.

42. Correa, P. (1982). Precursors of gastric and esophageal cancer. *Cancer* 50, 2554–2565.

43. Shu, Y. J. (1983). Cytopathology of the esophagus: An overview of esophageal cytopathology in China. *Acta Ctyol.* 27, 7–16.

44. Hirayama, T. (1990). Lifestyle and mortality: A large scale census based cohort study in Japan. *Contributions to Epidemiology and Biostatistics*, Vol. 6, J. W. Heidelberg (ed.). Karger, Basel.

45. Crespi, M., Grassi, A., Amiro, G. et al. (1979). Esophageal lesions in northern Iran: A premalignant condition. *Lancet* 2, 217–221.

46. Daly, J. M., Karnell, L. H., and Menck, H. R. (1996). National cancer data base report on esophageal carcinoma. *Cancer* 78, 1820–1828.

47. Biot, W. J. (1994). Esophageal cancer trends and risk factors. *Sem. Oncol.* 21, 403–410.

48. Biot, W. J. and Fraumeni, J. F. (1982). Geographic epidemiology of cancer in the United States. *Cancer Epidemiology and Prevention*, D. Schottenfeld and J. Fraumeni (eds.). Saunders, Philadelphia, PA, pp. 179–193.

49. Pottern, L. M., Morris, L. E., Biot, W. J. et al. (1981). Esophageal cancer among black men in Washington, D.C. I: Alcohol, tobacco, and other risk factors. *J. Natl. Cancer Inst.* 67, 777–783.

50. Biot, W. J. (1997). Cancer of the esophagus: Its causes and changing patterns of occurrence. *ASCO* 159–163.

51. Endo, M., Takeshita, K., Yoshino, K. et al. (1994). Surgery of superficial esophageal cancer in high risk patients. *Shokaki-Geka* 17, 1341–1345.

52. Kouzu, T., Koide, Y., Armina, S. et al. (1995). Treatment of early esophageal cancer. *Stomach Intestine* 30, 431–435.

53. Nakamura T., Ide, H., Eguchi, R. et al. (1995). Pathology of esophageal cancer — Early esophageal cancer, superficial esophageal cancer. *Geka* 57, 1020–1022.

54. Makuuchi, H., Tanaka, Y., Tokuda, Y. et al. (1995). Endoscopic treatment for early esophageal cancer. *Geka* 57, 1027–1031.

55. Momma, K., Yoshida, M., Kota, H. et al. (1996). Clinical and pathological study on esophageal cancers reaching to the muscularis mucosae and the upper third of the submucosa. *Stomach Intestine* 31, 1207–1215.

56. International Agency for Research on Cancer: Tobacco smoking. IARC Monograph on Evaluation of Carcinogenic Risks to Humans, Vol. 38, Lyon, France, 1986.

57. International Agency for Research on Cancer: Alcohol. IARC Monograph on Evaluation of Carcinogenic Risks to Humans, Vol. 44, Lyon, France, 1988.

58. Biot, W. J. (1994). Esophageal cancer trends and risk factors. *Sem. Oncol.* 21, 403–410.

59. Tuyns, A. J., Pequiqnot, G., Jensen, O. M. (1977). Le cancer de esophageal an Ille et Vilane en function des niveaux de consomation d'alcool et de tabac: Des risques qui se multiplient. *Bull. Cancer* 64, 63–65.

60. Biot, W. J. (1992). Alcohol and cancer. *Cancer Res.* 52, 2119s–2123s (suppl).

61. Biot, W. J., Li, J. Y., Taylor, P. R. et al. (1993). Nutrition intervention trials in Linxian, China: Supplementation with specific vitamin/mineral combinations, cancer incidence, and disease specific mortality in the general population. *J. Natl. Cancer Inst.* 85, 1483–1492.

62. Mark, S., Liu, S. F., Li, J. Y. et al. (1994). The effect of vitamin and mineral supplementation on esophageal cytology: Results from the Linixian dysplasia trial. *Int. J. Cancer* 57, 162–166.

63. Correa, P. (1982). Precursors of gastric and esophageal cancer. *Cancer* 50, 2554–2565.

64. de Jong, U. W., Breslow, N., Goh, J. et al. (1974). Etiologic factors on esophageal cancer in Singapore Chinese. *Int. J. Cancer* 13, 291–303.

65. Kolicheva, V. J. (1974). Data on the epidemiology and morphology of precancerous changes and of cancer of the esophagus in Kazakhstan USSR. Thesis. Alma Ata.

66. Correa, P. (1982). Precursors of gastric and esophageal cancer. *Cancer* 50, 2554–2565.

67. Rose, E. F. (1981). A review of factors associated with cancer of the esophagus in the Transkei. *Cancer Among Black Populations*, C. Mettlin and G. Murphy (eds.). Alan Liss, New York, pp. 67–75.

68. Mukada, T., Sato, E., and Sasano, N. (1976). Comparison studies on dysplasia of esophageal epithelium in four prefectures of Japan with reference to risk of carcinoma. *J. Exp. Med.* 119, 51–63.

69. Crespi, M., Grassi, A., Amiro, G. et al. (1979). Esophageal lesions in Northern Iran: A premalignant condition. *Lancet* 2, 217–221.

70. Watanabe, H., Jass, J. R., and Sobin, L. H. (1990). *Histological Typing of Esophageal and Gastric Tumors*, second edition. Springer-Verlag, Berlin.

71. Rose, E. F. (1981). A review of factors associated with cancer of the esophagus in the Transkei. *Cancer Among Black Populations*, C. Mettlin and G. Murphy (eds.). Alan Liss, New York, pp. 67–75,

72. Postlethwait, R.W. and Wendell, A. (1974). Changes in the esophagus in 1,000 autopsy specimens. *J. Thorac. Cardiovasc. Surg.* 68, 953–956.

73. Kitamura, K., Kuwano, H., Yasuda, M. et al. (1996). What is the earliest malignant lesion in the esophagus? *Cancer* 77, 1614–1619.

74. Yang, C. S. (1980). Research on esophageal cancer in China: A review. *Cancer Res.* 40, 2633–2644.

75. Jacob, P., Kahrilas, P. J., Desai, T. et al. (1990). Natural history and significance of esophageal squamous cell dysplasia. *Cancer* 65, 2731–2739.

76. Daly, J. M., Karnell, L. H., and Menck, H. R. (1996). National cancer data base report on esophageal carcinoma. *Cancer* 78, 1820–1828.

77. Shu, Y. J. (1983). Cytopathology of the esophagus: An overview of esophageal cytopathology in China. *Acta Ctyol.* 27, 7–16.

78. Berry, A. V., Baskind, A. F., and Hamilton, D. G. (1981). Cytologic screening for esophageal cancer. *Acta Cytol.* 25, 135–141.

79. Silverberg, E. and Lubera, J. A. (1988). Cancer statistics. *Cancer* 38, 5–22.

80. Lightdale, C. J. (1994). Staging of esophageal cancer 1: Endoscopic ultrasonography. *Sem. Oncol.* 21, 438–446.

81. Mori, M., Adachi, Y., Matsushima, T. et al. (1993). Lugol staining pattern and histology of esophageal lesions. *Am. J. Gastroenterol.* 88, 701–705.

82. Dawsey, S. M., Fleischer, D. E., Wang, G. et al. (1998). Mucosal iodine staining improves endoscopic visualization of squamous dysplasia and squamous cell carcinoma of the esophagus in Linxian, China. *Cancer* 83, 220–231.

83. Yoshida, T., Inoue, H., Kawano, T. et al. (1999). The ultrasonic tactile sensor: In vivo clinical application for evaluation of depth of invasion in esophageal squamous cell carcinomas. *Endoscopy* 31, 442–446.

84. Vo-Dinh, T., Panjehpour, M., Overholt, B. F. et al. (1995). In vivo cancer diagnosis of the esophagus using differential normalized fluorescence indices. *Laser Surg. Med.* 16, 41–47.

85. Takeshita, K., Tani, M., Innoue, H. et al. (1997). Endoscopic resection of carcinoma in situ of the esophagus accompanied by esophageal varices. *Surg. Endosc.* 5, 182–184.

86. Endo, M., Takeshita, K., Yoshino, K. et al. (1994). Surgery of superficial esophageal cancer in high risk patients. *Shokaki-Geka* 17, 1341–1345.

87. Kouzu, T., Koide, Y., Armina, S. et al. (1995). Treatment of early esophageal cancer. *Stomach Intestine* 30, 431–435.

88. Dimick, J. B., Pronovost, P. J., Cowan, J. A., and Lipsett, P. A. (2003). Surgical volume and quality of care for esophageal resection: Do high volume hospitals have fewer complications? *Ann. Thorac. Surg.* 75, 337–341.

89. Aly, A. and Jamieson, G. G. (2004). Reflux after esophagectomy. *Br. J. Surg.* 91, 137–141.

90. Dicken, B. J., Bigam, B. L., Cass, C. et al. (2005). Gastric adenocarcinoma. *Ann. Surg.* 241, 27–39.

91. Elder, J. B. (1995). Carcinoma of the stomach. *Bockus Gastroentology*, 5th edition, W. S. Haubrich and F. Schaffner (eds.), W. B. Saunders, pp. 854–874.

92. Blot, W. J., Devesa, S. S., Kneller, R. W. et al. (1991). Rising incidence of adenocarcinoma of the esophagus and gastric cardia. *JAMA* 265, 1287–1289.

93. Cady, B., Rossi, R. L., Silverman, M. L. et al. (1989). Gastric adenocarcinoma: A disease in transition. *Arch. Surg.* 124, 303–308.

94. Fuchs, B. M. J. and Mayer, R. J. (1995). Gastric carcinoma. *N. Engl. J. Med.* 333, 32–41.

95. Albert, C. (1995). Clinical aspects of gastric cancer. *Gastrointestinal Cancers: Biology, Diagnosis, and Therapy*, A. K. Rustgi (ed.). Lippincott, Wilkins, and Williams, pp. 197–216.

96. Gore, R. M. (1997). Gastric cancer: Clinical and pathologic features. Radiology clinics of North America. 35, 295–308.

97. Catalano, V., Labianca, R., Beretta, G. D. et al. (2005). Gastric cancer. *Crit. Rev. Oncol. Hematol.* 54, 209–241.

98. Karpeh, M. and Brennam, M. (1998). Gastric carcinoma. *Ann. Surg. Oncol.* 5, 650–656.

99. Zuccaro, G., Gladkova, N., Vargo, J. et al. (2001). Optical coherence tomography of the esophagus and proximal stomach in health and disease. *Am. J. Gastroeneterol.* 96, 2633–2639.

100. Poneros, J. M., Brand, S., Bouma, B. E. et al. (2001). Diagnosis of specialized intestinal metaplasia by optical coherence tomography. *Gastroenterology* 120, 7–12.

101. Jackle, S., Gladkova, N., Feldchtien, F. et al (2000). In vivo endoscopic optical coherence tomography of the human gastrointestinal tract-toward optical biopsy. *Endoscopy* 32, 743–749.

102. Braunwald, E., Fauci, A. S., Kasper, D. L. et al. (2001). *Harrison's Principles of Internal Medicine*, 15th edition. McGraw-Hill, New York.

103. Kiemeney, L. A., Witjes, J. A., Verbeek, A. L., Heijbroek, R. P., and Debruyne, F. M. (1993). The clinical epidemiology of superficial bladder cancer. *Br. J. Cancer* 67, 806–812.

104. Fleshner, N. E., Herr, H. W., Stuart, A. K., Murphy, G. P., Mettlin, C., and Menck, H. R. (1996). The national cancer data base report on bladder carcinoma. *Cancer* 78, 1505–1513.

105. Ozen, H. (1997). Bladder cancer. *Curr. Opin. Oncol.* 9, 295–300.

106. Fleshner, N. E., Herr, H. W., Stuart, A. K., Murphy, G. P., Mettlin, C., and Menck, H. R. (1996). The national cancer data base report on bladder carcinoma. *Cancer* 78, 1505–1513.

107. Walsh, P., Gittes, R., Perlmutter, A., and Stamey, T. (1986). *Campbell's Urology.* W. B. Saunders Co., Philadelphia, PA, p. 1374.

108. Barentsz, J. O., Witjes, A., and Ruijs, J. H. J. (1997). What is new in bladder cancer imaging? *Urol. Clin. North Am.* 24, 583–602.

109. Liu, J. B., Bagley, D. H., Conlin, M. J., Merton, D. A., Alexander, A. A., and Goldberg, B. B. (1997). Endoluminal sonographic evaluation of ureteral and renal pelvic neoplasms. *J. Ultrasound Med.* 16, 515–521.

110. Koraitim, M., Kamal, B., Metwalli, N., and Zaky, Y. (1995). Transurethral ultrasonographic assessments of bladder carcinoma: Its value and limitation. *J. Urol.* 154, 375–378.

111. Koenig, F., McGovern, F. J., Enquist, H., Larne, R., Deutsch, T. F., and Schomacker, K. T. (1998). Autofluorescence guided biopsy for the early diagnosis of bladder carcinoma. *J. Urol.* 159, 1871–1875.

112. Jesser, C. A., Boppart, S. A., Pitris, C., Stamper, D. L., Nielsen, G. P., Brezinski, M. E., and Fujimoto, J. G. (1999). High resolution imaging of transitional cell carcinoma with optical coherence tomography: Feasibility for the evaluation of bladder pathology. *Br. J. Radiol.* 72, 1171–1176.

113. Zagaynova, E. V., Streltsova, O. S., Gladkova, N. D. et al. (2002). In vivo optical coherence tomography feasibility for bladder disease. *J. Urol.* 167, 1492–1496.

114. Pan Y. T., Xie, T. Q., Du, C. W. et al. (2003). Enhancing early bladder cancer detection with fluorescence guided endoscopic optical coherence tomography. *Opt. Lett.* 28, 2485–2487.

115. Peto, R., Lopez, A. D., Boreham, J. et al. (1994). Mortality from smoking in developed countries 1950-2000. Indirect estimates from National Vital Statistics.

116. National Cancer Institute, Surveillance Epidemiology and End Results Program (2002). http://www.seer.cancer.gov.

117. Lubin, J. H. (1994). Invited commentary: Lung cancer and exposure to residential radon. *Am. J. Epidemiol.* 140, 323–331.

118. Hirayama, T. (1981). Nonsmoking wives of heavy smokers have a higher risk of lung cancer: A study from Japan. *BMJ* 282, 183–185.

119. Trichopoulos, D., Kalandidi, A., Sparros, L. et al. (1981). Lung cancer and passive smoking. *Int. J. Cancer* 27, 1–4.

120. Smith, R. (2000). Epidemiology of lung cancer. *Radiol. Clin. North Am.* 38, 453–470.

121. Mountain, C. F. (1986). A new international staging system for lung cancer. *Chest* 89, 225s–233s.

122. Omenn, G. S., Goodman, G. E., Thornquist, M. D. et al. (1996). Effects of a combination of beta and vitamin a on lung cancer and cardiovascular disease. *N. Engl. J. Med.* 334, 1150–1155.

123. Edell, E. S. (1987). Bronchoscopic phototherapy with hematoporphyrin derivative for treatment of localized bronchogenic carcinoma: A 5 year experience. *Mayo Clin. Proc.* 62, 8–14.

124. Wardwell, N. R. and Massion, P. P. (2005). Novel strategies for the early detection and prevention of lung cancer. *Sem. Oncol.* 32, 259–268.

125. Sone, S., Takashima, S., Li, H. et al. (1998). Mass screening for lung cancer with mobile spiral CT scanner. *Lancet* 351, 1242–1245.

126. Wardell, N. R. and Massion, P. P. (2005). Novel Strategies for the early detection and prevention of lung cancer. *Sem. Oncol.* 32, 259–268.

127. Hirsch, F. R., Prindiville, S. A., Miller, Y. E. et al. (2001). Fluorescence versus white light bronchoscopy for detection of preneoplastic lesions: A randomized study. *J. Natl. Cancer Inst.* 93, 1385–1391.

128. Sutedja, G. (2003). New techniques for early detection of lung cancer. *Eur. Respir. J.* 21, (suppl 39), 57s–66s.

129. Tockman, M. S., Mulshine, J. L., Piantadosi, S. et al. (1997). Prospective detection of preclinical lung cancer: Results from two studies of heterogeneous ribonucleoprotein A2/B1 overexpression. *Clin. Cancer Res.* 3, 2237–2246.

130. Brezinski, M. E., Tearney, G. J., Bouma, B. E., Izatt, J. A., Hee, M. R., Swanson, E. A., Southern, J. F., and Fujimoto, J. G. (1996). Optical coherence tomography for optical biopsy: Properties and demonstration of vascular pathology. *Circulation* 93, 1206–1213.

131. Pitris, C., Brezinski, M. E., Bouma, B. E., Tearney, G. J., Southern, J. F., and Fujimoto, J. G. (1998). High resolution imaging of the upper respiratory tract with optical coherence tomography. *Am. J. Respir. Crit. Care Med.* 157, 1640–1644.

132. Tearney, G. J., Brezinski, M. E., Bouma, B. E., Boppart, S. A., Pitris, C., Southern, J. F., and Fujimoto, J. G. (1997). In vivo endoscopic optical biopsy with optical coherence tomography. *Science* 276, 2037–2039.

133. Tsuboi, M. et al. (2005). Optical coherence tomography in the diagnosis of bronchial lesions. *Lung Cancer.* 49, 390–398.

134. Broome, U., Olsson, R., Loot, T. et al. (1996). The natural history and prognostic factors in 305 Swedish patients with primary sclerosing cholangitis. *Gut* 38, 610–615.

135. Cameron, J. L., Pitt, H. A., Zinner, M. J. et al. (1990). Management of proximal cholangiocarcinoma by surgical resection and radiotherapy. *Am. J. Surg.* 159, 91–98.

136. Kosuge, T., Yamamoto, J., Shimada, K. et al. (1999). Improved surgical results for hilar cholangiocarcinoma with procedures including major hepatic resection. *Ann. Surg.* 230, 663–671.

137. Neuhas, P., Jonas, S., Bechstein, W. O. et al. (1999). Extended resections for hilar cholangiocarcinoma. *Ann. Surg.* 230, 808–819.

138. Tearney, G. J., Brezinski, M. E., Southern, J. F., Bouma, B. E., Boppart, S. A., and Fujimoto, J. G. (1998). Optical biopsy in human pancreatobiliary tissue using optical coherence tomography. *Dig. Dis. Sci.* 43, 1193–1199.

139. Poneros, J. M., Tearney, G. J., Shiskov, M. et al. (2002). Optical coherence tomography of the biliary tree during ERCP. *Gastrointest. Endosc.* 55, 84–88.

140. Lorincz, A. T., Reid, R., Jenson, A. B. et al. (1992). Human papillomavirus infection of the cervix. *Obstet. Gynecol.* 79, 328–337.

141. Wright, T. et al. (1994). Precancerous lesions of the cervix. *Blaustein's Pathology of the Female Genital Tract*, 4th edition, R. Kurman (ed.). Springer-Verlag, New York, p. 229.

142. Eddy, D. M. (1990). Screening for cervical cancer. *Ann. Intern. Med.* 113, 214–226,

143. Fahs, M. C., Mandelblatt, J., Schechter, J. et al. (1992). Cost effectiveness of cervical cancer screening. *Ann. Intern. Med.* 17, 520–527.

144. Ramanujam, N., Mitchell, M. F., Mahadevan, A. et al. (1994). In vivo diagnosis of cervical intraepithelial neoplasia using 337 nm excited laser induced fluorescence. *PNAS* 91, 10193–10197.

145. Ramanujam, N., Mitchell, M. F., Mahadevan, A. et al. (1994). Fluorescence spectroscopy: A diagnostic tool for CIN. *Gynecol. Oncol.* 52, 31–38.

146. Pitris, C., Goodman, A., Libus, J. L. et al. (1999). High resolution imaging of gynecological tissue using optical coherence tomography. *Obstet. Gynecol.* 93, 135–139.

147. Zuluaga, A., Follen, M., Boiko, I. et al. (2005). Optical coherence tomography: A pilot study of a new imaging technique for noninvasive examination of cervical tissue. *Am. J. Obstet. Gynecol.* 193, 83–88.

148. Felschtein, F. I., Gelikonov, G. V., Gelikonov, V. M. et al. (1998). Endoscopic applications of optical coherence tomography. *Opt. Express* 3, 257–270.

149. Escobar, P. F., Benlinson, J. L., White, A. et al. (2004). Diagnostic efficacy of optical coherence tomography in the management of preinvasive and invasive cancer of uterine cervix and vulva. *Int. J. Gynecol. Cancer* 14, 470–474.

150. Glickman, R. M. (1991). Inflammatory bowel disease. *Harrison's Prinicples of Internal Medicine*, J. D. Wilson, E. Braunwald, K. J. Isselbacher, R. G. Petersdorf, J. B. Martin, A. S. Fauci, and R. K. Root (eds.). McGraw-Hill Inc, New York.

151. Rosen, P., Baratz, M., Fefer, F., and Gilat, T. (1995). Low incidence of significant dysplasia in a successful endoscopic surveillance program of patients with ulcerative colitis. *Gastroenterology* 108(5), 1361–1370.

152. Kazadi-Buanga, J. and Jurado-Chacon, M. (1994). Etiologic study of 275 cases of endo-uterine hemorrhage by uterine curettage. *Rev. Fr. Gynecol. Obstet.* 89, 129–134.

153. Grimes, D.A. (1982). Diagnostic dilatation and curettage: A reappraisal. *Am. J. Obstet. Gynecol.* 142, 1–6.

154. Stovall, T. G., Solomon, S. K., and Ling, F. W. (1989). Endometrial sampling prior to hysterectomy. *Obstet. Gynecol.* 73, 405–409.

155. Stovall, T. G., Ling, F. W., and Morgan, P. L. (1991). A prospective, randomized comparison of the Pipelle endometrial sampling device with the Novak curette. *Am. J. Obstet. Gynecol.* 165, 1287–1289.

156. Chan, F.-Y., Chau, M. T., Pun, T.-C., Lam, C., Ngan, H. Y. S., Leong, L., and Wong, R. L. C. (1994). Limitations of transvaginal sonography and color Doppler imaging in the differentiation of endometrial carcinoma from benign lesions. *J. Ultrasound Med.* 13, 623–628.

157. Shipley, C. F., Simmons, C. L., and Nelson, G. H. (1994). Comparison of transvaginal sonography with endometrial biopsy in asymptomatic postmenopausal women. *J. Ultrasound Med.* 13, 99–104.

158. Goldstein, S.R. (1994). Use of ultrasonohysterography for triage of perimenopausal patients with unexplained uterine bleeding. *Am. J. Obstet. Gynecol.* 170, 565–570.

159. Boppart, S. A., Luo, W., Mark, D. L. et al. (2004). Optical coherence tomography: Feasibility for basic research and image guided surgery of breast cancer. *Breast Cancer Res. Treat.* 84, 85–97.

160. Welzel, J. (1997). Optical coherence tomography of the human skin. *J. Am. Acad. Dermatol.* 37, 958–963.

161. Strasswimmer, J., Pierce, M. C., Park, B. H. et al. (2004). Polarization sensitive optical coherence tomography of invasive basal cell carcinoma. *J. Biomed. Opt.* 9, 292–298.

162. Gladkova, S. G., Petrova, G. A., Nikulin, N. K. et al. (2000). In vivo optical coherence tomography imaging of human skin: Normal and pathology. *Skin. Res. Technol.* 6, 6–16.

163. Orengo, L. F. et al. (1997). Correlation of histologic subtypes of primary basal cell carcinoma and number of Mohs stages required to achieve a tumor free plane. *J. Am. Acad. Dermatol.* 37 (3 Pt 1), 395–397.

164. Wong, B. J., Jackson, R. P., Guo, S. et al. (2004). In vivo optical coherence tomography of the human larynx: Normative and benign pathology in 82 patients. *Laryngoscope* 115, 1904–1911.

165. Wong, R. K., and Elden, J. B. (2002). Propylene glycol as a contrast agent for optical coherence tomography to image gastrointestinal tissues. *Lasers in Surgery and Medicine.* 30, 201–208.

APPENDIX 17-1

NORMAL EPITHELIAL EXAMPLES

Chapter 17 deals primarily with neoplasms of the epithelium. However, the epithelium of the body varies considerably in composition. For those without a medical background, several examples of normal epithelium have been included below. It is the distortion of the epithelial architecture for a given region of the body that helps identify it as a neoplasm. For a more detailed discussion, histology or histopathology texts should be consulted.[1]

In the Figure 17A.1, simple columnar epithelium is shown with, among other structures, cells (C) and nuclei (N) identified. The epithelium cells are column shaped and the nuclei polarized at one side of the cell, near the basement membrane (B). This cell type is found generally in absorptive areas, such as the GI tract. Here, S is the surface.

The pseudostratified cells in the Figure 17A.2 are characterized by the appearance that cells are stacked. However, the cells actually extend from the basement membrane to the surface, which gives it the name pseudostratified rather than true stratification. This particular epithelium contains cilia (C) that protrude into the lumen. This particular cell type is found in the respiratory tract.

The epithelium show in Figure 17A.3 is truly stratified. It consists of layers of cells stacked on one another. The cell number is variable but in this case, the cells are

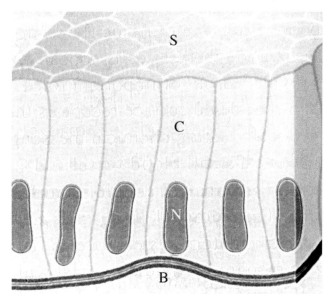

Figure 17A.1 Simple columnar epithelium. Courtesy Young, B., and Heath, J.W. (2000). *Functional Histology*, fourth edition, 251. [2]

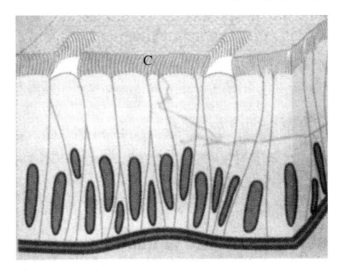

Figure 17A.2 Pseudostratified columnar ciliated epithelium. Courtesy Young, B., and Heath, J.W. (2000). *Functional Histology*, fourth edition, 251. [1]

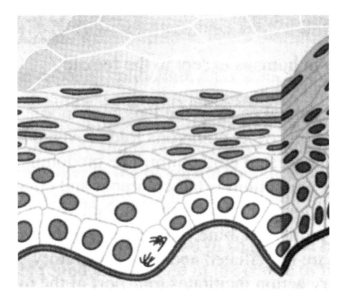

Figure 17A.3 Stratified epithelium. Courtesy Young, B., and Heath, J. W. (2000). *Functional Histology*, fourth edition, 251. [1]

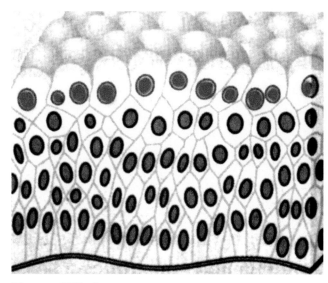

Figure 17A.4 Transitional epithelium. Courtesy Young, B., and Heath, J. W. (2000). *Functional Histology*, fourth edition, 251. [1]

cuboidal shaped at the base and more flattened at the surface. This epithelium is particularly resilient to abrasion. It is found in the oral cavity, esophagus, and vagina, and anus, among other sites.

The epithelium shown in Figure 17A.4 represents the epithelium of the bladder, where OCT may have a potentially important role. The epithelium allows both the distensibility of the bladder and prevents absorption of toxins. The epithelium's name is derived from the fact that it has characteristics of both cuboidal and squamous stratified epithelium.

REFERENCE

1. Young, B., and Heath, J. W. (2000). *Functional Histology, A Text and Color Atlas*, fourth edition, Churchill Living Stone, Edinburgh, Scotland.

18 OTHER APPLICATIONS AND CONCLUSIONS

Chapter Contents

18.1   GENERAL

This text focused on OCT imaging in nontransparent biological tissue, in particular its physical principles and applications. In Section I, a brief overview of relevant basic general physical and mathematical principles was discussed for the nonphysical scientist. In Section II, the physical principles behind OCT technology and imaging were discussed, and in Section III the potential clinical applications of OCT were examined. This book will conclude by discussing some other remaining potential applications that did not fit into the previous chapters. The topic of scattering and light in tissue will be discussed on the Elsevier web page, as this is almost a book in and of itself. The applications of OCT that will be discussed in this chapter are in dentistry, urology, nerve repair, vascular repair, and infertility.

18.2   DENTISTRY

18.2.1   General

OCT has potential applications in the field of dentistry where it has shown some feasibility, particularly the diagnosis of early dental caries and assessing the adequacy of restorations. Here, there is substantial interest to diagnose enamel disruption at earlier stages so that less aggressive interventions can be employed. However, the current costs of OCT engines may be prohibitive for use in most dental offices.

18.2.2   Anatomy and Histology

Adult humans have 32 teeth, which are divided into several categories based on shape and location, such as incisors and molars.[1] Each tooth has a section that

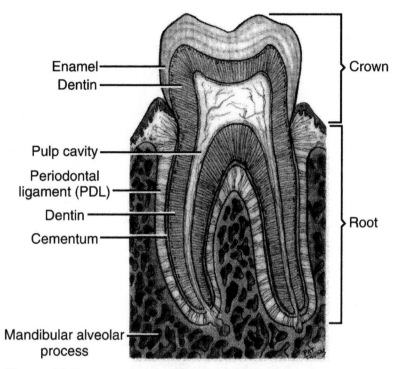

Enamel

Dentin

Crown

Pulp cavity

Periodontal
ligament (PDL)

Dentin

Cementum

Root

Mandibular alveolar
process

Figure 18.1 Schematic of tooth anatomy. Bath-Balogh, M. and Febrenbach,
M. J. (1997). *Illustrated Dental Embryology, Histology, and Anatomy.*

protrudes from the gum and at least one root below the surface. The roots are held in
place, under the gum, by alveolar bone. The layers of the tooth we are concerned
with are the outermost enamel, the dentin, and finally the pulp, often referred to
as the root. A cross section of the tooth is shown in Figure 18.1.

The enamel is the hardest tissue of the body. It consists almost exclusively of
calcium, mainly in the form of hydroxyapatite, and is only 0.5% organic in nature.
The organic enamel matrix is not composed of collagen but consists of predominately two classes of proteins, amelogenins and enamelins. The enamel is organized
into hard rods or prisms that are 4–6 μm in diameter, with heavy interprismatic
material in between. A drawing and electron microscopic image of the enamel rods
are shown in Figure 18.2. These rods begin upright on the surface of the dentin,
with a heavy inclination to the side. From the dentin, the rods run perpendicular to
the surface, spiral in the middle, and perpendicular to the surface near the surface.[2]
Shown in the figure are groups of adjacent prisms organized in bands with similar
direction and orientation, the Hunter-Schreger bands, which may contribute to the
birefringence of enamel. It should be noted that enamel is avascular, so there is no
inflammatory reaction like there is in the dentin.

Dentin also consists of both organic (20%) and inorganic components (80%).
About 90% of the organic components are collagen with significant amounts

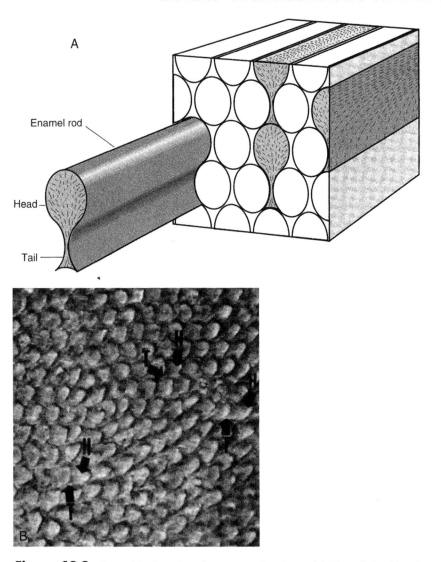

Figure 18.2   Enamel basic unit and corresponding SEM.   (A) The relationship of enamel rods in the tooth.   (B) Scanning electron microscope (SEM) showing the head (H) and tail (T). Courtesy of Bath-Balogh, M. and Febrenbach, M. J. (1997). *Illustrated Dental Embryology, Histology, and Anatomy.*

of glycoaminoglycan. The inorganic component is crystals in the form of hydroxy-apatite. The organic matrix is derived from odontoblasts on the inside of the dentin surface, separating it from the pulp cavity. Cytoplasmic processes or Tomes' fiber from the odontoblasts penetrate in the dentin through the dentinal tubules. An electron micrograph of dentin rods is shown in Figure 18.3. In a cross section of the tooth, the dentin appears to have a radial striated pattern due to these

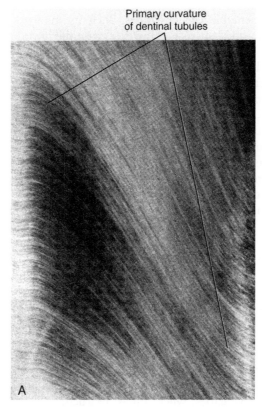

Figure 18.3 Primary curvature of dental tubules. Courtesy of Bath-Balogh, M. and Febrenbach, M. J. (1997). *Illustrated Dental Embryology, Histology, and Anatomy.*

tubules. Between the system of tubules are the collagen bundles embedded between glycoaminoglycan. The course of the collagen is along the access of the tooth and perpendicular to the dentinal tubules. Dentin, which does not contain nerves, is sensitive to touch and cold. Some suggest that the odontoblasts may be capable of transmitting pain to the pulp.

The pulp is a highly vascular structure with an extensive nerve component. It is supported by loose connective tissue. Blood vessels and nerves enter through the root canal.

18.2.3 Dental Caries

Dental caries is a chronic destruction of the tooth, characterized by demineralization, which can ultimately lead to mastication malfunction.[3] The disorder is present throughout the world, but until recently, it has traditionally been worse in industrialized countries and among higher socioeconomic classes due to the high intake of fermentable carbohydrates, such as monosaccharides (glucose, fructose) and disaccharides (sucrose). However, with the introduction of fluoride, the progression of the disease has changed dramatically. Approximately 30 billion dollars are spent annually in the United States on dental care.

Teeth, in general, have a high resistance to caries, as demonstrated by the low incidence of caries in primitive populations. The chemoparasitic theory of dental caries says that the microorganisms on the teeth metabolize carbohydrates and produce acid. A wide range of organic acids are produced. The acid results in dissolution of the mineral phase of the teeth. In the second phase, the enamel and dentin are broken down.

Caries is a dynamic process that does not necessarily result in the formation of frank cavitation. The tooth may undergo cycles of demineralization and remineralization. It is the net loss of minerals that ultimately determines the extent of caries. The earliest changes are dissolution of the enamel leading to pathways where diffusion can occur. On a microstructural level, caries begins with an enlargement of the surface intercrystalline spaces. As the process continues, cracks and microfissuring develop, opening the surface for diffusion. If, over a period of months to years, the surface weakens sufficiently then cavitation may result. Early lesions cannot be detected with current clinical techniques due to lack of resolution. With some exceptions, the first clinically detectable amount of caries is the white spot lesion, where demineralization has progressed to at least 300 to 500 μm. The white spot can be a reversible stage and the lesion may ultimately revert to normal looking enamel. There are two noted situations where a white spot may not occur on the progression to caries. The first is in rapidly progressive caries, where rapid destruction of the enamel may occur. The second is hidden or occult caries, where progression into the dentin occurs with the enamel surface appearing normal. This may occur because fluoride has resulted in healing of the surface. It should also be noted that fluoride has changed where caries is occurring. Occlusal surfaces are now affected more often than smooth or approximal surfaces.[4] It is estimated that over the next 10 years, over 90% of all caries in children will be on the occlusive surface. Occlusal caries are more difficult to detect and monitor than smooth surfaces, particularly in the early stages, using conventional dental exploring or bitewing X-rays. It is localized to the surface of the dentin and cementum. For patients over the age of 44, root caries is the principal reason for tooth loss.[5] It to is difficult to detect by conventional techniques.

As caries proceeds into the subsurface, demineralization becomes greater in the subsurface than in the enamel. Risk factors for caries include metabolizable carbohydrates, lack of fluoride use, dental plaque, genetics, virulence of the bacteria, and salivary flow rate. When present in the highly stable crystalline form of fluorapatite, this fluorinated compound is more stable than hydroxyapatite.[6] The effectiveness of fluoride has largely been attributed to its ability to enhance the remineralization process.

The composition and flow rate of saliva can effect dental caries. Among the ways saliva can serve to protect the tooth is by diluting acids, buffering acids, producing a mechanical cleansing action, and having antimicrobial properties.

Bacteria that can survive in the tooth environment at very low pH will be more likely to lead to tooth decay.[7] Some strains are more virulent then others, such as *Streptococcus mutans* that is common among patients with caries.

The presence of plaque, polysaccharides, and glucans can lead to the progression of plaque. When dietary sugar is no longer present, the microorganisms can still produce acid through intra- and extracellular storage of polysaccharides. The presence of water-insoluble glucans, which can form on the surface of teeth, can affect the flow within the tooth and also lead to dental caries. It should be noted that caries is a highly localized and complex process that can occur on one tooth and not the adjacent tooth. Because of the factors described, each tooth experiences a distinct environment. Occlusal surfaces of molars in general have the greatest risk of caries, due likely to the pits and fissures, whereas the lingual surface of the lower anterior teeth have the lowest risk.

With the use of fluoride, the progression of caries has changed dramatically. In the pre-fluoride era, progression of the disease was rapid so that dentists focused primarily on frank caries. The tendency was to restore, rather than monitor questionable areas. Now, with the slower progression, a need exists for earlier detection so that minimally invasive procedures can be used. The process is slow enough that if early lesions are detected, various interventions are available which stabilize or reverse the process.

18.2.4 Minimally Invasive Treatments

It is now generally believed that in the era of fluoride use, dentistry should move toward a more minimally invasive approach. Lesions should be treated at earlier stages and the use of large amalgam deposits should be avoided. Some of the treatments are discussed below. The preventative application of sealants to the general population is not cost-effective and currently are only recommended for high risk groups.

Early lesions can be largely cured with the ingress of salivary calcium, phosphate, and fluoride ions. Patients at high risk or with active caries should be put on antibacterial (e.g., chlorhexidine), buffering agents (e.g., baking soda), sugarless gum (to increase salivary flow), fluoride application, diet modification, and sealants.

When it is determined that the lesions need to be restored, removal of decay with maximal preservation of normal tissue is critical. This avoids the cycle of increased restoration size, tooth fracture, and crown.

When caries is present, small composite restorations referred to as "preventative resin restorations" offer a more conservative preparation than the large amalgam preparation, which requires more healthy tooth to be removed.

Sixty percent of approximal caries with radiographic evidence of penetration into the outermost portion of the dentin have not yet cavitated on the enamel and attempts should be made to remineralize prior to restoration approaches.

To conserve teeth air abrasion has recently become popular to treat caries and prepare teeth.[8] It uses a stream of aluminum oxide particles under compression. The particles strike the tooth at high velocity and remove small amounts of tooth structure. Air abrasion may not be suitable for removing large amalgam because of the mercury generated.

18.2.5 Diagnosing Caries

Current clinical technologies lack the sensitivity to identify early lesions. Clinical criteria, such as color and softness, occur at a relatively late stage. The sharp explorer, which for a long time was used to assess caries, is now believed to be a source of caries.[9] Conventional radiographic techniques cannot identify early lesions and miss more than half of deep dentinal lesions, in addition to poor performance on occlusive surfaces. Four new techniques are currently being evaluated, but are associated with significant limitations. They are quantitative light-induced fluorescence, electrical conductance measurements, direct digital radiography, and digital fiber-optic transillumination.

Several investigators have noted differences in the fluorescent properties between normal and carious tissue.[10] In general, the tooth is exposed to light in the blue-green region while the fluorescence measurements are made in the yellow-green region. Areas of demineralization appear as dark spots. Recently, the KaVo DIAGNOdent system has been introduced which uses excitation at 655 nm and emissions are in the near-infrared. In artificially created lesions without plaque, the sensitivities with and without dye were 0.76 and 0.54, respectively. The specificities were 0.64 and 0.29, respectively. When plaque was present, the sensitivities were 0.91 versus 0.43 and the specificities were 0.5 versus 0.55. Other disadvantages include the fact that the tooth surface needs to be kept clean as the presence of stains or deposits (including plaque) can register as caries and changes in normal tooth development may register as abnormal. Furthermore, the angle of the beam with the surface should be less than 2 degrees, and the limit of penetration is 100 μm. It generally only detects lesions large enough to reach the dentin, and it does not provide information about lesion depth or severity.

The theory behind electrical conductance measurements (ECM) is that normal teeth should have little or no measurable conductivity while demineralized or carious teeth will.[11] Moisture from saliva must be removed, typically with air, from the tooth since it is capable of conducting current. In a recent study comparing ECM (LODE Diagnostics) with bite wing radiographs the sensitivity was 93% and the specificity was 77% while bitewings had a specificity of 77% and a sensitivity of 62%. Currently, ECM is recommended to be used in conjunction with other modalities in the diagnosis of occlusal caries in which noninvasive intervention is indicated. In a recent study comparing visual inspection, fiber-optic transillumination (FOTI), and ECM, ECM was demonstrated to have a small but significant advantage in predicting occlusal caries. A limitation of ECM is the requirement for dry tooth surfaces. It has been demonstrated that the minimum flow rate required is 7.5 L/min. Other limitations include the need for direct contact, a conducting medium, and the need to isolate with cotton wool.

Recently, the advantages of digital radiography have been examined. Traditional radiography has significant disadvantages, which include the time/labor required for processing, variability related to chemicals, and inability to manipulate the images post-processing. Two types of digital radiography exist, CCD based and storage phosphor based. In the storage phosphor-based system, a reusable image plate is exposed to X-rays. The plate is then exposed to a laser scanner to store information. Digital radiographic images are obtained either by video recording with digitization or by direct digital radiography. This allows both image processing and the use of less radiation (up to 50% less). However, radiography has difficulty in detecting noncavitating caries, occlusal caries, root surface caries, and secondary caries.

FOTI has the advantage of using nonionizing radiation.[12] Light is transmitted through the tooth via fiber optics and is detected at the opposite side of the tooth. Modeling is used to assess the presence of caries. FOTI has led to mixed results due to the high level of intra- and interexaminer variability. To overcome these limitations, CCD image capture and digitization have been introduced (DIFOTI). One study demonstrated that DIFOTI was twice as sensitive as conventional radiography in detecting approximal lesions and three times as sensitive for occlusal lesions. For buccal-lingual lesions, the technology was $10\times$ more sensitive. However, even though it has been shown to be superior to conventional X-ray, sensitivities remain low (approximal 0.56, occlusal 0.67, smooth 0.43, and root 0.38).

Some groups are evaluating microbial testing to predict caries. They are probably most reliable when measured initially to establish baseline, then measured periodically to detect changes associated with increased risk for caries. However, limitations to the test due to the varied nature of organisms in the mouth may ultimately prevent its use.

One study looked at the sensitivities and specificities for the various technologies and the results were as follows[13]:

Test	Sensitivity	Specificity
Clinical exam	0.13	0.94
FOTI	0.13	0.99
Radiography	0.58	0.66
Fissure discolor	0.74	0.45
Electrical resistance	0.96	0.71

Ultimately, all of these techniques are limited in their effectiveness and for this reason, OCT is being evaluated for this application.

18.2.6 OCT and Dental

The first OCT images of teeth were generated by Colston et al. and Feldchtein in 1998.[14–16] In Figure 18.4, an *in vivo* OCT image of the anterior portion of a human tooth is shown.[16] In this figure, E is the enamel, D is the dentin, EP is the epithelium (gingival), CT is a connective tissue layer, AB is alveolar bone, and S is the sulcus. DEJ is the dentin-enamel junction which is sharply defined. In Figure 18.5,

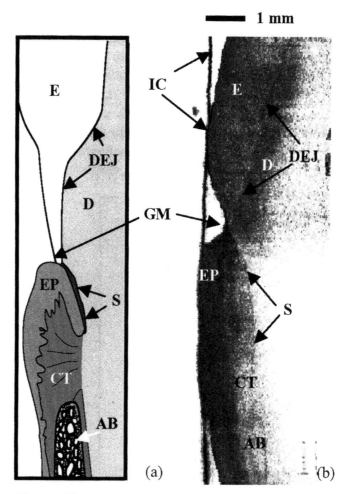

Figure 18.4   Sketch of tooth anatomy and the corresponding OCT image. Structures noted are the enamel (E), dentin (D), junction (DEJ), epithelium (EP), sulcus (S), and alveolar bone (AB). Courtesy of Colston, B.W., et al. (1998). *Optics Express.* 3, 230–239.

a composite restoration is imaged (CR) and in the enamel (E), polarization effects are noted which are represented by a banding pattern.[16] In Figure 18.6A, a caries lesion in the dentin below a restoration is seen.[15] In Figure 18.6AB, an air gap is seen below a restoration. In this study, an L-shape hand-held probe was used which allowed all sides of the teeth to be imaged. Caries tends to be highly scattering and lacks polarization sensitivity.[15]

PS-OCT has also been applied to imaging of the teeth. Polarization sensitivity was noted in the original papers examining OCT imaging of teeth.[16] More detailed studies with specially designed OCT systems followed which improved the

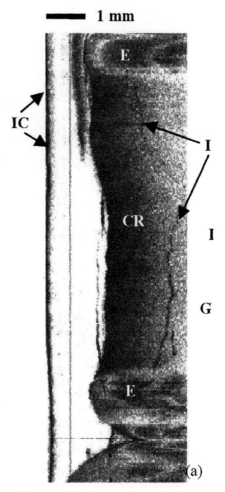

Figure 18.5 OCT image of a dental restoration. The composite restoration (CR) is clearly identified. Courtesy of Colston, B.W. et al. (1998). *Optics Express.* 3, 230–239.

characterization of caries.[17–19] As with cartilage and coronary arteries, future studies are needed to establish how this technology can be of diagnostic value and how additional information can be gained from PS-OCT data. In addition to diagnosing caries, OCT has the potential to be applied to periodontal disease and malignancies of the oral cavity.[15,20]

18.2.7 Limitations

There are several limitations to implementing OCT for dental imaging. First, the cost may be prohibitive for dental offices, where instruments generally do not exceed

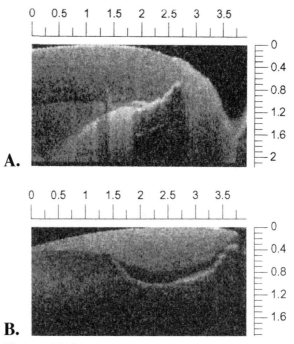

Figure 18.6 In vivo OCT image of caries lesion and gap below a CR. Courtesy of Feldchtein, F.I., et.al. (1998). *Optics Express.* 3(6), 239–250. A is the caries lesions while B is a gap below the composite restoration (CR). Courtesy of Colston, B.W. et al. (1998). *Optics Express.* 3, 230–239.

$30,000. Second, in screening the oral mucosa for carcinoma, scanning the entire surface of the mouth may be impractical. Third, large scale clinical trials are needed to establish if it ultimately affects patients' morbidity.

18.3 PROSTATE SURGICAL GUIDANCE

18.3.1 General

OCT can potentially be applied to prostatic resection to reduce erectile dysfunction and retrograde ejaculation associated with both endoscopic and open surgical procedures, in particular, resection for benign prostatic hypertrophy (BPH). The prostate is a fibromuscular and glandular organ that adds secretions to the seminal fluid during ejaculation. It surrounds the bladder neck and the first part of the urethra.[21] It is covered by a fibrous capsule, although the portion on the anterior and apical surfaces is often called the anterior fibromuscular stroma. Older classifications divide the prostate into five lobes: anterior, middle, posterior, right lateral, and left lateral. The prostate can also be divided into the transitional zone, which surrounds the proximal urethra; the central zone, which surrounds the urethral ducts; the peripheral zone, which makes up the largest part of the gland; and the anterior fibromuscular stroma, as described above. The transitional zone causes most cases of BPH. Histologically, the prostate consists of branched tubulo-acinar glands embedded in a stroma. The stroma consists of dense collagen, fibroblasts, and smooth muscle cells.

18.3.2 Benign Prostatic Hypertrophy

BPH or nodular hyperplasia is an enlargement of the prostate gland into the urethral lumen. It is estimated that by age 60, half of all men have histological evidence of BPH and all men have it by age 80.[22] Symptoms attributed to BPH fall into two broad categories, obstructive and irritative. Obstructive symptoms are mani-fested as a weak urinary stream, hesitancy, intermittency, and sensations of incomplete emptying. Irritative symptoms include frequency, urgency, urge incon-tinence, and nocturia. There is a static component to the obstruction from the physical presence of the tissue and dynamic due to increased sympathetic tone, producing tonic contraction of the muscular elements within the stroma of the prostate.

Resection is a very common procedure with an estimated number of surgical repairs in excess of 400,000 per year.[23] It is characterized by hyperplasia of the glands and stroma along with fibrosis, forming relative discrete nodules around the prostatic urethra and bladder neck. During prostatic hypertrophy, the enlargement of the median and lateral lobes of the gland produce elongation and lateral compression and distortion of the gland. Histologically, the tissue can consist of glandular proliferation, dilatation, muscular proliferation, fibrosis, or a combina-tion. There will also be areas of focal metaplasia and infarctions. Enlarged prostates generally weight 60 to 100 g. The etiology of prostatic hypertrophy is linked to sex hormones, specifically dihydrotestosterone (DHT), a metabolite of testosterone, which stimulates prostate growth. It is 10 times more potent at stimulating the prostate than testosterone.

There are a variety of technologies used to treat prostatic hypertrophy, all with their respective limitations. Among the most important complications of many of these procedures is the production of sexual dysfunction. Sexual dysfunction can be divided into impotency (erectile dysfunction), retrograde ejaculation, and loss of libido. Therefore, careful history is required to distinguish the source of sexual dysfunction both when evaluating a patient and assessing the results of clinical studies. The incidence of erectile dysfunction after a transurethral resection of the prostate (TURP) is debated but is probably between 4 and 35%.[24,25] Reflexogenic erections are mediated through the parasympathetic cavernous fibers (pelvic plexus) from S2–S4 while psychogenic erections are regulated by the thalamic and limbic system through nerves T11–L2 and S2–S4.[26] This is most likely secondary to transection of unmyelinated nerves near the capsule, particularly the cavernous nerves. The cavernous fibers run on the posterolateral base to the apex of the prostate between the capsule and the endopelvic fascia. These are the nerves most susceptible to injury during prostate resection. They are only a few millimeters from the capsule. OCT guidance of prostate resection can potentially lead to reduced impotence rates.

The incidence of erectile dysfunction varies with the procedure and rates remain controversial.

Retrograde ejaculation occurs when during ejaculation, some sperm is directed toward the bladder rather than forward out the penis. It occurs when the bladder's sphincter is compromised. Again, OCT guidance may be useful in preserving sphincter function.

18.3.3   Treatment

18.3.3.1 Medical treatment

The treatment for BPH may be either medical, minimally invasive, or invasive. Medical treatments can be either with alpha inhibition or 5-α reductase inhibition. The inhibition of the alpha receptor is believed to lead to relaxation of dynamic contraction of the prostate and improved urine flow. The inhibition of the 5-α reductase works by preventing conversion of testosterone to dihydrotestosterone, the androgen in the prostate that promotes hyperplasia.

18.3.3.2 Invasive procedures

Invasive modalities include TURP and open surgical resection. TURP has been the most common procedure for BPH for the last 30 years. It uses a probe that generates current to both cut and coagulate prostate tissue. Alternating current is applied at frequencies between 300 kHz to 5 mHz, which serves to heat the tissue. Direct current or reduced frequencies would stimulate muscles and nerves, which is why the relatively high frequency stimulation is used. If straight sine wave stimulation is used, the probe acts as a cutting device. When short bursts of sine waves are used, coagulation results. If the generator produces a mixture of pulses, then a combination cutting and coagulation current can be generated. During the cutting mode, a "chip" is produced in the shape of a canoe. While this may be the most effective treatment in terms of improving urinary flow, it has several serious complications other than impotence that have led to the search for newer methods of treatment.[27–29] Up to 13% of patients will require blood transfusions. Up to 60–80% will develop retrograde ejaculation while as many as 5–15% will have erectile dysfunction. However, some question exists whether significant erectile dysfunction actually is a result of the TURP and that retrograde ejaculation is a more significant problem.[30] About 10% will need re-operation in 5 years while 5% will develop bladder neck stenosis or urethral stricture. In addition, TURP syndrome is a serious complication. It is absorption of irrigation fluid through opened venous sinuses exposed during the surgery resulting in hypovolemia, hyponatremia, seizures, and hemolysis.

Open surgery, specifically retropubic and suprapubic prostatectomy, is the last option for men with BPH. It is used when the prostate is more than 50–75 g, the patient cannot be placed in the dorsal lithotomy position, or symptomatic bladder diverticulum is present. These procedures are the most invasive, requiring 3–4 days of hospitalization, risks of open surgical procedures, 3–5% erectile dysfunction, and almost 100% retrograde ejaculation.

18.3.3.3 Minimally invasive procedures

Minimally invasive procedures include transurethral electrovaporization (TUVP), visual laser ablation, transrectal high intensity focused ultrasound (HIFU), and transurethral needle ablation (TUNA). While these procedures claim reduced complication rates with success rates similar to TURP, minimal long-term data are available. It has even been suggested by some that the short-term benefits may simply be due to compression by the catheter and symptoms return over a period of time. Since long-term data are not available, the techniques will be described but their ultimate efficacy, particularly relative to TURP, will not be commented upon.

Electrovaporization — Simultaneous resection, vaporization, and coagulation of the prostate can be achieved with TUVP that uses a band electrode coupled to a high electrocuting energy source.[31,32] With electrovaporization, a wire loop electrocautery wire is pulled through the prostate tissue. About 1–3 mm of prostate tissue is vaporized with each pass. The procedure uses standard equipment used with TURP. Since desiccation occurs, it should reduce fluid loss and electrolyte abnormalities. TURP uses a standard wire electrode. However, the removal of tissue is slow and it is more difficult to vaporize with each pass.

TUIP — Transurethral incision of the prostate is another technique that is used to treat BPH.[33] In the original procedure, using a continuous flow resectoscope equipped with a knife electrode, the bladder is maintained at 50 cm. Incisions are made at the five and seven o'clock position within the ureteral orifice. These particular sites are chosen because they promote median lobe atrophy by disrupting blood flow. To avoid damage to the internal sphincter, the procedure was modified so that manipulation of the bladder neck is prevented. There are variations in the number and positions of the incisions. The indications for TUIP include (1) men with significant obstructive symptoms despite a "normal size" or moderately enlarged prostate (less than 30 g), (2) younger patients in whom sexual performance and antegrade ejaculation remain significant issues, and (3) debilitated patients who are a significant surgical or anesthetic risk. Advantages are that TUIP is useful for cost containment and leads to minimal incidence of bladder neck contractures. It also may be relatively good for prevention of impotency and antegrade ejaculation. Disadvantages are that it is not effective for treatment for significant median lobe disease, biopsies are not obtained as they are with TURP, and it can only be performed in relatively small glands (30 g or less).

Laser — Laser techniques include laser vaporization, laser resection-enucleation, and coagulation. The ND:YAG laser was originally the source of choice, but it has been largely replaced by the holmium laser. This is because of the reduced penetration of the latter. Holmiun laser prostatectomy by resection-enucleation has predominately replaced laser vaporization.[34] Its use is proposed for larger prostates and as an alternative to open surgical procedures. The theory behind interstitial laser coagulation is to shrink the prostate by generating necrosis within the prostate without damaging the urethra or leading to tissue sloughing. With this technique, the prostate lobes are punctured individually with a laser fiber without the use of a trocar. Laser irradiations are made at 45 degrees. Laser irradiation is performed at 5 minutes per fiber placement. Shrinkage takes place within 6 to 8 weeks of irradiation. Preliminary results suggest the main advantages are reduced blood loss, quicker catheter removal, and shorter hospital stay. Drawbacks of interstitial laser are the prolonged postoperative catheterization and the delay in symptom improvement, particularly irritative syndromes. Reintervention within 1 year may be as high as 8–15%. Adverse events and disadvantages are similar to those found with TURP, including a high incidence of retrograde ejaculation and erectile dysfunction.

TUNA — The transurethral needle ablation device (TUNA) uses radiofrequency (approximately 500 kHz) energy to heat and breakdown the tissue. With this technique, two needles are at the end of the scope and placed into the prostate parenchyma.[35] The urethral mucosa is preserved so that tissue sloughing is minimal. If the tissue is heated too rapidly there will be less water available for heating. The tissue is therefore heated for 3 to 4 minutes. The area of heating is approximately 6 mm from the tip. Tissue is reabsorbed after 8 weeks and patients generally note improvement in symptoms after 2–3 weeks. Advantages of the procedure include the fact it can be done in the office without general anesthetic and the morbidity is less. Incontinence, impotence, and retrograde ejaculation are reported as rare. In addition, since the adenoma contains less blood supply than the capsule, damage to the capsule is limited. There are several disadvantages. First, it cannot be used in the median lobe. Second, it should only be used with prostate glands 60 g or less. Finally, the long-term efficacy has not been determined.

TUMT — Many different systems for the delivery of transurethral microwave therapy (TUMT) have been developed in the past few years.[36,37] The process can be performed without the need for regional or general anesthesia. Coagulative necrosis occurs deep within the tissue by achieving temperatures greater than 45°C for not less than 20 minutes, which produce thermal coagulation, microvessel vascular thrombosis, and destruction of stromal smooth muscle. The current technique uses water cooling to protect the urethra and bladder. This prevents urethral sloughing.

Rectal temperatures are used to monitor the degree of ablation. The different prostate sizes lead often to variability in heating and therefore variable effectiveness. However, studies assessing the prostate histopathology and MRI show urethral destruction, which likely leads to destruction of the sensory nerves and therefore some of the relief in symptoms. New onset erectile dysfunction is less than 5%.

18.3.4   OCT and BPH

OCT has demonstrated considerable potential for guiding the treatment of prostatic hypertrophy.[38,39] In Figure 18.7, an image of the prostate, its capsule, and an unmyelinated nerve is shown.[38] The upper arrow demonstrates the boundary

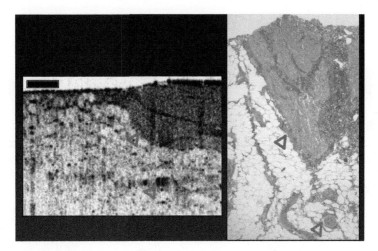

Figure 18.7 OCT image of prostate capsule. The prostate parenchyma is in the upper right. The capsule is to the left. The upper arrow demonstrates prostate-capsule border. The lower arrow is an unmyelinated nerve. It is transection of these unmyelinated nerves that leads to many of the complications of TURP. Courtesy of Brezinski, M.E., et al. (1997). *J. Surg. Res.* 71, 32–40.

between the capsule and prostate parenchyma (right). The lower arrow is an unmyelinated nerve vulnerable to transection.

Real-time monitoring of prostate ablation has also been monitored with OCT.[39] Ablation was performed with lasers and radiofrequency (RF) energy. An example image is shown in Figure 18.8. In this figure RF ablation was followed as a function of time. In addition to removal of prostate tissue, coagulation (ct) is also noted. The corresponding histology is included.

There are several significant challenges associated with OCT guidance of prostate resection. First, due to the penetration of only a few millimeters, guided interventional procedures are likely to take significantly longer. Second, blood and carbonization can reduce imaging penetration. However, in spite of the difficulties, the application of OCT for guiding BPH resection needs further examination.

18.4 NERVE REPAIR

18.4.1 General

OCT has the potential of improving the outcome of surgery to repair transected nerves. It can both image through the re-anastomosed nerve and distinguish motor from sensory nerves. The heart of the nervous system is the nerve cell. Nerve cells tend to have very long axons that are extensions that leave the cell and transmit signals over large distances. Nerves are generally collections of axons or nerve fibers. There are generally two types of nerve fibers, myelinated and non-myelinated. The axons of non-myelinated axons generally are simply enveloped by the cytoplasm

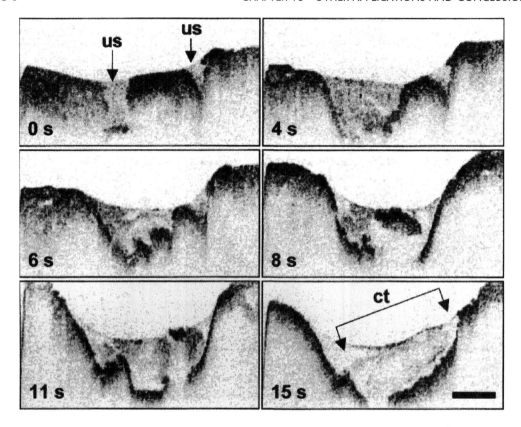

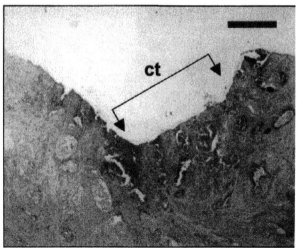

Figure 18.8 Laser ablation of human prostate. Courtesy of Boppart, S.A., et al. (2001). *Comput. Aided. Surg.* 6, 94–103.

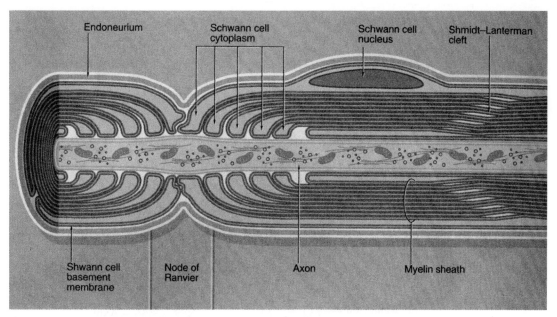

Figure 18.9 Diagram of myelinated nerve fiber. Courtesy Young, B. and Heath, J.W. (2000). *Functional Histology,* fourth edition. [100]

of a specialized cell known as a Schwann cell. The larger myelinated fibers are wrapped with concentric layers of Schwann cell plasma membranes forming what is known as the myelin sheath, as shown in Figure 18.9. The myelin is characterized by gaps along its length known as the Node of Ranvier. A cross section of a peripheral nerve is shown in Figure 18.10.[40] The peripheral nerves we are interested in contain both non-myelinated and myelinated nerve fibers, grouped together in what are known as fascicles. Fascicles are bundles of nerve fibers or axons within a nerve. Generally, the fascicles are either motor, containing almost exclusively fibers which direct muscles, or sensory, which contain fibers that perceive information from the outside world such as pain. Within a fascicle, the nerves are separated by supportive tissue known as endoneurium. A condensed layer of robust vascular supportive tissue known as perineurium surrounds each fascicle (the outer sheath of the fascicle). In peripheral nerves containing more than one fascicle, a layer of loose connective tissue called the epineurium is present and condenses peripherally to form a strong cylindrical sheath (the outer casing of the nerve).

Nerve injury, particularly of the extremities, is very common during trauma. Important nerves subject to traumatic injury, which are amenable to repair, are both large (cranial, paraspinal, brachial plexus) and small (distal upper and distal lower extremity) nerves. Unlike vascular repair, nerve repair does not need to be immediate, although this is optimal.[41] Repair can be performed in days rather than hours after the injury but before significant tension begins to develop. Nerve injury has now replaced vascular injury as the leading cause of amputation, as was evident in the first Gulf War.[42]

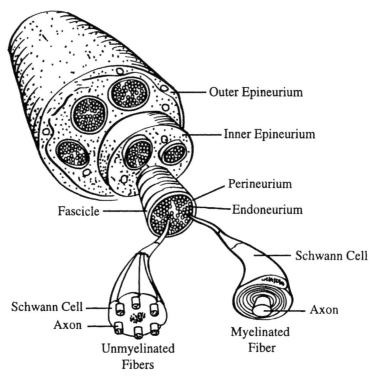

Figure 18.10   Diagram of subdivisions of a nerve. Courtesy of Diao, E., and Vannuyan, T. (2000). *Hand Clinics.* 16, 53–66.

18.4.2 Repair

The two predominate methods of repairing injured nerves are epineural (with or without interfascicular stitches) and perineural.[43] The epineural repair connects the outermost protective sheath around the nerve ends together. Epineural suturing was long considered a good choice for nerve repair since it provides reliable anchoring. Temporarily supplanted by fascicular repair (perineural), it regained popularity in the late 1970s and early 1980s. Combining an interfascicular stay suture with epineural repair is probably the conventional method that provides the best results. The problem with epineural repair is that when the epineurium is re-anastomosed, it cannot be determined whether the fascicles are correctly aligned. Perineural suturing is often called fascicular suturing.[43] Transfixing the perineurium (directly connecting the fascicles) is difficult because this sheath is not easy to see, is both elastic and fragile, and handling the fascicle often produces mushrooming (retraction of the nerve sheath). An additional problem associated with perineural repair is that it interferes with the rich network of blood vessels that run along and between the perineurium and epineurium, destroying them or producing extensive microthromboses, which result in significant peri- and intraneural fibrosis.

Finally, perineural repair results in longer tourniquet times, increasing the probability of ischemic injury.

The key factor of the success or failure of nerve repair seems to be alignment. When the fascicles are misdirected, the probability of failure increases dramatically. Of particular importance is that transected sensory nerves are not re-anastomosed with transected motor nerves, which leads to significantly poorer outcomes. In two sequential studies, satisfactory motor results were obtained in 84% of patients with fascicular alignment and only 53% without.[44] For sensory, 68% were satisfactory with alignment and only 42% without. Nerve contusions from the initial injury, which are poorly assessed by current imaging modalities, may also play an important role in procedure success.

18.4.3 Diagnostics

There is no current method for imaging through a nerve to assess the position of fascicles within the re-anastomosis site once closure has taken place, one of the areas we are examining with OCT. There are methods that have been used in an attempt to distinguish sensory and motor fascicles prior to reanastomosis, with significant limitations. These methods include anatomical landmarkers, electrical stimulation, and acetylcholinesterase levels. The position of some sensory and motor nerves can be determined from certain anatomical consistencies.[45,46] Examples are the large motor fascicle of the ulnar nerve and the anterior central motor fascicle of the median nerve. However, these are exceptions to the rule.

Another method for assessing motor versus sensory nerves is direct electrical stimulation.[47] The voltage used is generally between 1 and 5 μV, pulse rate of 30 per second, and duration of 2 msec. It is most useful for identifying motor nerves which cause muscle movement under stimulation. Using direct stimulation has several limitations. First, the injury must be new enough so that peripheral portions of the nerve have not degenerated. Second, the severance of the nerve should be clean. Third, it is time-consuming for the surgeon. Fourth, the level of anesthesia must be carefully regulated, too much and the patient cannot participate in the test, too little and the patient will be in pain. Therefore, this method has only been of limited use.

A test that is commonly used is the acetylcholinesterase assay.[48,49] With this test, myelinated fibers stain positive for acetylcholinesterase with motor nerves and negative with sensory nerves. While the test has a high specificity, it takes over an hour for the surgeon to get back the results, a difficult situation for the operating suite.

A technique capable of imaging below the tissue surface, assessing at high resolution both the exactness of re-anastomosis and identification of fascicle type, could be a powerful tool in the repair of injured nerves.

18.4.4 OCT and Nerve Repair

We initially demonstrated imaging of peripheral nerves in a publication in 1997.[50] In Figure 18.11, an OCT image of a peripheral nerve is seen. The individual fascicles

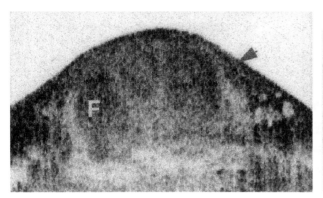 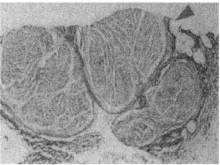

Figure 18.11   OCT image from nerve and corresponding histology. The individual fascicles (F) are identified through the nerve. Courtesy of Brezinski, M.E., et al. (1997). *J. Surg. Res.* 71, 32–40.

(F) are clearly delineated through the nerve wall and their presence confirmed by histology. Figure 18.12 is from a later work where the three-dimension reconstruction of a nerve is shown.[51] The individual fascicles are tracked through the nerve and its fascicles and a bifurcation (b) is easily identified. We are currently examining speckle and polarization techniques to distinguish motor from sensory fascicles. Recently, another group has demonstrated changes in scattering associated with the electrical activity of the abdominal ganglion of Aplysia.[52] It is unclear if this technique can be implemented on intact human nerves.

18.5   VASCULAR REPAIR

18.5.1   General

Three roles are envisioned for OCT in imaging for microsurgery of vessel repair. They are vessel re-anastomosis post-laceration, vessel re-anastomosis for flap procedures, and monitoring flap viability. Only vessel re-anastomosis post-laceration will be discussed here as an example.[53,54]

18.5.2   Vessel Re-anastomosis

The most common civilian injury to the vasculature is penetrating injury while second is blunt trauma. When motor vehicle accidents are the major cause, blunt trauma increases in prevalence. Most arterial injuries are repaired with end-to-end anastomosis followed by interposition with a saphenous vein graft. The latter has the higher reocclusion rate. In addition to arterial repair, vein repair is also often attempted.[55] Success rates are better for arterial rather than venous repair. The problem is that veins are flimsy and apposition of the edges tends to often be inaccurate. A technology is needed that can see into the vessel lumen and identify structural abnormalities such as size mismatch, flaps, and clots.

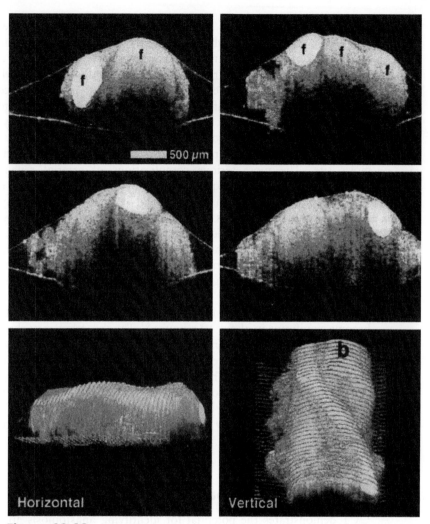

Figure 18.12 Three dimensional OCT imaging of a nerve. Courtesy of Boppart, S.A. et al. (1998). *Radiology*. 208, 81–86.

Injuries seen at military medical units during troop engagements are predominately injuries of the limbs, with almost universal vascular and nerve involvement.[56] Limb injuries have exceeded the sum of head/neck, abdomen, and chest injuries in WWI, WWII, the Korean War, the Vietnam War, the Israel-Arab Wars (1975, 1982), and the Falkland Island conflict.[56] The lower extremity is injured more frequently than the upper by over a 3:1 ratio.[57] Of all injuries, 75% are from explosive ammunition while 25% are from bullets from small arms, thus multiple site involvement is common.[58]

The complications associated with vascular surgery have steadily improved since WWII, where vascular insufficiency was the leading reason for amputation.[59]

During WW II, 50% of patients with major vessel involvement underwent amputation and of those who underwent vascular repair, 36% ultimately received an amputation. Today, many vascular procedures have success rates in excess of 99%. However, for some arteries, such as the popliteal and vessels less than 3 mm in diameter (including those of free tissue transfer or grafting), complication rates remain high. Amputation after popliteal artery repair continues to be in the range of 29.5%. For forearm arteries (ulnar and radial) in modern day, the 1-week and 5-month patency with microscope guidance were 86% and 68%.

It is known that vascular failure in these vessels generally occurs at the anastomosis site. The most serious complications of arterial repair are thrombosis and infection, the latter frequently associated with disruption of the arterial repair with bleeding. If infection occurs, ligation of the artery may be mandatory. Thrombosis typically results from misalignment at the anastomosis site, with intimal flaps and adventitial inversions leading to a local hypercoaguable state. The hypercoaguable state results from either the introduction of thrombogenic material into the lumen (i.e., adventitia) or the development of turbulent flow. Size mismatch, particularly when a vein graft is used, can also lead to turbulence and acute occlusion.[60]

In experimental lesions in microvessels, it was found that the thrombus can become significant in about 10 minutes.[61] The thrombus may shrink and grow over the next few days. However, since the surgeon cannot see within the vessel, he/she will be unaware unless thrombus forms. The vessel may then subacutely occlude within the next 72 hours. The current method for monitoring the site is the surgeon waiting at least an hour for signs of occlusion before closing the site, which is cost-ineffective. An imaging technique that can image through the vessel wall during the procedure could allow identification of intimal flaps, adventitial inversion, or thrombus.

18.5.3   OCT and Arterial Repair

OCT demonstrates the potential for identifying anastomosis abnormalities during and after repair. In Figure 18.13, a superior mesenteric artery is shown with the corresponding histology.[50] The intima, media, and adventitia are well differentiated in addition to the coverslip below (c). In Figure 18.14, a vessel is imaged after re-anastomosis.[51] An intima flap can be seen in Figure 18.14C while in Figure 18.14F, a complete occlusion is noted (O), which is not obvious on the previous image.

With OCT imaging, the optimal imaging is likely along the vessel axis as in Figure 18.14F, which shows the obstruction. The most significant problem associated with OCT evaluation of re-anastomosis would be the presence of blood. It is possible that, since the vessel diameter is small, penetration through the small amount of blood may not significantly influence imaging. If this is not the case, the surgeon would have the option of either occluding the vessel proximally and/or injecting saline.

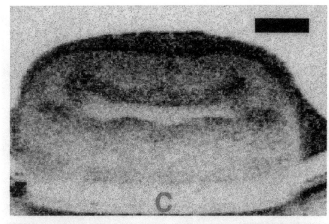

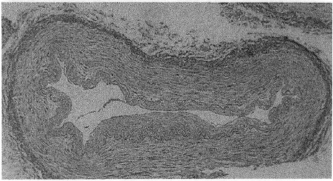

Figure 18.13 External imaging of a vessel demonstrating the three layered appearance. The top is the OCT image and the bottom is the histology. The coverslip (c) is seen below. Courtesy of Brezinski, M.E., et al. (1997). *J. Surg. Res.* 71, 32–40.

18.6 NEUROSURGERY

18.6.1 General

Nowhere in the body are vital structures more closely packed then in the central nervous system. Therefore, surgery is often complex and the unnecessary destruction of important neurologic pathways is always a concern for the neurosurgeon. Among the most important applications for OCT are guiding brain/spinal cord tumor resections, preventing nerve injury during spinal surgery, assisting in vascular repair (such as aneurysms), and guiding endoscopic procedures. This section will focus on guiding the resection of 'benign' brain tumors.

18.6.2 Brain Tumors

Metastatic tumors to the brain — usually lung, breast, melanoma, and kidney — will not be considered here. This is because most metastatic tumors generally

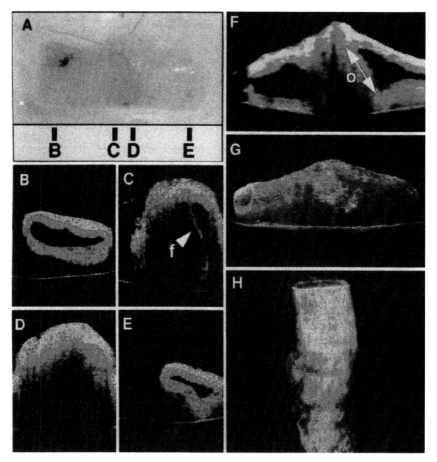

Figure 18.14   Three dimensional reconstruction of a reanastomosed vessel. The obstruction is best noted in F. Courtesy of Boppart, S.A. et al. (1998). *Radiology.* 208, 81–86.

are widespread and local resection is not likely to have a significant impact. Primary brain tumors represent 1.4% of all tumors and 2.4% of tumor deaths. "Benign" brain tumors represent 40% of all primary brain tumors and approximately 20,000 new cases are diagnosed each year.[62] The quotation marks around benign indicate that while brain tumors generally do not metastasize, many will infiltrate the surrounding adjacent tissue and the patient will ultimately succumb to the disease. There are a large number of brain tumor types, but this section will focus on the two most common, meningiomas and gliomas (specifically astrocytomas).

18.6.2.1 Meningiomas

Meningiomas are tumors of the meninges, the protective membranes around the brain and spinal cord. Most do not infiltrate or metastasize, and they represent almost 20% of primary brain tumors. They are most common in middle age patients and the elderly. Seizures are the presenting symptom 50% of the time. Additional symptoms such as headaches, visual problems, and changes in personality are from increased intracranial pressure. Surgery is almost always curative, but when the tumor is in the cavernous sinus, resection may not be possible. In this case, stereotactic radiosurgery may be used, where hundreds of beams of radiation are aimed at the tumor in a single session.

The overall recurrence rate of meningiomas is 13–40%, and is a function of prior resection, which in turn is a function of the difficulty in resection. OCT guidance of tumor resection could allow more complete removal of the neoplasm, therefore reducing this recurrence rate.

18.6.2.2 Gliomas

Over half of all brain tumors are gliomas, of which astrocytomas are the most relevant. The astrocytomas are generally graded from type I to type IV. Type IV is often referred to as glioblastoma multiforme. Low grade gliomas make up 25% of all brain tumors, and there are about 17,000 newly diagnosed cases each year.[62] Grade I or II is described as benign, but cells infiltrate the surrounding brain tissue in clusters too small to be detected or resected by the surgeon. The terms low grade and benign therefore reflect histopathological appearances and do not imply a benign clinical course.[63,64] While they are slow growing tumors, median survival is approximately 93 months. In addition, the tumor has a high incidence of transformation to a malignant cell type.[64] Type III has a median survival of about 12 months. Type IV is extremely aggressive and has a 5-month median survival which can be increased to approximately 1 year with therapy. With all forms, the greater the ability to resect the tumor with minimal damage to normal tissue, the better the survival. Therefore aggressive resection should be performed whenever possible as viable tumor cells may exist outside areas of traditional definitions of tumor infiltration.[65,66]

However, the benefits of aggressive surgery must be weighed against potential permanent neurological damage. In some patients, unless improved prognosis is achievable, more conservative therapy may be indicated. Widely accepted indications for neurosurgical intervention include increased intracranial pressure, neurological deficit due to mass effect or hemorrhage, and symptomatic epilepsy unresponsive to antiepileptic drug treatment.

18.6.3 Imaging Brain Tumors

Within the white matter the distinction between tumor and edematous brain may be even more difficult to detect. Texture and color at the margin is also less reliable for low grade tumors. Stereotactic techniques have therefore been used in an attempt to better localize the tumor.[67,68] The imaging is performed with X-ray, MRI, or ultrasound. External reference points are used and they can be either a frame attached to the patient or marked points on the patient's surface anatomy (frameless). These techniques can also be used for taking biopsies. Neuronavigation can be aided by using sensors that track the instruments.

However, shifts in the brain due to surgical retraction, the extent of the resection cavity, and leakage of cerebrospinal fluid can occur when the surgeon is relying on preoperative data. This may lead to misorientation. Alternatively, intraoperative imaging with CT or MRI can provide images of the resection that reduce the problem of brain shift by providing real-time images during surgery. Among the limitations affecting image quality are artifacts which may be due to electrical environment in the operating room or the instability of the magnetic field. But the main limitations of the intraoperative use of MRI is its considerable economic cost, which includes the purchase of the equipment, modifications of the operating room, and operation costs.[69] The ultimate impact on patient outcome in terms of survival or quality of life has yet to be established.

Intracranial ultrasound can also be used in situations where there is a question if the lesion has shifted after preoperative MRI.[70] Intracranial ultrasound is local imaging and at a higher resolution than MRI or CT. However, images are often difficult to interpret because echogenic structures cannot reliably differentiate normal from abnormal tissue. In addition, blood products in the surgical field may cause misinterpretation of ultrasound images.

Another approach used in the awake patient is intraoperative neurophysiological monitoring with direct cortical stimulation to precisely define cortical regions. For example, stimulation of the language region is performed with the patient reciting, counting, naming objects, or filling in missing words to preserve language skills. This is more difficult with the white matter where fibers are very susceptible to damage from surgical manipulation. Also, arterial or venous damage can also result in deficits.[71] Other technologies that have been used to try to assess tumor margins on the cellular level include diffuse optical tomography, Raman spectroscopy, fluorescence, and confocal microscopy.[71–74] However, only limited studies are available on these technologies.

18.6.4   OCT and Brain Tumor Resection

OCT has the potential to guide tumor resection on a micron scale. In Figure 18.15, an OCT image of the cerebellum is seen with the corresponding histology. The molecular and grandular layers are clearly defined.[50] Figure 18.16 shows a series of OCT images obtained parallel to the surface of the cerebrum of a metastatic melanoma.[75] The en face images were obtained at different depths and the corresponding histology is included. At 1200 μm, as seen by the arrow, a small cluster of cells less than 50 μm is seen.[75] Recently, OCT imaging was performed of meningiomas and gangliogliomas, but at 800 nm using a solid state laser, which limited penetration to a few hundred microns.[76] The resolution was approximately 5 μm. Figure 18.17 shows normal brain tissue where the leptomeningi are noted by the white arrows, but deep tissue remains relatively homogeneous. Among the features noted in tumors were the enlarged nuclei and clustering of the tumor cells as shown in Figure 18.18. The white arrows show a strip of clustered tumor cells.

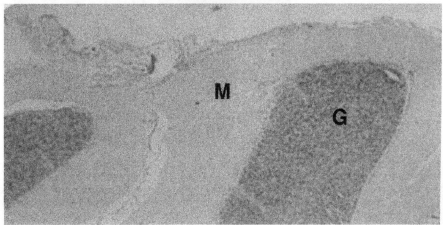

Figure 18.15 OCT image of human cerebellum. The gray matter (M) and white matter (granular) (G) are identified in the OCT image and its corresponding histology. Courtesy of Brezinski, M.E., et al. (1997). *J. Surg. Res.* 71, 32–40.

The difficulty with applying OCT to brain tumor resection is the limited penetration. The imaging depth of only a few millimeters would make it difficult to continuously monitor the surgical resection, particularly with large tumors. Likely, OCT would be used at particular points during the procedure to assess margins. It can either be performed through a hand-held probe or imaging device on the surgical instrument.

In addition to assessing tumor margins, OCT may potentially play an important role in spinal surgery, for example, during disc repair to reduce nerve injury. While this topic will not be discussed in detail, relevant OCT images are seen in Figure 18.19.[50] In Figure 18.19a, an unmyelinated nerve is demonstrated less than 500 μm from an intervertebral disc (right arrowhead). In Figure 18.19b, the

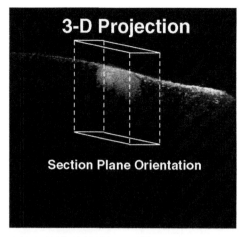

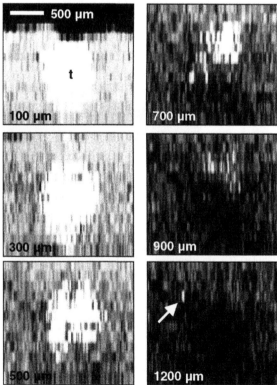

Figure 18.16 Serial imaging of cranial melanoma. In the final image, an isolated cluster of tumor cells is noted in the image at 1200 μm. Courtesy of Boppart, S. A. et al. (1998) *Neurosurgery*, 43, 834–841.

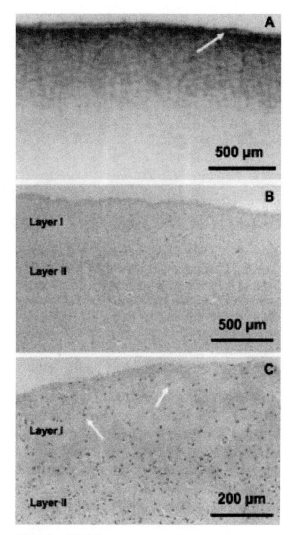

Figure 18.17 Ex vivo imaging of healthy brain tissue. The top of the OCT image of the leptomeninges. The bottom is the corresponding histology. Arrows represent neuron and glial cells. Courtesy of Bizheva, K. et al. (2005). *J. Biomed. Optics.* 10(1), 011006-1 to 011006-7.

corresponding histology is shown. A large image of an intervertebral disc is shown with the characteristic fibrocartilage pattern (C) and supportive tissue (S). In Figure 18.19d, a small unmyelinated nerve is noted embedded in the retro-peritoneum. In Figure 18.19e, the corresponding histology is shown. These images suggest that OCT may have a role in reducing the morbidity associated with spinal surgery.

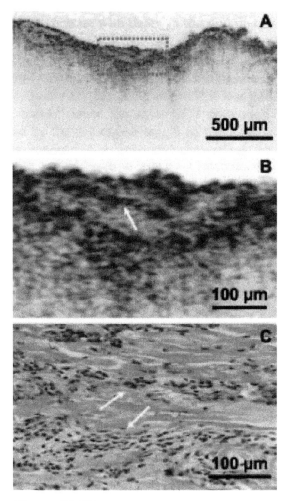

Figure 18.18 OCT image of a ganglioglioma. The tumor is identified in the OCT images (A and B) and the corresponding histopathology. Courtesy of Bizheva, K. et al. (2005). *J. Biomed. Optics.* 10(1), 011006-1 to 011006-7.

18.7 THE ROLE OF OCT IN ASSESSING SUBFERTILITY

18.7.1 General

According to the Centers for Disease Control, approximately 5% of all women between the ages of 15 to 44 have sought out infertility treatment such as ovulation induction, surgery, or assisted reproduction.[77] About 20% of all women have their first child after the age of 35. By age 35, a woman has half the chance of becoming pregnant when compared with age 25. The common causes of subfertility are pelvic

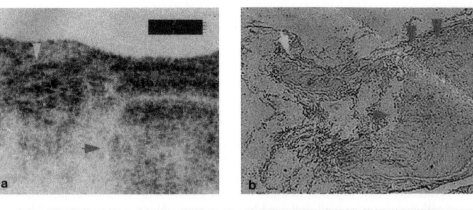

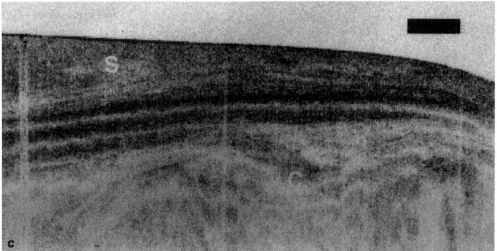

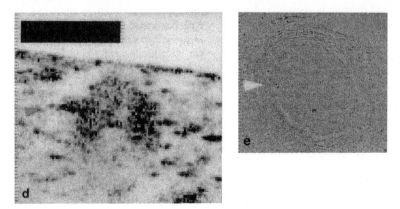

Figure 18.19 Imaging of unmyelinated nerves. (A) An unmyelinated nerve (upper arrow) in close proximity to an intervertebral disc (lower arrow). (B) The corresponding histology to A. In C, the intervertebral disc is shown where C is the cartilage and S is the supportive tissue. In D, a retroperitoneal unmyelinated nerve is shown with the corresponding histology in C. Courtesy of Brezinski, M.E., et al. (1997). *J. Surg. Res.* 71, 32–40.

disease (tubal disease and endometriosis), semen abnormalities, ovulation disorders, and idiopathic infertility. The most common cause of subfertility, the reduced ability to bear children, is abnormality of the oviduct (the fallopian tube).

18.7.2   The Oviduct

The oviduct is a structure that transmits the ovum from the ovary to the uterus.[78] It is about 14 cm long and muscular with one end opening into the peritoneal cavity near the ovary and the other into the uterine cavity. It has four segments that gradually phase into one another. The first section is the intramural section, which is within the uterine wall, followed by the isthmus. Next, a section more dilated than the isthmus is the ampulla. The final section is near the ovaries and is termed the infundibulum, which contains finger-like projections toward the ovaries known as fimbriae. The oviduct consists of three layers: the mucosa, muscularis layer, and external serous coat (that is visceral peritoneum). The epithelium consists predominately of simple columnar cells of two types, secretory cells and cells with cilia that beat toward the uterus. The mucosa of the ampulla is thick and has large numbers of branching folds that appear in cross section as a labyrinthine system with little space in between. The folds are much less highly branched and numerous in the isthmus and almost nonexistent in the intramural section. The oviduct receives the ovum expelled by the ovary, and moves it toward the uterus through rhythmic contractions. Its secretions provide nutrition to the embryo and fertilization typically takes place at the ampullar-isthmic junction. If over 50% of the ampulla is removed or destroyed, pregnancy will be extremely unlikely.

18.7.3   Oviduct Pathology

Tubal disease is recognized as the cause of infertility in as many as half the cases.[79] About 40% of these have a previous history of pelvic inflammatory disease (PID) or infection, which is usually the result of sexually transmitted diseases.[80] Pathologic changes with PID of the oviduct include deciliation, flattening of the mucosal folds, and degeneration of the secretory cells. Sources of infection include *Chlamydia trachomatis*, gonorrhea, genital tuberculosis, post-pregnancy sepsis, and intrauterine contraceptive devices. The next most common cause of tubal subfertility is endometriosis. Endometriosis is characterized by the presence of endometrial glands and stroma outside of the uterus.[81] Salpingitis isthmica nodosa (SIN) is another common cause of infertility.[82,83] SIN is characterized by hypertrophy and hyperplasia of the muscularis surrounding pseudogland-like structures. SIN is thought to be the cause for as much as 5% of all cases of female infertility. It is identified in 20–60% of cases of proximal tube occlusion. The documentation of SIN within the fallopian tube has implications for both fertility and ectopic pregnancy.

Fallopian tube disease can be either occlusive or nonocclusive. Occlusions in the oviduct can be described as distal, midtubal, and proximal.[84] A very common lesion is distal tube occlusion that leads to distention of the ampulla, referred to as a hydrosalpinx. It is typically associated with tubal damage and peritubal scarring. Midtubal occlusion is the site least likely for occlusion. Proximal tube occlusion can occur from SIN, PID, endometriosis, and previous ectopic pregnancy.

18.7.4 Infertility Testing

The standard noninvasive tests for infertile couples in general are semen analysis, measurement of FHS/estradiol levels or clomiphene challenge test, and assessment of ovulation via progesterone levels.[85] The most common invasive techniques for assessing the oviduct are hysterosalpingography, laparoscopic dye studies, sonohysterography, selective salpingography, and falloposcopy.[86] All these techniques have limitations and currently, multiple techniques are necessary to evaluate the fallopian tubes appropriately.

Hysterosalpingography is the injection of radiopaque contrast material into the uterus and fallopian tubes through the cervix.[87,88] While it is a relatively simple procedure to perform, it has several significant problems. First, it has a high false positive rate for obstruction due to the inability to achieve sufficient pressure to open the orifice of the oviduct. Second, it does not have sufficient resolution to identify the microstructure of the tube. Third, debris or spasm can result in a false positive. Fourth, it is reliable for proximal occlusion, substantial for distal obstruction, and poor for adhesions.

Some physicians view laparoscopic evaluation of the fallopian tube with dye injection as the current gold standard evaluating the oviduct.[89] It is the best method for assessing peritubal disease such as external endometriosis and adhesions of the fimbria and ovaries. However, it has several important disadvantages. First, it generally requires the use of general anesthesia. Second, dye pressure must be sufficient to overcome the opening pressure of the oviduct orifice, otherwise a false positive results. Third, its value for evaluating the luminal surface of the oviduct is limited. Fourth, there is a risk of vascular, bowel, or urologic injury

Sonohysterography is the performance of ultrasound after the injection of sonolucent fluid (air with saline or air-filled albumin microspheres) into the uterus and fallopian tubes to improve contrast.[90,91] While it is easy to perform, it has several significant disadvantages. First, it has a high false positive rate for obstruction because insufficient intrauterine pressure is achieved to overcome the physiologic opening pressure of the oviduct orifice. Second, debris in the oviduct or spasm can present as an obstruction even though no obstruction exists. Third, it has a limited resolution so it is not useful for identifying intraluminal adhesions. Fourth, it is not particularly useful for assessing peritubal disease.

Direct injection of dye into the fallopian tubes is referred to as selective salpingography.[92] This allows the problem of spasm and the development of sufficient pressure to be overcome. Selective salpingography is a better diagnostic tool for proximal tube occlusion than the use of laparoscopic dye. There was no difference for distal tubular occlusion. Laparoscopy is better for assessing peritubal disease. The disadvantage of selective salpingography is that it requires technical expertise, using more expensive equipment, and more radiologic table time is required.

The incidence of perforation is 3–11%. Of note, selective salpingography by itself can be therapeutic and has been suggested to improve fertility rates in women with proximal disease.[93]

Falloposcopy is the direct visualization of the lining of the oviduct via an endoscope introduced through the cervix.[94] The falloposcopy scope is 0.5 mm in diameter and is introduced through a 1.2–1.3 mm outer diameter Teflon cannula. Its ability to give information about the surface structure of the oviduct makes it superior as a diagnostic modality for assessing tubal causes of infertility. In one study, with falloposcopy, management was changed in 52.4% of the patients in the series.[95] The spontaneous pregnancy rate was higher (27.6%) for those with tubes normal by falloposcopy compared with mild (11.5%) or severe (0%) disease.[96] However, falloposcopy does not give as good a view of the tubal folds past the isthmic region. In addition, there is little distention of the lumen and the catheter has poor image quality. Finally, it gives little information about microstructure below the surface of the lumen.

18.7.5   Treatment

There are a variety of approaches to the treatment of tubal infertility, including mechanical lavage, balloon dilatation, tubal transplantation, laparoscopic surgery, guidewires, microsurgery, and *in vitro* fertilization (IVF). The appropriate treatment varies from patient to patient.

In some instances, the mechanisms behind fallopian tube malfunction include partial adhesions, debris, mucus plugs, or endometrial fragments. Therefore, simple lavage of the tube may result in restoration of fertility. In one study, in patients who underwent elective salpingography with previously documented patent ducts, the total subsequent pregnancy rate (48.6%) within 12 months was significantly higher than that of control groups (11.6%).[97] Of these, 33% of the pregnancies occurred within the first month of the procedure. In women with nonpatent fallopian tubes, the patency rate from salpingography alone was 35–75% with a pregnancy rate between 25 and 35%.

If selective salpingography does not open the tube, other options exist which can result in patency. One option is transcervical balloon tuboplasty.[98] In one study of patients undergoing balloon dilatation, 66.2% achieved patency and the spontaneous pregnancy rate was 33%, avoiding surgery for the patients.

There is substantial controversy whether microsurgery or IVF should be used to treat subfertility.[99] In young patients who are interested in having multiple children, microsurgery should be considered, particularly when the disease is proximal. Common indications for microsurgery include lysis of adhesions, fimbrioplasty, salpingostomy, tubal re-anastomosis, endometriosis removal, ovarian cystectomy, and management of ectopic pregnancy. As stated above, if the oviducts are severely diseased, unless new techniques are developed, IVF should be strongly considered. IVF is associated with a much higher expense and many patients cannot afford or are unwilling to undergo the procedure. Ultimately, high resolution imaging is needed to distinguish those patients who need microsurgery versus those who should undergo IVF, in addition to those who potentially need minimally invasive procedures.

18.7.6   OCT and Imaging the Fallopian Tubes

The potential of OCT for imaging the oviduct was initially demonstrated in 1997.[50] In Figure 18.20, the distal (a) and proximal (p) oviduct is imaged. The mucosal

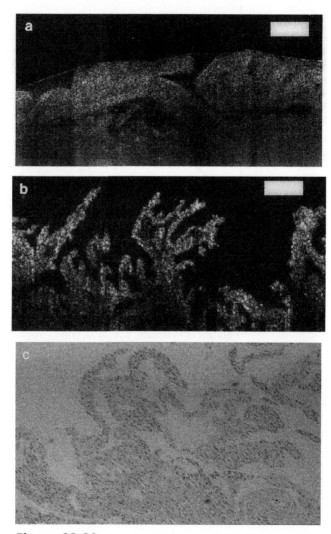

Figure 18.20   Imaging of a human oviduct. A and B are OCT images of the oviduct. A is the distal oviduct with less branching while B is the proximal portion. Image C is the corresponding histology for B. Courtesy of Brezinski, M.E., et al. (1997). *J. Surg. Res.* 71, 32–40.

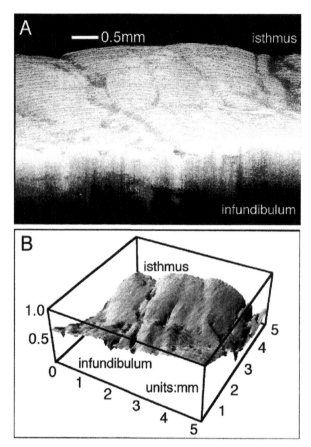

Figure 18.21   Three dimensional images of an oviduct. Courtesy of Herrmann, J.M., et al. (1998). *Fertility and Sterility.* 70, 155–158.

folds are noted in (b) of the ampulla (proximal) in this figure. In Figure 18.21, a three-dimensional image the isthmus is shown. Since the smallest current OCT guidewires are 0.017", they should easily be able to track up the oviduct, performing OCT imaging in real time at micron scale resolution. This should allow both the identification of stenosis and guidance of subsequent interventions. In addition, more subtle disease such as fibrosis, SIN, and endometriosis should be more easily identified. However, studies are still limited in this area and future work is needed.

18.8   CONCLUSION

In conclusion, the basic scientific principles of OCT were initially discussed in Section I. This was intended to help the nonphysical scientist have a better intuitive understanding of OCT physical principles. Then OCT physics and technology were described in detail. As this is a relatively new modality, it is anticipated that significant advances in the technology beyond those discussed in Section II are likely in the near future. In addition, this section discussed some topics unavailable in the OCT literature, such as the importance of quantum noise, which are in part derived from other areas of science. Finally, the potential OCT applications in clinical medicine were discussed in Section III. This section began with chapters aimed at aiding investigators in performing basic and clinical research. The applications discussed in Section III should not be considered exhaustive. The focus was on those applications that are likely to reach clinical utility the earliest. Examples of other potential applications include examining neoplastic regions after therapy (such as PDT or radiation therapy) for viable tumor cells or guiding solid tumor biopsies. Also, no attempt was made to study transparent tissue, specifically the eye which is covered in a separate text, or developmental biological and tissue engineering, as the focus is clinical medicine. In conclusion, OCT represents an attractive new technology for biomedical imaging. Substantial future research is necessary to establish its ultimate utility.

REFERENCES

1. Bath-Balogh, M. and Febrenbach, M. J. (1997). *Illustrated Dental Embryology, Histology, and Anatomy.* W. B. Suanders Company, Philadelphia, PA.
2. Hanaizumi, Y., Kawano, Y., Ohshima, H. et al. (1998). Three dimensional direction and interrelationship of prisms in cuspal and cervical enamel of dog tooth. *Anat. Rec.* 252, 355–368.
3. Zero, D. T. (1999). Dental caries process. *Dent. Clin. North Am.* 43, 635.
4. Harris, N. and Garcia-Godoy, F. (1999). *Primary Preventive Dentistry*, 5th ed. Appleton & Lange, Stamford, CT.
5. NIH (2001). Diagnosis and management of dental caries throughout life. NIH Consensus Statement 18.
6. LeGeros, R. Z. and Tung, M. S. (1983). Chemical stability of carbonate and fluoride containing apatites. *Caries Res.* 17, 419–429.
7. van Houte, J. (1994). Role of micro-organisms in caries etiology. *J. Dent. Res.* 73, 672–681.
8. White, J. M. and Eakle, S. Rationale and treatment approach in minimally invasive dentistry. *JADA* 131, 13S–19S.
9. Wenzel, A. (1993). New caries diagnostic methods. *J. Dent. Educ.* 57, 428–431.
10. Shi, X. Q., Welander, U., and Angmar-Mansson, B. (2000). Occlusal caries detection with Ka DIAGNOdent and radiography. *Caries Res.* 34, 151–158.
11. Ashley, P. F., Ellwood, R. P., Worthington, H. V., and Davies, R. M. (2000). Predicting occlusal caries using the electronic caries monitor. *Caries Res.* 34, 201–203.
12. Scheiderman, A., Elbaum, M., Shultz, T. et al. (1997). Assessment of dental caries with digital imaging fiber-optic transillumination. In vitro study. *Caries Res.* 31, 103–110.
13. Verdonschot, E. H., Wenzel, A., and Bronkhorst, E. M. (1993). Assessment of diagnostic accuracy in caries detection: An analysis of two methods. *Community Dent. Oral Epidemiol.* 21, 203–208.

14. Colston, B. W., Everett, M. J., Da Silva, L. B. et al. (1998). Imaging of hard and soft tissue structure in the oral cavity by optical coherence tomography. *Appl. Opt.* 37, 3582–3585.
15. Feldchtein, F. I., Gelikonov, G. V., Gelikonov, V. M. et al. (1998). In vivo imaging of hard and soft tissue of the oral cavity. *Opt. Express* 3, 239–250.
16. Colston, B. W., Sathyam, U. S., DaSilva, L. B. et al. (1998). Dental OCT. *Opt. Express* 3, 230–239.
17. Baumgartner, A., Dichtl, S., Hitzenberger, C. K. et al. (2000). Polarization sensitive optical coherence tomography of dental structures. *Caries Res.* 34, 59–69.
18. Fried, D., Xie, J., Shafi, S. et al. (2002). Imaging caries lesions and lesion progression with polarization sensitive optical coherence tomography. *J. Biomed. Opt.* 7, 618–627.
19. Wang, X. J., Milner, T. E., de Boer, J. F. et al. (1999). Characterization of dentin and enamel by use of optical coherence tomography. *Appl. Opt.* 38, 2092–2096.
20. Otis, L. L., Colston, B. W., Armitage, G., and Everett, M. J. (1997). Optical imaging of periodontal tissues. *J. Dent. Res.* 76 (SI), 383.
21. Snell, R. S. (1981). *Clinical Anatomy for Medical Students*, second edition, Little, Brown, and Company, Boston, MA.
22. Berry, S. J., Cofffey, D. S., Walsh, P. C., and Ewing, L. L. (1984). The development of human benign hyperplasia with age. *J. Urol.* 132:L, 474–479.
23. Groner, C. (1988). Desktop resource: Top 25 most frequently performed surgeries. *Health Week*, June 23.
24. Lefaucheur, J. P., Yiou, R., Salomon, L. et al. (2000). Assessment of penile nerve fiber damage after transurethral resection of the prostate by measurement of penile thermal sensation. *J. Urol.* 164, 1416–1419.
25. Bieri, S., Iselin, C. E., and Rohner, S. (1997). Capsular perforation localization and adenoma size as prognostic indictors of erectile dysfunction after transurethral prostatectomy. *Scan. J. Urol. Nephrol.* 31, 545–548.
26. Krane, R. J., Goldstein, I., and Tejada, I. (1989). Impotence. *N. Engl. J. Med.* 321, 1648–1659.
27. Schatzl, G., Madersbacher, S., Djavan, B. et al. (2000). Two year results of transurethral resection of the prostate versus four less invasive treatment options. *Eur. Urol.* 37, 695–701.
28. Schatzl, G., Madersbacher, S., Lang, T. et al. (1997). The early postoperative morbidity of transurethral resection of the prostate and of 4 minimally invasive alternatives. *J. Urol.* 158, 105–111.
29. Naspro, R., Salonia, A., Colombo, R. et al. (2005). Update of the minimally invasive therapies for benign prostatic hyperplasia. *Curr. Opin. Urol.* 15, 49–53.
30. Kassablan, A. S. (2003). Sexual function in patients treated for benign prostatic hyperplasia. *Lancet* 361, 60–62.
31. Talic, R. F. and Al Rikabi, A. C. (2000). Transurethral vaporization resection of the prostate versus standard transurethral prostatectomy. *Eur. Urol.* 37, 301–305.
32. Poulakis, V., Dahm, P., Witzsch, U. et al. (2004). Transurethral electrovaporization versus transurethral resection for symptomatic prostatic obstruction. *BJU* 94, 89–95.
33. Kletscher, B. A. and Oesterling, J. E. (1992). Transurethral incision of the prostate: A viable alternative to transurethral resection. *Sem. Urol.* 10, 265–272.
34. Tooher, R., Sutherland, P., Vostello, A. et al. (2004). A systematic review of holmium laser prostatectomy for benign prostatic hypertrophy. *J. Urol.* 171, 1773–1761.

35. Hill, B., Belville, W., Bruskewitz, R. et al. (2004). Transurethral needle ablation versus transurethral resection of the prostate for the treatment of symptomatic benign prostatic hyperplasia: 5 year results. *J. Urol.* 171, 2336–2340.

36. Trock, B. J., Brotzman, M., Utz, W. J. et al. (2004). Long term pooled analysis of multicenter studies of cooled thermotherapy for benign prostatic hyperplasia, results at three months through 4 years. *Urology* 63, 716–721.

37. Blute, M. L. (1997). Transurethral microwave thermotherapy: Minimally invasive therapy for benign prostatic hyperplasia. *Urology* 50, 163–166.

38. Tearney, G. J. Brezinski, M. E., Southern, J. F. et al. (1997). Optical biopsy in human urologic tissue using optical coherence tomography. *J. Urol.* 157, 1915–1919.

39. Boppart, S. A., Herrmann, J. M., Pitris, C. et al. (2001). Real time optical coherence tomography for minimally invasive imaging of prostate ablation. *Comput. Aided Surg.* 6, 94–103.

40. Diao, E. and Vannuyan, T. (2000). Techniques for primary nerve repair. *Hand Clin.* 16, 53–66.

41. Toby, E. B., Meyer, B. M., Schwappach, J. et al. (1996). Changes in the structural properties of peripheral nerves after transection. *J. Hand Surg.* 21A, 1086–1090.

42. Burke, F. M., Newland, C., Meister, S. J. et al. (1994). Emergency medicine in the Persian Gulf War. *Ann. Emerg. Med.* 23, 755–760.

43. Culbertson, J. H., Rand, R. P., and Jurkiewicz, M. J. (1990). Advances in microsurgery. *Adv. Surg.* 23, 57–88.

44. Deutinger, M., Girsch, W., Burggasser, G. et al. (1993). Peripheral nerve repair in the hand with and without motor sensory differentiation. *J. Hand Surg.* 18A, 426–432.

45. Sunderland, S. (1945). The intraneural topography of the radial, median, and ulnar nerves. *Brain* 68, 243–299.

46. Watchmaker, G. P., Gumucio, C. A., Crandall, E. et al. (1991). Fascicular topography of the median nerve. *J. Hand Surg.* 16A, 53–59.

47. Hakstian, R. W. (1968). Funicular orientation by direct stimulation. *J. Bone Joint Surg.* 50, 1178–1186; Gual, J. (1986). Electrical fascicle identification as an adjunct to nerve repair. *Hand Clin.* 2, 709–722.

48. Engel, J., Ganel, A., Melamed, R. et al. (1980). Choline acetyltransferase for differentiation between human motor and sensory nerves. *Ann. Plast. Surg.* 4, 376–380.

49. Yunshao, H. and Shizhen, Z. (1988). Acetylcholinesterase: A histolochemical identification of motor and sensory fascicles in human peripheral nerve. *Plast. Reconstr. Surg.* 82, 125–131.

50. Brezinski, M. E., Tearney, G. J., Boppart, S. A. et al. (1997). Optical biopsy with optical coherence tomography: Feasibility for surgical diagnostics. *J. Surg. Res.* 71, 32–40.

51. Boppart, S. A., Bouma, B. E., Pitris, C. et al. (1998). Intraoperative assessment of microsurgery with three dimensional optical coherence tomography. *Radiology* 208, 81–86.

52. Lazebnik, M., Marks, D. L., Potgieter, K. et al. (2003). Functional optical coherence tomography for detecting neural activity through scattering changes. *Opt. Lett.* 28, 1218–1220.

53. Disa, J. J., Cordeiro, P. G., and Hidalgo, D. A. (1999). Efficacy of conventional monitoring techniques in free tissue transfers. *Plast. Reconstr. Surg.* 104, 97–101.

54. Suominen, S. and Asko-Seljavaara, S. (1995). Free flap failures. *Microsurgery* 16, 396–400.

55. Meyer, J., Walsh, J., Schuler, J. et al. (1987). The early fate of venous repair after civilian trauma. *Ann. Surg.* 206, 458–464.

56. Ryan, J. M., Milner, S. M., Cooper, G. J. et al. (1991). Field surgery on a future conventional battlefield: Strategy and wound management. *Ann. R. Coll. Surg. Engl.* 73, 13–20.

57. Trouwborst, A., Weber, B. K., and Dufour, D. (1987). Medical statistics of battlefield casualties. *Injury* 18, 96–99.

58. Zajtchuk, R. and Sullivan, G. R. (1995). Battlefield trauma care: Focus on advanced technology. *Mil. Med.* 160, 1–7.

59. Spencer, F. C. and Grewe, R. V. (1955). The management of arterial injuries in battle casualties. *Ann. Surg.* 141, 304–313.

60. Harria, J. R., Seikaly, H., Calhoun, K. et al. (1999). Effect of diameter of microvascular interposition vein grafts. *J. Otolaryngol.* 28, 152–157.

61. Arnljots, B., Dougan, P., Weislander, J. et al. (1994). Platelet accumulation and thrombus formation after microarterial injury. *Scand. J. Plast. Reconstr. Hand Surg.* 28, 167–175.

62. Central Brain Tumor Registry of the United States data, 1995–1999; CBTRUS, Central Brain Tumor Registry of the United States, 2002.

63. Peipmeier, J. and Baehring, J. M. (2004). Surgical resection for patients with benign primary brain tumors and low grade lymphomas. *J. Neuro-Oncol.* 69, 55–65.

64. Peipermeier, J., Christopher, S., Spencer, D. et al. (1996). Variations in the natural history and survival of patients with supratentorial low grade astrocytomas. *Neurosurgery* 38, 872–878.

65. Croteau, D., Scarpace, L., Hearshen, D. et al. (2001). Correlation between magnetic resonance spectroscopy imaging and image guided biopsies: Semiquantitative and qualitative histopathological analysis of patients with untreated glioma. *Neurosurgery* 49, 823–829.

66. Nicolato, A., Gerosa, M. A., Fina, P. et al. (1995). Prognostic factors in low grade supratentorial astrocytomas: A univariate statistical analysis in 76 surgical treated adult patients. *Surg. Neurol.* 44, 208–211.

67. Slavin, K.V., Anderson, G. J., and Burchiel, K. J. (1999). Comparison of three techniques for calculation of target coordinates in functional stereotactic procedures. *Stereotact. Func. Neurosurg.* 72, 192–195.

68. Wolfsberger, S. et al. (2002). Anatomical landmarkers for image registration in frameless stereotactic neuronavigation. *Neurosurg. Rev.* 25, 68–72.

69. Hall, W. A., Kowalik, K., Liu, H. et al. (2003). Costs and benefits of intraoperative MRI guided tumor resection. *Acta Neurochir.* Suppl. 85, 115–124.

70. Kelles, G., Lamborn, K. R., and Berger, M. S. (2003). Coregistration accuracy and detection of brain shift using intraoperative sononavigation during resection of hemispheric tumors. *Neurosurgery* 53, 556–564.

71. Vives, K. P. and Piepmeier, J. M. (1999). Complications and expected outcome of glioma surgery. *J. Neurooncol.* 42, 289–302.

72. Mizuno, A., Hakashi, T., Tashibu, K. et al. (1992). Near infrared Raman spectra of tumor margins in a rat glioma model. *Neurosci. Lett.* 141, 47–52.

73. Poon, W. S., Schomacker, K. T., Deutsch, T. F., and Artuza, R. L. (1992). Laser induced fluorescence: Experimental intraoperative delineation of tumor resection margins. *J. Neurosurg.* 76, 679–686.

74. Haglund, M. M., Hochman, D. W., Spence, A. M., and Berger, M. S. (1994). Enhanced optical imaging of rat gliomas and tumor margins. *Neurosurgery* 35, 930–941.

75. Boppart, S. A., Brezinski, M. E., Pitris, C., and Fujimoto, J. G. (1998). Optical coherence tomography for neurosurgical imaging of intracortical melanoma. *Neurosurgery* 43, 834–841.

76. Bizheva, K., Unterhuber, A., Hermann, B. et al. (2005) Imaging ex vivo healthy and pathological human brain tissue with ultrahigh resolution optical coherence tomography. *J. Biomed. Opt.* 10, 011006-1–011006-7.

77. http://www.cdc.gov

78. Seibel, M. M. (1997). *Infertility, A Comprehensive Text*. Simon & Shuster, London.

79. Darwish, A. M. and Youssef, A. A. (1998). Screening sonohysterography in infertility. *Gynecol. Obstet. Invest.* 48, 43–47.

80. Khalaf, Y. (2003). Tubal subfertility. *BMJ* 327, 610–613.

81. Marcoux, S., Maheuxx, R., and Berube, S. (1997). Canadian collaborative group on endometriosis. Laparoscopic surgery in infertile women with minimal or mild endometriosis. *N. Engl. J. Med.* 337, 217–222.

82. Honore, L. H. (1978). Salpingitis isthmica nodosa in female infertility and ectopic pregnancy. *Fertil. Steril.* 29, 164–168.

83. Creasy, J. F., Clark, R. L., Cuttino, J. T., and Groff, T. R. (1985). Salpingitis isthmic nodosa: Radiologic and clinical correlates. *Radiology* 154, 597–600.

84. Valle, R. F. (1995). Tubal cannulation. *Obstet. Gynecol. Clin. North Am.* 22, 519–540.

85. Falcome, T. (2001). What the interested needs to know about infertility. *Cleve. Clin. J. Med.* 68, 65–72.

86. Papaioannou, S., Bourdrez, P., Varma, R. et al. (2004). Tubal evaluation in the investigation of subfertility: A structured comparison of tests. *BJOG* 111, 1313–1321.

87. Mol, B. W., Swart, P., Bossuyt, P. M. et al. (1996). Reproducibility of the interpretation of hysterosalpingography in the diagnosis of tubal pathology. *Hum. Pathol.* 11, 1204–1208.

88. Glatstein, I. Z., Slepper, L. A., Lavy, Y. et al. (1997). Observer variability in the diagnosis and management of the hysterosalpingogram. *Fertil. Steril.* 67, 233–237.

89. (2004). *National Institute for Clinical Excellence, NHS Fertility; Assessment and Treatment for People with Fertility: Assessment and Treatment for People with Fertility Problems. Full Guidelines*. RCOG Press, London.

90. Hamilton, J. A., Larson, A. J., Lower, A. M. et al. (1998). Evaluation of the performance of hysterosalpingo contrast sonography in 500 consecutive patients, unselected, infertile women. *Hum. Reprod.* 13, 1519–1526.

91. Dijkman, A. B., Mol, B. W., van der Veen, F. et al. (2000). Can hysterosalpingo contrast-sonography replace hysterosalpingography in the assessment of tubal infertility? *Eur. J. Radiol.* 335, 44–48.

92. Lang, E. K., Dunaway, H. E., and Roniger, W. E. (1990). Selective osteal salpingography and transvaginal catheter dilatation in the diagnosis and treatment of fallopian tube obstruction. *Am. J. Roentgenol.* 154, 735–740.

93. Honore, G. M., Holden, A. E., and Schenken, R. S. (1999). Pathophysiology and management of proximal tubal blockage. *Fertil. Steril.* 71, 785–795.

94. Kerin, J., Daykhovsky, L., Grundfest, W. et al. (1990). Falloposcopy: A microendoscopic transvaginal technique for diagnosing and treating endotubal disease incorporating guide wire cannulation and direct balloon tuboplasty. *J. Reprod. Med.* 35, 606–612.

95. Surrey, E. S. (1999). Microscopy of the human fallopian tube. *J. Am. Assoc. Gynecol. Laparosc.* 6, 383–389.

96. Dechaud, H., Daures, J., and Heron, B. (1998). Prospective evaluation of falloposcopy. *Hum. Reprod.* 13, 1815–1818.

97. Kamiyama, S., Miyagi, H., and Kanazawa, K. (1999). Therapeutic value of selective salpingography for infertile women with patient fallopian tubes: The impact on pregnancy rate. *Gynecol. Obstet. Invest.* 49, 36–40.

98. Valle, R. F. (1995). Tubal cannulation. *Obstet. Gynecol. Clin. North Am.* 22, 519–539.

99. Penzias, A. S. and DeCherney, A. H. Is there ever a role for tubal surgery? *Am. J. Obstet. Gynecol.* 174, 1218–1223.

100. Young, B., and Heath, J. W. (2000). *Functional Histology, A Text and Color Atlas*, fourth edition, Churchill Livingstone, Edinburgh, Scotland.

INDEX

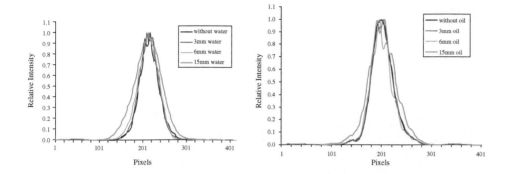

Figure 5.7 Data obtained from the autocorrelation function off a mirror after light has passed through different amounts of water and oil. The water is shown on left; the oil on the right. It can be seen that water results in a substantially greater broadening of the PSF even at depths as small as 5 mm. Courtesy Liu, B., et al., 2004. *Phys. Med. Biol.* 49, 1-8 [5].

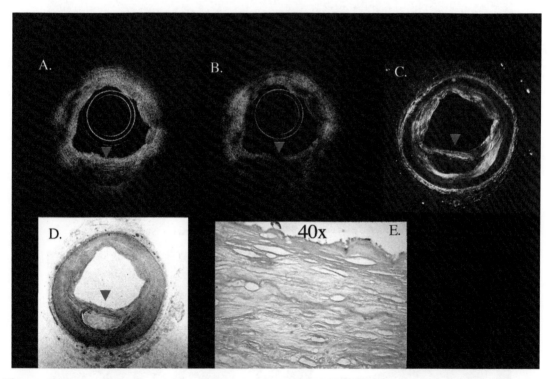

Figure 8.5 Picrosirus staining for identifying organized collagen. In this figure, polarization sensitivity of a human atherosclerotic coronary artery imaged with single detector PS-OCT is performed (A and B), as well as picrosirus staining (C) and trichrome blue (D and E). It can be seen that organized collagen is detected by OCT and picrosirus but not trichome blue. PS-OCT images were taken with the reference arm controllers in two different positions. Courtesy of Giattina, S.D. et al. (2006). *Intern. J. Cardiol.* 107, 400-409.

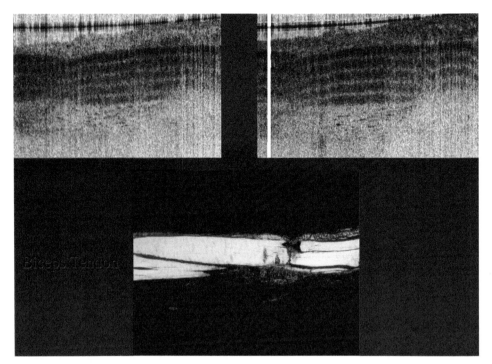

Figure 8.9 Birefringence assessment of a biceps tendon using SDPS-OCT. This slide illustrates the ability of SDPS-OCT to identify organized collagen in biceps tendon. It can be seen that the high collagen concentration is represented by the banding pattern that is noted at the two different controller positions in the reference arm. Courtesy of Martin, S.D., et al. (2003). *Int. Orthoped.* 27, 184-189.

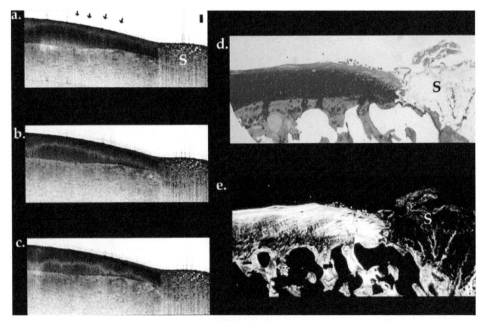

Figure 8.20 SDPS-OCT of normal articular cartilage. In a, b, and c, a band moves evenly through the cartilage as the polarization paddles are rotated in the reference arm. In the picrosirus stained section, the cartilage is bright identifying organized collagen. Courtesy of the Drexler, W., et al. (2001). *J. Rheum.* 28, 1311-1318.

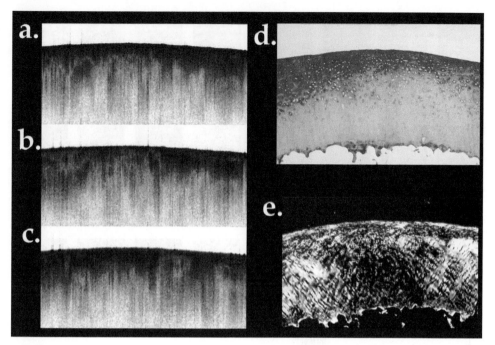

Figure 8.21 SDPS-OCT of osteoarthritic cartilage of relatively normal thickness. In a, b, and c, no significant banding patterns are noted. In e, the picrosirus stained section, no evidence of organized collagen is noted. In d however, the cartilage appears relatively normal. PS-OCT has therefore detected early osteoarthritic changes prior to cartilage thinning. Courtesy of the Drexler, W., et al. (2001). *J. Rheum.* 28, 1311-1318.

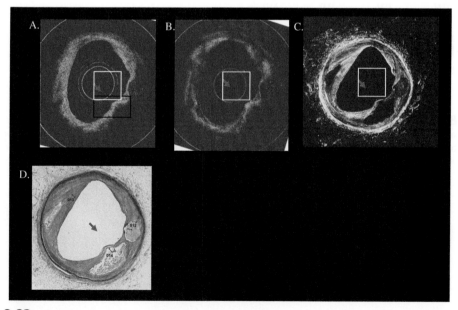

Figure 8.23 Polarization sensitive imaging of human atherosclerotic coronary arteries. This is a thin walled plaque with minimal organized collagen. In parts A and B, at two different controller positions, the white box shows an area of thin walled plaque. This is confirmed in the picrosirus stained section C, where brightness is minimal. Courtesy of Giattina, S.D. et al. (2006). *Intern. J. Cardiol.* 107, 400–409.

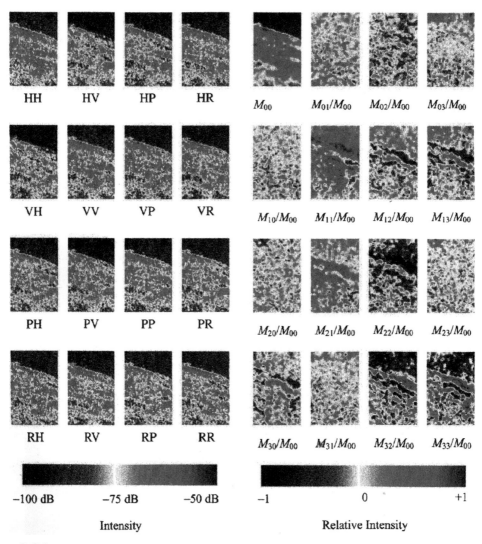

Figure 8.26 Mueller matrix of a tissue sample. The left are the raw OCT images; the right are the corresponding Mueller matrix parameters. Details of each image are not discussed here, but can be found in the original publication of this. The image is provided as an example of the current difficulty of interpreting data from a single Mueller's matrix. Courtesy of Jiao, S., et al. (2000). *Applied Optics.* 39, 6318-6325.

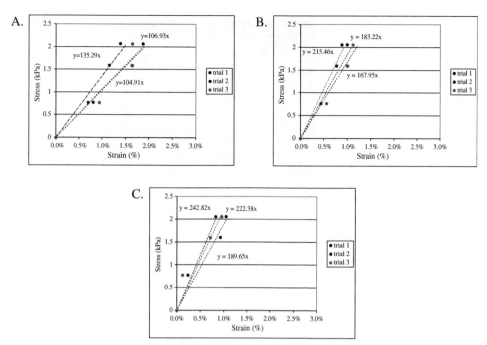

Figure 9.4 Strain-stress curves for (A) 1%, (B) 2%, and (C) 3% phantoms using three caliper measurements (trials 1-3). Young's modulus estimators are displayed as slope values of linear regression equations. Courtesy Rogowska, J. et al. (2002). *OSA Technical Digest*, PD20.1–20.3. [32,33]

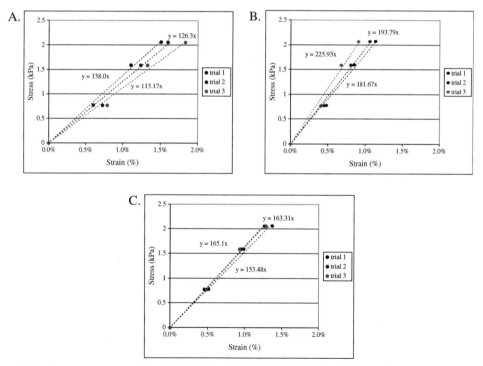

Figure 9.5 Strain-stress relationship for (A) 1%, (B) 2%, and (C) 3% phantoms using three OCT elastography measurements (trials 1-3). Young's modulus estimators are displayed as slope values of linear regression equations. Courtesy Rogowska, J. et al. (2002). *OSA Technical Digest*, PD20.1–20.3. [32]

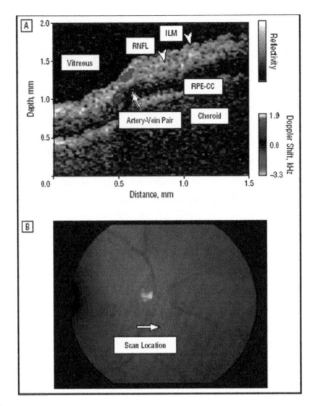

Figure 10.7 (A) *In vivo* color Doppler OCT (CD-OCT) image (2048 axial × 100 lateral pixels) of bidirectional flow in the human retina acquired in 10 seconds. The axial dimensions indicate optical path. CD-OCT is able to distinguish various layers of tissue and to quantify blood flow magnitude and direction in sub-100-μm diameter retinal blood vessels. (B) Fundus photograph marked to illustrate the position of the linear scan inferior to the optic nerve head. ILM, inner limiting membrane; RNFL, retinal nerve fiber layer; and RPE-CC, retinal pigment epithelium-choriocapillaris complex. (Courtesy of Dr. J. A. Izatt, *Arch Ophthamol.* 2003, 121, 235–239.)

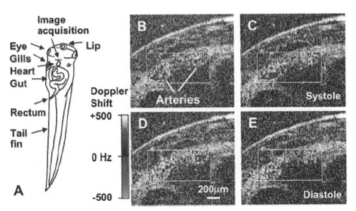

Figure 10.8 Demonstration of *in vivo* real-time Doppler flow imaging. (A) Schematic ventral view of a 17-day-old tadpole, *Xenopus laevis*. The imaging plane is slightly superior to the heart, transecting two branches of the ductus arteriosus. (B–E) Subsequent frames acquired at 8 frames per second. The blood pulses first through the left artery (B–C) and then, slightly delayed, through the right artery (D). (Courtesy of Dr. J. A. Izatt, *Opt. Lett.* 2002, 27, 34–36.)

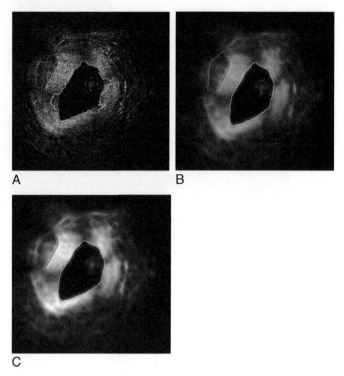

Figure 11.11 Results of segmentation using a Sobel/watershed algorithm applied to (A) the original image, (B) the image processed RKT technique (11 × 11 pixels kernel), and (C) the image processed with the RKT technique (21 × 21 pixels kernel). Red and yellow outlines represent the lipid lesion and lumen, respectively.

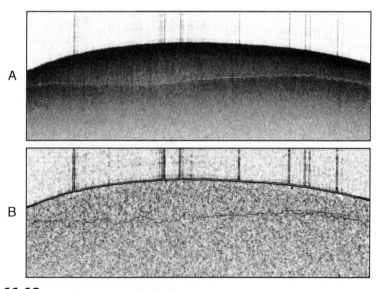

Figure 11.12 Results of the Sobel edge detection and graph searching algorithm applied to the original image; (A) two cartilage borders: outer (green), and inner (red), and (B) the corresponding cost image used in graph searching. Reprinted with permission from J. Rogowska, C.M. Bryant, and M. E. Brezinski, *J Opt Soc Am A*, 20(2):357–367, 2003.

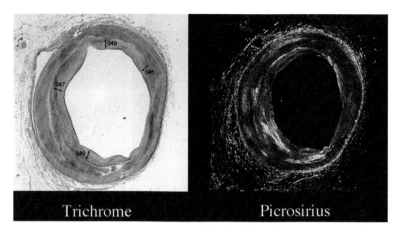

Figure 12.2 Picrosirus staining. Using the appropriate staining techniques is critical for hypothesis testing. One the left is a trichrome stained section of coronary artery, a technique which is sometimes used by investigators to 'quantitate' collagen concentrations. On the right is a picrosirus stained section that demonstrates organized collagen by bright areas, which is clearly more informative than the trichrome image. Courtesy of Stamper, D., et al., 2006. Plaque characterization with optical coherence tomography. *J. Am. Coll. Cardiol.* 47(8), 69–79. [25]

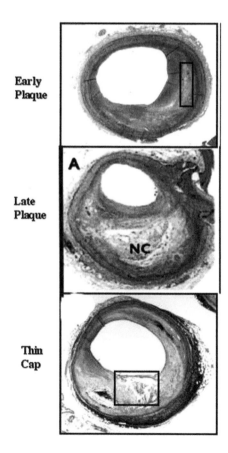

Figure 15.4 'Vulnerable Plaque' classified by in terms of thin capped fibroatheroma (TFCA). Squares demarcate diseased area and NC designated necrotic core. Courtesy Kolodgie, F.D., et al. (2003). *N.E.J.M.* 349, 2316-2325. [25] (modified)

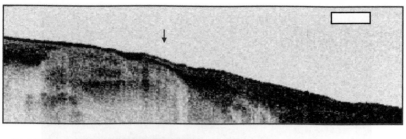

Reflectance

Figure 15.5 Thin intimal cap. The top is an OCT image of a heavily calcified plaque with minimal plaque while the bottom is the corresponding histopathology. The plaque is on the left and the relatively normal tissue on the right. The arrow denotes a thin intimal cap less than 50 μm in diameter at some points. Bar represents 500 μm. Image courtesy Brezinski, M.E., et. al. (1996). *Circulation*. 93(6), 1206–1213.

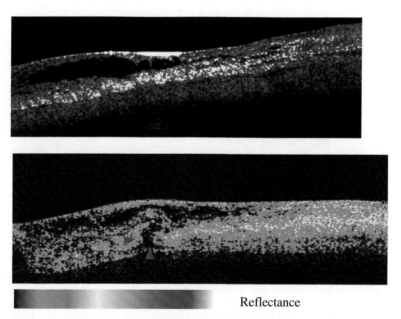

Reflectance

Figure 15.6 Unstable plaque. The top image is an unstable plaque with the black area on the left being lipid and the yellow arrow is the thin intimal cap. The bottom image is fissuring below the plaque surface shown by the arrows. Image courtesy Brezinski, M. E., et. al. (1996). *Circulation*. 93(6), 1206–1213.

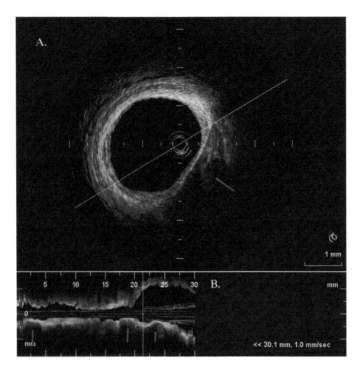

Figure 15.14 In vivo human imaging. In this image, a plaque is seen which covers almost two quadrants in cross section (line in A). In the axial image (b), the lines demonstrate plaques viewed along the catheter length. Image courtesy of Professor Ishikawa (Kinki University Japan) and Joseph Schmitt PhD, Lightlab Imaging. Published in Brezinski, M.E. (2006). *Intl. J. of Cardiol.* 107, 154–165.

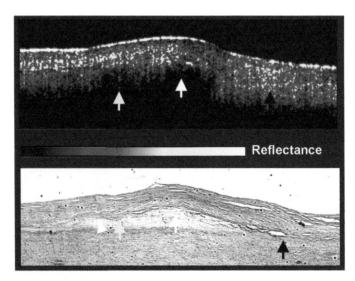

Figure 15.21 Examining intima boundary sharpness as a marker for lipid. In this figure, the yellow arrows demonstrate the cap-lipid interface that is diffuse and is covered by a highly scattering cap (yellow reflections). The white arrow, which is also over lipid, identifies cap with lower scattering and the cap-lipid interface is sharply defined. Finally, the black arrow shows the intimal-elastic layer interface (no lipid present) that is diffuse with an intima that is highly scattering. The reason for the diffuse nature of the intimal-elastin border is likely that multiple scattering is obscured due to high scattering within the intima. It is unclear, therefore, whether the lipid interface is diffuse because of the core comp[osition, as previously suggested, or multiple scattering from the cap. *Courtesy of Circulation* (1996). 93, 1206–1221 (modified).

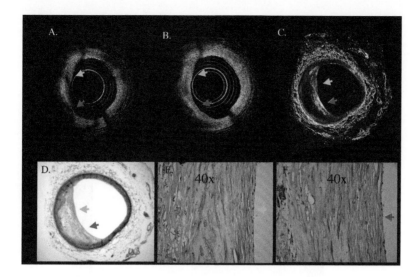

Figure 15.22 Assessing plaque collagen with PS-OCT. Single Detector PS-OCT changes the polarization state in the reference arm and allows collagen assessment. A and B demonstrate OCT images taken at two different polarization states. The image changes more in some regions than in others due to the higher concentration of collagen in the former. In C, the picrosirus section shows reduced collagen with the upper arrow but increased collagen at the lower arrow. In D, E, and F, trichrome stained sections are shown which is notable for a lack of inflammatory cells, the significance of which is discussed in the text. Unpublished data based on Giattina, S. D., et al. (2006). *Int. J. Cardiol.* 107, 400–409.

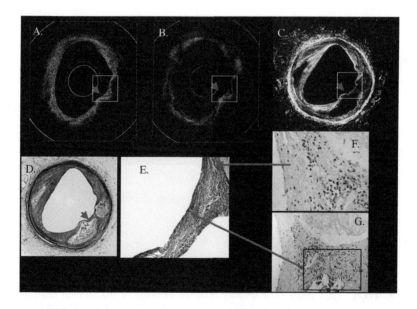

Figure 15.23 Thin walled plaque with minimal organized collagen. In the figure, a plaque with a wall thickness at points less than 60 çm in diameter is imaged. In parts A and B, the white box demonstrates an area of thin intima with almost no organized collagen. This is confirmed in the picrosirus stained section. Some inflammatory cells are noted in deeper sections, but have minimal presence near the surface. Courtesy Giattina, S.D., et al. (2006). *Int. J. Cardiol.* 107, 400–409.

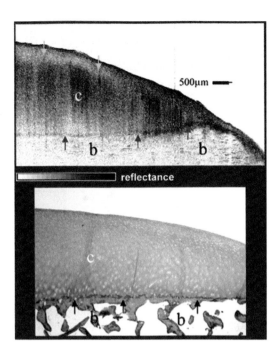

Figure 16.1 OCT imaging of normal human cartilage. The top is the OCT image, the bottom is the histopathology. Cartilage (c) and bone can be clearly identified in vitro along with the bone cartilage interface in this normal patella. Figure courtesy Herrmann, J. et al. (1999). *J. Rheumatolog.* 26(3), 627–635. [41]

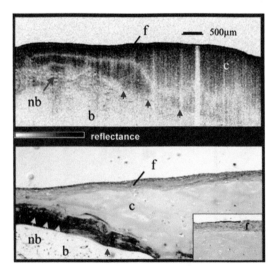

Figure 16.4 OCT imaging of severely diseased cartilage of the femoral head. In this image, severe factors are noted. First, the cartilage is thinned on the left. Second, on the left identified by nb is disruption of the bone-cartilage interface. Third, a fibrous band is noted at the surface of the cartilage. Image courtesy Herrmann, J. et al. (1999). *J. Rheumatolog.* 26(3), 627–635. [41]

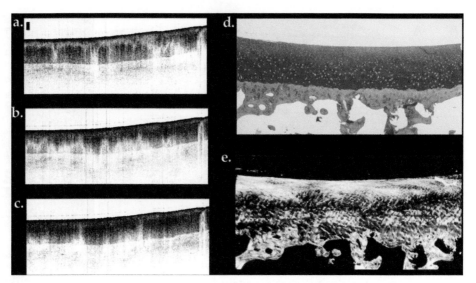

Figure 16.8 PS-OCT imaging of mildly diseased cartilage. In the OCT images (a, b, and c), the banding pattern is present but not as organized as in figure 16.8. In e, organized collagen is present by picrosirius but areas of disorganization are also noted. Image courtesy Drexler, W., et al. (2001). *J. Rheumatol.* 28, 1311–1318. [42]

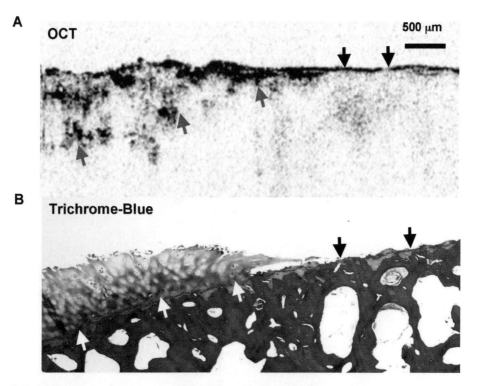

Figure 16.11 Severely osteoarthritic in vivo knee. In this section of tissue, cartilage remains on the left but is grossly inhomogeneous with poor resemblance to the normal cartilage seen above. On the right, no significant cartilage exists. Image courtesy Li, X., et al. (2005). *Arthritis Res. Ther.* 7, R318–323.

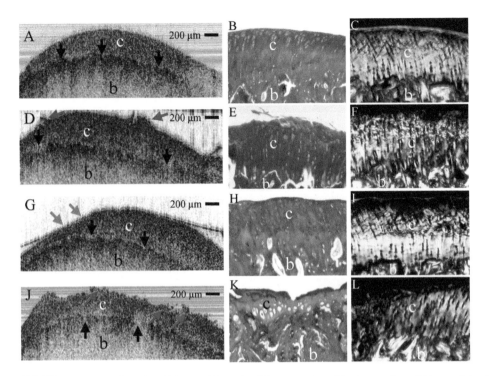

Figure 16.15   Mechanically induced OA. In this study, OA was produced by transection of the medial collateral ligament and full thickness artificial meniscal tear. The left are the OCT images at 0 (A), 1 week (D), 2 weeks (G), and 3 weeks (J) respectively. The second column is the trichome stained section while the third is the corresponding picrosirius. Progressive degeneration is noted in each. Courtesy Roberts, M.J. (2003). *Anal. Bioanal. Chem.* 377, 1003–1006.

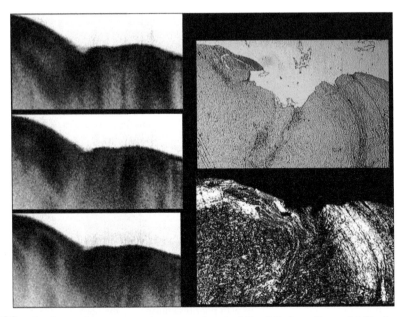

Figure 16.22   The figure demonstrates the ruptured end of the ACL from Figure 16.21. It can be seen that polarization sensitivity has been completely lost in the OCT images, and in the picrosirius stained region, it can be seen that collagen organization is almost completely lost. Courtesy Martin, S. D., et al. (2003). *Int. Orthop.* 27, 184–189.

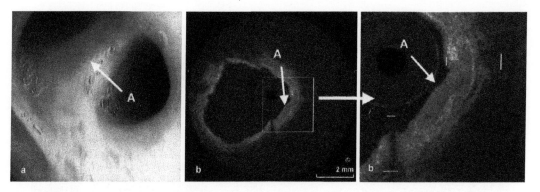

Figure 17.11 In vivo imaging of squamous cell carcinoma. In a, subtle changes in the bronchus are noted. In b (and its enlargement), the tumor in the mucosa and submucosa are seen, as well as loss of submucosal glands and structural definition. Courtesy of Tsuboi, M. et al. (2005). *Lung Cancer.* 49, 387–394.

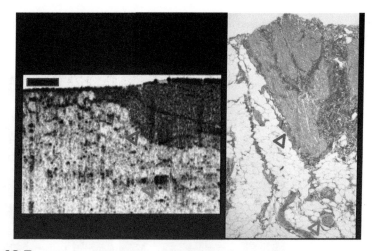

Figure 18.7 OCT image of prostate capsule. The prostate parenchyma is in the upper right. The capsule is to the left. The upper arrow demonstrates prostate-capsule border. The lower arrow is an unmyelinated nerve. It is transection of these unmyelinated nerves that leads to many of the complications of TURP. Courtesy of Brezinski, M.E., et al. (1997). *J. Surg. Res.* 71, 32–40.

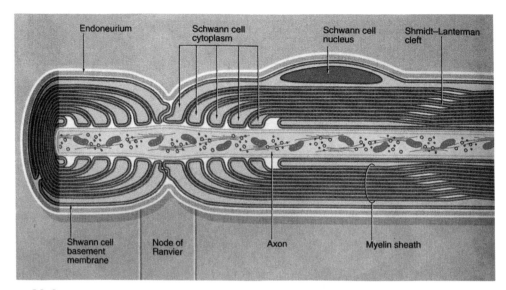

Figure 18.9 Diagram of myelinated nerve fiber. Courtesy Young, B. and Heath, J.W. (2000). *Functional Histology*, fourth edition. [100]

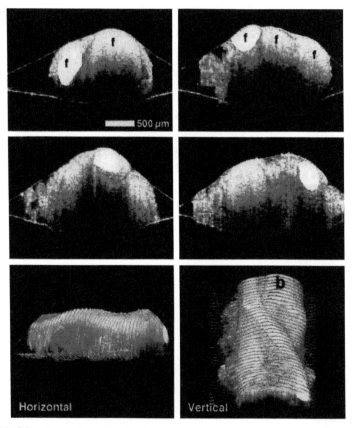

Figure 18.12 Three dimensional OCT imaging of a nerve. Courtesy of Boppart, S.A. et al. (1998). *Radiology*. 208, 81–86.

Printed and bound by CPI Group (UK) Ltd, Croydon, CR0 4YY

08/05/2025

01864865-0001